ASTRONOMY AND ASTROPHYSICS LIBRARY

Series Editors: G. Börner, Garching, Germany
A. Burkert, München, Germany
W. B. Burton, Charlottesville, VA, USA and
   Leiden, The Netherlands
M. A. Dopita, Canberra, Australia
A. Eckart, Köln, Germany
T. Encrenaz, Meudon, France
M. Harwit, Washington, DC, USA
R. Kippenhahn, Göttingen, Germany
B. Leibundgut, Garching, Germany
J. Lequeux, Paris, France
A. Maeder, Sauverny, Switzerland
V. Trimble, College Park, MD, and Irvine, CA, USA

R. E. Gershberg

Solar-Type Activity in Main-Sequence Stars

Translated by S. Knyazeva

With 69 Figures and 17 Tables

 Springer

Professor Roald E. Gershberg
Crimean Astrophysical Observatory
Crimea Nauchny 98409, Ukraine

Dr. Svetlana Knyazeva
Department of International Programmes
Siberian Branch of RAS
17, Prosp. Akademika Lavrentieva
Novosibirsk 630090, Russia

Cover picture: Flare on AD Leo of 18 May 1965. (Gershberg and Chugainov, 1966)

Library of Congress Control Number: 2005926088

ISSN 0941-7834
ISBN-10 3-540-21244-2 Springer Berlin Heidelberg New York
ISBN-13 978-3-540-21244-7 Springer Berlin Heidelberg New York

This work is subject to copyright. All rights are reserved, whether the whole or part of the material is concerned, specifically the rights of translation, reprinting, reuse of illustrations, recitation, broadcasting, reproduction on microfilm or in any other way, and storage in data banks. Duplication of this publication or parts thereof is permitted only under the provisions of the German Copyright Law of September 9, 1965, in its current version, and permission for use must always be obtained from Springer. Violations are liable to prosecution under the German Copyright Law.

Springer is a part of Springer Science+Business Media

springeronline.com

© Springer Berlin Heidelberg 2005
Printed in The Netherlands

The use of general descriptive names, registered names, trademarks, etc. in this publication does not imply, even in the absence of a specific statement, that such names are exempt from the relevant protective laws and regulations and therefore free for general use.

Typesetting by the authors
Final Layout: TechBooks India
Cover design: *design & production* GmbH, Heidelberg

Printed on acid-free paper   SPIN: 10992593   55/3141/BM - 5 4 3 2 1 0

*In memory of Solomon Borisovich Pikelner,
a marvelous person and eminent astrophysicist*

Preface

Over the last decades our insight into the realm of stars was deepened considerably by many outstanding discoveries: pulsars, neutron stars predicted in the 1930s; discrete X-ray sources, stellar systems with much stronger X-ray emission than optical luminosity; maser emission related to the early stage of stellar evolution and to some oscillating supergiants that have reached an advanced evolutionary stage; intense outflow of matter from hot stars, etc. Parallel to these discoveries, which are mainly due to progressing methods of astrophysical observations, the comprehension of many stellar events improved markedly. The spectra of supernovae were successfully interpreted and the hydrodynamic theory of flares on them was elaborated. The theory of stellar evolution in binary systems yielded the concept of mass exchange as a decisive factor of their evolution and gave a key to understanding the activity of a wide family of novae, symbiotic and cataclysmic systems. As a result of the development of the evolution theory of early spectral-type stars, oscillating cepheids – the lighthouses of the Universe that were used to determine distances to different stellar systems – had become space-time references of the development of such systems.

As compared to these impressive findings for extremely interesting but rather rare types of stars, which attract the attention of astronomers of all specialties and not only astronomers, the study of very faint and cool variable dwarfs – UV Cet-type flare stars – might seem too narrowly specialized. However, successful investigation of these stars changes essentially the general concepts of the "stellar kingdom". During the last decades it became clear that, first, flare dwarfs make up the most abundant type of stars and, secondly, their activity, being of the same physical nature as solar activity, plays a major role among lower main-sequence stars.

Among several tens of thousands of variable stars known today, only several hundreds are classified as UV Cet-type variables in the solar vicinity. However, the relatively low number is due to low luminosities, which prevent detection of the stars at large distances from the Sun. Indeed, variables of this type total about half of all stars within a distance of several parsecs from the Sun. One cannot be sure that all flare stars have been detected in this vicinity, only the lower limit has been estimated so far. Moreover, the study of the nearest stellar clusters has shown that flare stars form an overwhelming majority of populations of these systems. Finally, according to the spectral classification of all field red dwarfs, a significant part of M4 and later stars have strong chromospheres, which suggests their flare activity. Thus, one can state that 40 to 90% of all stars are UV Cet-type objects. In other words, if there was a detector capable of recording stars "apiece" without regard to their luminosity, the image of the Galaxy would essentially differ from normal photos of extragalactic systems. In addition to hundreds of millions of practically constant stars of middle-spectral types, which determine the integral luminosity of galaxies, tens of thousands of regularly oscillating cepheids and semiregular late supergiants, hundreds of novae stars that appear every year and grandiose flares of supernovae occurring once a decade, the detector first of all would find billions of UV Cet-type stars flaring irregularly with characteristic time intervals between flares of about one or several hours. The prevalence of flare stars in the stellar population not only raises general interest in this basic form of stellar matter, but also necessitates the revision of conventional notions about the dynamics of the interstellar medium, the ways of formation of its chemical composition, and the sources of high-energy particles.

The similarity of the solar and UV Cet-type activity was noted in the early 1950s, but even in the late 1960s this fact was only an occasion for discussing one of many physical models of such variable stars. This situation was thoroughly considered in my first monograph "Flares on Red Dwarf Stars" (Gershberg, 1970a). However, further intense photometric, spectral, and radio-astronomical observations of UV Cet-type stars and impressive advances in the study of solar activity supported the idea on the identity of physical nature of active red dwarfs and the Sun, as well as an approach to these variables as objects with maximum surface magnetic activity (Gershberg and Pikelner, 1972; Mullan, 1975a). Many researchers of flare stars could not immediately accept the analogy with solar activity. But now that X-ray emission of strong stellar coronae and ultraviolet emission line spectra of transition regions have been directly detected, ultraviolet and X-ray radiation of stellar flares has been recorded, statistical studies of several thousand stellar flares have proved that energy and time characteristics of solar flares are commensurable with those of stellar flares, and the spectra of solar and stellar chromospheres have been theoretically calculated using identical programs, qualitative similarity of the general picture of solar and stellar spottedness has been detected, long-term cycles of stellar activity similar to the 11-year solar cycle have been detected in observations of stellar chromospheres, and very young and very old G dwarfs have enabled reconstruction of the evolution of solar activity in the past and predictions for the future billions of years – the thesis about physical identity between the solar activity and that of red dwarf stars sounds trivial. However, this "trivial statement" acquires increasingly constructive substance.

Indeed, the results of high time, spectral, and spatial resolution observations of active processes on the solar surface are initial data for the working models of transient flares, dark spots, active regions, coronal loops, and other local stellar structures. On the other hand, flares on UV Cet-type stars are accompanied by an energy release sometimes exceeding that of the strongest solar flares by 100–1000 times. This ensures a strict fidelity criterion of theoretical constructs for solar flares: a theory of solar flares is correct only if, under the appropriate change of essential parameters, it can present 2–3 orders of magnitude stronger and shorter flares than the strongest and shortest flares on the Sun. Finally, the dependences of the intensity of permanent chromospheric emission and thermal radiation from the corona on rotational velocity and stellar age are initial experimental data for constructing the theory of the magnetism of stars with convective envelopes, in general, and the evolutionary theory of solar activity in particular. In essence, the statement about physical identity between the activity of the Sun and red dwarf stars underlies quickly developing topics of solar-stellar physics.

By now, general outlines of evolutionary and physical bases of the activity of medium- and low-mass stars have been clarified: the initial angular momentum of a protostellar cloud produces a rotating star → in a gravitationally bound gaseous sphere, rotation becomes differential and convection transfers the energy to the outer layers → the combination of differential rotation and convection generates a magnetic field, which eventually causes a whole variety of phenomena considered in this book; it forms structures on the stellar surface and participates in generation and outward transfer of nonradiative energy fluxes from subphotospheric layers; its dissipation is responsible for the existence of the chromosphere, corona, and stellar wind; the magnetic field forms spatial structures on the whole stellar atmosphere and has a decisive impact on nonstationary processes occurring there. Since the final energy source of this chain is stellar rotation, this sequence inevitability results in evolutionary delay of rotation. Each link of the chain involves many diverse processes and phenomena.

Comparative research of flare stars and the Sun has become increasingly fruitful. It attracts the ever-broadening audience of researchers – observers and theorists, stellar and solar physicists, experts in plasma physics, high-energy astrophysicists and solar–terrestrial relationships. The goal of this book is to provide the researchers with an overview, as complete as possible, of observational results for UV Cet-type stars and related objects, as well as with qualitative, though not always unambiguous, physical interpretation.

***

I am deeply grateful to all those who helped me in the preparation of this book. M.M. Katsova, M.N. Lovkaya, A.V. Terebizh, and N.I. Shakhovskaya actively contributed to the compilation of the Crimean Database of Flare Stars that was then updated by M.N. Lovkaya. The Catalog of Flare Stars in the

Solar Vicinity compiled on the basis of this database is reproduced in the Appendix by courtesy of Dr. C. Bertout, Chief Editor of Astronomy and Astrophysics. L.S. Liubimkov made useful remarks to the chapter "Quiescent Photospheres", M.A. Livshits and M.M. Katsova – to "Models of Stellar Flares." I wish to thank E.I. Zhigalkina for preparing the illustrations, N.P. Bannova, T.A. Granitskaya, E.V. Kostyleva, and M.I. Taran for the provision of access to publications. This book would not have appeared without financial support from ASTROPRINT Publishing Co. and its director Dr. G.A. Garbuzov.

December, 2001 *Nauchny, Crimea*

Preface to the English edition

The English-language edition includes results obtained during recent years. The most important of them are the new data of far ultraviolet observations from the Hubble Space Telescope, the recently launched Far Ultraviolet Spectroscopy Explorer, and the new-generation X-ray telescopes XMM-Newton and Chandra. As a result, about 100 publications were added to the list of references. This edition does not contain The Catalog of Flare Stars in the Solar Vicinity, which is accessible in Astronomy and Astrophysics Supplement Series, Vol. 139, p. 555, 1999. I am very grateful to Drs. L.A. Pustil'nik and M.M. Katsova for useful remarks, to L.I. Filatova and A.V. Terebizh for active help in preparing figures and the list of references, respectively. My special thanks go to the translator, Dr. Svetlana Knyazeva, for fruitful collaboration.

October, 2004 *Nauchny, Crimea*

Contents

Introduction 1

1 **Flare Stars in a Quiescent State** 9
   1.1 General Characteristics of Flare Stars 9
   1.2 Photospheres 13
      1.2.1 Quiescent Photospheres 13
      1.2.2 Starspots 17
          1.2.2.1 Estimates of the Parameters of Individual Starspots 20
          1.2.2.2 Estimates of the Parameters of Spotted Regions within the Zonal Model ... 31
          1.2.2.3 Some Problems of Starspot Physics 37
      1.2.3 Magnetic Fields at the Photospheric Level 40
          1.2.3.1 Zeeman Spectropolarimetry 42
          1.2.3.2 Robinson's Spectrophotometry 44
          1.2.3.3 Other Magnetometric Methods 54
   1.3 Chromospheres and Transition Regions 57
      1.3.1 Optical and Ultraviolet Spectra of Chromospheres and Transition Regions 60
          1.3.1.1 Calcium Emission in the H and K Lines 60
          1.3.1.2 Hydrogen Emission 68
          1.3.1.3 Other Emission Lines in the Optical Range ... 78
          1.3.1.4 Ultraviolet Spectra 81
      1.3.2 Models of Stellar Chromospheres 103
          1.3.2.1 Semiempirical Homogeneous Models 104
          1.3.2.2 Surface Inhomogeneities at the Chromospheric Level 112
   1.4 Coronae and Stellar Winds 121
      1.4.1 Soft X-ray Emissions: X-ray Photometry and Colorimetry 124
      1.4.2 EUV and X-ray Spectroscopy 152

| | | 1.4.3 | Microwave and Shortwave Emissions | 163 |
| | | 1.4.4 | Stellar Winds and Far-IR Emission | 177 |

 1.5 Heating Mechanisms for Stellar Atmospheres 181

2 Flares 191
 2.1 General Description of Stellar Flares 194
 2.2 Temporal Characteristics of Flares 205
  2.2.1 Time Scales of Flares 205
  2.2.2 Time Distribution of Flares 211
 2.3 Flare Energy 214
  2.3.1 Energy of Optical Flare Emission 214
   2.3.1.1 Spectrum of Maximum Flare Brightness 217
   2.3.1.2 Spectrum of Flare Energy 220
   2.3.1.3 Total Energy of Flare Emission 234
  2.3.2 Estimates of Total Energy of Flares 237
 2.4 Dynamics and Radiation Mechanisms of Flares
     in Different Wavelengths 240
  2.4.1 X-ray Emission of Flares 241
  2.4.2 Radio Emission of Flares 280
  2.4.3 Ultraviolet Emission of Flares 302
  2.4.4 Optical Emission of Flares 323
   2.4.4.1 Light Curves 324
   2.4.4.2 Colorimetry 335
   2.4.4.3 Polarimetry 341
   2.4.4.4 Spectral Studies 342
 2.5 Models of Stellar Flares 360

3 Long-Term Variations in Activity of Flare Stars 381
 3.1 Activity Cycles 381
 3.2 Evolutionary Changes in the Activity of Stars 394
  3.2.1 Evolution of Stellar Activity 396
  3.2.2 Evolution of Solar Activity 409

4 Conclusion 419
 4.1 Activity and Subphotospheric Magnetic Fields 421
 4.2 Activity and Magnetism of Stellar Atmospheres 426

References 433

Index 489

Introduction

Brief History of Investigation of Flare Red Dwarfs

In 1924, Hertzsprung noticed that one of the faint stars in Carina was brighter by 2^m on one of the photographs taken on the night of January 29. The rate of brightness increase suggested that this star did not belong to novae or oscillating RR Lyrae-type stars. Thus, Hertzsprung decided that the effect could be produced by the fall of an asteroid on star. Apparently, it was the first recorded flare on stars of this type.

Studying the spectra of faint stars in Orion in December 1938, Wachmann (1939) found an unusual variable star with an abnormal spectrum. The spectrum was obtained using an objective prism, the spectra being discretely broadened in one direction at 0.02 mm every 6 min during an hour. During the first third of the exposure the stellar spectrum resembled that of a nucleus of a planetary nebula of the WR type: on a background of continuous radiation of the type of continuum of B or A stars one could see strong emission lines H_γ, H_δ, H_ε, and H_ζ. Then, the brightness of the object decreased by at least one and a half stellar magnitudes, and in the band corresponding to the other part of the exposure mainly emission lines became visible. At the same time, the spectrum of the adjacent star displayed uniform darkening throughout its width. In the images obtained a month later, the star had a regular K spectrum without emission. Apparently, Wachmann was the first to record the spectrum of a flare on a UV Cet-type star.

In 1940, while examining the parallaxes of faint stars van Maanen (1940) noticed that the brightness of the M6e star Lalande 21258 (= WX UMa) was about 16^m on more than 20 plates, but in two images obtained on 11 May 1939 with an interval of about half an hour it was $14\overset{m}{.}2$ and $14\overset{m}{.}5$ (Fig. 1). Several years later, van Maanen (1945) detected a similar phenomenon in measuring the parallax of Ross 882 (= YZ CMi). Reporting this fact, he noted that both variables had low luminosity, belonged to a late spectral class, and should be objects of the same type.

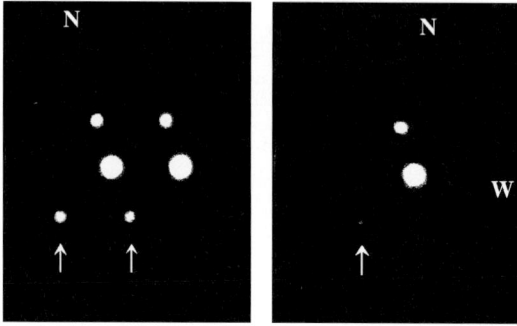

Fig. 1. Photos of WX UMa obtained by van Maanen (1940): stellar brightness of $14^m\!.2$ and $14^m\!.5$ (*left*) and normal brightness of about 16^m (*right*)

In 1947, Luyten found very large proper motion – $3''3$ per year – of the 14^m star L 726-8 and invited astronomers from several observatories to study this object. Carpenter found that the star was very red and determined its parallax. Page and Struve obtained spectrograms of the star, which appeared to be an M6 star with hydrogen and calcium emission lines. Joy and Humason observed L 726-8 at the 100″ reflector and established its binarity. They estimated the difference of brightness ($\sim 0^m\!.5$) and angular separation ($\sim 1''5$). Van Biesbroeck and van den Bos carried out micrometric observations of the system. Determining the parallax within this cooperative study, on 7 December 1948 Carpenter obtained a plate on which one of the five images of the star was much brighter than the other four. Upon studying this image, Hughes and Luyten suggested that there was a strong flare on the weak component of the system: over 3 min its brightness increased by more than 12 times. Having collected preliminary communications of observers, Luyten (1949a) found that L 726-8 was a binary system with the smallest mass components. He estimated the energy release in the flare as 4×10^{31} erg and noted its explosive character. He remembered similar observations by van Maanen. Then, Joy and Humason (1949) reported additional important data obtained at the 100″ reflector. In August–September 1948, they obtained spectrograms for each component of the system. In all spectrograms, except for the one of 25 September 1948, the spectrum of the M dwarf had very strong emission lines of hydrogen and calcium. The spectrum of 25 September 1948 strongly differed from the others (Fig. 2): absorption lines and bands were almost flooded by strong continuous radiation, which was the most explicit in the range of wavelengths shorter than H_δ, bright hydrogen lines were strengthened as compared to calcium lines, also the lines λ 4026 Å and λ 4471 Å of neutral helium and λ 4686 Å of ionized helium were seen in the emission, but there were no forbidden lines. Joy and Humason were sure that on 25 September 1948 they observed a similar phenomenon to that recorded by Carpenter in the direct image of 7 December 1948. But unlike Luyten, who assumed that the flare

Fig. 2. Spectra of UV Cet obtained by Joy (1960): stellar spectrum in the normal state (*top*) and during the flare of 25 September 1948 (*bottom*)

was due to the occurrence of emission lines in the weak component of the system, Joy and Humason concluded that the increased brightness was primarily caused by continuous radiation. Later, the weak component of the system L 726-8 was named UV Cet.

Cooperative research of UV Cet by American observers stimulated a thorough examination of dMe variable stars. The next decade was a period of fast accumulation of data. Over these years, the first photoelectric records of the light curves of flares were obtained (Gordon and Kron, 1949; Liller, 1952; Roques, 1953, 1954), numerous collections of negatives obtained in the course of long-term monitoring at different observatories were investigated (Luyten, 1949b; Shapley, 1951, 1954; van de Kamp and Lippincott, 1951; Lippincott, 1952, 1953; Hoffleit, 1952; Luyten and Hoffleit, 1954; Gaposhkin, 1955; Petit and Weber, 1956), series of visual observations were performed, and the first attempts to determine statistical characteristics of flare activity were undertaken (Petit, 1955, 1957; Oskanian, 1953, 1964), slit spectrograms of two stellar flares of the considered type were obtained (Herbig, 1956; Joy, 1958), the data on sporadic changes of spectra were collected for some red dwarf stars that suggested affiliation of these objects to flare stars (Joy, 1960; Thackeray, 1950; Popper, 1953; Münch and Münch, 1955; Bidelman, 1954). It should be emphasized that these studies were not focused on the identification of the nature of flares on the UV Cet-type stars. They mainly dealt with the description of the objects that were definitely or presumably attributed to this type or with the examination of general characteristics of this class of stellar variability and the place of UV Cet-type stars among other eruptive variables of late spectral classes. Numerous publications by Petit (1954, 1955, 1957, 1958, 1959, 1961), who initiated an international cooperation of observers of variable stars aimed at studying flare objects, made an important contribution to this field. Results of this decade of "initial accumulation" of data on the activity of red dwarfs were thoroughly described in my small monograph (Gershberg, 1970a).

In 1958, at the X General Assembly of the International Astronomical Union UV Cet-type stars were ranked as a special type of eruptive variables in the general classification of variable stars. They were defined as "dMe stars subjected to flare with the amplitude of 1^m to 6^m. Maximum brightness is attained in seconds or several dozen seconds after the commencement of the flare, the star returns to its normal brightness after several minutes or dozens of minutes. A typical representative – UV Cet." This purely photometric definition is reproduced in all editions of the "General Catalog of Variable Stars".

Further advances in studying flare red dwarfs were to a great extent due to the development of new astrophysical methods and updating of traditional methods, which treated transience and irregularity of stellar flares as essential properties of the objects. Thus, since the 1960s photoelectric monitoring has become the basic source of data on photometric properties of flare stars: light curves of flares, their energy and time distribution, the properties of the quiescent state beyond flares. The photographic method of multiple exposures has become the basic way of detection of flare stars in the nearest stellar clusters and made it possible to estimate the prevalence of such phenomena in the stellar realm. At the same time, the first successful observations of stellar flares in the radio range were carried out at the Jodrell Bank Observatory. In the mid-1960s, the first high time resolution slit spectra of flares on UV Cet-type stars were recorded with the Shajn reflector in the Crimea and the Struve telescope in Texas. The analysis of these spectra together with the available data on radio emission of flares and features of chromospheres and photospheres of such stars led to a conclusion on the physical identity between the activity of flare stars and the Sun. Started in the 1970s, extraterrestrial studies of flare stars were advanced in the 1980s by the IUE satellite that recorded ultraviolet spectra of the chromosphere and the transition region to the corona in quiescent and active states. Using the Einstein Observatory, X-ray emission of coronae of many flare stars was discovered. Finally, by the mid-1980s, the Very Large Antenna in New Mexico (USA) recorded microwave radio emission of flares and quiet coronae of flare stars. X-ray emission of many UV Cet-type stars was recorded by EXOSAT. In the early 1990s, the EUVE satellite provided the data on the far ultraviolet of stellar flares. However, the results of these and later studies from the X-ray telescopes ROSAT, BeppoSAX, ASCA, Chandra, XMM-Newton and from the Hubble Space Telescope (HST) and Far Ultraviolet Spectroscopy Explorer (FUSE) should not be considered as a history. They are included in the database of modern multiwavelength astrophysical observations of flare stars, and will be thoroughly considered in the following chapters.

The data on flare stars will be theoretically interpreted in the context of general notions on the activity of lower main-sequence stars eventually related to stellar magnetism. The current level of these ideas is still rather far from an ideal strict deductive theory, and experimental studies are still governing the research of these variable stars. This fact determines the structure of the book: as yet the list of experimental confirmations of a strict and complete physical theory cannot be compiled, but we can already deviate from the structure of two previous small monographs devoted to this subject (Gershberg, 1970a, 1978), in which we consistently considered the results of various observational methods and on this basis developed and substantiated a general concept of activity of UV Cet-type stars. Here, reasoning from the accepted general concept of physical identity of the activity of the Sun and the lower main-sequence stars of middle and late spectral types, we will consider various phenomena and structures on such stars using the whole body of relevant observational data.

Terminological Remarks

The systematics of variable stars divides these objects according to the character of brightness changes with time into three large groups: oscillating, eclipsed, and eruptive. The brightness of oscillating stars changes continuously and strictly periodically or almost periodically. The physical nature of these changes is explained by the features of their internal structure: transparency of the outer layers or maximum possible efficiency of convective transfer on such stars never provides an outward carryover of all the energy released in their interior. Such objects can exist only in the mode of periodic pulsations, when a part of the superfluous energy is spent for the expansion of a star and at maximum phase radiation increases appreciably. Eclipsed variables are binary stars, whose orbital planes are oriented so that a ground-based observer periodically records an eclipse of one component of the system by the other. It is obvious that the observed brightness changes in such systems can also occur at constant brightness of each component. (Notwithstanding their name, pulsars should not be attributed to oscillating stars. Pulsars are eclipsed systems, whose variability depends only on changing visibility conditions: the accessibility of directed radio-emitting areas above the stellar surface to ground-based observers is determined by the rotation of a degenerated star.)

Variables with noncyclic high-amplitude brightness changes belong to the immense class of eruptive stars. As a rule, individual "eruptions" can be separated on the light curves of these variables, i.e., emissions, abrupt increase and fading of brightness. Eruptive variables are rather diversified. It suffices to say that they include both supernovae that release a flare of $10^{49}-10^{52}$ erg once in a lifetime and red dwarfs of the UV Cet-type with frequent weak flares within $10^{27}-10^{32}$ erg and less. Qualitative similarity of the light curves of eruptive stars means only that all eruptive processes suggest a break in the smooth development of a star as a whole or its separate parts, though the mechanism of instability, energy source, and the triggering mechanism of the break can be quite different. Except for rather few in number variable types of R Corona Borealis and UX Orion, whose brightness sometimes weakens abruptly, the overwhelming majority of eruptive stars experience fast brightness increases, i.e., flares. However, sometimes, mainly in the popular science literature, all eruptive variables are called flare stars. But researchers of variable stars use this term only for dwarfs of middle and late spectral types of medium to low masses. In the 1960s, the term "flare stars" was applied mainly to designate the nearest stars in the solar vicinity, and stars with similar photometric features in more remote stellar clusters were referred to as flash stars. Now this term has practically disappeared, since the objects it labeled were found to be young Orion variables or brighter flare UV Cet-type stars of slightly earlier spectral type.

The term "flare UV Cet-type stars" appeared after the detection of transient flares on red M dwarf stars. Today, it is clear that the low luminosity of the stars creates favorable conditions for the detection of the UV Cet-type activity but does not favor the occurrence of this activity, which, being physically identical to the solar activity, takes place on dwarf stars in a wide spectral range from G to M stars. Transformation of the phenomenological term designating the observed phenomenon into an astrophysical term reflecting the physical essence of the process evidently illustrates the progress of astronomical knowledge. In this book, the term "flare UV Cet-type stars" (as well as reduced variants "flare stars" and " UV Cet-type stars") will be used for physical variables of the lower main sequence displaying solar-type activity. Finally, the same meaning will be put into the traditional term "flare red dwarfs", though, strictly speaking, "flare UV Cet-type stars" also include less numerous orange dwarfs and even less frequent yellow G dwarfs, similar to our Sun.

Manifestations of the activity of lower main-sequence stars, as on the Sun, are rather numerous and diverse. Certainly, first of all these are the "principal" local nonstationary phenomena, sporadic flares involving all layers of the stellar atmosphere. Other manifestations are cool spots observed on the stellar surface and the variability of large-scale structures of stellar atmospheres, chromospheres, and coronae. Since there are experimental evidences of structural arrangement of chromospheres, while the observations of coronae are represented within the loop model, the upper layers of stellar atmospheres should also be considered in the context of stellar activity. Finally, the data on long-term cycles of stellar activity, similar to the 11-year solar cycle, are accumulated.

The physical identity of the final cause of all phenomena of stellar activity of the considered type is doubtless. But observations reveal regular differences of average spectral types of dwarf stars that are characterized by different manifestations of this activity: flares and quiescent chromospheres of the coolest M stars are the most thoroughly investigated, early M and late K stars prevail among spotted dwarfs, long-term activity cycles are found mainly for early K and late G dwarfs. This regularity reflects observational selection: flares and chromospheric emission are better seen on the background of weak photospheric radiation of the coolest dwarf stars. To detect spots, i.e., to record slight distortions of the light curve of a star, the level of its regular brightness should be determined with high accuracy, which requires a brighter photosphere. Radiation of the photosphere will hide chaotic brightness bursts caused by flares. Finally, long-term activity cycles are found mainly in high-resolution spectrometric observations, which are applicable only to even brighter stars. X-ray emission of coronae, for which photospheric radiation is not a hindrance, is recorded from early F to late M dwarfs. The question on the degree to which observational selection masks the real dependence of the total level of various manifestations of stellar activity on their luminosity, mass, and age can be solved upon special analysis of the observational data.

Thus, we propose a wider astrophysical definition of UV Cet-type stars as compared to the above purely photometric definition: UV Cet-type stars are lower main-sequence objects that display the phenomena typical of solar activity.

1

Flare Stars in a Quiescent State

1.1 General Characteristics of Flare Stars

Flare stars of the UV Cet type are known in the solar vicinity and in several closest open clusters. Being absolutely weak or very weak objects, as a rule, these stars can be studied in detail at distances of not more than 20–30 parsecs. However, they represent a significant part of the stellar population: the spatial density of flare stars in the solar vicinity is 0.056 stars/pc^3, whereas the total density of the stars is only approximately twice as high (Shakhovskaya, 1995). The rapid decrease in the density of flare stars with increasing distance from the Sun is due to the observational selection of low-luminosity objects. This statement was confirmed by the special statistical study of Mirzoyan et al. (1988), who disproved the alternative statement on the existence of a real clustering of flare stars in the solar vicinity. Thus, the known UV Cet-type stars are a random sample of the stellar population of the Galaxy determined by the position of the Sun within our stellar system.

The list of known flare stars in the solar vicinity is presented in the Catalog by Gershberg et al. (1999). The previous catalog contained 226 objects including about 240 flare stars, since it listed the components of some binary and multiple systems separately, while other similar systems were presented as a single whole. The distribution of 168 stars, whose spectral types are known, is shown in the first row of Table 1. On the basis of spectral classification of about 2000 M dwarfs close to the Sun, Hawley et al. (1996) recently found that 105 of them were M0–M3 emission dwarfs and 208 of them are M4–M8 emission stars. The second row of the table shows the distribution of the objects according to the data of Hawley et al. (1996). Petit (1961) and Joy and Abt (1974) had already discovered the fast increase of the share of emission objects while proceeding from early M dwarfs to late dwarfs. Thus, Joy and Abt concluded that all dwarfs later than M5.5 belonged to the emission type. But Giampapa (1983) discovered that nonemission stars prevailed among M6 and later dwarfs. A more detailed consideration of the problem revealed that late Me dwarfs were young objects with low spatial velocities, while late M

Table 1. Distribution of flare stars over spectral types

Spectra	G0 – G9	K0 – K3	K4 – K8	M0 – M3	M4 – M8
Number of previously known UV Cet-type stars	9	18	24	53	64
Number of UV Cet-type stars in the Catalog by Gershberg et al. (1999)	10	19	25	146	212

stars were mainly objects with kinematic characteristics inherent in the stars of the old disk and halo (Giampapa and Liebert, 1986; Reid et al., 1995). According to Shakhovskaya (1995), the share of flare stars among dwarfs of the appropriate spectral types increases from 3% for early G to 30% for late M stars. According to Hawley et al. (1996), the share of emission objects among K6 is 1% and for M0–M3 it is close to 10%. The percentage monotonically increases up to 60% for M6 and then decreases; emission is observed only in 13 of 32 dwarfs of later subclasses. Such a complicated dependence is due to the combination of observational selection – flares and the strong chromospheres are more easily detected on spectrally later stars – with probable longer duration of the emission phase in lower-mass stars (Herbst and Miller, 1989; Hawley et al., 1996) and real decreasing activity of the coolest M stars. Here we would like to make some comments. First, Herbig (1956) already found a flare on the low-mass star VB 10 that was orders of magnitude stronger than the strongest solar flares. Later Linsky et al. (1995) recorded an ultraviolet flare on this star. Then, Fleming et al. (2000) recorded an X-ray flare on it. The variable H_α emission is suspected on the very cool and rapidly rotating dwarf BRI 0021–0214 classified as > M9.5 (Tinney et al., 1997), while on the dM9.5e star 2MASSW J0149090+295613 a flare was recorded with an H_α amplitude of equivalent width near 30 (Liebert et al., 1999). Secondly, according to Tinney (1995), as to the latest M dwarfs, the Catalog of the Nearest Stars CNS3 (Gliese and Jahreiss, 1991) is essentially incomplete. Thirdly, the discovered spottedness of very low mass stars (Terndrup et al., 1999; Krishnamurthi et al., 2001b) and X-ray flares on LHS 2065 (Schmitt and Liefke, 2002) and on LP 944-20 (Rutledge et al., 2000) evidence that this activity takes place on stars till the end of the main sequence and on brown dwarfs. This is confirmed by the existence of chromospheres on the components of the multiple system LHS 1070 (Leinert et al., 2000). Gizis (1998) discovered two M subdwarfs, Gl 781 A and Gl 455, with H_α emission: both were the components of binary systems and the properties of their chromospheres and coronae were similar to those of dMe dwarfs with solar metallicity. Two other M6.5 active dwarfs were found by Gizis et al. (2000a) in the near multiple systems. Hall (2002) detected variable H_α emission in L3 dwarf 2MASSI J1315309-264951. Finally, Gizis et al. (2000b) systematically studied the activity of the weakest main-sequence stars. Having examined 60 M7–L dwarfs, they found that near M7 the occurrence rate of H_α in the emission reached 100%, then it decreased to 60% for L0 and to 8% for L4 dwarfs, but the luminosity ratio $L_{H_\alpha}/L_{\text{bol}} = R_{H_\alpha}$ for these ultracool stars did not reach the values characteristic of earlier M dwarfs, and the ratio started decreasing for M6 and continued for late L dwarfs. From the H_α variations in the considered sample Gizis et al. (2000b) concluded that flare activity was common for M7–M9.5 dwarfs and up to half of their fluxes in H_α could be due to flares. On the other hand, Landini et al. (1986) discovered a strong X-ray flare on the single G0 V star π^1 UMa that was younger by an order of magnitude than the Sun, but its quiescent luminosity within 0.2–4 keV exceeded the appropriate solar value by two orders of magnitude.

The inhomogeneity of the above sample is primarily due to the accidental proximity of the stars of different origin and age to the Sun, which is evidenced by their heterogeneous kinematic characteristics (Kunkel, 1975a; Shakhovskaya, 1975; Hawley et al., 1996). Using these characteristics, Veeder (1974) ranked 15 of 33 flare stars as the old disk population, while Stauffer and Hartmann (1986), 9 of 25, respectively. Kinematic estimates of the age of different groups of UV Cet-type stars vary within 6×10^6–10^{10} years. Apparently there are members of different young kinematic groups among flare stars in the solar vicinity: the Hyades, the Pleiades, and Sirius (Poveda et al., 1995). The heterogeneity of the sample of flare stars is also due to the fact that the stars are attributed to the UV Cet-type variables using several independent criteria: optical, ultraviolet, and X-ray flares, powerful quiet chromospheres and coronae, and noticeable effects of spottedness. The efficiency of different selection criteria varies for stars of different spectral type.

Over 30% of the known UV Cet-type stars are members of binary and multiple systems, with the rate of occurrence increasing toward lower-luminosity objects (Pettersen, 1991). But among all spectral types there are definitely single-flare stars, thus the binarity is not a necessary condition for the considered activity type. Observational selection leads to the conclusion that the overwhelming majority of known binary systems comprising the UV Cet-type stars consist of pairs of flare red dwarfs: it is more difficult to find a flaring component in the systems comprising red dwarfs and higher luminosity stars, if the pair is not very wide. Sometimes, flare components in such pairs are found only during flare events (Arsenijevic, 1985; Shakhovskoy, 1993; Hünsch and Reimers, 1995).

Quiet M dwarfs are the slowest rotators among main sequence stars; their axial rotational velocity usually does not exceed 2 km/s (Marcy and Chen, 1992). But the axial rotational velocity of flare stars, as a rule, noticeably exceeds this value: for 40% of single UV Cet-type stars it is close to or slightly higher than 10 km/s. Thus, 11 of 29 dMe stars considered by Stauffer and Hartmann (1986) had a rotational velocity above 10 km/s, while of 170 dM stars only one or two had equally high rotational velocity. However, rotational velocities can be significantly higher in nonsynchronized pairs. For the dM1.5e star Gl 890, whose binarity has not been established, $v\sin i = 70$ km/s (Pettersen et al., 1987); $v\sin i$ of the K2 dwarf BD+08°102 that probably forms a pair with the white dwarf is about 90 km/s (Kellett et al., 1995). According to Bopp and Fekel (1977a) and Bopp and Espenak (1977), the strong chromospheric emission and spottedness occur at rotational velocities above 5 km/s. (This criterion is consistent with the opinion of Young et al. (1987a,b) that all M dwarfs involved in binary systems with rotational periods of less than five days are emission stars.) But this criterion is not rigorous: there are emission dwarfs with a rotation rate of about 4 km/s (Torres et al., 1985). The stars whose rotational velocity is lower than 1–2 km/s do not flare and have a very low activity level (Pettersen, 1991). However, if in G and K dwarfs the activity level correlates fairly well with the rotational velocity (Hartmann and Noyes, 1987), among the M stars this correlation vanishes. Thus, for the flare star Proxima Cen $v\sin i = 0.5$ km/s, for Gl 890 it is 70 km/s, while the values of $\log(L_{H_\alpha}/L_{\rm bol})$ that characterize the emission of the quiescent chromospheres of these stars differ only by 0.4. We note that the low $v\sin i$ of Proxima Cen is not due to the smallness of the angle i, but is actually conditioned by its slow rotation, since, according to Benedict et al. (1993), the period of its axial rotation is about 42 days. Based on high-resolution spectra of 99 studied red dwarfs, Delfosse et al. (1998) found that for 24 stars $v\sin i > 2$ km/s, all these stars appeared to be later than M3.5 and in kinematic characteristics were younger than the old-disk population. According to Basri (2001), after M5 the portion of rapidly rotating stars grows and approaches 100% for M9.5 and for L dwarfs. This circumstance may be due to longer deceleration of initial rotational velocities for lower-mass stars. The decisive role of rotation in the considered activity of lower main-sequence stars becomes obvious under systematic comparison of $v_{\rm rot}$ with the level of permanent chromospheric and coronal emission, which will be proved later in the book. It should be noted that, based on the analysis of M dwarfs in the Pleiades and Hyades, Terndrup et al. (2000) concluded that in the Hyades the average rotational velocity of the stars with masses of about 0.4 solar masses was lower by a factor of 2.5 than in the Pleiades, and this distinction was responsible for the lower luminosity of the chromospheres and coronae in the Hyades.

The distinctions between G0 dwarfs, whose structure is the most similar to the solar one, and the latest M dwarfs are very significant. Within this range of spectral types the effective temperatures vary from 6000 to 2500 K, stellar masses vary from 1 to 0.06 solar masses, the radii from 1 to 0.1 solar radius, and the luminosity from 1 to 0.0008 of the solar luminosity. In addition to these significant quantitative external distinctions, there are major qualitative distinctions of inner structure. G dwarfs are characterized by thermonuclear burning of hydrogen and by radiative transfer of the energy in the central part, while the convective zone occupies only the outer layers (about 30% of the stellar radius). On the other hand, in the stars close to the spectral type M5 with a mass of $0.3\,M_\odot$ and an effective temperature of about

3200 K, there is a transition to completely convective structures accompanied by the disappearance of the boundary between the radiative nucleus and the convective envelope on which, according to the modern concepts, the effective generation of the toroidal magnetic field occurs following the $\alpha - \omega$ dynamo mechanism. For even later dwarfs close to M9 the main sequence terminates: red dwarfs are replaced by brown dwarfs with a mass of about $0.07\,M_\odot$, the central stellar temperatures are not sufficient to sustain thermonuclear fusion of hydrogen, therefore they have no thermonuclear energy sources and "live" at the expense of gravitational compression. However, the transition from main-sequence stars to brown dwarfs can occur without significant change in the activity level, since all the diverse internal structures have convective transfer, at least in the outer regions of stars, and noticeable axial rotation, which result in the generation of magnetic fields of low- and medium-mass stars and all relevant activity phenomena. The recent discovery of the X-ray radiation from young brown dwarfs apparently supports these considerations (Neuhauser and Comeron, 2001) .

1.2 Photospheres

Because of the immense interstellar distances the only stellar surface we can observe directly is that of the Sun. To reconstruct the properties of "visible stellar surface" from the total observed radiation one should take into account two main factors: systematic limb darkening and irregular surface inhomogeneities due to starspots. The regions of the stellar surface free of spots are called quiescent photospheres.

1.2.1 Quiescent Photospheres

The structure, and consequently the radiation, of a stationary quiescent stellar photosphere are determined by three main independent factors: the effective temperature, gravity, and chemical composition of the radiating substance. These parameters are determined by means of spectral analysis of stellar spectra and are involved in theoretical calculations of the models of stellar atmospheres. Thus far, both analytical and synthetic operations have disregarded stellar activity. Otherwise, if the three parameters of flare and nonflare stars are equal, it is suggested that their structures are identical. Of course, this suggestion is valid only in a certain approximation, and recent studies reached the limits of its applicability.

The wide range of effective temperatures of flare stars produces considerable differences both in the spectral composition and the intensity of continuous radiation within the deep subphotospheric layers, and in the continuous and selective absorption coefficients governing the form of the observed absorption spectrum of photospheres. Thus, the intensities of the absorption lines of hydrogen and metals are comparable in the spectra of G stars. In the spectra of K stars, the lines and line blends of metals prevail. In the spectra of the latest K or M0 stars, the bands of TiO molecules arise, which soon strengthen in the later spectral subtypes. In the late M stars, the optical spectrum of the photosphere is determined by the molecular bands of TiO and VO, which bind the excess oxygen remaining after the binding of almost all the carbon in the CO molecule, while the infrared spectrum depends on the bands of water vapor, which binds hydrogen not involved in the H_2 molecules. All these molecules practically do not leave the spectral intervals with an intrinsic continuum of the outgoing radiation (Allard et al., 1997).

Spectral classification of active and nonactive dwarfs, i.e., the estimates of effective temperatures, is performed using the same spectral criteria, which evidences at least the essential similarity of the temperature structure of the two stellar types. It is known that initially the set of criteria of spectral classification was developed for the photographic range of the spectrum. Then, for cool stars the system was expanded to the near infrared region. Recently, Reid et al. (1995) proposed an algorithm for spectral classification of cool dwarfs based on the strongest band of TiO λ 7050 Å, which makes it possible to determine the spectral type of K7–M6 dwarf stars from the depth of the band to the accuracy of a half-subtype. This algorithm cannot be applied to hotter stars, since they do not have this band, while for cooler stars this band is saturated and weakens due to the increasing absorption of VO. However, the applicability range of this algorithm covers the vast majority of known UV Cet-type stars.

For G0–M5 main-sequence stars, the logarithm of gravity smoothly changes from 4.4 to 4.8 and there is no reason to expect systematic distinctions in this parameter for UV Cet stars and quiescent dwarfs of the same spectral types.

Finally, the strong differences in the chemical composition as on hot chemically peculiar stars are not known in flare stars (Hartmann and Anderson, 1977; Mould, 1978). Probably, the only well-established exclusion is the abnormally high content of lithium in the atmosphere of the flare star Gl 182 (= V 1005 Ori) (Bopp, 1974a; de la Reza et al., 1981). In the strict sense, probably this fact is not an exclusion but an extreme case, since recently it was suggested that there was a correlation between the content of lithium in the atmospheres of K dwarfs and the level of their activity (Zboril et al., 1997; Favata et al., 1997). A similar correlation for 110 F–G stars was established by Cutispoto et al. (2003). But such anomalies of individual elements with low absolute content do not affect the structure of the star and its atmosphere on the whole. Observations in the X-ray range (see Sect. 1.4.2) revealed a noticeably reduced, up to an order of magnitude, content of heavy elements in the coronae of a number of active dwarfs as compared to the Sun and the photospheres of these stars. The physical sense of this effect, the so-called FIP effect, is still not clear.

As to the general level of metallicity of the atmospheres of the stars under consideration, there are distinct spectral criteria for estimating the factor: as the metallicity decreases, the intensities of hydrid bands intensify as compared to oxide bands. This effect is easily explained: as the metallicity of a substance decreases, so does the relative content of both Ca and O that determines the selective absorption in the bands of CaO and H^-, which to a considerable degree governs the continuous absorption. Thus, the depth of the CaO oxide line, which in the first approximation is proportional to the expression $n(\text{Ca}) \times n(\text{O})/n(\text{H}^-)$, decreases. But the relative content of H at least does not decrease with the decrease in metallicity, thus the absorption in CaH bands that is proportional in the first approximation to the expression $n(\text{Ca}) \times n(\text{H})/n(\text{H}^-)$ at least does not decrease. These considerations are equally applicable to active and nonactive dwarfs. They formed a basis for the successful search for photometric effects of metallicity in the Hertzsprung–Russell diagram and in two-color infrared diagrams.

In the region of M dwarfs, the main sequence has a dispersion along the luminosity axis of about one and a half stellar magnitudes and the exact photometry shows that the stars with the metallicity of the young disk are the brightest in the infrared bands, the stars with metallicity of the old disk ($[M/H] < -0.5$) are weaker by $0\overset{m}{.}3$–$0\overset{m}{.}6$ and the dwarfs with $[M/H] < -1$ are weaker by $0\overset{m}{.}8 - 1\overset{m}{.}2$ (Leggett, 1992). (From CO bands in the near infrared region Viti et al. (2002) estimated $[M/H] = -1$ in the binary system CM Dra.) Although active and nonactive dwarfs are mixed, dMe stars are, on average, brighter by $0\overset{m}{.}34$ than dM stars in the diagram $(M_V, R-I)$ (Stauffer and Hartmann, 1986) and by $0\overset{m}{.}66$ in the diagram $(M_K, V-K)$ (Hawley et al., 1996). This small shift against the background of considerable dispersion was noted by Kuiper in 1942 and by Veeder in 1974. It can only partially be explained by the fact that some dMe stars have not reached the main sequence or by the unaccounted binarity, but it is most likely due to the increased metallicity of some dMe stars. Specially selected two-color diagrams support this conclusion (Stauffer and Hartmann, 1986; Leggett, 1992).

It should be noted here that kinematic characteristics largely account for the collective properties of stellar groups, while metallicity is a characteristic of an individual star, and there is no unique dependence between them. Nevertheless, while proceeding from the young disk to the old disk and further to the halo, we observe an increase in the dispersion of velocities accompanied by decreasing metallicity as the stellar age increases.

As stated above, the data on photospheric criteria of the activity of red dwarfs were obtained relatively recently. Hawley et al. (1996) discovered that the fine structure of the TiO bands in the infrared region of the spectrum correlates with the degree of activity of a star: TiO bands arising near the temperature minimum or deeper in the photosphere appear to depend on $L_{H_\alpha}/L_{\text{bol}}$. This spectral peculiarity cannot be related to the metallicity fluctuations. These fine effects can provide useful limitations for the models of dwarf stars.

For approximate estimates for the atmospheres of dwarf stars of the lower main sequence, one can use the values of effective temperatures and absolute

Table 2. Effective temperatures, luminosities and radii of stars

Spectral Class	Effective Temperature (K)	$L_{bol}/L_\odot$	$R_*/R_\odot$
G0	5940	1.25	1.06
G2	5790	1.07	1.03
G4	5640	0.92	1.00
G8	5310	0.66	0.96
K0	5150	0.55	0.93
K1	4990	0.46	0.91
K3	4690	0.32	0.86
K4	4540	0.26	0.82
K5	4410	0.22	0.80
K7	4150	0.14	0.72
K9	3940	0.096	0.66
M0	3840	0.077	0.63
M1	3660	0.050	0.56
M2	3520	0.032	0.48
M3	3400	0.020	0.41
M4	3290	0.013	0.35
M5	3170	0.0076	0.29
M6	3030	0.0044	0.24
M7	2860	0.0025	0.20
M8	2670		
M9	2440		

luminosities of stars reproduced from the study of de Jager and Nieuwenhuijzen (1987) (Table 2) and stellar radii calculated using these data and the relation

$$R_*/R_\odot = (L_{bol}/4\pi\sigma T_{eff}^4)^{1/2} = 3.33 \times 10^7 (L_*/L_\odot)^{1/2} T_{eff}^{-2} . \qquad (1)$$

In the atmospheric models of cool dwarfs, the dissociative equilibrium of more than one hundred of two- and three-atom molecules and the ionization equilibrium of a dozen atoms and ions are usually calculated. In addition to spontaneous and stimulated recombination and free–free transitions of tens of atoms, millions of atomic and molecular lines are taken into account (Allard and Hauschildt, 1995). These calculations are made in hydrostatic and LTE approximations and involve two principal difficulties: insufficiency of data on absorption coefficients of numerous molecules and dust particles in the photospheres of cool stars and the approximateness of the theory of convective energy transfer in subphotospheric layers. Over recent years, an amount of certain progress was achieved in both directions (Allard et al., 1997). Nevertheless, the estimates of effective temperatures of the coolest M stars in different modern models still differ by hundreds of degrees: for the dM9e star LHS 2924 the model of Kirkpatrick et al. (1993) for the red region of the photospheric spectrum yields 2750 K, for the best presentation of the infrared region it is 2625 K, while the molecular-dust model of the atmosphere by

Tsuji et al. (1996) yields 2130 K and their purely molecular model, 2110 K. According to Jones and Tsuji (1997), dust formation in stellar atmospheres should be taken into account already in constructing synthetic spectra within the range 6500–7500 Å for stars with an effective temperature below 3000 K. The recent discovery of the dust disks around active red dwarfs necessitates greater attention to the role of dust in such objects (Schütz et al., 2004; Liu et al., 2004). However, based on two- and three-dimensional radiative and hydrodynamic calculations, Ludwig et al. (2002) came to a fairly unexpected conclusion concerning the expected absence of qualitative differences between the granulation patterns on the Sun and on a dwarf with effective temperature of 2800 K and quantitative distinctions of the patterns in the intensity contrast, the horizontal scale of inhomogeneities, and the characteristic velocity of motion, which are consistent with the predictions of the mixing length theory.

A general representation of the optical spectra of the photospheres of flare red dwarfs is provided in the publications of Pettersen and Hawley (1987, 1989): they presented low-dispersion spectra of about 30 G9–M5 stars obtained from observations using the 2.1-m Struve telescope at the McDonald Observatory and measured intensity jumps at the heads of molecular bands of TiO, MgH, CaOH, and CaH.

Although stellar surfaces cannot be observed directly, using astroseismology methods Kjeldsen et al. (1999) presumed a granulation effect on α Cen A stars, whose physical parameters are close to solar ones. Indirect information on the inhomogeneity of the photospheres of active stars is provided by polarimetric observations (Alekseev and Kozlova, 2002, 2003a).

1.2.2 Starspots

Dark spots on the Sun, the first manifestation of solar activity, were probably detected by the priests of ancient Babylon with the naked eye. Sunspots were mentioned in Chinese, Japanese, and Korean medieval chronicles. Scientific investigation of sunspots was started with the invention of telescopes in the early 17th century. Mullan (1992) summed up the basic characteristics of sunspots known thus far.

Sunspots are the regions where strong local magnetic fields up to several kilogauss emerge on the solar surface. These fields substantially suppress convective transfer, the basic mechanism of upward heat transfer within the convective zone of the Sun. As a result, the effective temperature in the spot center is 3800 K, which is 2000 K lower than that of the quiescent photosphere, therefore sunspots look dark on its background.

The spots are surrounded by facular areas, brighter regions of the photosphere with weaker magnetic fields. The disturbed chromosphere, the regions of enhanced brightness in the CaII H and K emission lines, is located above the spots and the faculae. These features, already found during the initial optical observations of the Sun, are called active regions. They demonstrate noticeable variation with time. The subsequent radio, UV, and X-ray observations of the Sun have shown that active regions are related to the sources of increased emission in these wavelength ranges as well. In other words, active regions include the solar atmosphere throughout its height.

Due to the occurrence of dark spots, the summary flux deficit of the solar photosphere radiation reaches $0\overset{m}{.}002$, but there is some excess in overlapping with the additional emission of facular areas. Thus, the maximum bolometric luminosity of the Sun occurs at the epoch of the activity maximum, i.e., under the heaviest spottedness. However, a decrease in the luminosity of the visible solar hemisphere was recorded with confidence within the ACRIM experiment (Willson et al., 1981), when a group of spots passed over the solar disk.

At the beginning of the 11-year solar cycle the spots appear at heliographic latitudes about 35°, by the end of the cycle the regions of their appearance drift toward approximately the 5° latitude. The dynamo theory relates this visual effect to the differential rotation of the Sun, which generates a toroidal magnetic field and provides buoying of newly formed magnetic tubes upward to the solar surface.

The size of the smallest spots, pores, is of the order of 1000 km, which corresponds to the characteristic size of photospheric granules. Since the granules are convection cells near the solar surface, their properties are closely related to the convection characteristics. Large spots can be an order of magnitude larger, while the greatest spot groups cover up to 0.006 of the area of the visible solar hemisphere and their diameter achieves 10° in the heliocentrical coordinate system. The number of spots quickly increases as their sizes decrease. The lifetime of pores is several hours, while large spots can exist for up to several months.

Large spots have distinct photometric structure: they consist of a darker central part, the umbra, and a somewhat lighter periphery, the penumbra. The degree of reduction of the flux intensity $I_{\mathrm{umbra}}/I_{\mathrm{photosphere}}$ varies considerably with the wavelength: from 0.02 at about 4000 Å to 0.6–0.7 at about 2 µm. As a spot approaches the disk edge, this ratio decreases within the whole wavelength range due to a steeper temperature gradient in the spot as compared to the quiescent photosphere.

In the spot center, the magnetic field is vertical, but as observation moves away from the spot axis, a transverse field component arises. It is still not clear which factor determines the main characteristic of a spot, its magnetic flux. The strength of magnetic field and the spot size are not correlated, the field strength satisfies the condition of hydrostatic equilibrium of the magnetic flux tube and the quiescent photosphere surrounding the spot.

In 1949, Kron (1952) discovered spots on red dwarfs from the slight distortions of the light curve of the YY Gem binary system, which differed from the curve typical of eclipse systems. Both components of the YY Gem system are dMe flare stars.

Hall (1994) stated that the idea of stellar spottedness was as old as searches for the cause of photometric variability of stars. The point is that spottedness

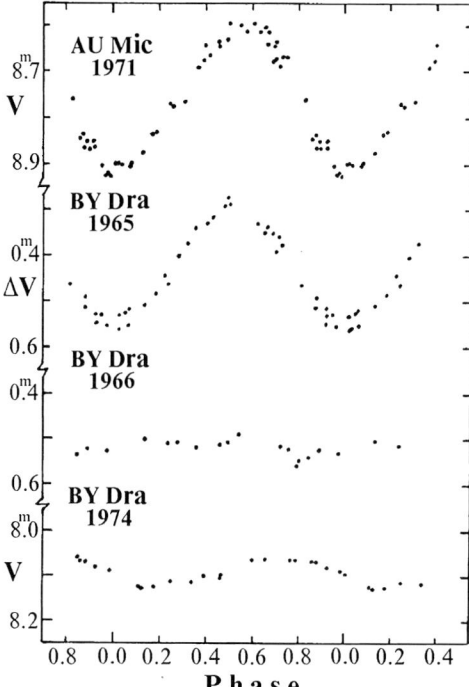

Fig. 3. Brightness variations of two red dwarfs due to spottedness (Rodonò, 1980)

is the first explanation of brightness variation offered by the discoverers of various types of variable stars starting from long-period variables. Pickering (1880) even proposed to explain all types of stellar variability by spottedness. Further studies soon disproved this global assumption and discredited the idea of stellar spottedness to the extent that when Kron did find the spottedness of the YY Gem system, this discovery did not impress variable-star investigators. Only 15 years later, when the interest in flare-emission dwarfs had become high enough and Chugainov (1966) found spottedness in the BY Dra binary system with similar components, did the studies of this phenomenon became widespread. The concept of starspots on red dwarfs had indisputable advantages over the competing models of eclipses and stellar oscillations and soon became conventional. Typical light curves of spotted stars are shown in Fig. 3. The term "BY Dra-type stars" has been introduced for spotted stars, though now it is clear that it does not mean a new variability type, but a new manifestation of the activity of red dwarfs. Hall (1972) was one of the first to relate the spottedness of cool stars to the phenomena similar to solar activity.

Since stellar surfaces cannot be studied directly, the analogs of the solar-activity regions and activity complexes are primarily used to analyze time variations of observed parameters and their correlations. It is obvious that in the case of surface inhomogeneities such time variations can occur only when the rotation axis of a star deviates noticeably from the line of sight. The surface inhomogeneities are investigated using spectral and photometric methods. The former requires dense series of high-resolution spectrograms for axial rotation period and a high signal-to-noise ratio (S/N). Thus, the method is applicable only to relatively bright objects, variable stars of the RS CVn type. Flare red dwarfs are too weak for the fine spectral analysis and, as a rule, are studied using less informative photometric methods.

Thus, the main photometric effect of red dwarf spottedness is periodical brightness changes in the optical range of wavelengths with amplitudes varying from one hundredth to two–three tenths of the stellar magnitude, which during 10–100 and more rotations of a star occur strictly synchronously with the axial rotation, while at larger time intervals the average brightness and its oscillation amplitude change sporadically, and phase shifts and slight variations of the period of these small-amplitude brightness variations occur. (Periodical brightness variations with an amplitude of $0^{m}_{.}01$ were recorded during the high-precision observations of Proxima Cen with the Hubble Space Telescope (Benedict et al., 1993). Two-day ground-based observations of BD+22°4409 with an axial rotation period of about ten hours detected a change in the brightness minimum depth by $0^{m}_{.}03$ (Robb and Cardinal, 1995).) Obviously all these effects can be related to the appearance, development, and decay of starspots, as well as to their drift over the surface of a differentially rotating star. At the same time, it is obvious that determining the characteristics of the surface inhomogeneity from the variations of the observed integral flux is a rather ambiguous task.

As a first step in solving this problem, Krzeminski (1969) "gathered" all inhomogeneities in one spot, "decreased" its temperature by 350 K, which corresponds to two spectral subtypes, and estimated the spot size required for the observed photometric effect: the obtained value was about 10% of the stellar surface. His result gave an impetus to constructing a number of spottedness models for red dwarfs.

It should be noted that shortly before the discovery of spots on YY Gem Kron (1947) found spottedness in the AR Lac binary system. The systems have many differences: YY Gem consists of two red dwarfs of the main sequence, while AR Lac, one of the most studied of the RS CVn-type systems, consists of enhanced-brightness stars that had undergone a noticeable evolution and left the main sequence. RS CVn-type objects will not be thoroughly considered in this monograph, but since the methods for analyzing the spottedness of both types of variables were developed in parallel, they have much in common.

1.2.2.1 Estimates of the Parameters of Individual Starspots

Quantitative characteristics of starspots from the observed light curve of cool stars have been the object of intense studies for 30 years: about three hundred publications are devoted to this subject. To avoid the situation when "an extensive literature substitutes research by cross-referencing to contradicting opinions of numerous authors rather than provides answers to thrilling questions" (L.N. Gumilev), we shall mention only those that were important for conceptual progress of the studies.

Evans (1971) constructed the first theoretical light curve for a star with a spot at the equator for the case when the rotation axis of the star lies in the picture plane, while the spot boundaries are set by the sections of meridians and parallels. Reducing the problem to the two-parametric statement, he successfully presented observations of the CC Eri spotted star for each of the three observational seasons using the spots stretched along the equator for many tens of degrees.

Bopp and Evans (1973) proposed a computational layout for the light curves of spotted stars with arbitrarily oriented rotation axes and with allowance for limb darkening. Within this model, they successfully presented all the above observations by Chugainov, Krzeminski, Evans, and the data of Torres et al. (1972). To present the light curves with a continuous brightness change throughout the entire axial rotation period, Bopp and Evans used an hypothesis on large nonsetting high-latitude spots, while to explain slight variations of the color index $(B - V)$ they used an hypothesis on the low temperature of spots. As a result, the light curves of CC Eri for three seasons and those of BY Dra for two seasons were successfully presented by 2000 K spots stretching up to the latitude of $65°$, the total area varying from 4 to 20% of the stellar hemisphere. Later, using the formalism developed in this publication, Bopp and Fekel (1977b) analyzed the observational data for another flare-spotted star, FF And, and concluded that these observations could be presented by a dark-spot model ($\Delta T \sim 500\,\text{K}$) with a longitudinal length of about $125°$ or by a "black" spot model ($\Delta T \sim 1800\,\text{K}$) with a length of up to $70°$.

Torres and Ferraz Mello (1973) developed the theory of light curves and color indices of stars with blackbody spots, reproduced the above observations of BY Dra and CC Eri, and employed the second dark spot to present the asymmetrical light curve of AU Mic, noting that the observations in the B and V bands were insufficient for unequivocal selection of the parameters of the two-spot model. Later, the two-spot approach was used to present a number of asymmetrical light curves of the RS CVn-type stars (Bopp and Noah, 1980).

Vogt (1975) made a correction to the Torres–Ferraz Mello model for the second component of a binary system and used the energy distribution in cooler star spectra as the energy distribution in the spot spectrum. (It is noteworthy that using comparative spectrophotometry of sunspots and late-type stars Badalyan and Obridko (1984) found that the sunspot should be attributed to M0, K5 or G8–K0 spectral types, respectively, depending on spectral details being compared, such as continuum and weak or strong lines.) Vogt developed Mullan's idea (1974) that the large longitudinal Evans spots were, in fact, the sequences of spot groups resulting from the dynamo mechanism.

He noted that within this model the stellar brightness variability could be governed by differential changes in the spot belt.

Friedemann and Gürtler (1975) performed extensive calculations for theoretical light curves of a blackbody star with one blackbody round spot for various combinations of the spot radius, its position on the star, and orientation of the stellar rotation axis with respect to an observer. Using this model, they found that at equiprobable spatial orientation of rotation axes and equiprobable location of spots on the stars only for 72% of stars with the spot of radius 45° and only for 44% of stars with a spot of radius 25° should the photometric effect exceed $\Delta V = 0\overset{m}{.}1$.

In 1970, the light and dark hemispheres of BY Dra had the maximum brightness over 10 years of observations. Assuming that during this season the light stellar hemisphere had no spots and, following the Torres–Ferraz Mello formalism (1973), Chugainov (1976) analyzed its brightness and color index $(B-V)$ variations over the decade and concluded that usually both hemispheres were spotted and at minimum stellar brightness the spots covered up to 60% of the area of the dark hemisphere. He explained the relatively small temperature differences between the quiescent photosphere and spots of about 400 K by the predominance of penumbrae in the spot areas.

Oskanyan et al. (1977) analyzed the two-decade photometry of BY Dra and concluded that for a period of about two years with dense observational series the stellar brightness could be presented by a train of six sinusoids with abrupt changes: average brightness of up to $0\overset{m}{.}3$, the amplitudes of periodical variations within $0\overset{m}{.}01$–$0\overset{m}{.}02$, the photometric period of up to 0.2 day, and the phase up to 130°. These features could be explained by disappearing old spots and new spots appearing at different latitudes, including the near-polar ones. Oskanyan et al. noticed that the estimate of the brightness of a spot-free star was extremely important for constructing the spottedness pattern, however, the values could be determined only under long-term photometric observations. It is not inconceivable that the underestimated level of the maximum brightness could sometimes lead to erroneous conclusions on the existence of hot spots.

Davidson and Neff (1977) carried out the first multicolor $BVRI$ observations of BY Dra and concluded that their data agreed with the model of the dark spot 200–300 K cooler than the photosphere and stretched longitudinally to 200–220°.

Following Kopal's formalism proposed for the light curves of eclipse systems, Budding (1977) developed an analytical theory for constructing the light curves for stars with one round spot. Adding an automated iteration procedure, he reprocessed Evans' observations for CC Eri and together with his own observations for YY Gem presented the brightness curves of these spotted stars by estimating the optimal spot parameters and probable errors. Using the Budding formalism, Oláh (1986) calculated the expected $B-V$ color indices for the M0 dwarf considering the temperature differences $T_{\text{photosphere}} - T_{\text{spot}}$, the spot size, the inclination of the rotation axis with respect to the line of sight, spot latitudes, and limb-darkening coefficients. Later, the Budding ideology was further developed by Banks et al. (1991).

Vogt (1981) noted that the above models could present the observed photometric effects only, but did not state and solve the strict inverse problem of mapping of inhomogeneous stellar surfaces. He proposed an algorithm that uncoupled the determination of temperature and spot geometry using the observations in the V and R bands and applied the Barnes–Evans relation associating the surface brightness and the color index. The Vogt algorithm yields the difference of color indices $\Delta(V-R)$ of the quiescent photosphere and a spot, as well as a certain function of the other parameters: the inclination of the rotation axis with respect to the line of sight, the limb-darkening coefficient, the size and form of the spot. Given a number of additional assumptions, the function can be used to estimate the spot geometry. One of the assumptions implies a spot-free bright hemisphere, i.e., precise determination of the absolute maximum of stellar brightness. Assuming the simplest geometry of a single round spot, Vogt used the Budding technique (1977) to determine the optimal dimensions and positions of such spots on BY Dra and constructed the first pattern of its spottedness evolution. According to his calculations, in 1965 a dark spot of radius 30–50° was at a latitude of about 40°, then it drifted by 20–30° toward the pole, in 1970–75 it disappeared leaving small hot spots in its place, then in 1977 a somewhat smaller dark spot appeared again at about 40° latitude. From the variations of the axial rotation period Vogt estimated the velocity of differential rotation of the stellar photosphere, which was almost coincident with the appropriate value for the Sun. The estimated temperature of dark spots on BY Dra was 600 K lower than the photospheric temperature. According to Stauffer's statistical colorimetric study (1984), for the spotted stars in the Pleiades $\Delta T > 700$ K.

It should be noted that Oskanyan et al. (1977), Vogt (1981), and Poe and Eaton (1985) considered photometric effects for both cool dark and hot light spots. However, the subsequent observations steadily pointed to reddening of stars during the phase of maximum spottedness, which unequivocally evidenced the occurrence of dark and cool spots. A long-living "blue" spot found in the Walraven photometric system on one of the most active red dwarfs AU Mic is apparently an active region with strong Balmer emission or a result of numerous weak flares (Byrne, 1993a). The anticorrelation of brightness in the V band and the $U-B$ color index of YZ CMi found by Amado (1997) could be due to the same cause.

A proper program for estimating the parameters of two spots with account of the energy distribution in the spectra of stars of different spectral types was developed at Villanova University (USA). The program was used to analyze a number of RS CVn-type stars with asymmetrical light curves (Dorren et al., 1981; Guinan et al., 1982; Dorren and Guinan, 1982b).

La Fauci and Rodonò (1983) developed a program to calculate theoretical light curves for stars with two different arbitrarily located blackbody round spots. They successfully presented the recorded light curves of the

RS CVn-type star II Peg; the curves were often asymmetrical and appreciably changed from one season to another. First, the program calculates a dense grid of theoretical curves in the total space of the sought parameters, then using an iteration procedure it restricts the parameter range, and thus selects the best model. A pattern of spottedness symmetric about the equator, i.e., including four spots instead of two, is considered.

Based on the Budding formalism and the Barnes–Evans relationships for a star with one or several round spots, Poe and Eaton (1985) developed a computer program for calculating theoretical light curves and color indices. The program accounted for the contribution of a secondary component to the total brightness of the system. It revealed a marked dependence of the calculated color indices on the limb-darkening coefficient. The analysis of BY Dra observations in 1965–80 with this program led to a conclusion that the observed photometric history of the system could be presented by the evolution of one cool spot, whose radius and longitude varied from 17° to 58° and from 50° to 83°, respectively, and the area varied from 2 to 23% of the stellar surface.

To analyze the photometry of some red dwarfs and RS CVn-type stars within the joint programs involving IUE, Rodonò et al. (1986) elaborated an interactive calculation technique for two-spot models based on the Friedemann and Gürtler formalism (1975). The calculation involved two stages: first, using the trial-and-error method the parameters of one spot were selected with respect to the χ^2 criterion, then the parameters of the second spot were estimated from remaining discrepancies. The large number of varying parameters provided a high-accuracy presentation of observations. Later, the program had become widely used for analyzing both symmetrical and asymmetrical light curves for both types of spotted stars (Fig. 4). On the assumption of equal temperature of spots, the observations shown in Fig. 4 were presented by the models with the following parameters: for BY Dra $\Delta T = 600$ K, the spot radii of 39° and 22°, the latitudes of 77° and 0°, the 100° difference of longitudes, the spot areas of 11 and 4% of the stellar surface, respectively; for AU Mic $\Delta T = 850$ K, spot radii of 37° and 22°, latitudes of 81° and 5°, a 90° longitude difference, and areas of 10 and 4% of the stellar surface.

Summing up the calculation results for starspot parameters obtained in some earlier studies and using the described technique, Rodonò (1986b) noted that within the brightness amplitude of $0\overset{m}{.}05$–$0\overset{m}{.}35$ the radii of typical spots varied from 10° to 32° and the ratio $T_{\text{spot}}/T_{\text{photosphere}}$ varied from 0.70 to 0.86. These ratios lie between the values characteristic of sunspot penumbrae and umbrae and closely approach the latter. Recording the effects of smaller spots requires much more accurate photometry.

Budding and Zeilik (1987) developed an iterative algorithm for determining the parameters of the two-spot model that enabled a reliability assessment of the resulting solution and a possibility of searching for minimum χ^2. This algorithm was applied to a number of systems of the RS CVn type. As a result,

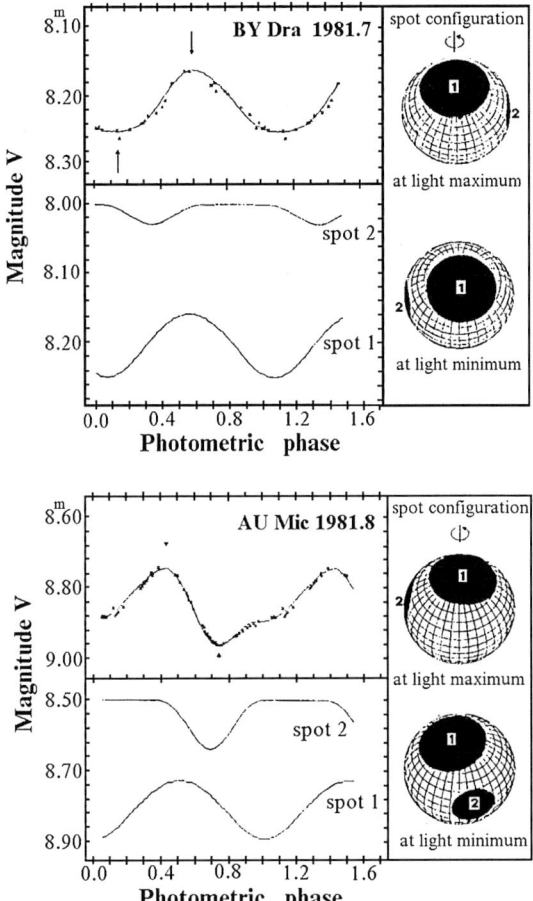

Fig. 4. The light curves of two red dwarfs presented within the two-spot model (Rodonò et al., 1986)

active longitudes and rather high latitudes were suggested for relatively small spots (Eaton, 1992).

Strassmeier (1988) developed an algorithm for simultaneous calculation of theoretical light and color curves and spectral-line profiles in the spectra of single and binary stars with any number of spots of arbitrary geometry from the digitized stellar surface map. The algorithm enables interactive adjustment of the spot map to photometric and spectral observational data.

Kjurkchieva and Shkodrov (1986) and Kjurkchieva (1987) elaborated an analytical theory of the light curves of spotted stars for the case when a round spot was completely hidden behind the stellar limb at a certain phase interval of axial rotation and the light curve had a horizontal section. If there is more than one clear minimum on the brightness curve, the algorithm makes it possible to attribute a proper spot to each minimum and to estimate the acceptability of the obtained set of parameters from the proximity of independent estimates of the inclination of the rotation axis. Later, Kjurkchieva tailored this procedure for the nonlinear limb-darkening coefficient and for a certain family of temperature profiles of spots (Kjurkchieva 1989, 1990).

Using the Kopal–Budding analytical approach, Vetesnik and Esghafa (1989) proposed an algorithm for calculating the light curves and the curves of radial velocities for different spottedness patterns: for a single round spot, for a finite number of round spots, and for spots with brightness varying along the radius. Somewhat earlier, Lodenquai and McTavish (1988) developed an algorithm for determining the parameters of the two-spot model with regard to the photometric umbra-penumbra structure of each spot and applied it to the five-year observational series of one of the RS CVn-type stars. Their model reflected the variations of light curves based on the migration of spots of invariant size over the star.

Reasoning from the suggestion that for a single spot passing over the stellar disk the brightness should be described by a negative sinusoid branch, Hall et al. (1989) proposed splitting of the observed light curves into sections and separated 7 to 9 individual spots with lifetimes from 3.5 months to 2 years in the light curve of the G8 V star V 478 Lyr. For RS CVn the radii of such spots appeared to be close to $14°$, i.e., much less than in the relevant two-spot models.

The international symposium held in 1990 in Armagh (Byrne and Mullan, 1992) was focused on up-to-date photometric and spectral methods for studying surface inhomogeneities on cool stars and astrophysical results obtained using the methods. In the introductory paper, Eaton (1992), in particular, reported the results of analyzing the spottedness of several RS CVn-type eclipse systems, for which the photometry of eclipses provided additional information about the spots.

Of particular interest were the results obtained under uniform processing of long-term series of uniform observations. Thus, Cutispoto analyzed the observations of the BY Dra star between 1981 and 1987 and concluded that within the two-spot models the light curves of the star could be presented by means of a large near-polar spot, which over the years drifted from $75°$ via $73°$ to $82°$ latitude and made a complete revolution over the longitude, and a 2–3-times smaller spot that over the same period drifted from $40°$ via $19°$ to $36°$ latitude and made three complete revolutions over the longitude (Rodonò, 1992).

Based on the long-term observational series, the latitudes of the spots found from the stellar spottedness models were compared with the axial rotation periods during the appropriate seasons. This enabled the estimations of differential rotation for a number of stars. As opposed to Vogt's estimate (1981) for the differential rotation of BY Dra that practically coincided with the solar value, the new estimates for this star appeared to be much lower (Panov and Ivanova, 1993). In addition, the values decreased toward the later spectral types, which contradicted the expectations based on the dynamo theory. In one case, the angular velocity of the near-polar zone appeared to be even higher than in the equatorial zone. Different ad hoc schemes were used to explain these features (Rodonò, 1986c; Lanza et al., 1992; Mullan, 1992).

The progress of the two-spot presentation of the photometry of spotted stars had a decisive effect on the determination of the character of spottedness from the spectral line bisectors: Toner and LaBonte (1992) sought the characteristics of two spots that would provide the observed variations of the bisectors.

Kang and Wilson (1988) developed an iterative algorithm for estimating the parameters of two spots as applied to the RS CVn-type binary systems. The algorithm based on the least-squares method accounts for the effects of tidal and radiation interaction of the components of the systems.

Strassmeier and Bopp (1992) analyzed photometric observational series of the chromospherically active star VY Ari and found that over two seasons its light curve was subject to changes at weekly intervals and could be presented by the following model of spottedness of the main component: there are always one large polar spot and from 2 to 4 large spots between 0° and 50° latitudes approximately at the same longitudes, some of them appeared over several days and their total area was equal to 10–15% of the stellar surface.

Having considered the long-term observational series of the BY Dra and EV Lac stars, Pettersen et al. (1992b) concluded that large near-polar spots existed on both stars for 14 and 10 years, respectively, but all spots on EV Lac were mostly on one hemisphere.

Scheible and Guinan (1994) studied the dG0 star EK Dra, which was in many respects similar to the very early Sun, and found that the observed light curve of the star could be presented by the two-spot model, the spots with a radius of about 20° being placed at middle latitudes and spaced in longitude for 140° with $\Delta T \sim 500$ K.

Eker (1994) developed a new analytical algorithm for calculating the light curves and the color indices of spotted stars with an arbitrary number of round spots and in numerous model calculations proved the dependence of theoretical curves on the accepted law of limb darkening and the size of spots, as well as on the inclination of the rotation axis to the line of sight and the spot latitude. He proved that the limb darkening should be taken into account, in particular, for the color-index curves, and that the linear approximation was sufficient for the darkening coefficient. Furthermore, Eker thoroughly considered the models of uniform filling of the equatorial band by different numbers of spots of identical size. On the basis of this experience using the trial-and-error method, within the two-spot model he presented the light curves of BY Dra and HK Lac, while for the latter he considered both the pure two-spot model and that supplemented by the equatorial spottedness band. In all cases, two spots under consideration were approximately symmetrical about the equator, approximating the solar situation, and the deviation of the theoretical light curve from the initial observational data did not exceed $0\overset{m}{.}0025$.

Later, Eker (1995, 1996) improved the algorithm introducing the search for the best solution of spot parameters using automatic iterations and studied the unambiguity of the determination of the parameters of one spot from the light curves in the V and R bands.

Butler et al. (1996) considered the observations of the YY Gem system in four photometric bands and using the Budding–Zeilik algorithm (1987) for eclipse systems found that, on average, $\Delta T \sim 600$ K.

Using the analytical Budding approach (1977), Kövári and Bartus (1997) studied the stability and reliability of the spottedness models obtained using this algorithm. They found that the required accuracy of the initial photometric data was $0\overset{m}{.}002$–$0\overset{m}{.}005$. The error in the estimate of the brightness of an unspotted star by $0\overset{m}{.}03$–$0\overset{m}{.}05$ resulted in an essential distortion of the distribution of spots, whereas the error in the inclination of the rotation axis by $10°$ resulted in considerable errors in the latitude and the size of spots and for an inclination of less than $20°$ it was impossible to get reliable values of the parameters.

Kövári (1999) showed that the light curves and $B - V$ color indices of BY Dra over 30 years could be satisfactorily presented if the spottedness of both components of the systems was considered.

Let us sum up the results.

The current analytical and numerical calculation methods for theoretical light curves and color indices of spotted cool stars integrate the intensity over the surface of stars of variable brightness due to accidental spots and systematic limb darkening. They solve the direct problem of photometric mapping. The fact that analytical algorithms deal with round spots, while numerical algorithms usually consider rectangular spots is not a principal distinction. When solving the direct problem, one can take into account any type of limb-darkening coefficient and the spot temperature profile. Both algorithms present observations with high accuracy.

As to the inverse problem on the restoration of a stellar surface map from the observed integral flux, despite the availability of the algorithms varying from the simple trial-and-error method to the image restoration developed by Banks et al. (1991) and the matrix operations proposed by Wild et al. (1994), the current situation leaves much to be desired. Certainly, there is no doubt that starspots are cooler by hundreds of degrees than the photosphere and occupy tens of per cent of the stellar surface. (Saar and Neff (1990) proposed a purely spectral method of estimating the temperature and size of starspots. The method is based on the comparison of the jumps of the heads of two molecular bands λ 7100 Å and λ 8860 Å with different temperature dependences in the spectrum of an active star with those in the spectra of presumably spotless stars. For BD+26°730 this photometry-independent method yielded $\Delta T = 750$ K and the spotted region of 20% of the disk. According to similar later studies, for BD+26°730 $\Delta T = 1330$ K and the surface filling factor by the spots f = 0.51, for EQ Vir $\Delta T = 830$ K and f = 0.43, and for LQ Hya

$\Delta T = 1540\,\text{K}$ and $f = 0.50$ (Saar et al., 2001).) But the results of traditional photometric mapping give rise to serious doubts.

The point is that despite the diversity of the applied computational algorithms, all the above studies of the spottedness of red dwarfs were focused on solving the problem stated by Krzeminski: to present the observed light curves by a minimum number of separate surface structures with determined individual parameters. But it is not obvious that the way to mapping a spotted stellar surface lies through solving the Krzeminski problem. Moreover, one can mention some principal shortcomings of the above models of spottedness of red dwarfs.

First, all considered solutions apply the idea of high-latitude or polar spots to interpret the light curves without noticeable horizontal section with maximum stellar brightness, which is valid for practically all observed light curves. However, there are no such spots on the Sun. Schüssler and Solanki (1992) and Granzer et al. (2000) showed that at fast rotation of a star, when the Coriolis force exceeded the buoyancy force of the magnetic flux tube, the spots should shift to the pole. But only a few red dwarfs are fast rotators, though on one of them, Gl 890 = HK Aqr with $v \sin i = 70\,\text{km/s}$, spots probably exist at high latitudes (Young et al., 1990). The results of the Doppler mapping of the RS CVn-type stars are usually cited in support of the existence of high-latitude and polar spots. In fact, many such variables are fast rotators and the close binarity of these systems can result in phenomena that differ qualitatively from those on the Sun. However, the reality of polar spots obtained with the Doppler mapping of such stars also raises doubts, since the rotational modulation of the radial velocity underlying this mapping method in the near-polar regions is negligible by definition (Byrne, 1996). Thus far, the Doppler mapping results have been published for several dwarfs only.

According to Hatzes (1995), the Doppler mapping of YY Gem comprising two dMe stars leads to the conclusion of a noticeable spottedness of both components at the latitudes of about $45°$, on weak equatorial spots, and the absence of spots at latitudes higher than $60°$.

Saar et al. (1992) constructed the surface temperature map of the single K2 dwarf LQ Hya, discovered dark spots with $\Delta T \sim 300\,\text{K}$ at middle latitudes, but did not find high-latitude spots. Strassmeier et al. (1993) and Rice and Strassmeier (1996) considered again the LQ Hya star on which Jetsu (1993) found differential rotation. They found that each of nine spectral lines used in mapping yielded a dark spottedness band within the range of latitudes from $-10°$ to $+35°$ with a mean temperature difference $\Delta T = 400\text{--}500\,\text{K}$, while for the existence of spots at high latitudes different spectral lines gave contradictory results. Probably the shift of the middle of the spottedness band from the stellar equator can be caused by the inclination of the stellar rotation axis to the sky plane, and in reality the spottedness regions form a structure symmetrical about the equator. In observations run two years later they found again a wide equatorial spottedness band between $-10°$ and $+35°$ latitudes (or two lines symmetrical about the equator with a small gap) with $\Delta T < 700\,\text{K}$

and slight polar spots even less distinct than before (Rice and Strassmeier, 1998). Using a new technique for image restoration, Berdyugina et al. (2001) revised the observational data for LQ Hya in 1993–99 and from nine spectral lines found that the star always had large active regions at latitudes of 50–70° and small spots of about 40° with $\Delta T \sim 900$ K. Recently, Kövári et al. (2004) studied LQ Hya using 28 temperature maps for 1996–2000. All the maps evidence the prevailing spottedness at low and middle latitudes from −20° to +50° and an absence of polar spots. This confirmed the conclusions of Berdyugina et al. on active longitudes on the star.

Earlier, Strassmeier and Rice (1998) analyzed 12 lines in the spectrum of EK Dra, which was very close in many parameters to the Sun, but its age was only about 70 million years and $v \sin i \sim 17$ km/s. In all lines they found with confidence the availability of spots near the equator and at middle latitudes in both hemispheres, while polar spots were found in nine of the 12 lines. The MHD calculations yielded the expected latitudes of the exit of the flux tubes of the toroidal field only within 25–65°.

Stout-Batalha and Vogt (1999) constructed the images of the surfaces of two rapidly rotating K dwarfs in the Pleiades. For both of them, spots at the high latitudes of 70–77° at $\Delta T \sim 800$–900 K and the absence of low-latitude spots were found. Earlier similar results were obtained by Ramseyer et al. (1995) for the K2 dwarf V 471 Tau, which makes up the binary system with a white dwarf and has an axial rotational velocity of 91 km/s. On the K star, they found a wide spottedness band in the range of 60–80° latitudes and small spotted regions near 40°.

On the other hand, in analyzing two rapidly rotating G dwarfs in the α Per cluster to study the surfaces of these rather faint ($\sim 11^{m}\!.5$) stars, Barnes et al. (1998) and Jeffers et al. (2002) developed the method of Doppler mapping using many spectral lines of echellograms. On each they found spottedness regions at low latitudes and near-polar spots. Lister et al. (1999, 2001) used a similar method and about 2000 photospheric spectral lines to map the rapidly rotating K dwarf LO Peg. They also made a conclusion on the existence of an intensely spotted band on the latitude $25 \pm 10°$, the absence of spots at middle latitudes and the availability of high-latitude spots that did not cover the pole, since an appropriate flat section was not observed in the center of the line profiles.

Near-polar spots on rapidly rotating red dwarfs and solar-type stars are being discussed (Bruls et al., 1999; Linsky, 1999; Schrijver and Title, 2001).

Secondly, the above photometric mapping algorithms often involve the search for hierarchical models, i.e., the surface structures, in which one can pick out a main spot and estimate its parameters, subtract its photometric effect from the observed light curve, and estimate the parameters of a second-rank spot – see, for example, Rodonò et al. (1986), Berrios-Salas et al. (1989), Banks et al. (1991). In some cases this procedure was continued till the third spot. Each step of the procedure requires 3–4 additional free parameters, thus it is obvious that these approximations can lead to a rather high reproduction of observations. However, as in the case of high-latitude spots, the resulting pattern does not resemble the solar spottedness even remotely. This is not only an oversimplification of the pattern of solar spottedness, it is not even reduced to this pattern in principle. Otherwise, the ideology of the above algorithms does not account for the a priori information that is of major importance for image restoration: it does not involve the requirement of the class of solutions, which includes the observed pattern of solar spottedness.

Thirdly, all one- and two-spot models have a pure photometric shortcoming. If a spot is placed at a not very high latitude, it is obvious that it will determine two measured photometric values: total attenuation of the stellar brightness, i.e., its maximum brightness in the epoch under consideration, and the amplitude of the rotational brightness modulation. Thus, there should be an unequivocal relation between the two values. An increase in the number of spots and limb darkening should somehow wash away the expected dependence. But the observations do not justify the expectations. Thus, Alekseev and Gershberg (1996c) collected the data on 140 observation seasons of 13 red dwarfs and established that in 127 successive seasons the signs of variations of maximum brightness and of the amplitude of periodical oscillations coincided only in 32 cases.

To overcome the shortcomings of the existing algorithms of photometric mapping of the spottedness of red dwarfs, a fundamentally new approach to the problem was proposed and developed by Alekseev and Gershberg (1996a–c, 1997b). The approach suggests a search for common properties of spotted regions instead of determining the parameters of individual starspots.

1.2.2.2 Estimates of the Parameters of Spotted Regions within the Zonal Model

In the simplest model of zonal spottedness, an ensemble of starspots is approximated by two dark bands symmetric about the equator with varying filling factor along longitudes. This model is described by four free parameters: the distance between dark bands and the equator, the band width, the ratio of the surface brightness of spots to undisturbed photosphere, and variation of the filling factor along longitudes. It is assumed that at a certain longitude the filling factor is equal to unity, i.e., spots occupy practically the entire band width, and at the opposite longitude it achieves a certain minimum, whose value is the fourth parameter of the model. Between these extremes the changes of the filling factor are assumed to be linear. In addition to the analogy with a washed-away pattern of the solar spottedness, the pure stellar arguments in favor of this model are the conclusions by Evans (1971), Bopp and Evans (1973), Vogt (1975), Bopp and Fekel (1977b) and Davidson and Neff (1977) on the extension of starspots along the equator for tens of degrees, as well as the reasoning of Mullan (1974) on the formation of such a structure through the dynamo mechanism. A geometrically similar scheme was considered by Eaton and Hall (1979) in analyzing the spottedness of the components of the RS

CVn-type systems: in two bands parallel to the equator, they specified such a distribution of small numerous dark spots that a summary effect of darkening along the bands was described by one cosinusoid period shifted along the ordinate axis for a unity. The photometric equivalence of a nonsetting high-latitude or near-polar spot and a nonuniform equatorial band of small spots was noted by Vogt (1981), Rodonò et al. (1986), Pettersen et al. (1992b), and Panov and Ivanova (1993). As they failed to find the rotational modulation of brightness of the dMe star FK Aqr in 1983, when its brightness was somewhat lower than in 1979, Byrne et al. (1990) formulated an hypothesis on uniform distribution of spots along the stellar longitude. Such a uniform distribution of spots along the equator and two active longitudes with large spotted regions were found by Messina et al. (1999b) on the young G dwarf HD 134319.

The computational algorithm for the direct problem within the model of zonal spottedness based on general correlations of Dorren (1987) was developed to analyze the observations performed in the $BVRI$ photometric bands. Assuming that the spots radiate as cooler stars, the expected relationships between the intensity ratios $\beta_\lambda = (I_{\mathrm{spot}}/I_{\mathrm{photosphere}})_\lambda$ were determined from the stars with well-known photometric data and radii

$$\beta_B = \beta_V^{1.7}, \quad \beta_R = \beta_V^{0.70}, \quad \text{and} \quad \beta_I = \beta_V^{0.40} \, . \tag{2}$$

The listed exponents are valid to the accuracy of one–two hundredths for G8–M6.5 stars. These relationships replace the Barnes–Evans relationships or the assumption on the Planck energy distribution in the radiation of spots that were used in the above algorithms. The limb-darkening coefficients obtained by van Hamme (1993) were used in the calculations. The measurements for brightness maxima and minima were taken in four photometric bands, the four parameters of the zonal spottedness model were determined from the excessive system of 8 equations using the least squares method.

Table 3 presents the calculation results for the parameters of the zonal spottedness models of EV Lac for 23 observational seasons (Alekseev and Gershberg, 1996c; Alekseev, 2000): the difference of the seasonal maximum stellar brightness and the absolute brightness maximum, for which its maximum brightness over all considered observation seasons is taken, the amplitude of periodical brightness variations ΔV, the distance along the latitude φ_0 from the equator to the spottedness band, the width of the spottedness band $\Delta \varphi$, the minimum filling factor within the band $f_{\min}$, the ratio of intensities $(I_{\mathrm{spot}}/I_{\mathrm{photosphere}})_V$ in the V photometric band and the spottedness of darker and lighter hemispheres in per cent of the area of the stellar hemisphere. The formal calculation accuracy of φ_0 and $\Delta \varphi$ is tenths of a degree. The same result $\varphi_0 = 0$ for all observations evidences that for EV Lac the sought four-parameter spottedness model degenerates to a three-parameter model: two sought bands symmetric about the equator merge into one equatorial band. The values of β_V found vary from 0.37 to 0.62 with an average

Table 3. The models of zonal spottedness of EV Lac after Alekseev (2000)

Epoch	ΔV_{max}	ΔV	φ_0	$\Delta\varphi$	f_{min}	β_V	S_{max}	S_{min}
1971.6	$0\overset{m}{.}13$	$0\overset{m}{.}11$	$0°$	$20\overset{°}{.}0$	0.42	0.54	29.2	19.4
1972.6	$0\overset{m}{.}05$	$0\overset{m}{.}14$	$0°$	$16\overset{°}{.}5$	0.07	0.53	21.8	8.6
1973.7	$0\overset{m}{.}00$	$0\overset{m}{.}12$	$0°$	$13\overset{°}{.}6$	0.00	0.57	17.6	5.8
1974.6	$0\overset{m}{.}11$	$0\overset{m}{.}07$	$0°$	$13\overset{°}{.}9$	0.46	0.51	20.8	14.2
1975.6	$0\overset{m}{.}30$	$0\overset{m}{.}01$	$0°$	$16\overset{°}{.}4$	0.90	0.37	27.6	26.0
1976.6	$0\overset{m}{.}04$	$0\overset{m}{.}02$	$0°$	$6\overset{°}{.}1$	0.54	0.62	9.4	7.0
1979.6	$0\overset{m}{.}08$	$0\overset{m}{.}08$	$0°$	$13\overset{°}{.}4$	0.33	0.53	19.2	11.6
1980.7	$0\overset{m}{.}06$	$0\overset{m}{.}08$	$0°$	$12\overset{°}{.}5$	0.25	0.55	17.6	9.4
1981.7	$0\overset{m}{.}10$	$0\overset{m}{.}06$	$0°$	$12\overset{°}{.}8$	0.47	0.52	19.2	13.2
1983.7	$0\overset{m}{.}08$	$0\overset{m}{.}06$	$0°$	$11\overset{°}{.}9$	0.41	0.54	17.6	11.4
1984.7	$0\overset{m}{.}06$	$0\overset{m}{.}11$	$0°$	$14\overset{°}{.}7$	0.16	0.53	20.0	9.4
1985.5	$0\overset{m}{.}11$	$0\overset{m}{.}05$	$0°$	$12\overset{°}{.}6$	0.55	0.52	19.4	14.4
1986.7	$0\overset{m}{.}06$	$0\overset{m}{.}12$	$0°$	$15\overset{°}{.}5$	0.14	0.53	21.0	9.4
1987.7	$0\overset{m}{.}09$	$0\overset{m}{.}06$	$0°$	$12\overset{°}{.}4$	0.44	0.53	18.4	13.0
1988.8	$0\overset{m}{.}13$	$0\overset{m}{.}00$	$0°$	$9\overset{°}{.}9$	0.94	0.53	17.0	16.4
1990.0	$0\overset{m}{.}10$	$0\overset{m}{.}08$	$0°$	$14\overset{°}{.}2$	0.39	0.51	20.8	13.2
1991.7	$0\overset{m}{.}11$	$0\overset{m}{.}02$	$0°$	$10\overset{°}{.}3$	0.76	0.54	16.8	14.6
1992.7	$0\overset{m}{.}11$	$0\overset{m}{.}06$	$0°$	$13\overset{°}{.}2$	0.50	0.51	20.0	14.2
1993.7	$0\overset{m}{.}12$	$0\overset{m}{.}02$	$0°$	$10\overset{°}{.}9$	0.76	0.53	17.8	15.6
1994.7	$0\overset{m}{.}09$	$0\overset{m}{.}02$	$0°$	$9\overset{°}{.}3$	0.71	0.56	15.0	12.6
1995.7	$0\overset{m}{.}12$	$0\overset{m}{.}04$	$0°$	$13\overset{°}{.}9$	0.66	0.58	22.0	17.8
1996.7	$0\overset{m}{.}10$	$0\overset{m}{.}06$	$0°$	$14\overset{°}{.}4$	0.51	0.58	21.8	15.6
1998.8	$0\overset{m}{.}13$	$0\overset{m}{.}04$	$0°$	$14\overset{°}{.}5$	0.68	0.57	23.2	19.2

of 0.53. Following Pettersen (1976), we take the efficient temperature of EV Lac equal to 3300 K and obtain for β_V the following differences between the photospheric temperature and the estimates of blackbody spot temperatures: from 370 K to 190 K with an average of 240 K.

Using the above calculation technique for the parameters of the zonal spottedness model, Alekseev (2001) uniformly analyzed the observations for 25 dwarfs over more than 340 observational epochs. The total number of known objects with the evidences of spottedness was over one hundred, thus, the sample analyzed was sufficiently representative. It covered dG1e-dM4.5e spectral types with rotational velocities from a few to 25 km/s.

The results prove that the observed variety of the light-curve parameters of spotted red dwarfs at maximum and minimum, i.e., the amplitudes of the rotational brightness modulation and the ratio of amplitudes $\Delta B/\Delta V$, $\Delta R/\Delta V$, and $\Delta I/\Delta V$, can be presented within the simplest four-parameter model of zonal spottedness with the differences O–C that do not exceed the observation errors at $\varphi_0 = 0$–$55°$, $\Delta\varphi = 0\overset{°}{.}5$–$34°$, $f_{min} = 0.00$–0.95, and $\beta_V = 0.03$–0.58. The models describe the general stellar spottedness, which covers from 2 to 50% of the total stellar surface. (It should be noted that the spottedness parameters of LQ Hya obtained within the framework of the zonal model do not contradict the results of the above Doppler mapping.)

The zonal model presents the observations without applying the hypothesis on large cool near-polar spots, although when calculating the model parameters only the natural restriction $\varphi_0 + \Delta\varphi \leq 90°$ was used and the near-polar spots were not excluded in advance.

Calculation of the parameters of the zonal model does not involve any hierarchical considerations.

For 10 of the 25 considered stars the photometric mapping was performed before using traditional models. Comparison of the mapping results for spot area and temperature calculated with the values obtained using the zonal models shows satisfactory agreement.

Since the average stellar brightness and the amplitude of periodical oscillations within the zonal spottedness model are determined by two parameters – the width of the spottedness band $\Delta\varphi$ and the degree of its uniformity f_{min} – that vary independently, the increase in the average brightness of a star from one season to another can be accompanied by an increase or a decrease of the amplitude of the rotational variation of its brightness observed in reality.

The obtained individual and average parameters of the zonal models were compared with the global characteristics of the stars: absolute magnitudes M_V, rotational velocities v_{rot} and the Rossby numbers Ro that are equal to the ratio of the axial rotation period of the star to the characteristic circulation time of convective vortices. Although the mass of a main sequence star is the main factor governing its absolute luminosity and the circulation velocity of convective vortices, and therefore the Rossby number functionally depends on M_V and v_{rot}, it is appropriate to compare the found parameters of the zonal models with these global characteristics independently.

Figure 5a shows six plots outlining some physically meaningful correlations with the global characteristics of the zonal model parameters: the latitude φ_0, the average spot latitude $\langle\varphi\rangle = \varphi_0 + (\Delta\varphi)/2$, the spot temperature, the temperature difference of the photospheres and spots, and the average degree of spottedness S (%); the encircled dot marks the middle position of the Sun in these diagrams, the vertical sections in the two upper charts represent the ranges of the solar values φ and $\langle\varphi\rangle$.

The following conclusions were drawn from the data presented in Fig. 5a. For the stars with $M_V > 7^m$ the calculations revealed merging of two spottedness bands assumed symmetric about the equator. For half of the hotter stars the calculations yielded two bands, as on the Sun. The stars with split spottedness bands demonstrate a tendency to the growth of φ_0 for brighter stars, including the Sun. The star BE Cet, the most similar to the Sun, displays the least value of $\Delta\varphi$, which corresponds to the situation that is the most similar to the solar one.

Comparison of $\langle\varphi\rangle$ and M_V shows that on all the stars the spotted regions are localized at low and middle latitudes. One can clearly see that on cooler stars spots tend to approach the equator: for late K and M dwarfs the average

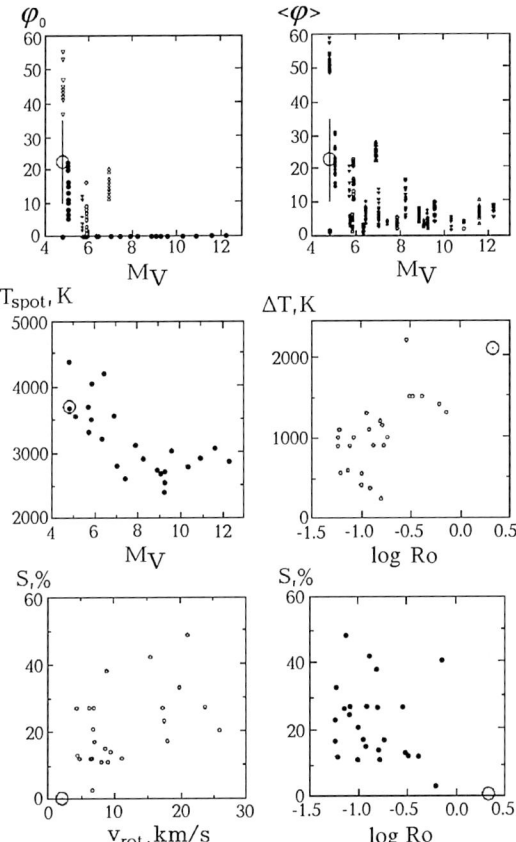

Fig. 5a. Comparison of the calculated parameters of the zonal models of red dwarfs (the latitude φ_0, the average spot width $\langle\varphi\rangle = \varphi_0 + (\Delta\varphi)/2$, the spot temperature $T_{\rm spot}$, the difference between the temperature of photospheres and spots ΔT, the degree of spottedness of stellar surfaces S) with the global stellar parameters (the absolute stellar magnitude M_V, the rotational velocity $v_{\rm rot}$, and the Rossby numbers) (Alekseev, 2001)

latitude of spots does not exceed 20°. On hotter stars, spots are shifted to middle latitudes and occupy a greater latitude range.

The calculation results for the zonal models show that the temperatures of spotted regions vary from 4000 K for solar-type stars to 2500–3000 K for the coolest M dwarfs. The correlation coefficient of the values on the relevant diagram is $r(T_{\rm spot}, M_V) = 0.69 \pm 0.08$.

Comparison of the temperature differences of the photosphere and the spot ΔT with global stellar parameters showed that, on average, this difference achieved 2000 K for hot and 300 K for cool stars. In addition, one can suggest the statistical growth of ΔT with the growth of the Rossby number: the correlation factor is $r(\Delta T, \log \mathrm{Ro}) = 0.67 \pm 0.05$.

The maximum areas of the spotted regions tend to grow with increasing stellar rotation rate and with decrease of the Rossby number. However, taking into account the above dependence $\mathrm{Ro}(v_\mathrm{rot})$, the two latter plots in Fig. 5a cannot be considered independent.

Thus, in all the plots of Fig. 5a the solar spottedness parameters fall into the regions occupied by the parameters of the stellar zonal spottedness models or into the natural extension of these regions. Otherwise, there is a tendency to join the parameters of the calculated stellar zonal models with the solar spottedness characteristics. This fact suggests that the zonal spottedness models actually reflect the essential properties of the surface inhomogeneity of red dwarfs.

The subsequent multicolor photometric observations in the Crimea were also successfully represented within the zonal spottedness model (Alekseev and Kozlova, 2001, 2002, 2003a,b). We would like to emphasize the qualitatively new result obtained by Livshits et al. (2003) and Katsova et al. (2003): for several most heavily spotted stars, for which the zonal models yielded separate spottedness belts, they found systematic shifts of the lower borders of the belts to the equator as the activity cycles developed simultaneously with the increase of the belt areas (see Fig. 5b). Otherwise, stellar analogs of the Maunder solar butterflies were found on the active stars. The rate of decrease of φ_0 is 2–3 times lower than on the Sun.

The zonal model requires a certain revision of the above results on the differential rotation of red dwarfs: now the observed axial rotation periods should be related to certain middle latitudes of the spotted bands, and since all spotted structures in the zonal models have a narrower range of latitudes than in the models for individual large spots, one should expect greater values of stellar differential rotation, i.e., the values will approach the solar values. Naturally, this remark is not valid for the direct estimates of differential rotation rates obtained using a recently proposed method for the analysis of the absorption spectrum of a star; but in this method some difficulties are concerned with symmetrically distributed spots (Reiners and Schmitt, 2002a,b).

The zonal model is free of the above fundamental shortcomings of the models with individual large spots, but it currently yields to them with respect to the mathematical formalism used. Being a model with 3–4 free parameters, it can present only symmetrical light curves, as the models with one large spot. This disadvantage can be removed in specifying a more complicated change of filling factor along longitudes: for the two-peaked change of filling factor, as for adding a second spot in the above algorithms, three additional varied parameters should be added. The transition from the estimates by the least squares method to the modern techniques of image restoration should increase the reliability of the estimates of the model parameters. Probably, the first work in this direction was the publication by Messina et al. (1999b) on the spottedness of the young G5 dwarf HD 134319 based on the algorithm of Lanza

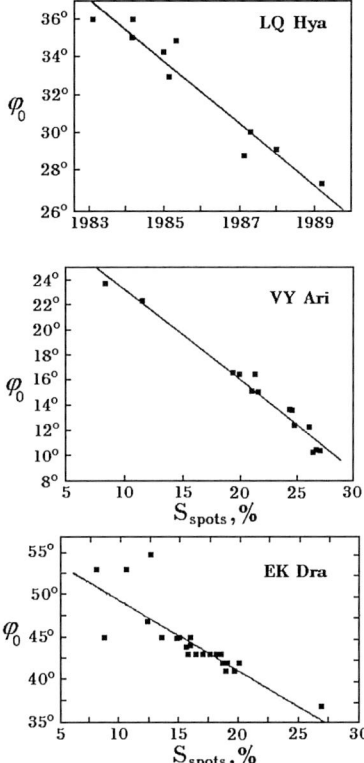

Fig. 5b. Stellar cycle evidences: the analogs of the Maunder solar butterflies obtained using the zonal spottedness models (Livshits et al., 2003) and Katsova et al. (2003)

et al. (1998). Using the methods of image restoration, the maximum entropy principle and Tikhonov's regularization on the basis of five-year observations with an automatic photometric telescope, they discovered a uniformly spotted equatorial region and two opposite active longitudes with large regions of spots preserved over the whole observation period.

1.2.2.3 Some Problems of Starspot Physics

Starpots are concerned with two fundamental problems of stellar physics: stellar magnetism and the radiation deficit of a spotted photosphere. Experimental data for the first problem will be considered further in this chapter. For many decades, the Parker hypothesis (1955a) on the emergence of the tubes of toroidal magnetic field was widely used in solving the problem of sunspot formation, though there is another point of view on the decisive role of convective motions (Getling, 2001). As to the second fundamental problem, its acuteness grows sharply on proceeding from the Sun, on which the area of spots hardly reaches 0.5% of the surface, to red dwarfs, on which spotted regions cover tens of per cent of their surfaces and the radiation deficit in the optical range reaches $0^{m}\!.5$ and more (Phillips and Hartmann, 1978; Hartmann et al., 1981; Bondar, 1996). Expanding the considerations of de Jager (1968) who stated that the radiation deficit of a sunspot could cover the energy requirements of a strong solar flare, Mullan (1975b) advanced an idea on the removal of the radiation deficit of stellar photospheres by Alfven waves. Hartmann and Rosner (1979) considered several opportunities for the energy channel of the photospheric radiation deficit: the transmission into another spectrum region or into an unobserved energy form carried out upward (Alfven waves?); pumping-over to the adjacent photospheric regions; conservation inside and redistribution with time. However, they were unable to choose one of them. Spruit (1992) concluded that this energy remained in the subphotospheric layers and was spent for a slight heating of the convective zone. According to his estimates, such a heating does not lead to detectable changes in the stellar structure. Besides, it is not clear for which values of the deficit this conclusion is valid.

The uniform construction of zonal spottedness models carried out in the Crimea for 24 red dwarfs during more than 340 epochs made it possible to estimate the bolometric deficit of the radiation of spotted photospheres of dwarf stars (Alekseev et al., 2001)

$$\Delta L_{\text{bol}}/L_{\text{bol}} = S(T_{\text{phot}}^4 - T_{\text{spot}}^4)/T_{\text{phot}}^4 , \qquad (3)$$

where S is the portion of the stellar surface covered by spots, and T_{phot} and T_{spot} are blackbody temperatures of the photosphere and spots, respectively. The calculations showed that for the most active dKe stars this deficit achieved 30% of bolometric luminosity and in absolute units varied from 3×10^{29} to 5×10^{32} erg/s.

The problem of radiation deficit is related to another problem of sunspots and starspots: the uncertainty in the equation of the energy of matter in spots does not allow one to precalculate their effective temperature and construct the model of internal structure (Mullan, 1992). On the other hand, the convection theory is related to lifetimes and minimum sizes of spots. If it is assumed that the lifetimes of spots are determined by the diffusion decay of magnetic flux tubes and the diffusion is determined by the convective turbulence, the expected and observed lifetimes of sunspots and starspots are rather close. This circumstance supports the validity of the existing phenomenological models of stellar convection.

In addition to the above fundamental problems of starspots, a number of special questions are still to be answered. Are large spotted regions indeed giant spots (or close pairs of spots of different polarity) that cover up to tens of per cent of a stellar surface or groups of smaller spots as the largest groups of sunspots? Having considered the stabilization conditions of sunspots, Mullan (1983) concluded that large spotted regions on red dwarfs should be formed by a great number of small spots that practically had no penumbra. Which factor determines the maximum size of starspots: total depth of the convective zone or giant convective cells that for whatever reason did not degenerate into granulation cells or the size of supergranules? Can large spots coexist with smaller spots? How small are the latter? How large are the smallest spots on red dwarfs?

It is believed that the magnetic flux tubes in the solar-type shell dynamo are formed on the interface of the convective zone and the radiative core. If a distributed dynamo starts acting on low-mass stars, do the spots of different size emerge from the same depth? How great is the depth?

Does differential rotation of a star destroy large starspots? If so, what pieces do they fall into? Can small spots merge into large spots?

What do the determined spot temperatures correspond to? Do the giant spots have an umbra-penumbra-type structure?

Although the list of unsolved questions is long, the fact of considerable spottedness of flare stars is sufficient to make an important physical conclusion: the magnetohydrodynamic situation on such stars should, as a rule, differ significantly from that on the Sun, where, as mentioned, the total area of the spotted surface does not exceed 0.5% of the surface. Due to the close location of starspots they form more often the so-called δ-configurations (Zirin and Liggett, 1987) and in their vicinity strong flares should occur more often than on the Sun. This is confirmed by observations. It is not clear whether isolated active regions are preserved at such a high density of spots and whether under these conditions the sympatic flares become typical rather than a rare phenomenon, as occurs on the Sun.

We note two qualitatively new observational results.

Apparently, the first spectral evidences of cool spots on a solar-type star were obtained by Campbell and Cayrel (1984). Using the cross-correlation, they found the molecular lines of TiO and CaH (typical of sunspots) in the spectrum of the G2 dwarf HD 1835 (= BE Cet) recorded with a signal-to-noise ratio $S/N \sim 1000$.

On the basis of 4-season spectral monitoring of the G8 dwarf ξ Boo A Toner and Gray (1988) found variations of the asymmetry and equivalent width of absorption lines with the axial rotation period of the star. First, they interpreted the data obtained as a result of passing of a surface inhomogeneity, the so-called "starpatch" covering about 10% of the disk, over the stellar disk, the temperature of the inhomogeneity being 200 K lower and the dispersion of granulation velocities 1.5–2 times as high as in the undisturbed photosphere. Later, Toner and LaBonte (1991) considered another interpretation of the observed spectral features of ξ Boo A assuming horizontal flows of matter analogous to the Evershed flow in sunspots. According to the calculations, the starpatch covering 10% of the disk at a latitude of about 30° is darker by 10–20% than the surrounding photosphere with a characteristic flow velocity of about 8 km/s and the same dispersion of the velocity explains the observed asymmetry and broadening of spectral lines. Different ratios of umbra and penumbra in the starpatch structure were considered. However, there is no certainty that the starpatches were really a new type of surface inhomogeneity and not a new manifestation of ordinary starspots or activity centers revealed by the new research technique.

Finally, it is appropriate to note that the long time span of the constant spotted region phase can be due both to the considerably long starspot lifetime and to the existence of active longitudes with regenerating starspots (Rojzman and Lorents, 1991).

1.2.3 Magnetic Fields at the Photospheric Level

The structure of the solar magnetic field at the photospheric level is very complicated. Apparently, it concentrates in small discrete flux tubes with diameters of 100–200 km that are irresolvable in ground-based observations. The tubes come to the solar surface and form different visible structures: thick ropes of tubes form dark sunspots, small groups of bundles are responsible for such regions of increased brightness as light faculae and the knots of the chromospheric network; the field strength in the tubes is 1–2 kG. The total magnetic field of the Sun is rather weak, its strength is of the order of 1 G. Main magnetic structures are local fields of sunspots, with a strength of 1.5 to 3–4 kG. Since sunspots are located at middle and low latitudes, the contribution of the regions with latitudes of above 50° to the total solar magnetic flux does not exceed 10%. Sunspots are surrounded by facular areas, the regions with total radiation higher by several per cent than the radiation of the quiescent photosphere. The facular areas are tens of times larger than the spots and live 2–3 times longer; therefore facular areas are observed in the absence of spots as well. The main contribution to the total magnetic flux on the Sun is produced by fields of 1–2 kG and the total area of the magnetized surface is 1–2%, in the faculae the filling factor increases to 5–25%. The structure of the magnetic field in the solar photosphere changes over a day or even faster. Apparently, the described complicated pattern is only a section on the photospheric level of the three-dimensional structure of the solar magnetic field. All these circumstances should be taken into account in discussing the magnetism of red dwarfs.

Three dwarfs were mentioned in Babcock's Catalog of Magnetic Stars (1958): ε Eri and 61 Cyg A among the stars with narrow lines showing a low Zeeman effect or its absence, and HD 88230 among the stars, in which a magnetic field is probable, but have not been confidently established. After the publication of the Catalog the magnetism of the main sequence cool dwarf stars was not studied; the interest in this problem resumed only 15 years later, after the discovery of spottedness of the stars.

The effect of sunspots on the solar luminosity is negligible. They are of particular interest to heliophysics as places where strong magnetic fields are localized. Therefore, when huge (with respect to solar scales) spottedness areas were found on red dwarfs, at which flare activity was stronger by orders of magnitude than on the Sun, a number of estimates of expected properties of starspots was published long before the magnetometric techniques reached a level sufficient for experimental study of the structures.

Assuming that the deficit of radiative energy due to spots converts into the magnetic field energy that is then radiated in the flares, Evans and Bopp (1974) estimated the characteristic field strength as tens of kilogauss.

It is known that, the sizes of sunspots correspond to the characteristic sizes of supergranules (Švestke, 1967). Assuming that a similar situation occurs on red dwarfs, Mullan (1973, 1974) calculated the size of such structure on stars and concluded that spots on them should achieve 50–60°. Rucinski (1979) estimated the size of granules and supergranules in a wide range of spectra and found that the size of supergranules decreased monotonically from G2 to M stars, the size of granules was maximum for K4 dwarfs, and by M8 both characteristic sizes decreased to 100–140 km. According to Mullan's calculations within his theory of cellular convection and the hypothesis on the removal of the radiation deficit from spots by Alfven waves, one should expect magnetic fields with a strength of up to 20 kG and a temperature below 2000 K on the surface of such spots. Given such strong magnetic fields in starspots, one should expect magnetic fields of 5 kG and higher in the stellar active regions. Such fields could ensure the heating of rather dense chromospheres by Alfven waves and finally the existence of dMe as opposed to dM stars (Mullan, 1975a).

Assuming that there was Biermann's battery effect in dMe stars, Worden (1974) estimated the expected strength of the magnetic field on such stars as 10^4–10^5 G. Accepting these estimates, Mullan (1975c), however, showed significant advantages of the dynamo mechanism over the battery effect.

The first magnetometric observations of F–G dwarfs with calcium emission after Babcock was undertaken by Boesgaard (1974). Among 8 objects, for which Zeeman spectrograms were obtained, only for the two coolest (G8 star ξ Boo A and K0 star 70 Oph A) was a longitudinal magnetic field of 140 and 115 G found at the level of 4σ and 9σ, respectively. However, the observation results for ξ Boo A were not confirmed later (Boesgaard et al., 1975).

Zeeman observations of the H_α region in the spectrum of the BY Dra star suggest that the found profile splitting can be associated with a magnetic field of up to 40 kG, although the absorption spectrum did not evidence a field of over several kilogauss (Anderson et al., 1976). These results were not justified later (Anderson, 1979; Vogt, 1980).

Mullan and Bell (1976) developed the theory of polarization of stellar radiation due to a large number of magnetesensitive absorption lines falling into the $UBVR$ photometry bands. Applying these calculations to the observations of BY Dra, they estimated the photospheric field as 10 kG; however, later the initial polarimetric results were not reproduced.

1.2.3.1 Zeeman Spectropolarimetry

Vogt (1980) observed about twenty red dwarf stars with the sensitivity of Zeeman spectropolarimetry increased by two orders of magnitude using the Reticon detector, but did not reveal an effective magnetic field with a strength confidently exceeding the measurement error of 100–160 G either in the absorption lines, or in H_α emission. Then he thoroughly studied all theoretical predictions for strong fields on the stars under consideration and noted that they were insufficiently rigorous and rather ambiguous. Thus, he concluded that his result admitted the existence of general incoherent magnetic fields on the stars with a strength of up to one kilogauss and magnetic fields of spots of up to 10–15 kG. However, the concept proposed by Mullan for dMe stars as magnetic dM stars remains valid for this essential revision of the expected magnetic properties of red dwarfs.

The significant achievements in magnetometric polarized-light studies of hot magnetic stars cooled off the interest of observers of cool stars, and over almost two decades only a few studies were undertaken in this field.

Brown and Landstreet (1981) measured the longitudinal magnetic field using Zeeman polarimetry in many spectral lines. They installed a diaphragm in the focus of the Palomar coude spectrograph, which cut out the absorption lines of the spectrum of a K star, and the Fabry lens accumulated transmitted radiation on a photomultiplier. The spectrum was scanned by swinging a flat-parallel quartz plate. They observed ξ Boo A, 70 Oph A, two K5 dwarfs HD 131977 A and 61 Cyg A, and two dMe stars, but did not reveal a field over twenty gauss; according to their estimates for ξ Boo A, B = +1 ±12 G. Thus, they reduced the upper limit of the longitudinal field found by Vogt (1980) by an order of magnitude.

Borra et al. (1984) used the CORAMAG spectropolarimeter updated for a Cassegrain focus. The polarization optics was inserted in the parallel beam, and the right- and left-polarized radiation was recorded alternately with a switching frequency of 100 Hz. The device was able to record many lines simultaneously. Over 11 nights they made 112 observations of the longitudinal field of ξ Boo A. Once they recorded a field of 25.0 ±6.4 G, then four nights later, of 72 ±30 G; on the other nights no significant field was found. Comparing these results with the measurements of the field of this star from the Zeeman broadening, Borra et al. concluded that there should exist several large magnetic regions composed of several hundreds of smaller two-pole structures.

In the Crimea, a stokesmeter with a CCD matrix as a detector was produced and mounted on the 2.6-m Shajn telescope for spectropolarimetric studies. To avoid nonuniform pixel sensitivity, the spectra were recorded in two positions of a quarter-wave plate differing by a 90-degree turn. Using this device, ξ Boo A was observed for 10 nights, but only during two of them was the total magnetic field with a strength of 46 ± 14 and 55 ± 15 G found (Hubrig et al., 1994). Then Plachinda and Tarasova (1999) used this device to observe several F–G stars. The observations with a spectral resolution of 30 000 and a signal-to-noise ratio of 300–450 provided an observation accuracy of up to several gauss. But on β Com, the only G0 dwarf in the observational program, no significant magnetic field was found. The first real result was obtained by Plachinda and Tarasova (2000) when they found distinct periodical variations of the total field of ξ Boo A from $+30$ to -10 G with a period of about 6 days that coincided with the axial rotation period of the star. The results obtained by Brown and Landstreet (1981) and Borra et al. (1984) fit rather well the phase curve constructed by Plachinda and Tarasova. Then, Tarasova et al. (2001) measured the total magnetic field for three solar-type stars – ε Eri, χ^1 Ori, and 61 Cyg A – and obtained the following results: during two of nine observational nights for ε Eri they recorded a field of -10 and $+21$ G at the level of 4–5σ, during two of five observational nights for χ^1 Ori they recorded a field of -8 and $+11$ G at the level of 2σ, and during two successive nights of 15 observational nights for 61 Cyg A they recorded a field of about 13 G at the level of 4–5σ. It should be noted that close values of magnetic fields for ε Eri, ξ Boo A, and 61 Cyg A exceed the relevant solar value by an order of magnitude, although the first two of the three stars are younger and more active than the Sun, the last one is older and has a similar activity level to the Sun.

Plachinda et al. (2001) compared the above 15 magnetometric measurements of 61 Cyg A by the Shajn telescope with 10 additional observations. They noted a rather confident change of the total magnetic field of the star within $+4 - -13$ G that was synchronous with the stellar rotation. But against the background of a smooth sinusoidal curve they found 3 or 4 sharp spikes that decayed over 2–4 days. The authors interpreted these facts as a result of the appearance of a large magnetic structure on the visible stellar hemisphere. First, a structure similar to the leading spot of bipolar group with a noticeable magnetic flux emerged. Then, an analogous structure of the type of a tail spot of the group with magnetic flux of the opposite sign appeared tens of hours later. Numerical modeling has shown that to obtain the observed effect it suffices to get such structures at a low latitude with a field strength of up to 3–4 kG and a characteristic size of about 10^{10} cm.

Donati et al. (1997) applied a new Zeeman–Doppler imaging (ZDI) technique for magnetometric studies, in which circular polarization was measured at separate sections of the line profile that corresponded to different sectors of the surface of stars with considerable rotation. In principle, this method provides the data on the three-dimensional structure of the field, since it allows one to determine radial, meridional, and azimuthal components of the field using the profiles of a large number of spectral lines in circularly polarized light. This method enables direct investigation of the evolution of magnetic fields of stars and their cycles but is insensitive to the strongest fields, since it does not detect dark spots. The sought effect amounts to a few thousandths of the measured values, thus it can be applied only to rather bright stars, for which one can obtain a high signal-to-noise ratio over a short phase interval. Among about 20 objects studied by Donati et al. (1997) there were two K dwarfs: LQ Hya and CC Eri. Up to two thousand lines were recorded in their spectra. The analysis of the lines proved that magnetic fields on these stars were composed of separate magnetic regions of different polarity and orientation, and their surfaces definitely were not hotter than the quiescent photosphere, i.e., these were not faculae but spots. Later, using more complete data, Donati (1999) concluded that large-scale structures of the azimuthal field with a strength of several hundred gauss and with different signs at different latitudes existed on the LQ Hya surface for many years.

1.2.3.2 Robinson's Spectrophotometry

In the article that appeared simultaneously with the above Vogt publication (1980) Robinson (1980) proposed a new method based on magnetometric principle for studying cool stars from the Zeeman broadening of lines in non-polarized spectra, which significantly expanded the possibilities of the studies. This method presents the observed profile of the absorption line as a sum: $F_{observ} = fF_m(B) + (1-f)F_q(B=0)$, where f is the portion of the stellar surface occupied by the magnetic region, $F_m(B)$ is the profile of the line occurring in the magnetic area with the field strength B, and F_q is the profile of the line appearing in the undisturbed photosphere. Such an expansion of the observed line profiles is possible when considering the lines with different magnetic field sensitivity, i.e., different Lande factors. The paper, in which Robinson used the Fourier transforms of the profiles and the assumption on normal Zeeman triplet of optically thin lines, formed a basis for real measurements of magnetic fields on late dwarfs. But these observational techniques and the ideology of data processing and calculations of $F_m(B)$ have evolved greatly over the past years.

The first observations applying the Robinson method were carried out in the λ 6843 Å and λ 6810 Å lines of neutral iron that belong to the same triplet (Robinson et al., 1980). Observations of the chromospherically active G8V star ξ Boo A and the sunspot revealed the expected differences in the profiles of these lines, whereas on the quiet solar surface these profiles, as one would expect, were identical. According to these observations, the magnetic field of up to 2900 ±550 G occupied up to 45% of the surface of ξ Boo A, whereas on the K0V star 70 Oph A the field of strength of up to 1800 ±350 G occupied up to 10% of the surface. The estimates of the field strength in sunspots and faculae yielded results close to those of traditional magnetometric observations of the

Sun. The later data by Brown and Landstreet (1981) and Borra et al. (1984) on small longitudinal fields did not contradict the revealed values of kilogauss and provided good mutual compensation of circular polarization from numerous local stellar magnetic fields with different polarity. Since a high strength of the magnetic field is observed not only in relatively small sunspots, but also in more extensive facular areas of active regions and in the bright chromospheric network, at first it was not clear whether the magnetized regions of the stars should be identified with large starspots or extended structures of higher brightness.

Marcy's study (1981) of another couple of FeI lines – λ 6173 Å and λ 6241 Å – did not confirm the presence of a noticeable magnetic field on ξ Boo A, although using the observations of a sunspot he tested the validity of the measurement technique. Marcy concluded that the point was the variability of the magnetic field of ξ Boo A. This conclusion was supported by the known evidences of the variability of the emission of the CaII K line, and the absorption of HeI λ 10830 Å line, and the X-ray emission from the corona. Qualitative estimates showed that the detectable field had to disappear if its strength reduced to B = 1000 G or the filling factor f reduced to 0.06 of the stellar hemisphere surface.

Marcy (1984) compared different computational versions of the analysis of the Zeeman broadening, considered probable sources of errors in the estimates of the field parameters, and summed up the results of his magnetometric observations of the FeI lines λ 6173 Å and λ 6241 Å at the Lick Observatory. For 19 out of 29 program G0–K5 dwarfs he revealed magnetic fields within 600–3000 G and the filling factor of the visible hemisphere varying from 50% to 89%. Such a high factor is typical of sunspots, i.e., the stars should be considered as completely covered by spots. Marcy found that the regions with magnetic fields on K dwarfs were definitely larger than those on G stars, but he did not reveal a dependence between the field strength and the spectral type. The magnetic fields measured by Marcy on ξ Boo A and 70 Oph A were noticeably weaker but had a higher filling factor than those found by Robinson et al. (1980). For two chromospherically active dwarfs Marcy suggested field variations within a day. G dwarfs prevailed among the stars that did not show any evidence of a magnetic field. The magnetic flux proved a statistical dependence on the rotation rate of the star as $v_{rot}^{0.5-1.0}$.

Gray (1984) generalized the Robinson algorithm for simultaneous analysis of several spectral lines taking into account some effects of radiative transfer in the lines. Using the 2.1-m Struve telescope of the McDonald Observatory for magnetometric observations of 18 F–G–K dwarfs, he found on seven G6 and even later stars magnetic fields of 1.9–2.4 kG and a filling factor of 25–40%.

The brightest flare star AD Leo was the first dMe star on which the magnetic field was found and measured from Zeeman broadening. The magnetometric observations were performed by Saar and Linsky (1985) on a Fourier spectrometer mounted on the 4-m Mayall telescope at the Kitt Peak Observatory. The observations were performed in the infrared region of the spectrum

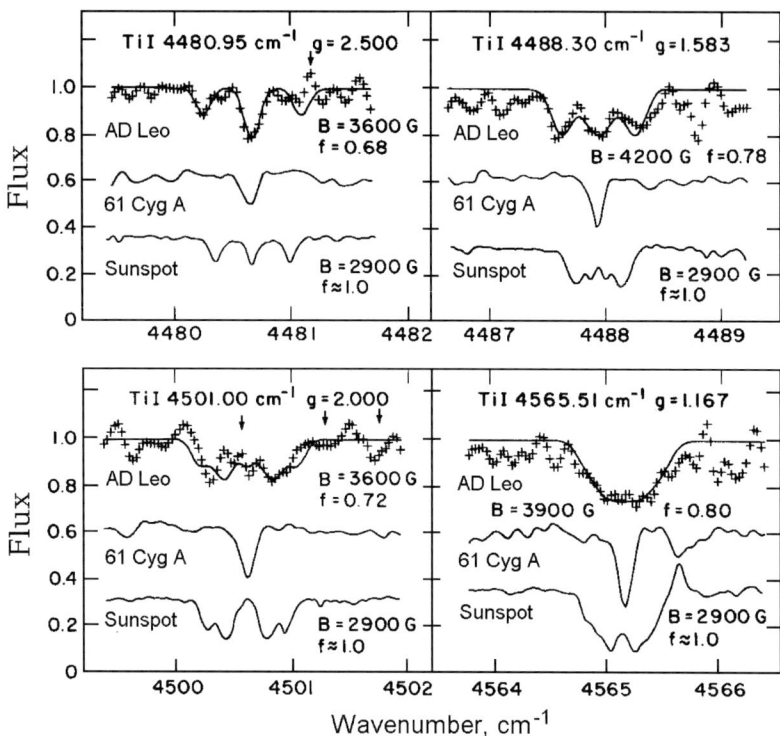

Fig. 6. The spectra of the AD Leo flare star, the nonactive star 61 Cyg A, and a sunspot in the region of IR lines of neutral titanium with different magnetic sensitivity (Saar and Linsky, 1985)

near 2.22 μm, where Zeeman splitting proportional to the squared wavelength was more noticeable. Saar and Linsky used five lines of neutral titanium and found magnetic fields with B = 3800 ±260 G and f = 0.73 ±0.06 (Fig. 6). The observational results for a sunspot and the nonactive star 61 Cyg A illustrate the noticeable magnetism of AD Leo (Fig. 6). Saar and Linsky showed that the revealed field strengths corresponded to the gas pressure of a quiescent photosphere, while the greater area of the magnetized region should be associated with high-efficiency generation of magnetic-flux from a rapidly rotating star through the dynamo mechanism.

Then, Saar et al. (1986a) published their magnetometric observations for the known flare and spotted dK5e star EQ Vir. Zeeman broadening was studied using the 4.5-m multimirror telescope from the magnetosensitive line FeI λ 6173 Å and nearly NaI, CaI, and FeI lines with lower Lande factors. For the first time, radiative transfer in the lines was taken into account, while the blends were thoroughly considered by comparing the lines with the spectra of the nonactive star 61 Cyg A. The calculations were performed for the LTE

Milne–Eddington atmospheric model on the assumption that the atmosphere parameters were identical in the magnetic and nonmagnetic regions. As a result, the strength of the field covering up to 80 ±15% of the stellar surface was estimated as 2500 ±300 G. These results combined with the earlier data on the field absence obtained from the polarized-light measurements evidenced the complicated topology of the magnetic field and numerous small bipolar structures. Similarly to AD Leo, the field strength corresponded to the gas pressure in the photosphere, while the greater average field $\langle B \rangle = fB$ corresponded to the greater rotational velocity of the star.

Using the above technique, Saar et al. (1986b) analyzed 11 spectra of the active K2 dwarf ε Eri obtained within two months in 1984. For the λ 6173 Å line, which is the most sensitive to the magnetic field, the comparison profile was obtained from the nonactive star spectrum. As opposed to considerable changes in the field strength and in the filling factor at approximate steadiness of the magnetic flux found by Marcy (1984) from 10 measurements made over two years, Saar et al. established variations of the field within 1.7–2.3 kG and the filling factor within 7–15%, but did not confirm the magnetic-flux constancy suggested by Gray (1984). They found that the changes of the measured parameters occurred at times that were shorter than the rotational period. When these measurements were supplemented by the results of magnetometric analysis for five other G–K dwarfs, the assumption on the proximity of measured field strength and their expected values, given equal field and gas pressure in the quiescent photosphere, was confirmed again (Saar and Linsky, 1986).

Gondoin et al. (1985) thoroughly studied the spectra of G8 stars in photometric H and K bands and found that only 4 nonblended lines of neutral iron near 1.56 µm matched the observations of the Zeeman broadening. But no manifestations of magnetic field were found on ξ Boo A. Explaining this result by the field variability, they noted that the observations in the optical and infrared ranges could result in incongruous conclusions, because the optical range was characterized by a low contribution of spots in the total radiation and fields dominating in the faculae, whereas in the infrared range the relative contribution of spots increased and for substantial spot fields they could become dominant in the total signal of the magnetic field.

Applying the Fourier spectrometer technique tested in the observations of AD Leo (Saar and Linsky, 1985), Saar et al. (1987) carried out magnetometric studies of a number of M dwarfs. The magnetic-field-insensitive FeI line was used to determine the parameters of nonmagnetic broadening of the lines and these parameters were used to calculate theoretical magnetosensitive profiles. Based on the observational results, Saar et al. established that five of the six dM stars showed no sign of the field, i.e., B < 700 G or f < 0.20, but they obtained B = 2.5 kG and f = 0.20 for the dM1 star Gl 229 and B = 2.5–5.2 kG and f = 0.6–0.9 for the flare dMe stars AD Leo, EQ Vir, BY Dra, AU Mic, and EV Lac. Combining these results with the measurements for magnetic fields on K dwarfs, Saar et al. confirmed the compliance of B with the gas pressure in the photosphere within 1–5 kG and the distinct growth of B for the stars of later spectral classes. The filling factor f correlates with the angular rotation rate of the star Ω rather than with the spectral type. The correlation of Ω with the averaged field fB is even closer. Probably, for axial rotation periods longer than 4 days f grows linearly together with Ω, then f is saturated at a level of 0.8. This relation and the observed strong fields on completely convective stars AD Leo and EV Lac did not fit the dynamo theory proposed by Durney and Robinson (1982). The saturation of f was suggested by the Skumanich and McGregor dynamo theory (1986), f being in a better correlation with the Rossby number than with Ω, whereas strong fields in completely convective stars required use of the distributed dynamo idea. Nevertheless, some important conclusions were based on the established correlations. First, if the strength of magnetic fields is governed by the condition of hydrostatic equilibrium of the field flux tubes and the ambient photosphere, the values of B are independent of the specific dynamo mechanism. Secondly, the efficiency of the acting dynamo mechanism and the variations of stellar magnetism are manifested in the changes of f (Saar, 1987). Later Saar (1994), using new observational data and improved analysis technique, determined more precisely the parameters of the magnetic field of active M dwarfs AU Mic, AD Leo, and EV Lac: B = 4.0–4.3 kG and f = 0.55–0.85.

Hartmann (1987) analyzed different estimates of magnetic-field parameters from the Zeeman broadening and concluded that since the lines with different relative depth were compared, both B and f could be systematically overestimated, which did not occur if the lines of one multiplet were compared. On the other hand, comparison of lines in the spectra of different stars can yield erroneous conclusions, since the effect of Zeeman broadening is influenced by the micro- and macroturbulence velocities and stellar rotation rates, which can hardly be identical for different stars. Finally, downward of the main sequence the lines of neutral metals become stronger and wider, therefore on M dwarfs only the fields that are stronger than those on the Sun can be found. Saar (1987) noticed some shortcomings of the first estimates of B and f of the field made using the Zeeman effect, which particularly disregarded weak blends distorting the profiles of most lines in the spectra of late stars. He believed that Marcy (1984) could discover the magnetic field in the nonactive star 61 Cyg A from the line λ 6173 Å because he disregarded CN and TiO blends. According to Saar (1987), the obtained high values of f evidenced that the magnetized regions were the analogs of the solar faculae and the bright network, rather than cool spots. This explained the significant variations of f at low photometric variability of many active dwarfs. By studying 29 G0–M4.5 dwarfs Linsky and Saar (1987) found an anticorrelation of B and T_{eff}, an independence of B from Ω and the Rossby number, as well as an independence of f from T_{eff}.

Using the 2.5-m telescope of the Mount Wilson Observatory, Bruning et al. (1987) studied seven late K and early M dwarfs. In six of them, magnetic fields varied from 750 to 2000 G and f varied from 0.20 to 0.65. Using the nonmagnetic line FeI λ 6180 Å they found the parameters of nonmagnetic broadening of the lines, then using the line λ 6173 Å and the criterion χ^2 they determined the optimal values of B and f.

Saar (1988) used and thoroughly analyzed the analytical solution of the transfer equation with the source function linearly depending on the optical depth and on the assumption of constant ratio of opacities in the line and in continuum throughout the atmosphere depth. Unlike Saar, Marcy and Basri (1989) developed the theory of magnetometric observations, introducing the numerical LTE stellar atmosphere model, in which all parameters could vary with depth, and the transfer equations for individual Stokes parameters. From the magnetic-field-insensitive line of neutral iron λ 7748 Å they determined the content of iron and the turbulent velocity in the stellar atmosphere and then analyzed the line λ 8468 Å sensitive to the field, adjusting the observed profile to the theoretical one by varying B and f. The observations were executed at the 3-m Lick telescope. Six of the 11 examined late G and K dwarfs distinctly displayed the Zeeman broadening, the strongest effect was found for ξ Boo A (B = 1600 G, f = 0.22) and ε Eri (B = 1000 G, f = 0.30). By the example of two other stars it was shown that among the late K dwarfs there were definitely stars with very weak fields, although the old and slowly rotating star 61 Cyg A, on which Saar and Linsky (1985) did not reveal any sign of magnetism, demonstrated again a clear presence of the field. A field of above 1600 G and a filling factor of over 30% were not found, though higher values of both parameters were measured before.

Saar (1991) performed magnetometric observations of two interesting K dwarfs. The rapidly rotating BD+26°730 star is of interest because its 60-year activity cycle was photometrically established and it is observed almost from the pole, thus the observed variations are primarily due to the proper evolution of the surface structures, not due to the rotation of the star itself. The analysis of the profiles of the magnetosensitive line λ 6173 Å and nearly low-sensitivity FeI, CaI, and NiI lines in the spectrum of the star and of the nonactive dwarf HD 32147 revealed a good agreement between the lines profiles with small Lande factor and noticeable broadening of the λ 6173 Å line, which can be attributed to a field with B = 2600 G and f = 0.5. Over 8 years, the brightness of this star decreased by $\Delta B = 0^{m}.15$, but the noticeable increase in its spottedness was not accompanied by any noticeable change of the intensity of the H_α line and ultraviolet lines of its chromosphere and transition region. For another interesting star HD 17925, the youngest star for which magnetometric observations were successful, Saar (1991) found B = 1500 G and f = 0.35.

Combining the principle of Doppler mapping with the magnetometric analysis based on nonpolarized spectra, Saar et al. (1992) studied the spectra of the rapidly rotating K2 dwarf LQ Hya during seven phases of the rotational period. They considered the profile variations of five FeI, CaI, and NiI lines, including the magnetosensitive FeI line λ 6173 Å and constructed temperature and magnetic maps. The maps demonstrated a rather good correspondence of dark and magnetic regions with fB = 1.0 kG. To match this result with the above pattern, Saar et al. suggested that light and nonmagnetic regions were the regular photosphere, while magnetometric observations showed photometrical suppression of light faculae by dark spots, in the center of which fB reached 2.5 kG.

The first evidences of the complicated distribution of the magnetic field strength were obtained by Saar (1992) in analyzing the infrared TiI lines in the spectrum of AD Leo using the technique described above (Saar and Linsky, 1985). The obtained line profiles could not be represented by one pair of f and B describing the two-component photosphere with uniform magnetic and nonmagnetic regions throughout the star: B = 3.5 kG and f = 0.30 in the 0.28 phase, B = 2.7 kG and f = 0.45 in the 0.60 phase, and B = 3.0 and f = 0.40 in the 0.91 phase (Saar et al., 1994c). The observations could be satisfactorily presented either by two types of magnetic regions with B_1 = 2.4 kG, f_1 = 0.45 for faculae and B_2 = 5.0 kG, f_2 = 0.30 for spots, or by a magnetic field with a vertical strength gradient, which was first considered by Grossmann-Doerth and Solanki (1990) in connection with the measurements of stellar fields. One should remember that large sunspots are characterized by high field strength and low vertical gradient, whereas individual thin flux tubes are distinguished by low dispersion of the strength at a fixed level and a considerable vertical gradient.

For the magnetic field of about 4 kG found for active M dwarfs by Saar (1994), the splitting of the infrared lines into the π and σ components should be distinctly seen in the high-resolution spectra. Reasoning from these considerations, Johns-Krull and Valenti (1996) observed two active M4.5e dwarfs EV Lac and Gl 729 by the 2.7-m Harlan J. Smith telescope at the McDonald Observatory. They observed the FeI λ 8468 Å line with a resolution of 120 000, the signal-to-noise ratio varied from 170 to 280, and an exposure of several hours. In parallel, they observed nonactive M dwarfs. The resulting spectra made it possible to "see" immediately the Zeeman splitting: beyond the magnetosensitive line λ 8468 Å the spectra of nonactive stars Gl 725 B and Gl 876 "lead" the EV Lac spectrum but the profile of the line itself differed sharply, its core was much shallower and σ components were clearly seen on both sides (see Fig. 7). For qualitative analysis of the spectra following the Allard and Hauschildt model (1995), theoretical profiles of this line for the atmosphere with and without field were calculated for T_{eff} = 3100 K and $\log g$ = 5.0 using the transfer equations for individual Stokes parameters. As a result, it was established that the parameters of magnetic fields of active stars were determined ambiguously: for EV Lac B = 4.2 kG and f = 0.4 from the profile wings and B = 3.4 kG and f = 0.6 for the entire profile; analogously for Gl 729 B = 2.8 kG and f = 0.4 from the line wings and B = 2.4 kG and f = 0.6 from the entire profile. As Saar (1992) did in studying AD Leo, Johns-Krull and Valenti concluded that the absence of common estimates suggested nonuniform magnetic fields on the stellar surface or noticeable vertical gradients.

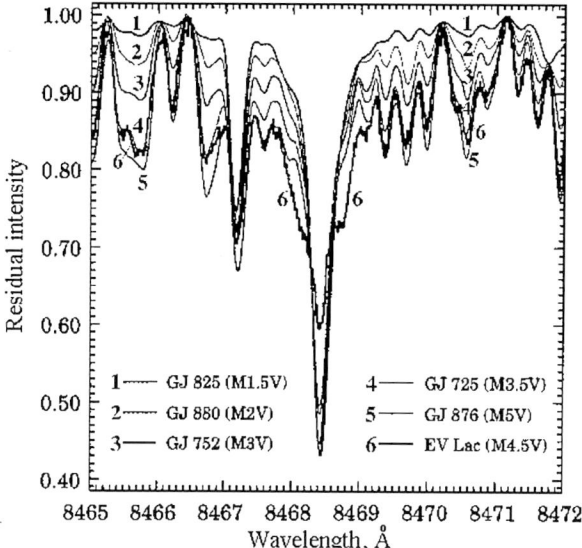

Fig. 7. The spectra of EV Lac flare star and nonactive M dwarfs in the region of magnetosensitive line λ 8468 Å (Johns-Krull and Valenti, 1996)

But the difference $\Delta B \sim 1$ kG for two flare stars of identical spectral type suggests that the idea of gas pressure in the photosphere completely governing the field strength should be defined more precisely. Probably, this is valid for G and K dwarfs, but for M stars where, with increasing stellar activity due to increasing depth of the convective zone and high rotation rate, f approaches unity, both f and B increase. Following the estimate of Kochukhov et al. (2001) made using the same method at the 3.6-m ESO telescope with a resolution of 140 000 and $S/N \sim 200$–250, for the M0 star GJ 1049 B ~ 5 kG and f is at least 0.5.

From 16 infrared lines of neutral iron near 1.56 μm Valenti et al. (1995) analyzed the photospheric magnetic field of the K2V star ε Eri from the observations with the 4-m Mayall telescope with a Fourier spectrometer. To present the recorded profiles, the Stokes parameters were calculated by varying five free parameters: effective temperature, relative abundance of iron, macroturbulent velocity, B, and f. The temperature variations in the stellar atmosphere were assumed identical for magnetic and nonmagnetic regions and were taken from the linearly scaled model of the solar atmosphere. As a result, Valenti et al. found that the parameters B = 1.44 kG and f = 0.088 ensure the best representation within the two-component model. The corresponding value of fB = 0.13 kG is much lower than the previous estimates, but the thorough analysis of probable errors showed that the last estimate was the most reliable. The high reliability of the result is primarily due to the fact that in the obtained spectra the σ component of the line λ 1.56 μm with the Lande factor equal to 3.0 is distinctly separated from the π component. Apparently, probable errors of the obtained parameters do not exceed 15% for B and 35% for f owing to the identical atmospheric model accepted for magnetic and nonmagnetic regions. Only somewhat lower estimates for the upper limits of B were obtained for 40 Eri and σ Dra.

Using space technologies, an echelle spectrograph was developed for magnetometric studies in the infrared range. The spectrograph had an InSb detector whose sensitivity was a hundred times higher than that of the Fourier spectrograph used before. The first observational results obtained using this device were reported by Saar (1996a). Analysis of the observations of the K5Ve star Gl 171.2 A (=BD + 26°730 = V 833 Tau) within the framework of the two-component model of its surface with regard to the entire pattern of the Zeeman splitting and the calculation of the radiative transfer considering the magnetooptical effect in the Milne–Eddington atmosphere yielded the estimates B = 2.8 kG and f = 0.60, which complied with the previous estimates based on optical lines (B = 2.6 kG and f = 0.50). This correspondence evidences that some earlier optical observations were sufficiently precise, but the significant spottedness of this star provides an ambiguous answer to the question on the regions to which the obtained measurements belong. Thus, if the contrast between the spot and photosphere is large in the optical range, the radiation of spots can be neglected and the magnetic signal is produced by faculae and the bright network, while in the infrared region this contrast is weak and the contribution of spots to the magnetic signal becomes substantial or even dominant. Thus, from the observations of LQ Hya Saar (1996a) obtained B = 3.5 kG and f = 0.7. However, the equivalent width of one of the infrared titanium lines appeared to be four times greater than in the less active ε Eri that belongs to the same spectral type. Such a significant amplification of the line can be explained only by the fact that it was formed at a lower temperature, i.e., the magnetic signal in this case is emitted by the spots and for them $f \sim 0.5$. These considerations require the three-component scheme: quiescent photosphere, faculae, and spots. For three dMe stars DT Vir, AD Leo, and YZ CMi, preliminary estimates yielded B > 3 kG and high values for f. The data on AD Leo and YZ CMi do not fit the two-component scheme either: there are a narrow π component and very broad σ components, which can be explained by the nonuniformity or appreciable vertical gradient of the magnetic field.

In 1995, Saar (1996b) summed up the results of 15 years of observations of magnetic fields on dwarf stars of the lower main sequence. Over these years new observational techniques were developed and data analysis became more complete, thus Saar ignored all observations that were performed with relatively low signal-to-noise ratios, without regard to the radiative transfer in the lines and the integration on the stellar disk, the results for K stars obtained for the line λ 8648 Å and inconsistent with infrared data, and the results on Zeeman intensification of lines that yield the low-accuracy product of fB^2 (see Sect. 1.2.3.3). The results of such a strict selection are presented in Table 4.

Table 4. Magnetic parameters of dwarf stars (after Saar, 1996b)

Star	Spectral Type	Axial Rotation Period, Days	f, %	B, kG	log F_X, erg/cm^2 s	log F_{CIV}, erg/cm^2 s
Sun	G2 V	25.4	1.5	1.5	4.8	4.0
HD 115383	G0 V	4.9	19	1.0	6.2	5.1
HD 20630	G5 V	9.4	20	1.8	5.9	4.8
ξ Boo A	G8 V	6.2	18	1.9	6.4	4.9
HD 131511	K1 V	9	6	1.7	5.8	4.7
HD 26965	K1 V	37	<2.7	1.7	5.3	3.7
HD 185114	K1 V	27.2	<1.9	1.36	5.0	4.0
HD 22049	K2 V	11.3	8.8	1.44	5.7	4.6
HD 17925	K2 V	6.76	35	1.5	6.1	5.0
LQ Hya	K2 Ve	1.60	70	3.5	7.3	5.6
Gl 171.2 A	K5 Ve	1.85	50	2.8	7.1	5.7
EQ Vir	K5 Ve	3.9	55	2.5	7.0	5.3
DT Vir	M2 Ve	1.54	50	3.0	6.9	
AD Leo	M3.5 Ve	2.6	60	4.0	6.6	5.5
YZ CMi	M4.5 Ve	2.78	67	4.2	6.8	5.3
EV Lac	M4.5 Ve	4.38	50	3.8	6.7	
Gl 729	M4.5 Ve		50	2.6	6.2	

These data confirmed and defined more precisely the statistical regularities derived earlier. Thus, for G–K stars and at least for the low-activity M dwarf Gl 729 there is a relationship B ≤ B_{eq}, where $B_{eq} = (8\pi P_{ph})^{1/2}$, where P_{ph} is the gas pressure in the quiescent photosphere, for the very spotted K2 star LQ Hya and active dKe and dMe dwarfs B > B_{eq}. As mentioned above, for LQ Hya this is associated with the contribution of spots to the magnetic signal, when it is natural to expect that $B_{spots} > B_{faculae} \sim B_{eq}$. Apparently, a similar situation is characteristic of emission K and M dwarfs.

Furthermore, the earlier proposed concept of the critical rotation rate remains valid. For slowly rotating stars, the relationship B ≤ B_{eq} is valid and f depends strongly on the rotation rate: it is proportional to $P_{rot}^{-1.8}$ for $P_{rot} > 3$ days and for faster rotation a saturation at the level of f ∼ 0.6 occurs. Instead of the axial rotation period as a global stellar parameter one can consider the Rossby number. In this case, the following statistical relations are valid: either f is proportional to Ro$^{-1.3}$ at Ro^{-1} < 8 and f ∼ 0.6 for Ro^{-1} > 8, or log f = −0.26−0.85Ro. In the latter relationship, the dependence of f on Ro is identical to that obtained by Montesinos and Jordan (1993) from an earlier sampling of magnetometric measurements, but the absolute value of f is approximately half as large. Saar explained this general tendency of a systematic decrease of B and in particular of f by the fact that an improved analysis technique could reveal more subtle effects responsible for some parts of the profile broadening that were first completely attributed to magnetic field.

A principal disadvantage of the Saar analysis (1996a,b) is the hypothesis on identical thermodynamic structures of the magnetic regions and normal photospheres.

To determine the magnetic field parameters, Rüedi et al. (1997) analyzed the optical spectra of 13 G1-K5 dwarfs of low and moderate activity and found fB = 165 ±30 G only for ε Eri. Their analysis was based on the numerical solution of the Unno–Rachkovsky equations for many lines with regard to magnetooptical effects. The results are in good agreement with the conclusions of Valenti et al. (1995), but Rüedi et al. doubted the significantly higher values of fB obtained before and the validity of separate estimates of f and B from optical spectra with a resolution of less than 100 000 and a signal-to-noise ratio less than 250. For ε Eri they also obtained restrictions on the temperature of magnetic regions: it should be the same as or higher than that in nonmagnetic regions, i.e., the obtained magnetic-field parameters were attributed to faculae. According to Valenti et al. (2001), the Zeeman broadening of molecular FeH lines provided a principal possibility for measuring magnetic fields on late M and brown dwarfs.

1.2.3.3 Other Magnetometric Methods

Stenflo–Lindegren Statistical Method. To study the magnetic properties of spatially irresolvable structures on the solar surface, Stenflo and Lindegren (1977) proposed the method of statistical analysis of many spectral lines. The essence of the method is that a measured parameter of the line profile, e.g., its depth, width at a certain fixed level or the equivalent width, is written as a sum of functions of known atomic constants and unknown thermodynamic and magnetic parameters in the region of line formation. From these relations one can find unknown parameters by studying a set of nonblended lines using the regression method. Mathys and Solanki (1987) and Solanki and Mathys (1987) used this approach to analyze magnetic fields of stars. As in the Robinson method, this statistical method provides an estimate of the effective strength of the magnetic field and the area it occupies on a star. In addition, it allows thermodynamic parameters of magnetized and nonmagnetized regions of the photosphere to be separated. Thus-obtained results depend weakly on the availability of weak blends.

The first results of studying the magnetic fields on lower main-sequence stars by the statistical method were reported by Mathys and Solanki (1987) and Solanki and Mathys (1987). They observed three stars with a resolution of 100 000 and a signal-to-noise ratio of several hundred and analyzed many tens of optical FeI lines. For ε Eri, the field strength determined with the greatest reliability was 1800–2500 G and the filling factor was 0.15–0.20. Strictly speaking, here, as in the Robinson method, it is not the value of f itself that is determined but rather the product $f\delta_c$, where δ_c is the continuum brightness contrast between the magnetic and nonmagnetic parts of the photosphere.

Analysis of these observations resulted in the conclusion: magnetic regions are hotter than nonmagnetic ones, i.e., they are similar to solar faculae.

Mathys and Solanki (1989) found fB and B for four stars and in two of them, also studied by Saar (1988), they found much higher values. They did not manage to find an unambiguous cause of the discrepancy but suggested that, if magnetic regions were hotter than nonmagnetic ones, the estimates of f obtained on the assumption of identical magnetic and nonmagnetic atmospheres were overestimated. Grossmann-Doerth and Solanki (1990) paid attention to the fact that in magnetometric studies different observers often used various lines of neutral iron λ 5250, 6173, 8468, and 15648 Å formed at noticeably different depths. If one considers that the vertical gradient of a stellar magnetic field is analogous to that occurring in the flux tubes of solar faculae, i.e., with the field pressure decreasing exponentially, in the estimates for B one can expect a discrepancy of up to 1 kG, but this circumstance should not distort an estimate of the magnetic flux fB.

Savanov and Savel'eva (1997) refined the Mathys and Solanki algorithm (1989) and estimated magnetic-field parameters for 10 G–K dwarfs. They estimated fB for four stars: for ξ Boo A fB = 990±250 G, for 70 Oph A fB = 1700±500 G, for ε Eri fB = 1200±500 G, and for HD 2047 fB = 3200±1100 G. For ξ Boo A two spectra were obtained with an interval of one hour, but only one of them provided a magnetic-field signal, which is probably explained by the motion of a magnetoactive region over the stellar disk due to rotation. However, it is not excluded that simple expansion of the absorption line parameters in the magnetic field into the sum of simple functions is not sufficient to replace the allowance for the radiative transfer effects and averaging over the stellar disk required strict solution of the problem. It does not provide the required precision of the estimate of stellar field parameters.

Zeeman Enhancement of Equivalent Widths of Lines. Another statistical method for estimating stellar magnetic fields was developed by Basri et al. (1992). The essence of the method is as follows. First, nonactive stars with temperature and iron abundance close to those of active stars under consideration are selected. Two to three dozen nonblended lines of neutral iron with different magnetic sensitivity are chosen in the spectra of the nonactive stars. The expected equivalent widths of these lines for nonmagnetic and magnetic atmospheres with a field of 2 kG are calculated from the known atmospheric models. The ratio of an equivalent width of a line in the magnetized atmosphere to the appropriate value in the nonmagnetic atmosphere is taken as the factor of Zeeman enhancement of the line. Then the factors are compared in the spectra of real active stars with those theoretically calculated, which enables estimation of the stellar field parameters. This method requires high-resolution spectra with high signal-to-noise ratio, the spectra of nonactive stars should be "rotated up" to the value of $v \sin i$ typical of an active star. Basri and Marcy (1994) applied this method to study five chromospherically active stars. The axial rotation period of four of them was less than seven days, 11.3 days for ε Eri, and for the compared stars the period was about 30 days. In observations at the 3-m Lick telescope with an echelle spectrograph and during 1–2 h exposures the spectra within 3800–9000 Å were recorded; 23 iron lines within 5320–8760 Å were used. As a result, with the highest confidence the field for EQ Vir was found, fB = 1.73 kG, which is in rather good correspondence with the estimates based on the Zeeman broadening. For ε Eri and HD 17925 fB < 0.5 kG, which is the detection limit of the method. For LQ Hya and Gl 171.2 A the data scattered considerably, which was probably caused by the significant spottedness of these red dwarfs. In principle, the method of Zeeman enhancement of lines should work for weaker stars better than the method of Zeeman broadening.

In addition to the above methods of estimating the parameters of magnetic fields on dwarfs, Ripodas et al. (1992) and Babel et al. (1995) proposed two new methods also based on the analysis of high-dispersion spectra. However, these methods did not yield specific results. Considerable attempts were undertaken to find stellar magnetic fields using broadband polarimetry (Huovelin and Saar, 1991; Saar and Huovelin, 1993), but they faced some principal difficulties concerned with the necessity of obtaining extremely high-precision observations and separating the magnetic-field effects from the Rayleigh and Thompson scattering.

Let us sum up.

The long-term solar studies showed that photospheric magnetic fields were the decisive factor for various nonstationary processes on the solar surface. This stimulated long-term efforts aimed at obtaining data on magnetic fields of the lower main-sequence stars. But magnetometric observations of cool stars are a very fine experimental study that requires reliable data and correct analysis. Nevertheless, some results of magnetometric studies can be considered reliable.

As for the Sun, lower main-sequence stars do not have global strong magnetic fields that are characteristic of some stars of early spectral types. The global fields of the stars under consideration are not stronger than one–two tens of gauss.

But many dwarfs of intermediate and low masses, the Sun as well, have local magnetic fields of numerous small different-polarity structures with the strength of hundreds of gauss to many kilogauss, though on some of the explored G–K–M dwarfs the fields are weaker than the detection threshold.

The interpretation of measured parameters of local magnetic fields is not evident and requires additional simulation. Apparently, the measurements in the optical region, in particular on G and early K dwarfs, yield the magnetic parameters of photospheric faculae, whereas in the infrared region, in particular for M stars, the fields of starspots start being revealed. Thus far, there is no generally accepted algorithm for calculating the three-component model including a normal photosphere, faculae, and spots.

In comparing the results of magnetometric observations in different spectral regions one should bear in mind not only the difference in the contrast between the spot and photosphere, but also the fact that the coefficients of continuous absorption of stellar photospheres noticeably vary in different regions: they are almost equal at 0.62 and 2.2 μm, higher at 0.85 μm and much lower at 1.56 μm. Thus, at 1.56 μm one can expect greater B and lower f, whereas at 0.85 μm B is lower and f is higher.

The strength of local magnetic fields of yellow and not too late red dwarfs is defined by the condition of equal pressures of the field and the ambient photosphere. Therefore, they systematically increase from many hundreds of gauss for G stars to many kilogauss for M dwarfs, "forget" the specific generation mechanism, and do not correlate with the surface filling factor and the rotation rate of the star.

The total efficiency of the field-generation mechanism is determined by the depth of the convective zone and the rotational velocity of a star. The total efficiency and the temporal variations of the magnetic flux are manifested in the filling factor that systematically increases from several hundredths for G stars and approaches unity for M dwarfs.

As the angular rotation rate increases, so does the total magnetic flux or the average field strength fB that is proportional to the flux.

The observational data thus far have not provided direct confirmation of the existence of magnetic fields of tens of kilogauss in starspots, though the facular areas of several kilogauss can be considered as an indirect argument in favor of such strong spot fields.

Messina et al. (2001, 2003) considered the reasons why fB should correlate with the maximum amplitudes of rotational modulation of stellar brightness in the V band. They found a correlation of fB with the rotational periods and Rossby numbers of active stars of the field and of six clusters of different age. At the photospheric level this result closes the earlier revealed rotation–activity correlations at chromospheric and coronal levels and in the transition region. The maximum amplitudes are the highest in K dwarfs of the age of the Pleiades and the α Per cluster and decrease as the age of the cluster increases. As for the activity in the upper layers of stellar atmospheres, the proposed photospheric activity index suggests the effect of activity saturation for the fastest rotators with the rotational periods of less than 0.35 days and Rossby numbers less than 0.02.

1.3 Chromospheres and Transition Regions

In the very first physical models, stars were considered as opaque gravity-bound formations of ideal gas. Under the conditions of radiative equilibrium the bodies should have maximum temperatures and densities in their centers and minimum temperatures and densities on the surfaces. However, solar studies showed that this model was incorrect: as matter comes up from the subphotosphere its temperature decreases, but upon reaching about 4200 K at a certain level it starts increasing again. Thus, at a height of about 300 km above the temperature minimum an effective ionization of hydrogen starts. These layers of the solar atmosphere were discovered first from the emission of the bright red line of hydrogen H_α during solar eclipses and got the name "chromosphere".

The contradiction between the initial stellar model and the observations of the Sun was removed when the convective energy transfer was found to play an important role in subphotospheric layers of intermediate and low-mass stars and to generate the nonradiative energy flux released in the atmosphere. This energy flux, as acoustic and/or hydromagnetic waves, interacts with atmospheric matter. The density of matter decreases with height following the barometric law approximately, while the density of the nonradiative energy flux decreases slowly enough because of the weak absorption and weak divergence of the flux. Thus, if the absorption coefficient changes slightly, the condition of the steadiness of the atmosphere, at which the amount of the absorbed mechanical energy should be equal to the amount of emitted radiative energy, can be realized only in the case of a noticeable increase of the temperature of matter with height. As a result, there is a temperature inversion: above the photosphere with an effective temperature of about 5700 K and a cooler temperature minimum there is a chromosphere with a temperature of 5000–20 000 K, then there are a very thin transition region with the temperature varying from 20 000 to 10^6 K and the corona with a temperature of above 10^6 K. The temperature jumps between the adjacent atmosphere layers are eventually due to the discreteness of states of hydrogen and helium atoms, which mainly govern the thermodynamic properties of the space plasma. Certainly, these simple considerations can provide only a general physical explanation of the temperature inversion in stellar atmospheres, but the real pattern is much more complicated: it should include different types of hydromagnetic waves, which interact with matter and with each other in different ways, the heat conduction of the hot corona toward the cooler underlying layers, variations at different spatial and temporal scales of the initial flux of nonradiative energy, and its significant deviations from the uniform distribution over the stellar surface owing to the complicated structure of the magnetic field.

Initially, the solar chromosphere was observed only in the limb during short-term solar eclipses. With the invention of out-of-eclipse coronagraphs and spectroheliographs a regular monitoring of optical characteristics of the chromosphere was started. Today, the solar chromosphere is observed in optical and radio ranges by a broad network of ground observatories and in ultraviolet from space vehicles, which provides simultaneous observations of the emission lines of the chromosphere and the transition region.

1.3 Chromospheres and Transition Regions

In brief, the results of studying the solar chromosphere are as follows.

The basic feature of the solar chromosphere is the extreme nonuniformity of all characteristics. In the lower layers of the chromosphere, to a height of about 1500 km, numerous emission lines are observed that correspond to absorption lines of the photosphere. On increasing height these lines weaken, but the emission intensity of ionized calcium and hydrogen increases and helium emission arises. In the upper chromosphere expanding from 4000 to 10 000 km, calcium emission disappears, hydrogen emission weakens, and helium emission strengthens. The density of matter in the chromosphere varies from 10^{15} cm^{-3} at the base to 10^9 cm^{-3} near the upper limit, but at the chromosphere base the ionization degree of matter is only 0.1%, near the upper limit it is complete, thus the range of electron densities is much narrower, 10^{12}–10^9 cm^{-3}. The systematic changes with height are superimposed by the considerable variations on the solar surface, since the chromosphere is not a continuous plane-parallel structure but an ensemble of numerous plasma loops. At the limb in the lower chromosphere these details are indistinguishable, these loops are visible as separate blobs in the midchromosphere, while the upper chromosphere contains only short-lived vertical 3000-km thick structures, spicules. Simultaneously, there are tens of thousands of spicules on the Sun, but they occupy less than 1% of the solar surface, and there is hotter coronal gas between them. Bright structures, floccules in the active regions and a bright chromospheric network, whose cells outline the supergranules, are seen in the lines of ionized calcium in spectroheliograms. During the epoch of the solar maximum activity in 1980, when the mean total area of sunspots was close to 0.2% of the solar disk area, the mean total area of calcium floccules was about 3% of the disk area. The floccules of the midchromosphere were clearly seen in calcium spectroheliograms, when passing to the upper chromosphere at the level of formation of the emission of Ly_α, grow by approximately 40% (Lean, 1992). In the center of the H_α line near the active regions one can clearly see dark filaments. Characteristic motions in the quiescent chromosphere are the rise of gas in spicules with a velocity of up to 20 km/s and a spread of matter from the centers of supergranules to their periphery with a velocity of about 0.5 km/s. Close neighboring active regions on the Sun form activity complexes. Usually these structures are placed on the solar disk in a nonrandom way; there are regions of preferable occurrence that persist for long periods, the so-called active longitudes. During the epochs of activity maxima the neighboring activity complexes are overlapped and are difficult to distinguish.

As stated above, the solar atmosphere comprises a thin transition region, wherein the temperature increases from chromospheric to coronal values. Its thickness is much less than the pressure height scale, only of the order of hundreds of kilometers, but this region provides strong emission in ultraviolet lines of the high ionization stages of carbon, nitrogen, silicon, and some other elements. Both photospheric spots and chromospheric floccules and the emission lines above the spots in the transition region depend on the solar-cycle phase. Thus, during the epoch of solar maximum the spots were not seen in the light of CIV lines, whereas a ten-fold increase in the intensity of the lines of the transition region was found in the vicinity of the solar minimum straight above the spot umbra (Mullan, 1992). From the Earth, the transition region of the Sun is studied in the range of millimeter and centimeter wavelengths using radio-astronomy methods.

It is obvious that stellar chromospheres and transition regions cannot be studied as thoroughly as those of the Sun, since only integral stellar characteristics can be observed, but the multiplicity of observed stellar chromospheres to a certain extent compensates for this limitation.

1.3.1 Optical and Ultraviolet Spectra of Chromospheres and Transition Regions

Stellar chromospheres were discovered with the beginning of spectral studies of stars of intermediate and late spectral types using the strong emission of the resonance CaII H and K doublet in the violet part of spectra. In addition to these bright lines, strong hydrogen emission was already found in the first spectra of flare stars. The notation dKe and dMe is used to notify the presence of hydrogen emission in the spectra of K–M dwarfs. The overwhelming majority of studies carried out thus far are based on the analysis of hydrogen and calcium emissions recorded from the Earth and ultraviolet spectra obtained from the IUE satellite. Few, but fundamentally new, ultraviolet data were obtained from the space telescopes HST and FUSE. The publications by Pettersen and Hawley (1987, 1989) provided a general view of optical spectra of the chromospheres of flare stars. Based on the observations of about thirty objects on the 2.1-m Struve telescope, they dealt with the energy distribution in the spectra of G9–M5 stars in the broad wavelength range from 3900 to 9000 Å, the Balmer decrement from H_α to H_9, and absolute surface fluxes in the H_β and CaII lines.

1.3.1.1 Calcium Emission in the H and K Lines

Chromospheric CaII H and K emission lines seen against the background of broad and deep absorption profiles are one of the most well-studied details of the solar spectrum. This is explained by their accessibility for ground-based observers and high sensitivity of photoemulsions in this wavelength region. Being excited by electron collisions, the lines provide important information on the temperature structure of atmospheres. Depending on the photospheric base – a quiescent photosphere, a spot or a facula – the profiles of these lines differ significantly, even being averaged over the solar disk they differ for different phases of the solar cycle.

Long lists of dwarf stars with the calcium emission were published in the 1940s–1950s (Joy and Wilson, 1949; Bidelman, 1954). Later Glebocki et al. (1980) made a compilation of the lists. The atlases by Pasquini et al. (1988)

and Rebolo et al. (1989) contain the records of several tens of spectra of F4–K5 dwarfs with different activity level in the vicinity of the CaII H and K lines with a resolution of 30 000–100 000 and high S/N ratio. The spectra were recorded by the echelle spectrograph located at the coude focus of the 1.4-m telescope of the South European Observatory and their quality is rather close to that of solar spectrum records.

In 1966, O. Wilson started a long-term program of systematic measurements of the intensities of emission cores of the CaII H and K lines in the spectra of about one hundred medium- and low-mass stars. Instead of the eye estimates of emission intensity used in his previous study (Wilson, 1963), in 1966 photoelectric measurements were started. A scanner with four sequentially opening entrance slits installed at the coude focus of the 2.5-m telescope of the Mount Wilson Observatory recorded the fluxes in line cores and in the broad bands of the adjacent continuum. Based on the results of 11-year spectral observations, Wilson concluded that all stellar chromospheres, from the weakest to the strongest, demonstrated significant variations of calcium emission fluxes for the times from one day to several months and these variations increased together with the fluxes (Wilson, 1978).

The studies initiated by Wilson were actively continued. His results gave an impetus to different astrophysical directions, from registration of individual flares, the estimates of axial rotation periods and characteristic lifetimes of active regions to the discovery of the analogues of the 11-year solar cycle and the conclusions on the decay rate of the stellar magnetism at evolutionary times. Below, we consider the results of spectral monitoring of calcium emission that are mainly related to the physics of stellar chromospheres, conclusions on cyclic and evolutionary changes will be presented in the third part of the book.

In 1980, a new four-channel HK-photometer was installed on the 1.5-m telescope at the Mount Wilson Observatory. Between July and October 1980 Wilson's team carried out intensive observations of 46 dwarfs (Fig. 8) and made the following conclusions. Calcium-emission intensities modulated with periods of days for 19 stars; the modulation was due to nonuniform distribution of the chromospheric emission over the surfaces of rotating stars. It is noteworthy that thus-obtained direct estimates of axial rotation periods are independent of the orientation of rotation axes and include slowly rotating stars for which the traditional spectroscopic method is inappropriate. In many cases, the phases of rotational modulation lasted over the whole observational period, which evidences the long-term asymmetric distribution of active regions along the longitude. The comparison of the found periods with the emission intensity made it possible to suggest that the level of chromospheric activity of stars depended mainly or even solely on their rotation (Vaughan et al., 1981; Baliunas et al., 1983). Stimets and Giles (1980) concluded that the analysis of nonuniformly distributed HK-photometer measurements through the autocorrelation method had certain advantages over the power-spectrum method, thus Vaughan et al. (1981) used the former.

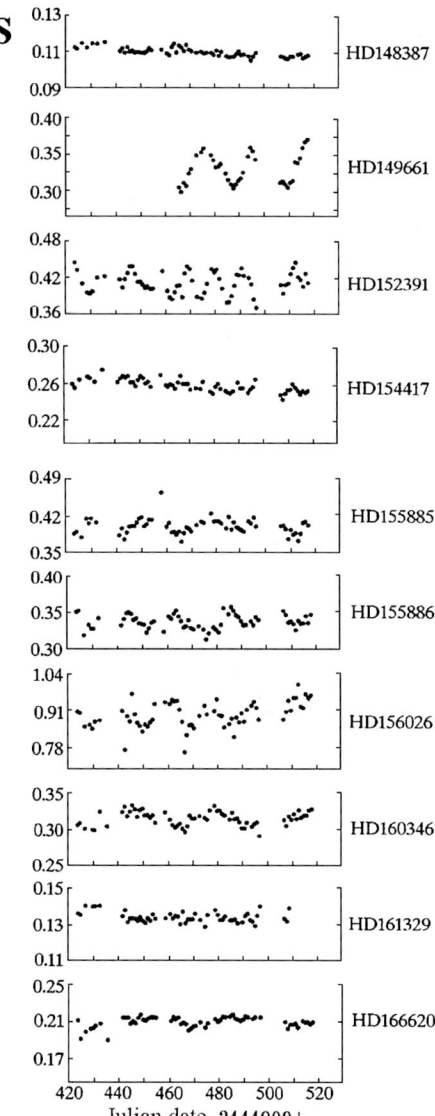

Fig. 8. Time variation of S, the ratios of fluxes in the cores of the CaII H and K emission lines to fluxes in the adjacent regions of the photospheric spectrum, measured in the course of nightly monitoring in 1980 using the HK-photometer for F–K dwarfs (Baliunas et al., 1983)

Baliunas et al. (1981) studied the intensities of the CaII H and K lines in the spectrum of the active star ε Eri and found that the variations of both lines correlated at the level of 5% from one night to another and at the level of 7% at a time interval of 15 min. The total energy of these fast variations was about 2×10^{30} erg and, since the localization of the source of the radiation variability on a small part of the disk was quite probable, one could not exclude that stellar flares were responsible for these variations.

On the basis of dense observational series of about 100 program stars obtained with the HK-photometer during three seasons by the team of O. Wilson, Baliunas et al. (1985) made spectral analysis of the time changes of the intensity of the calcium emission. They found that the periodicity of this intensity varied from one season to another and even within the same season for 12 stars, and that two different periods coexisted during at least one season for 10 stars. The result can be explained either by the existence of active regions at different latitudes of a star with noticeable differential rotation or by the phase shift due to the emergence of a new active region at the longitude where no active regions existed before. For four stars, for which the data were collected over 18 years, the effect of differential rotation was the greatest. On the G8 star HD 101501 cyclic changes of the calcium emission with the characteristic time of about 7.5 years and a differential rotation of 10% were found. On the G0 dwarf HD 114710 no cyclic changes were detected but the value of the differential rotation was determined as 21%. On the G1 star HD 190406 and the G0 star HD 206860, cyclic variations with the characteristic times of 2.6 and 5.3 years and differential rotations of 11 and 5% were established. During the third observational season two simultaneous periods of variation in the calcium emission were found on HD 190406, the second period coincided with the period found during the first observational season. On the whole, the pattern is consistent with the solar one when spots of a cycle coming to an end are at low latitudes and have one rotational period, while coexistent spots of a new cycle appear at higher latitudes and owing to the differential rotation for the Sun have another rotational period. On HD 206860 with an axial rotation period of about 5 days the active region was preserved during three observational seasons. It should be noted that the differential rotation for the Sun does not exceed 3%. On the basis of 10-year observations of HD 114710, Donahue and Baliunas (1992) concluded that the star had two longitudinal activity zones, but, as opposed to the Sun, the axial rotation period increased as the chromospheric emission in the cycle weakened. Considering the rotation character of 22 stars, for which rotational periods were determined during several seasons, Donahue and Baliunas (1994) found on 12 of them the characteristics of solar "butterflies", on six stars there was an inverse pattern, when one could expect a drift of active regions during one cycle to the pole, and for four stars they suspected a change in the direction of the drift of active zones during the cycle.

In 1991, 18 participants of the O. Wilson program published over 65 000 measurements for about 1300 stars obtained in 1966–83 (Duncan and 17 coauthors, 1991). In 1995, 27 participants of the program published the results of 25-year studies (Baliunas et al., 1995). Recently, Write et al. (2004) submitted for publication 18 000 measurements of S for 1200 F–M stars obtained from the archives of the Keck and Lick observatories.

According to Blanco et al. (1974), in the sample of stars of the same age the maximum ratio of the CaII K line luminosity to the stellar bolometric luminosity falls at the spectral type K0.

Linsky et al. (1979a) obtained at the 4-m telescope of the Kitt Peak Observatory the high-dispersion spectra of 17 G0–M2 dwarfs in the region of CaII H and K lines, calibrated them in absolute units, and found the ratios of fluxes in these lines to bolometric stellar luminosities – R_{HK}. These ratios were lower for the stars older than the Sun, α Cen A and α Cen B, were the same for 61 Cyg A and 61 Cyg B, and were greater for younger G dwarfs. Later, using the same equipment, Giampapa et al. (1981a) observed seven K–M dwarfs. The ratios of the fluxes in the calcium lines to the bolometric luminosities varied from 3×10^{-5} to 9×10^{-5} for dwarfs with hydrogen emission and are lower by an order of magnitude for dwarfs without the emission, the latter values being close to the solar value of 8×10^{-6}. Since the emissions in the CaII H and K lines are the first indicator of the existence of the stellar chromosphere, the search for stars without chromospheres is reduced to the search for stars without calcium emission. Out of four late M dwarfs without traces of the H_α line considered by Mathioudakis and Doyle (1989a), one had very weak calcium emission and three stars did not display any. However, to make a final conclusion, these observations carried out with a resolution of 1.4 Å should be repeated with a higher resolution.

Middelkoop (1982) calibrated the values of S measured by the team of the Mount Wilson Observatory to the absolute fluxes from a unit of the stellar surface F_{HK}. In so doing, he found the dependence of F_{HK} on rotation, which is common for stars of different spectral types, for single stars, and the components of binary systems. This suggests that the high chromospheric activity of the components is due to fast rotation determined by the tidal interaction in the systems rather than directly by the tidal heating of the atmospheres.

Based on the analysis of several hundreds of F–G–K stars, Mewe et al. (1981) and Schrijver (1983) showed that the radiation fluxes in the CaII H and K lines within each spectral interval of F–G–K stars overlapped by more than an order of magnitude, but at the same time they reliably recorded the lower level of such fluxes $F_{HK}^{\min}(Sp)$, the so-called basal chromospheric level, which was definitely higher than the detection threshold of this emission. The differences ΔF_{HK} between F_{HK} of specific stars and the appropriate $F_{HK}^{\min}(Sp)$ were more closely related to other indicators of the stellar activity than the fluxes F_{HK} themselves. Hence, Mewe et al. (1981) suggested that $F_{HK}^{\min}(Sp)$ was caused by nonmagnetic heating of atmospheres. The intensity of calcium emission on the solar surface with minimum magnetic activity in the center of supergranules far from the facular areas falls exactly on the basal level of the chromospheric emission. Rutten (1984) performed additional observations with the HK-photometer. The observational results made it possible to revise the Middelkoop calibration and conclude that the basal level was independent of metallicity. Later, Rutten (1986, 1987) found the dependence on the color index in the linear relation coefficient between F_{HK} and the logarithms of axial rotation periods. He also established an age-independent correlation between ΔF_{HK} and the axial rotation period or the angular velocity common for single dwarfs, members of binary systems, and most giants. The dependence of the activity level on rotation and its independence of age showed that the magnetic field responsible for the activity was generated by the dynamo mechanism and was not a relict field. Schrijver et al. (1989a) considered the Mount-Wilson measurements for the emission in the CaII H and K lines, the data on ultraviolet emission of MgII h and k, and the data on stellar variability and rotation. They concluded that the basal level of chromospheric activity was not linked to the magnetic activity and was relevant to the entire stellar surface, whereas the magnetic component was present in active regions and in the chromospheric network. The extrapolation of the relations between the activity level and stellar variability showed that such a variability disappeared near the basal level of chromospheric activity, i.e., the active regions and the chromospheric network did not exist on the lowest-activity stars. Stellar rotation was negligible near the basal level of chromospheric activity, which meant that it was not likely that the dynamo mechanism was active on the stars with basal chromospheric activity.

Based on the observations of 26 F5–K3 dwarfs in the CaII H and K lines Schrijver et al. (1992) determined more precisely the basal level of the chromospheric activity: it was much above the level provided by the purely radiative atmosphere, which suggests the existence of nonradiative heating even for the lowest-activity stars The basal level of chromospheric activity can cover the entire stellar surface, but one should not exclude the existence of time-variable cool and hot zones, if the generation of acoustic waves below the photosphere is nonuniform.

The close relation between the chromospheric activity and local magnetic fields, which is known from solar studies, formed a basis for the concept of the secular weakening of stellar magnetic fields caused eventually by the deceleration of rotation. However, in addition to rotation, the magnetic field depends on another independent parameter: stellar mass, the mass of the convective zone where the field is generated, the mass-dependent spectral type of a star, or the time of overturn of convective vortices in the convective zone that depends also on the spectral type. All this led to the idea on the dependence of the level of chromospheric activity on the Rossby number, the ratio of the axial rotational period to the theoretically calculated overturn time of convective vortices. Using the axial rotation periods of more than 40 stars found from the variation of their CaII H and K emission, Noyes et al. (1984a) found a correlation of such periods with the emission level averaged over 15 years. A particularly close correlation was established between the flux ratios in the CaII H and K lines revised to take into account the contributions of

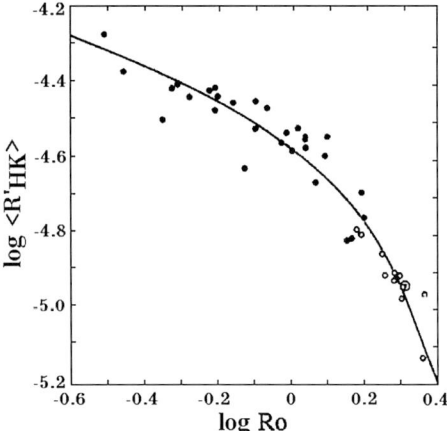

Fig. 9. Chromospheric luminosity in the CaII H and K lines related to bolometric luminosity versus the Rossby number (Noyes et al., 1984a)

the radiative atmosphere, to the bolometric luminosities of the stars, the values of R'_{HK} and the Rossby numbers (Fig. 9). The dependence turned out to be so close that it was used to estimate the axial rotation period from the observed luminosity of calcium emission. Quast and Torres (1986) made use of this fact: from the intensity of calcium emission they determined the axial rotation periods of several stars and then compared the periods and the color index $B-V$. The stars attributed to young and old objects according to their kinematic parameters are clearly distinguished in the constructed plots. Kim and Demarque (1996) refined the results of Noyes et al. (1984a) by estimating the time of passage of convective elements throughout the whole convective zone, while Noyes et al. used only the time of passage of the local height scale at the bottom of the zone. Within this approach the saturation effect remained: at low Ro, i.e., for fast rotators, the dependence of the activity level on the rotation rate disappeared (Montesinos, 2001).

Pasquini (1992) compared the high-accuracy profiles of the CaII K line in the spectra of G0-G5 stars with the profiles of this line obtained from the whole Sun and separate structures on its surface. General tendencies and relationships observed on the Sun throughout the cycle – line intensity, width, and asymmetry – take place on solar-type stars with differing chromospheric activity levels. Apparently, the stars cover a broader activity range than the Sun during the cycle. The spectrum of the most active stars suggests the presence of structures of the solar faculae type that should cover most of their surfaces. Despite the similarity of the spectra of solar faculae and active stars, in interpreting the latter it is impossible to choose the domination of the facula brightness or the filling factor of the stellar surface by faculae.

For four years, Garcia López et al. (1992) monitored the profile of the CaII H lines in the spectra of six G–K stars – ξ Boo A, α Cen B, ε Ind, ε Eri, κ Cet, and 70 Oph A – with a resolution of 80 000 and $S/N = 200$. They found that both the emission flux of the central component of the line and its width were variable. They revealed the shifts of the emission core corresponding to the velocities of hundreds of meters per second and of the absorption detail with lower velocities. The comparison of the wavelengths of H_3 and the emission core provided the first evidences of the vertical ascending and descending motions in the chromosphere, which were stable for several years on the star.

Using the Calar Alto 2.2-m telescope and the 2.5-m Isaac Newton telescope (Canary Islands), Montes et al. (1994) observed about 50 chronometrically active binary systems, including more than a dozen systems composed of active dwarfs with orbital periods of 0.7–6 days. They found that the K line at the level of half-intensity (W_0) and at its base (W_1) was wider than those corresponding to the Wilson–Bappu relationship: W_0 was more affected by the stellar rotation rate, while W_1 depended more on the emission intensity.

The qualitative relationships between the strengths of the magnetic fields of individual strictures on the solar surface and the brightnesses of the calcium floccules were established already in the late 1950s from the comparison of spectroheliograms and magnetograms of the Sun. Skumanich et al. (1975) found a linear relation between the CaII emission and the flux module of the local magnetic field for the quiet Sun. Schrijver et al. (1989b) found quantitative relationships between the intensity of CaII K and the magnetic flux density for a quiet Sun and its active regions that are valid for a broad range of the structures on the solar surface:

$$\Delta F_{CaII} = 0.055 \langle fB \rangle^{0.62 \pm 0.14} , \qquad (4)$$

where ΔF_{CaII} is the excess over the minimum nonmagnetic luminosity of the chromosphere. Later, Saar (1991) determined more precisely the exponent as 0.56 ± 0.03. The minimum luminosity level of the solar chromosphere measured in the centers of supergranules in very quiescent solar regions was equal to the minimum flux density of the solar-type stars and, on average, only by 30% lower than the minimum flux level in the sample of red dwarfs. For the chromospheric emission of CaII, this dependence is valid up to $\langle fB \rangle \sim 300$ G and saturation occurs at greater fluxes when the maximum efficiency of the magnetic heating is achieved or the filling factor approaches unity at the level where the optical thickness in the cores of the CaII lines achieves unity. The found saturation level coincides with the averaged magnetic flux from a solar-active region. This coincidence can be explained by the atmospheric physics: the point is the mechanism of the radiative transfer or magnetic heating rather than stellar age or rotation rate.

An example of synchronous change of the magnetic flux and the calcium emission on the scale of a long-term activity cycle was found for the K5V star κ Cet: according to Saar and Baliunas (1992b), from four points obtained in 1984–88 the value of ΔF_{CaII} is proportional to $\langle fB \rangle^{0.4 \pm 0.25}$, which does not contradict the correlation (4) with the exponent found by Saar (1991). Earlier,

Saar et al. (1986b) found the weak correlation of f and F_{HK} for the K2 dwarf ε Eri.

1.3.1.2 Hydrogen Emission

The hydrogen emission lines in the spectrum of a red dwarf are the first evidence of its affiliation to the UV Cet-type flare stars. This emission is seen even in low-resolution spectra and is the sufficient criterion for attributing a red dwarf to flare stars: the experience of many observers showed that, as a rule, 10–15-hour photoelectric monitoring in the U band is sufficient to record a flare on an emission dwarf. Furthermore, Torres et al. (1983) for several years ran photometric observations of 90 red dwarfs to find periodical brightness variations: 11 of 20 stars with H_α emission and 2 of 30 dwarfs with only calcium emission displayed the variations, whereas for none of the 40 nonemission stars were brightness variations observed. Thus, the existence of hydrogen emission to a high confidence ensures flare activity and spottedness of red dwarfs.

The emission of H_α is invisible on the disk in the spectrum of the quiet Sun. In the spectra of bright floccules, this strong absorption line is only slightly flooded in the center and the equivalent width of the flooding emission can be estimated as 0.02–0.03 Å. A similar effect is observed in the spectra of relatively similar to the Sun chromospherically active G–K stars (Zarro, 1983). The hydrogen emission in the solar-type stars was first studied by Herbig (1985). Using the coude spectrograph of the Lick telescope, he obtained the spectrograms of about 40 F8–G3 stars. Based on a high ratio $S/N \sim 200$–350, he established weak emission components with $W_{H_\alpha} \sim 0.1$ Å in the center of the H_α absorption lines, which were in good correlation with calcium-emission fluxes. According to his estimate, the ratios of fluxes in the main chromospheric channels of radiative losses of these stars are $F_{HK} : F_{IRCaII} : F_{H_\alpha} = 1 : 0.81 : 0.36$.

If, for the Sun and solar-type stars, W_{H_α} is usually a few hundredths of an angstrom, in the spectra of dKe–dMe stars it is of the order of several angstroms. The difference of two orders of magnitude is to a considerable extent due to the lower intensity of the photospheric continuum in the spectra of later stars, since for the effective temperatures presented in Table 2, for the wavelength H_α the ratios of the Planck functions $B(G2)/B(Sp)$ is 12 for the M2 spectra and 50 for M7 spectra. These qualitative reasonings can be refined using the results of Hawley et al. (1996) for spectral types and equivalent widths of 321 red dwarfs. Since during their spectral observations that underlie the measurements of W_{H_α}, no strict photometric monitoring was run, it is probable that some of the published values of W_{H_α} were obtained during flares; in this case W_{H_α} usually increased by a factor of 2–3. Reasoning from the average frequency and the average duration of stellar flares (see Sect. 2.2) one can estimate that among 321 the values from 1 to 5 could be obtained during flares. Thus, the exclusion of the 6 highest values from

Table 5. Characteristics of the H_α emission line in the spectra of late K and M dwarfs (measured by Hawley et al. (1996))

A. $W_{H\alpha}$ Distribution									
$W_{H\alpha}$ (Å)	< 2	2–3	3–4	4–5	5–6	6–7	7–8	8–9	9–10	> 10
N	54	55	59	55	37	20	15	6	7	13

B. $W'_{H\alpha}$ distribution								
$W'_{H\alpha}$ (0.08Å)	< 1	1–2	2–3	3–4	4–5	5–6	6–7	7–8	> 8
N	25	59	87	74	39	19	8	3	1

C. $\langle W'_{H\alpha} \rangle$ versus spectral type							
Spectral class	K5–M0.5	M1–M2.5	M3	M3.5	M4	M4.5	M5–M9
$W'_{H\alpha}$ (0.01Å)	32 ±10	23 ±12	25 ±10	26 ±13	24 ±11	22 ±11	19 ±11
N	25	9	30	58	70	53	50

consideration should nullify the distorting effect of flares on the estimates of parameters of quiescent stellar chromospheres.

Part A of Table 5 presents the distribution of 315 values of $W_{H\alpha}$ measured by Hawley et al. (1996): the distribution density of $W_{H\alpha}$ from the detection threshold of 1 Å to 5 Å is approximately constant, while the number of higher values smoothly decreases to a maximum of about 12 Å. Part B presents the distribution of the equivalent widths $W'_{H\alpha} = W_{H\alpha} B(Sp)/B(G2)$ reduced to the solar atmosphere. All $W'_{H\alpha}$ values of the sample under consideration are within 0.02–0.74 Å, and 70% of them are within 0.08–0.32 Å. The average for the sample $\langle W_{H\alpha} \rangle = 0.24$ Å corresponds to the absolute flux of 1.6×10^6 erg/cm^2 s. Part C presents the distribution of average values of $\langle W'_{H\alpha} \rangle$ along spectral types. The table shows that the dependence of $\langle W'_{H\alpha} \rangle$ on the effective stellar temperature is rather weak and there is only a small systematic decrease of $\langle W'_{H\alpha} \rangle$ with transition to later spectra, but within each spectral type the dispersion of the values is fairly large. Thus, the characteristic absolute surface brightness of stellar chromospheres in the H_α line averaged over the disk is 10 times higher than the appropriate value for solar floccules.

Using the echelle spectrograph of the 10-m Keck telescope Basri (2001) studied the region H_α of more than 60 late M and L dwarfs. He noticed clear emission from all M5–M9.5 stars of the sample and fast weakening of the emission after M9.5. He found that measured $W_{H\alpha}$ did not demonstrate the dependence on the rotation rate, while the ratios $L_{H\alpha}/L_{\text{bol}}$ systematically and rapidly decreased toward the late spectral types.

The investigation of physical conditions in the chromospheres of active dwarfs from hydrogen emission was started by Wilson (1961), who obtained the spectrum of EV Lac with the dispersion of 9 Å/mm using the Palomar telescope in 1960. During the 6-h exposure no appreciable brightening of the star was noticed, thus it was concluded that the resulting spectrum belonged to the quiescent state. The hydrogen emission lines in the spectrum were evidently broader than those of helium and metals. Thus, assuming a purely thermal mechanism of their broadening, Wilson estimated the upper temperature limit in the chromosphere as 14 000 K.

The first systematic consideration of hydrogen emission from the chromospheres of flare stars was performed in the Crimea (Gershberg, 1970b). The results of spectral observations of five emission red dwarfs with the 2.6-m Shajn telescope allowed the relative intensities and absolute luminosities of the Balmer lines and the width of the H_α profiles to be estimated. By comparing these data with the emission spectra of the solar chromosphere obtained at different phases of the solar eclipse it was shown that the density of matter in the layers of stellar chromospheres, in which the hydrogen emission was formed, was ten times higher than the appropriate values on the Sun. Since the structure of an isothermal atmosphere is determined by the ratio M_*/R_*, which for flare stars varies from 0.42 to 1.4 of the appropriate solar ratio, the general geometric structure of the chromospheres in both cases should be similar as well. Therefore, to heat a denser atmosphere to the chromospheric temperature, the power of the heating mechanism, the density of upward flux of nonradiative energy, and/or thermal conductivity from the corona (see Sect. 1.5) of flare stars should be higher at least by an order of magnitude than for the Sun.

Within the framework of the Sobolev concept of moving stellar envelopes, the theory of the Balmer decrement was developed for the case of purely collisional excitation of hydrogen lines in an isothermal gas free of external radiative excitation. Despite the considerable distinctions of this model from the stellar chromosphere, these calculations were applied to the Balmer decrements recorded in the quiescent states of nine flare stars. This resulted in the first estimates of the characteristic electron densities of stellar chromospheres: $(1–4) \times 10^{12}$ cm^{-3} at 10 000 K (Gershberg, 1974a). Then Grinin (1979) added photospheric radiation, but this almost did not influence the estimates of electron density. Though the concept of moving envelopes satisfactorily presented the observed decrement, it required noticeable internal movements in the radiating medium with velocities of up to 20–30 km/s, which had not been supported by observations. This fact raised doubts about the validity of the density estimates. The theory of the Balmer decrement elaborated later was free of the assumption on the substantial motion in the radiating medium: the quantum exit was achieved not at the expense of the differential motion of matter, but due to the quantum frequency drift at multiple scattering and their exit in the wings of profile lines. Application of the theory to the observed decrements yielded density estimates that exceeded the above estimates approximately by a factor of 3 (Katsova, 1990).

The observations of 9 flare stars performed in the Crimea resulted in the measurement of the absolute radiation fluxes of the M2–M6 chromospheres in the H_γ line. All the fluxes turned out to be close to 5×10^5 erg/cm^2 s to an

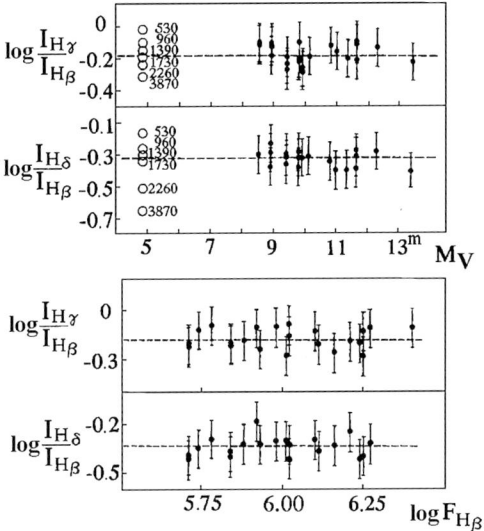

Fig. 10. The ratios of the intensities of the Balmer lines in the spectra of emission red dwarfs depending on the absolute magnitude M_V and absolute intensity of the H_β line. *Open circles* show the appropriate values in the spectrum of the solar chromosphere obtained at different phases of the solar eclipse (Shakhovskaya, 1974b)

accuracy of 2. The effective thicknesses of radiating layers estimated from the fluxes vary from several hundreds to one thousand kilometers.

Shakhovskaya (1974b) increased the number of studied spectra of emission red dwarfs at the 2.6-m Shajn reflector to 43. She found that for M0–M5 spectral types the ratios of the intensities I_{H_γ}/I_{H_β} and I_{H_δ}/I_{H_β} were practically independent of the absolute luminosity (Fig. 10).

The absolute flux in the H_β line for this sample also lies within a relatively narrow range of values and there is no systematic change of the relative intensities of the Balmer lines depending on the flux (see Fig. 10). However, later, Katsova (1990) suspected that the ratio I_{H_8}/I_{H_γ} grows systematically with the increase in the X-ray luminosity of the star. Shakhovskaya (1974b) found a systematic decrease of the ratio of the intensity of CaII emission lines to the intensity of the hydrogen emission for later M stars.

Worden and Peterson (1976) obtained the spectra of 8 dKe-dMe stars in the red part of the spectrum at the 4-m telescope of the Kitt Peak Observatory and found that the double-peaked profile of the H_α line was common for emission red dwarfs. Having considered various explanations of the profile formation, they concluded that the most probable model was the optically thick line in the center with the electron density of the radiating medium less than 10^{13} cm^{-3}. Pettersen and Coleman (1981) estimated the distance between the peaks of the double-peaked H_α profile in the spectrum of AD Leo as 0.6 Å for the profile half-width of 1.35 Å and its equivalent width of about

3.1 Å. The profiles of the H_α line and several other high-resolution Balmer lines and the estimates of their equivalent widths and FWHM values in the spectra of 17 red dwarfs were published by Worden et al. (1981). Pettersen et al. (1984b) studied the H_α profiles with the resolution of 0.45 Å in the spectra of EV Lac, EQ Peg A, and V 1054 Oph. They found that a self-absorption outlined in the EV Lac profile, which was not observed for EQ Peg A, while the profile of this line in the spectrum of V 1054 Oph could be presented as a superposition of two emission profiles with the self-absorption of up to 10–15% of the maximum level. According to their estimates, the electron density in the chromospheres varies from 10^{12} to 10^{13} cm^{-3}. Later, Pettersen (1989b) studied the H_α profiles of eight K5–M1 stars in the quiescent state with equal or even higher resolution (up to 0.20 Å): V 833 Tau, EQ Vir, BY Dra, 61 Cyg, DK Leo, V 1005 Ori, AU Mic, and YY Gem. For all single stars (AU Mic, V 1005 Ori, DK Leo, and EQ Vir) they revealed double-peaked profiles with FWHM = 1.4–1.5 Å, but for the components of YY Gem FWHM was 1.85 Å and rapid rotation ($v \sin i = 40$ km/s) flooded the central self-absorption. The resulting splits in the double-peaked profiles was equal to 0.55–1.0 Å. Using these values, the measured intensities, and the Cram-Mullan (1979) theoretical relations, Pettersen estimated the electron density of the chromospheres as 5×10^{11}–5×10^{12} cm^{-3}.

As stated above, the Balmer emission is a sufficient criterion for attributing a red dwarf to flare stars, but this is not a necessary condition. There are red dwarfs without hydrogen emission, at which flares characteristic of the UV Cet-type stars were recorded. One of them, GX And, is spectrophotometrically very close to AD Leo, but the level of its flare activity is lower approximately by a factor of 25. Pettersen and Coleman (1981) studied the star in the red region with high spectral resolution and found rather strong absorption of H_α. Cram and Mullan (1979) interpreted the existence of such objects as follows. In a cool dwarf with negligible nonradiative energy flux, the column density of neutral hydrogen atoms at the second level is very low, which results in weak hydrogen absorption lines in the photosphere spectrum with equivalent widths of H_α of less than 0.1 Å. For the dwarfs with growing temperature gradient, which leads to the formation of the chromosphere, the column density $n_2 l$ increases, and so does the absorption in the Balmer lines to $W_{H_\alpha} \sim 0.7$ Å. GX And is characterized by a similar situation. As the temperature gradient and the mass of the chromospheric layer further increase, the column density of collision-excited atoms grows at higher levels, and the Balmer emission arises first in the outer wings of the absorption profiles, then as a pure emission double-peaked profile. In this case, the electron density in the upper chromosphere exceeds 10^{11} cm^{-3}. (The prevalence of collision over photoionization in the formation of the H_α line for this electron density was noted by Fosbury (1974), this is a qualitative distinction of dMe stars from the Sun.)

1.3 Chromospheres and Transition Regions

The study of Cram and Mullan (1979) was an important step in understanding the chromospheres of cool stars. They proved that the difference between dMe and dM stars consisted not in the presence/absence of chromospheres, but in the fact that dMe stars had rather strong chromospheres, which could be characterized by a certain average increased density in the uniform models or by the filling of stellar surfaces by active regions.

Using the Lick telescope, Young et al. (1984) obtained the spectra of 77 M dwarfs in the region of the H_α line, among which they distinguished two single stars, Gl 410 and Gl 179, whose absorption profiles had an intermediate form between those characteristic of dM and dMe stars. Similar features were noticed for Gl 15 B and Gl 425 A, therefore Young et al. called them "marginal dMe stars." (Using the term "marginal BY Dra-stars" for the same group of K–M dwarfs, Saar and Bopp (1992) added a dozen potential members of this group.) These intermediary profiles can be presented as absorption profiles of dM stars flooded by the emission of 0.30–0.45 Å, which, according to Young et al., is formed in the active regions of the stars, and the stars are in transition from active dMe stars to nonactive dM stars. Giampapa (1985) estimated the surface filling factor of dM stars by faculae as 0.25. But the presence of a strong absorption of H_α evidenced noticeable nonradiative heating of the stellar atmosphere.

Stauffer and Hartmann (1986) thoroughly studied over 200 M dwarfs. Using the echelle spectra with a resolution of 0.15 Å, they measured the equivalent widths of the H_α line to an accuracy of up to 0.03 Å and the Gaussian half-widths to an accuracy of up to 10%. In their sample, the oldest and the lowest-activity stars had weak absorption H_α lines, then, according to Cram and Mullan (1979), as the chromospheric activity intensified, this absorption line increased and then transferred into emission. The majority of stars studied by Stauffer and Hartmann have H_α absorption lines with an equivalent width of not more than 0.7 Å systematically decreasing for cooler stars, while the average equivalent widths of the H_α emission lines grow toward cooler stars. Nine stars of the sample do not have any H_α lines. The dependence of W_{H_α} on the spectral type demonstrates a scatter that noticeably exceeds the measurement accuracy of the compared values. Stauffer and Hartmann associated this scatter with the difference in metallicity of the stars of different age: metal-poor stars had weaker chromospheres.

Cutispoto and Giampapa (1988) performed spectral monitoring of the H_α line of three dM stars with a resolution of about 40 000. They confirmed the Cram–Mullan conception and first found the low-amplitude variations W_{H_α} in the spectra of this type of stars (Gl 15 A, Gl 411, and Gl 526), which did not fit into the idea on stationary heating of the atmospheres of such stars and required a certain variable component. Later, the low-amplitude variations of the H_α absorption line in the spectra of several K–M dwarfs were suspected by Panagi et al. (1991) based on the observations at the 4.2-m William Herschel telescope with a resolution of about 10^5.

According to the calculations by Cram and Giampapa (1987), the observed profiles can appear both in summing up the radiation of heterogeneous sections of the stellar surface and when the general density of the chromosphere increases. Comparison of the H_α lines and the calcium emissions made them conclude that a chromosphere with a nonuniform surface was more probable. However, later Giampapa (1992) used the observations of Stauffer and Hartmann (1986) to revise the factors governing the observed broad activity range of red dwarfs, in particular, the considerable diversity of the H_α line: the variations of the filling factor of their surfaces by the active regions or the changing structure of chromospheres with height. He demonstrated that if any surface inhomogeneities of the chromospheres played the governing role, the widths of the H_α absorption line in the integral stellar spectra should either be independent of their equivalent widths, or increase as the equivalent widths decrease. But the data of Stauffer and Hartmann provide a definitely different pattern: systematic narrowing of the H_α absorption in the course of the decrease of their equivalent width. This result disproves the nonuniform model with the stellar surface including emission and absorption sections. This argument supports the decisive role of the changes of the chromosphere structure with height in the provision of the observed diversity of M star spectra. However, the obvious scheme of the stars with different degree of coverage by active region is still widely used (see Sect. 1.3.2.2). The recently discovered contribution of microflares to the Balmer emission (Alekseev et al., 2003) further complicates the situation.

Young et al. (1989) analyzed the observations of the H_α line in the spectra of M stars obtained by Stauffer and Hartmann (1986), Bopp (1987), and Young et al. (1987a) with a resolution of 0.15–0.5 Å. In all samples, the H_α emission widths were almost equal: FWHM = 1.3 Å. Selecting for dMe stars dM stars of appropriate spectral types with maximum absorption in the H_α line, Young et al. (1989) determined the excessive emission for each star and considered it as a measure of nonradiative heating of the chromosphere. This approach differs from the Cram and Mullan concept (1979), in which the chromosphere is considered responsible for the appreciable share of the absorption in the H_α line. But practically, the estimates of the excessive emission systematically differ in the two approaches by only a small amount that depends on the spectral type. The excesses of W_{H_α} found by Young et al. (1989) varied from 1 to 7 Å and the excessive absolute luminosities ΔL_{H_α} calculated from them systematically reduce for later stars. The most reliable correlation was found between this luminosity and the bolometric luminosity of a star, which suggests a very weak dependence of ΔL_{H_α}, that is equal to the product of the brightness of active regions and the filling factor, on stellar mass: $\Delta L_{H_\alpha} \sim M_*^{0.3}$.

Herbst and Miller (1989) measured the equivalent widths of the H_α lines of 118 K–M dwarfs using the filter technique at the Van Vleck Observatory. Considering the earlier spectral and photometric measurements of these values, they analyzed the chromospheric activity of 380 red dwarfs. In the diagram

W_{H_α}, $R - I$ one can clearly see the "main sequence" with the H_α absorption line and about 70 emission stars with a great dispersion of the values of W_{H_α} that, on average, grow toward later stars. The calculated luminosities L_{H_α} and the surface fluxes F_{H_α} of the emission stars systematically decrease toward cooler stars, whereas the upper limit of the ratio $R_{H_\alpha} = L_{H_\alpha}/L_{\rm bol}$ is practically constant and equal to $10^{-3.9}$ for the stars of all considered spectra. According to Delfosse et al. (1998), for M dwarfs this saturation level is $10^{-3.5}$.

Using a less precise method, Giampapa and Liebert (1986) and Liebert et al. (1992) continued the measurements of Stauffer and Hartmann (1986) for the weakest M dwarfs and found that the dispersion of the values of H_α equivalent widths increased with the increase of the absolute magnitude. The coolest star with H_α absorption, Gl 283 B, has $M_V = 16\overset{m}{.}8$, and stars with $M_V > 17^m$ have the greatest W_{H_α}. As opposed to the prevailing opinion, they found that among M dwarfs later than M5.5 there were stars without H_α emission and among the dwarfs with $M_V > 15^m$ the number of dMe and dM stars was comparable. The objects with the kinematics of the young disk prevailed among dMe stars, while among dM those with the kinematics of the old disk and halo dominated.

Robinson et al. (1990) used the 3.9-m Anglo-Australian telescope for high-resolution observations of H_α and CaII H and K in the spectra of 50 K5–M5 dwarfs. They published the records of the obtained spectra and determined more precisely the contribution of photospheres to the measured emission fluxes and the basal levels of chromospheric activity. Based on the observations of the H_α line they proved the existence of the lower and upper limits of W_{H_α} for different spectral types: the maximum absorption of this line decreased, while the maximum value of the emission increased toward later spectral types. They proposed qualitative estimates of the chromospheric fluxes for four typical levels of chromospheric activity:

$F_K = 5 \times 10^3$ erg/cm^2 s and $W_{H_\alpha} = -0.3$ Å,
$F_K = 5 \times 10^4$ erg/cm^2 s and $W_{H_\alpha} = -0.5$ Å,
$F_K = 10^5$ erg/cm^2 s and $W_{H_\alpha} = 0$,
$F_K = 10^6$ erg/cm^2 s and $W_{H_\alpha} = 5$ Å.

Pasquini and Pallavicini (1991) carried out observations with high S/N ratio using the echelle coude spectrograph at the 1.5-m telescope of the South European Observatory. They studied about 70 F6–K5 dwarfs with different activity level in the regions of H_α and CaII H and K lines. They compared the calculated surface fluxes, confirmed the results of Shakhovskaya (1974b) on systematic growth of the ratio F_{H_α}/F_{HK} toward cooler stars and established the equality of these fluxes in K3–K5.

Using the echelle spectrograph of the Keck telescope, Basri and Marcy (1995) obtained the spectra of 8 stars near the lower limit of the main sequence. They found that in the spectrum of the very late and rapidly rotating ($v \sin i \sim 40$ km/s) star BRI 0021-0214, a candidate for brown dwarfs, classified as M9.5+, the H_α emission and absorption lines were not seen. This can evidence fast weakening of magnetic heating of the stars with masses lower than 0.09 $M_\odot$ and a qualitative change of the mechanism of the loss of angular momentum for such stars, rather than a very cool atmosphere with negligible column density of hydrogen atoms at the second level. However, Tinney et al. (1997) suspected variable H_α emission on this star. With the same telescope Basri et al. (1996) carried out high-dispersion observations of 17 dwarfs of M6.5 and later spectral type with $v \sin i$ varying from < 5 to 40 km/s. In all cases, they detected the H_α emission with an equivalent width of 0.4–20 Å. The surface fluxes calculated using these values were noticeably weaker than for earlier M dwarfs, which conflicts with the already known systematic growth of F_{H_α} from F up to middle M dwarfs. Thus, rapid rotation of late M dwarfs does not guarantee stronger H_α emission, which is probably due to the conversion of the envelope dynamo into the distributed dynamo that is less sensitive to rotation.

With the 2.2-m telescope in Calar Alto and at the 2.5-m Isaac Newton telescope at the Canary Islands Fernandez-Figueroa et al. (1994) and Montes et al. (1995a,b, 1996) observed about 50 chromospherically active binary systems, including more than a dozen binary systems composed of active dwarfs with orbital periods varying from 0.7 to 6 days. The observations were performed in the regions of H_α and CaII H and K with a resolution of 0.1–1.0 Å. To estimate the level of chromospheric activity the researchers used the method developed by Young et al. (1989): they subtracted the spectra of nonactive stars of close spectral types and luminosities from the observed spectra of active systems. They concluded that the thus-obtained excess of the H_α emission was a preferable index of the chromospheric activity. They found a systematic decrease of the index with increasing axial rotational period and the Rossby number close to linear correlation of this H_α index with analogous indices with respect to CaII H and K lines, as well as nonlinear correlation of the index with the radiation in the CIV lines and in X-rays. For equal axial rotation periods of the components of binary systems, they obtained a stronger calcium emission than in single stars. In five systems, they revealed a strong emission of H_ε, which was probably pumped by the strong emission in the CaII H line.

Of particular interest are the observations of the Gl 890 (=HK Aqr) star, one of the fastest rotators among red dwarfs with $v \sin i \sim$ 70 km/s. Young et al. (1990) obtained a relatively dense observational series for this star and found an obvious asymmetry of the H_α line profile. Over the half-period of its axial rotation they revealed that the position of the bisector of the line changed as its equivalent width changed from 2.0 to 1.2 Å. They found the close correlation of the H_α emission with MgII h and k, as well as the localization of the minima of these emissions in the region of the broad minimum of stellar brightness in the V band, which meant noticeable mismatch of active and spotted regions on the star. The mismatch of the maximum of the equivalent width of the H_α emission and the minimum of stellar brightness was confirmed by Byrne and Mathioudakis (1993) in the course of special photometric and spectral monitoring of the star in August 1991. Byrne and McKay (1990) analyzed its ultraviolet spectra and did not detect a noticeable increase in the emission, which should have been expected for such a fast rotation. The fluxes in the MgII h and k lines and the FeII λ 2600 Å blends showed a distinct anticorrelation with the spottedness of the photosphere, and this phase relation preserved for at least two years. They interpreted the detected decrease in the brightness of the MgII h and k lines at a certain phase as the eclipse of the star by the low-mass component. However, Doyle and Collier Cameron (1990) noted simultaneous weakening of H_α, which was not observed in the broadband photometry, and concluded that the eclipse was caused by a cloud of neutral gas rather than by a small star. Later, this conclusion on the eclipse by the gaseous cloud was supported by Byrne et al. (1992b), who found that the emission of the CaII K line was also in antiphase with the stellar brightness. Further observations of the star required the combination of the bright active region in H_α preserved at least for several days and the flare activity of the star (Byrne et al., 1994). During three nights, Byrne et al. (1996) carried out spectral monitoring of the star with a resolution of 39 000 using the 3.9-m Anglo-Australian telescope. They obtained about 50 echelle spectra in the wavelength range from 4813 to 6810 Å with 5-minute exposures; H_α, H_β, and HeI D_3 were recorded simultaneously on the CCD matrix. Upon analysis of H_α profiles they concluded that from time to time absorption details passed along the chromospheric emission profile from the blue wing to the red one with velocities not higher than $v\sin i$, and this transition lasted for less than a half-revolution of the star. The simplest interpretation of these observations is that up to 7 structures of a type of low-latitude prominences pass in the stellar corona, they are preserved for at least 5 revolutions of the star and are projected on its disk. The fact that the star Gl 890 has one of the greatest axial rotation rates but does not demonstrate a noticeably increased activity level as compared to other dMe stars, according to Skumanich and McGregor (1986), can be caused by the saturation of magnetic activity.

By now, at least two late fast rotating dwarfs have been found on which prominence-type phenomena are suspected. These are the K5 stars BD+22°4409 with $v\sin i = 69$ km/s and RE 1816541 with $v\sin i = 61$ km/s, its axial rotation period is practically equal to that of HK Aqr, about 11 h (Byrne, 1997; Eibe et al., 1999). Probably, the electromagnetic forces retain such structures above the stellar equator at a distance of about two radii. According to Ferreira (2001), such structures can be maintained above the middle latitudes as well.

Comparing the spectral and photometric observations of DH Leo in 1984 and 1987, Newmark et al. (1990) noted that as the spot area decreased by 50%, the intensity of Balmer lines decreased only by 20%, which suggested that 60% of the chromospheric emission on the star occurred in the quiescent chromosphere not related to the spotted areas. The measured ratios of the fluxes in the Balmer lines are $F_{H_\alpha} : F_{H_\beta} : F_{H_\gamma} : F_{H_\delta} = 2.07 : 1 : 0.67 : 0.43$, which is close to the ratios of the solar chromosphere. The most complete Balmer decrement was measured for the flare star Gl 431 by Doyle et al. (1990b):

$$F_{H_\alpha} : F_{H_\beta} : F_{H_\gamma} : F_{H_\delta} : F_{H8} : F_{H9} : F_{10} : F_{11} : F_{12} : F_{13} : F_{14} =$$
$$= 4.7 : 1.7 : 1 : 0.57 : 0.34 : 0.21 : 0.14 : 0.13 : 0.10 : 0.06 : 0.03 \, .$$

Strassmeier et al. (1993) studied the chromosphere of one of the most active single stars, K2 star LQ Hya: the surface fluxes in the CaII H and K lines and the IR triplet and H_ε in it are close to those recorded in binary systems of the RS CVn type, and even the intensities of the emission calcium lines and the equivalent widths of emission averaged over the disk considerably exceed the appropriate values in the solar faculae. The H_α line profiles have two peaks and the "blue" peak is always higher than the "red" one, as on the flare dMe stars DK Leo, V 1005 Ori, and AU Mic. Apparently, this asymmetry can be stipulated by the proper velocity field in the chromosphere. Using the Cram–Mullan theory (1979) for an isothermal chromosphere, Strassmeier et al. estimated the density of matter on the level of formation of the H_α line from the splitting of the profile peaks of $(1–5) \times 10^{11}$ cm^{-3}. This value coincides with the boundary value of the density for the transition from radiative to collision-controlled emission of H_α line. At the phases of the maximum brightness of the star there is a minimum value of FWHM, minimal splitting of its peaks and consequently, above the spots the density of the chromosphere is lower than above the quiescent photosphere.

1.3.1.3 Other Emission Lines in the Optical Range

In the mentioned high-dispersion spectrogram of EV Lac obtained by Wilson (1961) in 1960 at the Palomar telescope, in addition to 22 absorption lines and the lines of hydrogen and calcium emission, there are the emission lines of FeI λ 3719.9 Å and λ 3859.9 Å, HeI λ 3888.6 Å and λ 4471.5 Å, SiI λ 3905.5 Å, and CaI λ 4226.7 Å. The lines of metals are narrow, those of helium are slightly broadened, and the lines of hydrogen are noticeably broadened. Worden and Peterson (1976) recorded the D_3 emission of the line of neutral helium in the spectra of several flare stars, in the course of one-week observations they recorded a five-fold slow increase of the equivalent width of the D_3 line of helium under simultaneous weakening of the H_α emission and emission cores of the D lines of sodium.

Giampapa et al. (1978) obtained, at the Kitt Peak Observatory, a series of spectrograms of the quiescent state of the star AD Leo in the red region with the resolution of 0.22 Å. In the summed spectrum corresponding to a 5-hour exposure they found the emissions of H_α, the lines of neutral helium λ 5876 Å and λ 6678 Å and D lines of sodium with equivalent widths of 1.4, 0.31, 0.058, and 0.7 Å, respectively. The ratio of intensities of triplet and singlet helium lines evidenced the collisional character of their excitation in the upper chromosphere at 20 000–50 000 K.

The D_3 helium absorption line is absent in the spectrum of the quiescent photosphere of the Sun but can be seen in active regions, and therefore is an indicator of local magnetic activity. Danks and Lambert (1985) obtained high-resolution spectra of 20 F–G–K stars in the D_3 region of the helium line and found a correlation of the equivalent widths of this absorption line with the chromospheric emission of CaII H and K and the coronal emission in soft X-rays. The correlations are close to linear, which can occur if they are stipulated by the surface filling by faculae. The rotational modulation of the D_3 line is found in the spectrum of χ^1 Ori and suspected for κ Cet and ε Eri. Cutispoto and Giampapa (1988) performed spectral monitoring of the helium D_3 line in the spectra of χ^1 Ori, ξ Boo A, and 70 Oph A with a resolution of 80 000 at the Kitt Peak. For χ^1 Ori and 70 Oph A they found the variation of the equivalent widths of the line, exceeding by 2–4 times the measurement errors, which supported the results of Danks and Lambert (1985).

The profiles of the D lines of sodium and the parameters of their emission cores in the spectra of seven red dwarfs are presented in the paper of Worden et al. (1981).

Shcherbakov (1979) obtained the first spectra of the flare stars AD Leo, BY Dra, EQ Peg, and EV Lac in the infrared region. He found a considerable weakening of the components of the absorption triplet CaII and associated it with their flooding by chromospheric emission. During higher spectral resolution observations Pettersen and Coleman (1981) directly revealed the emission cores in all components of the infrared triplet CaII in the spectrum of AD Leo. For earlier chromospherically active stars ξ Boo A, 70 Oph A, and ε Eri the flooding of the core of the line λ 8542 Å was found by Linsky et al. (1979b). Then, analogous data on chromospheric lines of helium, calcium, and sodium in the red region of the spectrum were obtained and described by Pettersen et al. (1984b). According to Pettersen (1988), for dKe and early dMe stars the surface fluxes in the CaII H and K lines exceed the fluxes in the infrared triplet, whereas for late dMe stars the ratio is inverse. Later, Pettersen (1989b) observed eight dK5–dM1 stars at the 2.7-m and 2.1-m telescopes of the McDonald Observatory, and the main McMath solar telescope with a resolution of 0.46–0.20 Å. He studied the regions of the D lines of sodium and helium, H_α, HeI λ 6678 A, LiI λ 6707 Å, and the infrared triplet CaII. The accompanying photometric monitoring identified the spectra related to the quiescent state of the stars. All considered stars, except for 61 Cyg, were known as spotted dwarfs, but in the above lines the rotational modulation was not noticed, though the monthly and annual changes of the line intensities were found. The weak emission of D lines of sodium was found in the spectra of V 1005 Ori and AU Mic, in the spectrum of BY Dra the absorption of D lines of sodium was to a great extent flooded by the emission, and the emission of HeI D_3 was seen in the spectra of V 1005 Ori and DK Leo. The HeI λ 6678 Å line was not found in either of the stars. The absorption line of lithium was very strong in the spectrum of V 1005 Ori: $W_{\text{Li}} \sim 0.3$ Å, by an order of magnitude weaker in the spectrum of V 833 Tau, but it was absent in the spectra of EQ Vir, DK

Leo, YY Gem, and AU Mic. Distinct emission of all the three components of the infrared triplet of CaII in the center of broad absorption profiles was recorded in the spectra of all considered stars, except for YY Gem, where it was flooded by fast rotation, and the nonactive 61 Cyg. The central emission in the components of the infrared triplet of calcium is the most obvious from the comparison of the spectra of pairs of stars of the same spectral type and different level of chromospheric activity. Foing et al. (1989) made this comparison using the observations of F9–K5 dwarfs with a resolution of 80 000 using the echelle spectrograph at the coude focus of the 1.4-m telescope of the South European Observatory. Using a similar procedure, Latorre et al. (2001) found the emission of the infrared triplet CaII, as well as H_α and H_β, in the spectra of each component of the OU Gem system.

Garcia López et al. (1993) studied the D_3 line of helium in the spectra of 145 F–G stars. This triplet is weaker than the infrared triplet of λ 10830 Å, but is more accessible for observations. Saar et al. (1997) studied the high-resolution spectra of 76 G and K dwarfs and the dependence of the D_3 helium line on rotation. They found that for $P_\mathrm{rot} > 4$ days the flux adsorbed in the line was $F_{D_3} \sim P_\mathrm{rot}^{-1.2}$, and at faster rotation its behavior depended on the spectral type: the flux was almost constant for G stars, decreased for K stars, and transformed into emission for late K dwarfs. For $P_\mathrm{rot} > 4$ days $F_{D_3} \sim \Delta F_{HK}^{1.5} \sim F_{CIV}^{0.7} \sim F_X^{0.6}$, which makes it possible to consider this line as formed in the upper chromosphere.

In the course of 12-year observations of 61 Cyg A Larson et al. (1993) noticed a clear variability of the core of the component of the infrared triplet CaII λ 8662 Å. These variations allowed the axial rotation period of the star to be determined as 36.21 days, which complied with the estimate based on CaII H and K lines. The stability of the 36-day period for many years suggests the long-term existence of active regions on the star or their regular recovery at the active longitudes.

Emission was found in the center of the broad absorption line CaI λ 4227 Å of the flare stars Gl 234 AB, Gl 375, Gl 431, and AD Leo (Fosbury, 1974; Doyle et al., 1990b; Mathioudakis and Doyle, 1991a).

Using the echelle spectrograph with a resolution of about 10^5 at the 4.2-m William Herschel telescope in the Canary Islands Panagi et al. (1991) observed the spectra of 11 K–M dwarfs. They obtained and published the profiles of the H_α line, D lines of sodium, and the components of the CaII infrared triplet. Comparison of the suspected variations of the depths of the D lines of sodium with the previous observations led to the assumption on the activity at the level of the temperature minimum. Andretta et al. (1997) noticed the perspectives of studying the D lines of sodium for the diagnostics of the lower and midchromospheres of M dwarfs. Then, Short and Doyle (1998) obtained and analyzed the high-dispersion spectra in the regions of H_α and D sodium lines of five M dwarfs with different activity level. They found a certain qualitative similarity in the behavior of sodium lines and H_α: as the activity level increases, first the absorption cores of D lines amplify, then emission of these lines develops. For dMe stars, the chromospheric models constructed on the basis of the lines of sodium and H_α are very close, while for low-activity dwarfs, the column density at the level of the transition region from sodium lines is higher by an order of magnitude than that obtained from the H_α line.

Thatcher and Robinson (1993) observed the spectra of early K stars with a resolution of 55 000 to determine a number of indicators of the chromospheric activity at different levels of the stellar atmosphere. Each of the stars was spectrographed during one or two successive nights in the regions of sodium D lines, CaI λ 4227 Å, the green triplet and λ 4571 Å MgI, the CaII H, K, and IR triplet lines, H_α and H_β. As the indicators, they considered the integral fluxes in the calcium and H_α lines calculated as the differences between the profiles of these lines for active and nonactive stars, the equivalent widths of the lines of H_α and H_β, and the depths of their cores. Except for the equivalent widths of the Balmer lines, all the above indicators correlated well.

Byrne et al. (1998) obtained high-dispersion spectra with high S/N ratio for 14 K4–M5 dwarfs with different activity level and found that the behavior of the HeI λ 10830 Å line was in principle similar to that of the H_α line: for low-activity stars this was absorption, with increasing activity, in particular, with growing L_X, the absorption weakened, while on AT Mic, one of the most active stars, it transformed into the emission, though there was no certainty about the absence of flares during the exposure.

Schmitt and Wichmann (2001) found the coronal forbidden line of Fe XIII λ 3388 Å in the optical spectrum of the M6 dwarf CN Leo.

1.3.1.4 Ultraviolet Spectra

The ultraviolet radiation of the quiet Sun comes from its chromosphere and the region transitional to the corona. But it does not form a uniformly luminous layer at a certain level of the solar atmosphere, it comes out mainly from the chromospheric network outlining the boundaries of supergranules: flows in the photosphere rake magnetic flux tubes toward the boundaries. The changes in the chromospheric emission averaged over the solar disk during the 11-year cycle are in close correlation with the portion of the surface covered by the faculae.

In early 1978, an American–European satellite IUE was orbited. It was the first efficient facility for studying ultraviolet spectra of weak stars and other celestial bodies. The primary mirror of the telescope was 45 cm in diameter; the Ritchey–Chretien optical design had a field of view of 16 arcmin. The telescope was equipped with two echelle spectrographs designed for observations in the alternative regime in the ranges of 1150–2000 Å and 1825–3200 Å. When the echelle and the cross-dispersion diffraction grating were used the spectrographs provided a resolution of 10^4. When the echelle was shut down by flat mirrors and only the grating was used the resolution was 6 Å. Each spectrograph had two entrance diaphragms: a round diaphragm with a 3″ diameter and an oval diaphragm with the axes of 10″ and 20″; the latter enabled several spectra to be obtained by sequential shifting of the image. The entrance diaphragms were drilled in the mirror plate that cast the image of the field of view onto the system of fine guiding. The spectrograph detectors, Uvicon television systems, accumulated images. In front of the detectors, multichannel converters of ultraviolet into visible radiation were installed.

IUE was used to study red dwarfs for more than 15 years and provided data for an extensive databank on the ultraviolet radiation of these objects.

Carpenter and Wing (1979) were among the first who obtained the ultraviolet spectra of flare stars at IUE. In the spectra of UV Cet, Proxima Cen, and YZ CMi, they noted the strong CIV resonance doublet and weak OI and SiII lines in the short-wavelength range, and the strong MgII and FeII emission in the long-wavelength range.

Hartmann et al. (1979) obtained the spectra in the range of 1215–1820 Å of the EQ Peg system composed of dM3.5e and dM4.5e stars, and the G8V star ξ Boo A. In their spectrograms, one can see the high-temperature lines of NV, SiIV, and CIV arising in the transition region, the chromospheric CI, OI, CII, and SiII lines and the line HeII λ 1640 Å, whose origin required a particular consideration. The high-temperature lines in the spectra of both objects had similar absolute fluxes per unit surface area, which implied a weak dependence of the properties of the transition region on the effective stellar temperature. These fluxes of emission lines averaged over stellar surfaces are comparable with the appropriate values in the active solar regions or should noticeably exceed them in the case of surface unevenness of the chromospheric structure and the transition regions.

In 1979, Butler et al. (1981) obtained the spectra of three M dwarfs with different levels of flare activity in the IUE short-wavelength range: one of the most active UV Cet-type stars AU Mic, the active star Gl 867 A, and the low-activity star Gl 825. During one exposure of Gl 867 A a flare occurred. In the spectra of AU Mic and Gl 867 A, the NV, OI, CII, SiIV, CIV, HeII+FeII, CI, and SiII emission lines were identified and their absolute fluxes were measured, but in the spectrum of Gl 825 no emission lines were found, though flare activity was recorded before.

In the long-wavelength part of the ultraviolet spectrum of Proxima Cen, Haisch and Linsky (1980) identified the strongest MgII h and k emission doublet and less confidently a number of other lines of metals – FeII, CrII, MnII, FeI, CuI, and TiII. With allowance for the earlier published observations of Carpenter and Wing (1979) and Hartmann et al. (1979), Haisch and Linsky suggested that increased NV emission as compared to the spectrum of the quiet Sun and weakened OI and SiII emissions were generally characteristic of active red dwarfs.

In considering the high-dispersion spectra of a number of G–K stars of different luminosity in the region of the resonance magnesium doublet obtained at IUE with the resolution of 0.20 Å, Basri and Linsky (1979) did not reveal self-absorption in the emission cores in the spectra of ξ Boo A and ε Eri.

Soon, Linsky et al. (1982) published an extensive study on ultraviolet spectra of cool dwarfs based on their observations of seven emission objects (AU Mic, EQ Peg, AT Mic, YZ CMi, Proxima Cen, UV Cet, and EQ Vir) and several nonemission K–M dwarfs within the whole IUE wavelength range. As in the previous observations at IUE, to eliminate the flare effect, the stellar images were shifted along the slit. Figure 11 illustrates a general idea of the character of the low-dispersion spectra obtained: in the wavelength region shorter than 1600 Å these are purely emissive spectra and in the longer-wavelength region the emissions are seen above the weak photospheric background. The main emissions are identified in Fig. 11. Almost simultaneously with the ultraviolet observations, the spectra of the considered stars in the optical region from H_α to CaII H and K were obtained. The following conclusions were made on the basis of analysis of all these data.

The emission lines of the h and k MgII resonance ultraviolet doublet are the strongest chromospheric lines on the Sun, i.e., their radiation is the main mechanism of radiative losses of the solar chromosphere. The fluxes in these lines averaged over the stellar surface systematically decrease toward cooler stars. Sometimes, the total flux in the lines of the CaII H and K resonance doublet and the blends of FeII λ 2610 Å are comparable to or even exceed the fluxes in the magnesium doublet. Nevertheless, the ratio of absolute fluxes in the MgII h and k lines to bolometric luminosity R_{hk} can be considered as a measure of total losses for the chromospheric radiation without regard to hydrogen emission. The data revealed the practical independence of this value from the effective temperature, although for active stars $R_{hk} \sim 10^{-4}$, whereas for nonactive stars $R_{hk} \sim 2 \times 10^{-5}$. But for dMe stars, the fluxes in the Balmer lines significantly exceed the fluxes in the MgII lines, by a factor of 5 for AU Mic and YZ CMi and by a factor of 17 for UV Cet, and exceed the fluxes in all other chromospheric lines taken together. Thus, the hydrogen emission of the stars is the main mechanism of radiative losses of their chromospheres.

Among the emission lines arising in the transition region, the strongest are the lines of the ultraviolet resonance doublet CIV λ 1548/51 Å, and the flux ratio in these lines to the bolometric luminosity R_{CIV} characterizes energy losses for the heating of the region. R_{CIV} is much higher for active stars than for nonactive dwarfs and noticeably increases for cooler stars. The different dependence of R_{hk} and R_{CIV} on the effective stellar temperature suggests different heating mechanisms of the chromospheres and the transition regions.

Figure 12 shows the ratios of fluxes on the surfaces of cool stars in ultraviolet lines arising at different temperatures to the appropriate fluxes from the solar surface. Depending on the value of R_{CIV} the cool dwarfs are divided into 6 groups: 3 groups of dMe stars, active G–K dwarfs ξ Boo A and ε Eri, dM stars 61 Cyg B and HD 88230, and the quiescent G–K dwarfs α Cen A and α Cen B. One can see that all fluxes on dM stars are approximately three times lower than those from the quiet Sun, whereas for quiescent G–K dwarfs they are the same as for the Sun. The fluxes on active G–K stars are the same as those in the active regions of the Sun. Finally, as a rule, active dMe stars have

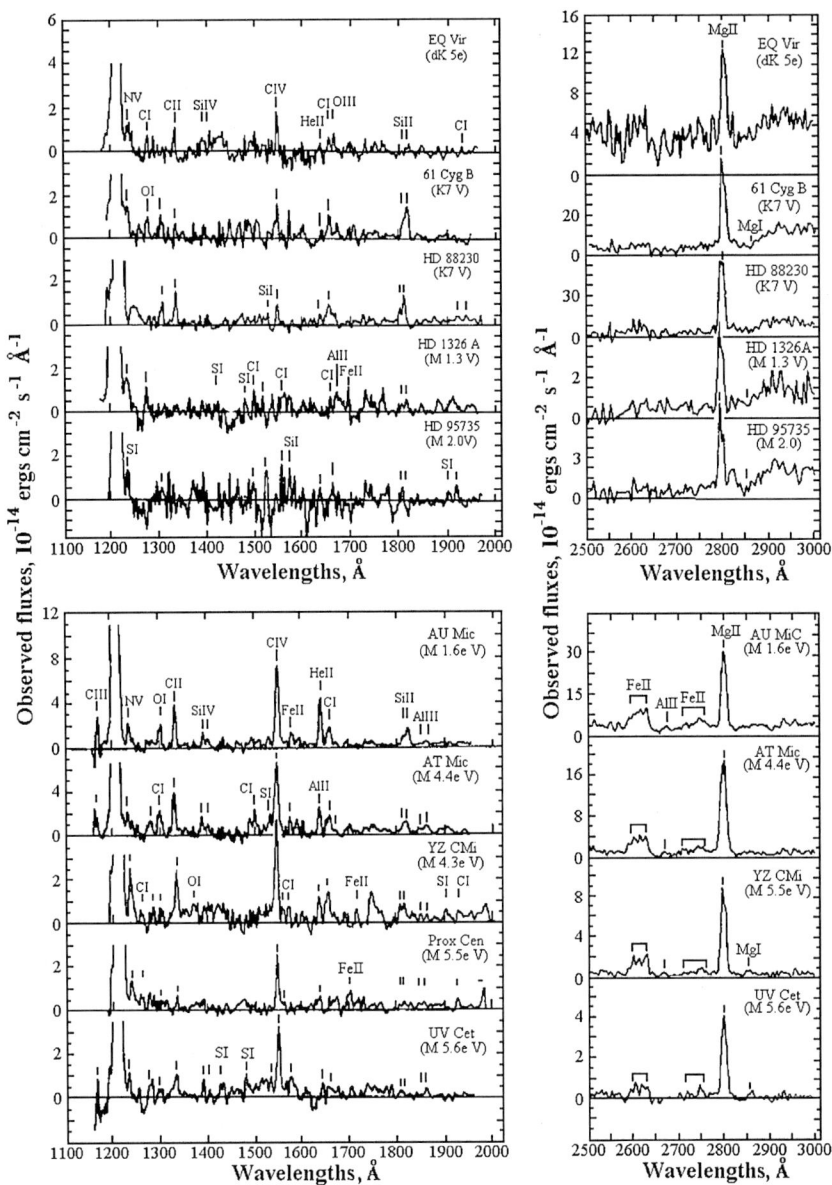

Fig. 11. The spectra of 10 K–M dwarfs obtained at IUE and calibrated in absolute units (Linsky et al., 1982)

even larger fluxes with qualitatively identical or even stronger dependence on the temperature of emission formation. Otherwise, the assumption that the stars are completely covered by structures similar to active solar regions is insufficient to explain their luminosity in these lines. Considering that some dMe stars have spotted chromospheres, the radiative intensities of their active

1.3 Chromospheres and Transition Regions

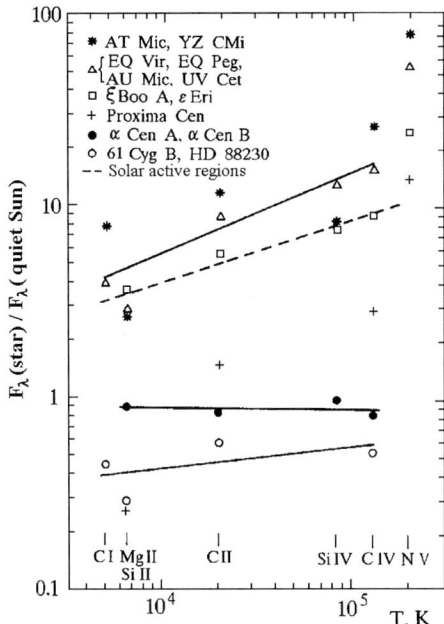

Fig. 12. The ratios of fluxes in ultraviolet lines per unit stellar surface to the appropriate solar fluxes versus their formation temperatures (Linsky et al., 1982)

regions should be even higher. Probably, the cause is denser packing of thin magnetic flux tubes, which occupy not more than 25% of the solar facular areas. The change of the emission measure with the temperature, at which different lines are formed, is similar in all cases, but the values of these measures for dMe stars are approximately 30 times larger and for dM dwarfs three times smaller than for the quiet Sun.

Among ultraviolet emissions, a particular place in the spectra of cool stars is occupied by the line He II λ 1640 Å, because, unlike the other lines of the transition region, it can be excited by recombinations after the second ionization of helium by soft X-rays and therefore reflects the conditions existing in the solar corona rather than those in the transition region. (Later, for solar-type stars Rego et al. (1983) found a rather close correlation of the fluxes in the line λ 1640 Å and in soft X-rays suggested already by Hartmann et al. (1979), which proved coronal excitation of the line. For stars differing from the Sun, this correlation is not valid due to the different contribution of the FeII emission to the blend λ 1640 Å.)

The ratio of HeII/CIV fluxes varies from 0.15 to 0.6 and is independent of the effective temperature of a star, the flux in the CIV lines, and stellar activity. As a convenient activity diagnostics parameter one should mention the SiII λ 1808 and λ 1817 Å lines: they are in good correlation with the magnesium doublet but are recorded simultaneously with the main lines of the transition region.

Table 6. Relative contributions of different emissions to total radiative losses of the chromospheres (after Pettersen, 1987)

Spectral Type	Star	Radiative Losses for			
		Balmer Lines	CaII HK+IR Triplet	MgII hk	Other Lines
dM6e	UV Cet	0.77	0.05	0.07	0.11
dM5e	YZ CMi, AT Mic	0.60	0.10	0.15	0.15
dM4e	EV Lac, EQ Peg	0.53	0.21	0.11	0.15
dM1e	AU Mic, YY Gem	0.43	0.29	0.14	0.14
dM0e	V 1005 Ori	0.30	0.36	0.25	0.09
dK5e	EQ Vir, BY Dra	0.20	0.40	0.20	0.20
dG2	Sun	0.13	0.64	0.23	

Along with the considerable similarity of the ultraviolet spectra of cool dwarfs, whose photospheric spectra differ noticeably, one should note the weakening of the lines SiII λ 1808/17 Å and HeII λ 1640 Å for active M stars, which can be caused by a sharper temperature gradient and/or higher atmospheric pressure. The appropriate line for estimating the density of matter is the intercombination line SiII λ 1892 Å.

Relative contributions of different emissions to total radiative losses of chromospheres are shown in Table 6.

Assuming the total radiation of the strong ultraviolet lines of OI and SiII proportional to all radiative losses of the chromosphere and the total radiation of the SiIV, CIV, and NV lines proportional to all radiative losses of the transition region, Oranje et al. (1982) compared these summary fluxes and found a rather close relation

$$F_{\rm tr} = 4.3 \times 10^{-3} F_{\rm chrom}^{1.44 \pm 0.02} . \tag{5}$$

Using a more complete sample of A–M stars, Oranje (1986) revealed close relations that could hardly be distinguished from the linear correlation between the intensities of different chromospheric lines and between the intensities of the transition region lines, but in comparing the emissions of these two types he found the relationships similar to (5). In particular, the fluxes in the ultraviolet chromospheric lines of CI λ 1657 Å, OI λ 1305 Å and SiII λ 1808/17 Å were proportional to F_{MgII} to the power of 1.14–1.22. The fluxes in the transition region lines of CII λ 1335 Å, SiIV λ 1400 Å, CIV λ 1548/51 Å, and NV λ 1240 Å are proportional to the flux in the MgII lines to the power of 1.6–1.7. All the found relationships had a correlation coefficient above 0.9. But the emission K–M dwarfs noticeably deviated from these general relationships, probably because the Balmer lines played the role of the main source of radiative losses in their atmospheres. There was a nonlinear dependence between coronal and chromospheric radiation similar to (5).

On studying ultraviolet spectra of close binary systems with solar-type components, Vilhu and Rucinski (1983) discovered the saturation effect in the "rotation–activity" dependence: for $P_{\rm rot} < 3$ days, the intensities of the transition-region lines did not depend on $P_{\rm rot}$ and the spectral type. This effect was later studied in different wavelength ranges. Thus, Vilhu et al. (1986) found the upper limits of the ratios $F_{MgII}/F_{\rm bol} < 2 \times 10^{-4}$ and $F_{CIV}/F_{\rm bol} < 2 \times 10^{-5}$ for the main-sequence stars of G0 and later spectral types. First, this saturation effect was discussed within purely geometric considerations on the filling of the entire stellar surface by active regions. But in this case one cannot expect surface fluxes that would exceed the solar values by more than one and a half or two orders of magnitude. Apparently, the found limits are due to the feedback between the rotation of stars with convective envelopes and magnetic fields generated in them: strong fields suppress differential rotation, which restricts the strength of magnetic fields.

Combining the IUE, HST, and EUVE observational results of six stars of late spectral types from the Sun to AU Mic and using the technique of emission measure as a temperature function, Doyle (1996a) found that within the range of 10^4–10^7 K all radiative losses of the chromosphere of a cool dwarf, except for those of the lines and continua of hydrogen, can be estimated practically as a linear function of the measured luminosity of the line CIV λ 1548 Å. Considering this fact, Doyle (1996b) doubted the validity of these saturation levels obtained before on the basis of individual lines. He preferred the estimates of total radiative losses from the measured fluxes of F_{CIV} and the ratio $F_{\rm tot}/F_{CIV}$ found for the Sun and a number of stars. For F–M dwarfs with axial rotation periods from 0.3 to 50 days, Doyle found quite a good correlation between the periods $P_{\rm rot}$ and thus found the fluxes $F_{\rm tot}$ as

$$\log F_{\rm tot} = 8.32 - 1.42 \log P_{\rm rot} \qquad (6)$$

with a correlation coefficient of 0.87. The fastest rotators did not demonstrate saturation, they had only a noticeable dispersion of points. An even better correlation was found between the ratio $R_{\rm tot} = F_{\rm tot}/F_{\rm bol}$ and the Rossby number

$$\log R_{\rm tot} = -3.82 - 1.45 \log {\rm Ro} \qquad (7)$$

with a correlation coefficient of 0.92. However, one can clearly see here the saturation at $\log {\rm Ro} < -1$, which corresponds to the axial rotation period of about two days. In this connection, Doyle suggested that the saturation should not be related to the surface filling by active regions or to the rearrangement of the dynamo mechanism mode, but only to the disregard for radiative losses for hydrogen, which should be large for fast rotators at the expense of an increased temperature minimum and increased density of the chromosphere.

Ayres et al. (1983a) compared high-dispersion ultraviolet spectra of three chromospherically active stars χ^1 Ori (G0V), ξ Boo A (G8V), and ε Eri (K2V) and two nonactive stars α Cen A (G2V) and α Cen B (K1V). Active stars displayed surface fluxes in the lines that exceeded the appropriate fluxes of quiescent stars by an order of magnitude, but the profiles of the main lines were practically identical. There was a similar difference in the spectra of the regions of the solar photosphere with strong and weak magnetic fields. This called into question the idea that the difference in the spectra of two star groups was due to the differences in filling factors, though the differing ratios of the intensities of the line SiII λ 1892 Å to the intensities of the lines of CIV λ 1548/51 Å and SiIV λ 1394 Å evidenced certain qualitative changes in the chromospheres. As in the spectra of the transition region of the Sun, the described stellar spectra illustrate a systematic and stable red shift of the SiIV and CIV lines by 4–8 km/s with respect to the chromospheric lines. Apparently, this is a result of a certain vibrational regime in small loops of magnetic flux tubes, which occurred under different conditions in the bases of such loops (Mariska, 1987).

Boesgaard and Simon (1984) observed the young G0 star χ^1 Ori by IUE and found increased surface fluxes as compared to the spectrum of the quiet Sun: in the MgII lines by a factor of 2.4, up to 6 in the other chromospheric lines, and up to 11–22 in the transition-region lines.

To study the profiles of ultraviolet emission lines, Ayres et al. (1983b) observed AU Mic with IUE and obtained a high-resolution 18-h exposure of the short-wavelength part and 4-h exposure of the long-wavelength part of its spectrum. Regular photometric monitoring did not reveal noticeable flares. Symmetric narrow emission lines without noticeable wavelength shifts were recorded. The values of FWHM were within 74–30 km/s, the latter value was close to the width of the instrumental profile. The widths of the transition-region lines were almost the same as for earlier dwarfs, which evidenced that the kinematics and heating mechanism of the stars with different luminosities, from solar-active regions to M dwarfs, were common. But the width of the MgII k line corrected for the instrumental profile was half that of G–K dwarfs.

Based on IUE observations of the flare star BD + 26°730 with small axial rotation period ($1\overset{d}{.}8$), Bopp et al. (1983) found that the absolute fluxes of the ultraviolet lines in its spectrum were greater than in all earlier studied dKe-dMe stars.

High-resolution observations at IUE made it possible to obtain MgII h and k line profiles and compare them with the characteristics of CaII H and K lines. Thus, Blanco et al. (1982) found this emission starting from the spectral type A7, which was much earlier than the calcium emission occurred. They found the independence of the ratio R_h from the effective stellar temperature, the exponential decrease in the flux in these lines with increasing axial rotation period, and decrease in the ratio of MgII/CaII fluxes with increasing radiative losses of stellar chromospheres. For the F2 dwarf 78 UMa Simon (1986) recorded high-temperature ultraviolet emission lines with the surface fluxes exceeding those from the Sun by a factor of 4.

Oranje and Zwaan (1985) found a clear linear dependence between the F_{hk} and F_{HK} fluxes. Later, Schrijver et al. (1992) compared the observations of 26 F5-K3 dwarfs in the CaII H and K and MgII h and k lines made at intervals of not more than 36 h and refined the linear relations between the appropriate surface fluxes. They determined the basal level of chromospheric activity in calcium emission and showed that in previous studies the points were dispersed due to the nonsimultaneity of the observations and the uncertainties in the absolute calibration.

Haisch and Basri (1985) compared the ultraviolet spectra of the Sun and 14 G dwarfs and found that the continuum in the region $\lambda > 1600$ Å provided information about the temperature minimum T_{min}. These values displayed noticeable dispersion for the stars of the same spectral types: more active stars had increased T_{min}. The temperature minimum on the Sun had spatial structures up to $1''$, the differences in the emission from this level could be explained by different filling factors of the solar photosphere by faculae.

Bromage et al. (1986) compared the ultraviolet spectra of the quiet Sun and the flare star AT Mic and found that in the stellar spectrum averaged over the surface fluxes in the transition-region lines and the HeII λ 1640 Å line were amplified by a factor of 10–40, whereas in the chromospheric lines, only by several times, while the MgII h and k doublet was weakened by a factor of 1.5.

From the set of lines and their widths determined during 48-h IUE monitoring of the flare star AT Mic with photometric support from South Africa, New Zealand, and Chili, Elgaroy et al. (1988) established that its spectra were very similar to those of AU Mic (Ayres et al., 1983b). In the spectra of AT Mic obtained definitely or hypothetically without flares, more than a dozen lines of ionized iron within the range of 2580–2632 Å were identified. For the range 1300–1820 Å they estimated the average surface fluxes in the lines that were stronger, as a rule, by a factor of 2–5, than in the very active regions on the Sun. The FWHM of the MgII, FeII, CII, SiIV, and CIV lines (in order of increasing formation temperatures) systematically increased from 36 to 88 km/s and their shifts with respect to the chromospheric lines systematically decreased from $+5$ to -12 km/s. Measured widths of the MgII lines in the spectra of AU Mic (Ayres et al., 1983b), AT Mic (Elgaroy et al., 1988), and AD Leo (Ambruster et al., 1989b) suggest that the Wilson–Bappu relation for the widths of the emission components of these lines and the stellar absolute luminosities is valid up to red dwarfs. However, later observations by Elgaroy et al. (1990) revealed that for a number of red dwarfs this relation was not valid: observed widths are too narrow, probably due to the small optical width.

On the basis of optical and ultraviolet observations of AD Leo Sundland et al. (1988) constructed a general spectrum of the quiescent star from the infrared triplet of calcium up to the NV line in the far ultraviolet. As follows from Table 7, the chromospheric lines add up to 4×10^6 erg/cm^2 s, while the lines of the transition region are 3×10^5 erg/cm^2 s. But this estimate disregards numerous lines of the transition region in the further ultraviolet. Using the solar spectrum in this region, from four measured lines of the transition region Sundland et al. estimated the total luminosity of the lines of the transition region of AD Leo as 10^6–10^7 erg/cm^2 s.

Table 7. Absolute radiative losses for different emissions from a unit surface of AD Leo of $10^4 \mathrm{erg/cm^2\,s}$ (Sundland et al., 1988)

CaII IR triplet	29
H_α	118
H_β	59
H_γ	24
H_δ	20
CaII H + H_ε	27
CaII K	18
MgII h+k	54
FeII λ 2751 Å blend	9
FeII λ 2610 Å blend	30
CIV λ 1548/51 Å	13
SiIV λ 1394 Å	2
CII λ 1335 Å	4
NV λ 1240 Å	3

Vilhu et al. (1989) measured the flux in the MgII h and k lines in the spectrum of the binary eclipse star CM Dra, whose components were dM4 stars, are apparently completely convective. Considering these measurements and published data, they determined more precisely the saturation levels of the CIV and MgII emissions in the range of spectral types F–M for normal stars with convective envelopes: in the CIV lines it was $\sim 10^6 \mathrm{\,erg/cm^2\,s}$ up to G0 and then decreased monotonically to $2 \times 10^5 \mathrm{\,erg/cm^2\,s}$ toward late M stars; in the MgII lines $\sim 10^7 \mathrm{\,erg/cm^2\,s}$ up to G7 and then decreased monotonically to $8 \times 10^5 \mathrm{\,erg/cm^2\,s}$ toward late M stars. For the latter, the saturation was achieved near the axial rotation period of about 5 days, for G stars of about 3 days. It was suggested that under saturation the whole stellar surface was filled with magnetic fields, whose strength was determined by gas pressure in the photosphere.

To estimate the total radiative losses of the upper stellar atmospheres at 2×10^5–2×10^6 K, Doyle (1989a) used the observational ratio of total losses to the luminosity of the CIV λ 1548 Å line on the Sun and measured fluxes in the line on stars. He obtained total losses of dKe-dMe stars in the region with the temperature above 10^4 K at the level of $(3–4) \times 10^7 \mathrm{\,erg/cm^2\,s}$, which was slightly higher than the appropriate value in the solar-active regions.

Stellar observations enable comparisons of the objects of different age and with different rotation in the broad range of effective temperatures, while the investigation of bright and extensive structures on the Sun enables observations with high spectral resolution, high S/N ratio, and in the regions with different levels of magnetic activity. Thus, Cappelli et al. (1989) compared eight main emissions in the short-wavelength part of the ultraviolet spectra of 45 F5–K5 stars and in 10 different structures of the solar surface – quiet regions, faculae of different strength, and flares – and found a considerable similarity. In particular, the solar data satisfied well the flux–flux ratios derived from the stellar data and covering three orders of magnitude: surface fluxes of slowly rotating stars were analogous to solar regions with low magnetic activity, while the fluxes of active rapidly rotating stars were similar to the strongest solar faculae and even flares. The weakest fluxes from solar structures were weaker than those from the least active stars, which suggested that even at the most nonactive stars several per cent of the surface could be covered by active regions. The filling factor of the most active objects was close to unity, or the active regions on them were brighter than the brightest solar analogs. But the continuous distribution of the solar structures over the activity level prevented unambiguous estimates of the surface filling factor by active regions.

Byrne and Doyle (1989) performed ultraviolet observations of two dM stars Gl 784 and Gl 825 and found that the surface fluxes in the chromospheric lines, in Ly_α and in the transition region lines were somewhat lower than on the quiet Sun. This means that the ratios $F^*/F_\odot$ are independent of the temperature of emission formation, whereas for dMe stars these ratios grow rapidly toward increased temperature of line formation. The fluxes in the MgII lines for these dM stars are much less than for dMe stars and the quiet Sun, but they continue to be the main coolers of the chromosphere surpassing the efficiency of Ly_α. The fluxes in nonhydrogen lines in the spectra of these dM stars are close to the appropriate fluxes in the spectra of nonactive K dwarfs 61 Cyg A and B. Thus, Byrne and Doyle concluded that the transition from dM to dMe stars was stipulated by the growth of the surface filling factor by active regions. Then, Doyle et al. (1990a) compared the fluxes in the MgII h and k and Ly_α lines in the spectra of 21 M dwarfs and found that the losses for Ly_α in dMe stars were almost twice as high as those in the MgII lines, they amounted to about 25% of the total radiative losses of the chromosphere and were comparable with the losses for H_α. However, such a situation can occur only if there is a temperature plateau in the lower part of the transition region or with different filling factors for Ly_α and H_α. In dM stars Ly_α and H_α provide a comparable contribution to the radiative losses. The fluxes in Ly_α of dMe stars are higher by almost an order of magnitude than those in dM dwarfs.

From ultraviolet spectra of the dM(e) stars Gl 105 B, Gl 447, and Gl 793 Byrne (1993c) measured surface fluxes in the Ly_α, CIV, and MgII lines and analyzed them together with analogous data for dM (Gl 784 and Gl 825), dM(e) (Gl 900), and dMe stars (YZ CMi and AU Mic). He found that F_{MgII} of the dM(e) star Gl 900 was actually intermediate between the appropriate values of the dM and dMe stars, but F_{MgII} of newly measured dM(e) dwarfs was lower than in dM stars, for example, for Gl 105 B by an order of magnitude. A similar situation occurs with the fluxes in two other lines. This means that dM(e) stars, Byrne's "zero H_α stars", cannot be considered as stars with nonuniform surfaces, partly analogous to dM and dMe stars, or as stars with uniform surface and nonradiative heating intermediary between the dM and dMe stars.

Landsman and Simon (1991) analyzed normalized fluxes in the Ly_α line, the ratios R_{Ly_α}, for more than 260 stars and found that the emission in this line appeared on late A and early F stars, though they had rather thin convective zones. For flare stars UV Cet, EV Lac, and AT Mic the values of R_{Ly_α} are among the highest, about 3×10^{-4}.

Haisch et al. (1990a) observed the YY Gem system in the ultraviolet and X-ray regions. The level of MgII emission appeared to be very stable over three years, which is apparently due to the uniform distribution of the regions radiating the lines along the longitude or even saturation, since, according to Doyle (1987), the stars with the short axial rotation period, $0\overset{d}{.}8$, should have a 100% surface filling factor for magnetic regions. However, the transition-region lines have rotational modulation and change over a greater time scale.

Applying the method of phase-dispersion minimization to analyze the time series of irregular sampling, Hallam et al. (1991) considered the IUE spectra of F–K dwarfs to find the rotational modulation of the major chromospheric and transition region lines. For the young active K2 dwarf ε Eri, they found modulation of the fluxes of ultraviolet lines with a period of about 2.8 days with preserved active longitudes during several hundred revolutions. This period is four times shorter than that found from the calcium monitoring from the Mount Wilson Observatory and could be its harmonics. Less certain evidence of rotational modulation of ultraviolet fluxes with a period of about 29 days was found for the old G2 dwarf α Cen A.

As in studying the fluxes in the CaII H and K lines, the basal level was found for the MgII, SiII, CII, SiIV, CIV lines, although in the transition-region lines this level was not as far from the detection threshold, as in the chromospheric lines (Rutten et al., 1991). Mathioudakis and Doyle (1992) extended the basal level of the chromospheric radiation of MgII to M dwarfs, and Byrne (1993c) measured the surface flux in the MgII h and k lines in the spectrum of Gl 105 B: the F_{hk} flux is lowest for M dwarfs – 8×10^3 erg/cm^2 s, which is 70 times lower than for the Sun.

Using GHRS HST, Peterson and Schrijver (2001) studied the MgII h and k lines for G dwarfs with metallicity 300 times lower than the solar metallicity. In all objects they noticed magnesium emission with split profiles, but the fluxes in the lines were within a narrower range than the metallicity range.

In 1986, Quin et al. obtained 40 ultraviolet spectra of one of the most active flare stars, AU Mic, for two full revolutions (Quin et al., 1993). The average surface fluxes calculated using the data are presented in Table 8. The accuracy of the estimate of absolute fluxes in the strongest lines was assessed as 20–30%. The fluxes in the chromospheric and transition region lines display a considerable dispersion, which is probably due to the frequent low-amplitude flares in the active regions with a filling factor close to unity. Against the background of the dispersion, no rotational modulation of the line intensity was found. From three density-sensitive intensity ratios of the lines (SiIV, CIII λ 1176 Å and AlIII to CIII λ 1908 Å), they obtained well-fitting lower-pressure limits in the region of formation of the lines. Using the

1.3 Chromospheres and Transition Regions

Table 8. Average surface fluxes of AU Mic (after Quin et al., 1993)

CIII	λ 1176 Å	8.8×10^4 erg/cm^2 s
Ly_α	λ 1215 Å	140 erg/cm^2 s
NV	λ 1240 Å	4.2 erg/cm^2 s
SiII	λ 1264 Å	1.1 erg/cm^2 s
OI	λ 1307 Å	6.0 erg/cm^2 s
CII	λ 1335 Å	8.8 erg/cm^2 s
SiIV	λ 1400 Å	7.0 erg/cm^2 s
CIV	λ 1550 Å	13.7 erg/cm^2 s
HeII	λ 1640 Å	10.5 erg/cm^2 s
CI	λ 1660 Å	5.3 erg/cm^2 s
AlII	λ 1670 Å	2.8 erg/cm^2 s
SiII	λ 1820 Å	9.7 erg/cm^2 s
AlIII	λ 1860 Å	1.4 erg/cm^2 s
CIII	λ 1908 Å	<0.7 erg/cm^2 s
MgII	λ 2800 Å	94.8 erg/cm^2 s
FeII	λ 2800 Å	50.8 erg/cm^2 s

method of differential emission measures, they estimated the total radiative losses within 10^4–10^7 K for the pressure $P_e = 3 \times 10^{15}$ cm^{-3} K and for solar content of the elements. According to Quin et al., the total radiative losses in the atmospheres of AU Mic and the Sun are as follows:

The Region of Stellar Atmosphere	Temperature Range	Radiative Losses (10^6 erg/cm^2 s)	
		on AU Mic	on the Sun
Chromosphere	$4.0 < \log T < 4.3$	34.7	2.0
Transition region	$4.4 < \log T < 6.1$	20.4	0.8
Corona	$6.2 < \log T < 7.6$	24.5	0.06
Total losses	$4.0 < \log T < 7.6$	80	3

The total radiative losses of the atmosphere are about 1% of AU Mic bolometric luminosity and 0.003% for the solar bolometric luminosity. The distribution of the losses among different atmospheric levels $F_{\text{chromosphere}}$: $F_{\text{transitionregion}}$: F_{corona} is 1:0.6:0.7 on AU Mic and 1:0.4:0.03 on the Sun.

Landsman and Simon (1993) compiled the catalog of measured fluxes in the Ly_α line for 275 stars, some of them were revised with respect to the interstellar absorption, which increased the values of the fluxes by a factor of 2–5. In almost all cases the radiative losses in Ly_α were comparable with those in the magnesium doublet, while the corrected relative fluxes R_{Ly_α} were in close correlation with the chromospheric emission CII λ 1335 Å formed in the upper chromosphere at a temperature of about 15 000 K.

In 1988, Butler et al. (1994) found for the YY Gem system a close similarity of the light curve in the V band and in the MgII h and k lines near the secondary minimum, which meant fairly complete and uniform filling of the stellar surface by the regions radiating in these lines. In the Ly_α line the eclipse was noticeably longer than the optical one, which can be caused either by greater extension of the radiating region of the eclipsed component or by greater effective part of the eclipsing component at the expense of circumstellar matter.

Doyle et al. (1994) obtained the spectra in the ultraviolet and visible wavelength ranges of four M dwarfs with very low chromospheric activity. Only Gl 821 had weak calcium emission with an absolute flux on the surface of 8×10^3 erg/cm^{2}s and only Gl 813 had weak magnesium emission with a flux of 4×10^3 erg/cm^2 s, half that from Gl 105 B that was considered before as the most nonactive dM(e) star. None of the stars had H_α emission and all stars had low rotation rate.

Simultaneous magnetometric and ultraviolet observations of ξ Boo A revealed strong modulation of the transition region lines in the phase with magnetic flux (Saar et al., 1988). Linsky et al. (1994b) compared the fluxes (IUE) in ultraviolet lines of CIV and MgII with the magnetic fluxes fB (McMath telescope) from the solar-type stars 59 Vir, ξ Boo A, and HD 131511. All three stars demonstrated appreciable variations of the line intensities, but the variations of 59 Vir had no rotational modulation. The comparison of the measured fluxes in the CIV lines and the fB value revealed a clear correlation $F_{CIV} \sim (fB)^{0.66 \pm 0.05}$, which is valid for the solar data presented by Schrijver (1990) and all considered stars, except for 59 Vir.

The next spacecraft that was successfully used to study stellar atmospheres after IUE was the Wide Field Camera (WFC) built in the United Kingdom and mounted together with the X-ray telescope ROSAT, the German–UK–USA satellite launched into orbit in June 1990. The camera was a "nested" telescope with grazing-incidence Wolter–Schwartzschild Type I mirrors and an aperture of 58 cm, a field of view of 5° in diameter, a spatial resolution of 1 arcmin in the center of the field of view and 3 arcmin at the edge, and a position accuracy from 20 arcsec to 1 arcmin. During sky surveys, the apparatus rotated about the axis perpendicular to the Sun, shifting the swath by one degree every day, thus each object was observed for 5 days every 96 min with an exposure of up to 80 s. In addition to survey observations, the apparatus was used to make more than 6000 long-duration exposures of individual pointing. The detectors were multichannel plates with a curved focal surface sensitized by cesium iodine. The filters were layered on a thin film and separating the bands 90–210 Å (S_1), 62–111 Å (S_2), 56–83 Å (P_1), and 17–24 Å (P_2) were mounted in front of the detector; in survey observations the filters were changed once a day. Thus, the apparatus was suitable for studying the upper transition regions and the lower coronae of cool stars.

WFC discovered the active system RE 0618+75, the binary system with a period of $0\overset{d}{.}54$ of two very rapidly rotating dM3e stars. Its orbit, spatial orientation, and global characteristics of the components were determined in optical observations (Jeffries et al., 1993). Though it is a fast rotator, the level of its activity does not exceed that of slower rotating dMe dwarfs, which supports the idea of the existence of the activity maximum stipulated either by the fact that the surface filling factor for active regions approached unity, or by the reverse negative feedback in the dynamo mechanism.

In the course of WFC observations, Jeffries and Bromage (1993) found strong EUV radiation of the late dwarf Gl 841 A. The subsequent optical observations revealed the binarity of the source with a period of about 1.12 days, the emission in H_α and CaII in each component and their flare activity.

Wood et al. (1994) made a statistical analysis of all nondegenerated stars within 10 pc recorded during the WFC all-sky survey. Among 220 objects known in these vicinities of the Sun, 41 objects were found with confidence and 14 were suspected at least in one photometric band. Luminosities of the objects were calculated in the S_1 and S_2 bands and compared with X-ray luminosities from the IPC Einstein Observatory data (see Sect. 1.4.1). The luminosity functions calculated for different spectral types demonstrated similarity for the stars of middle spectral types and a noticeable dispersion for M dwarfs: in EUV their luminosities were noticeably lower, which is probably due to the saturation effect of the ratio R_{EUV}. Most members of binary systems do not differ from single stars of the appropriate spectral types in L_{EUV}, but in the systems AT Mic, ξ UMa, FK/ FL Aqr, Wolf 630, ξ Boo, and EQ Peg there is an increased EUV radiation. Brighter sources have a higher ratio L_X/L_{EUV}, which can be explained by a higher coronal temperature. For AU Mic, the brightest coronal source in the radius of 10 pc from the Sun, this temperature achieves 4×10^7 K. The dependence L_{S1} and L_{S2} on rotation is steeper than $L_X(v_{\text{rot}})$.

The Hubble Space Telescope (HST) that started its operation in April 1990 opened new opportunities for obtaining ultraviolet spectra of flare stars with much higher resolution and S/N ratio than before.

In the course of cooperative observations of AD Leo in May 1991 Saar et al. (1994c) obtained spectral data from HST. The constancy of fluxes in the lines for the times 5, 10, and 30 s was analyzed. All lines showed variability at all these times. In a certain conventional definition of a flare, 35% of the flux in the SiIV lines can be attributed to the constant level, 40% to flares, and 25% to microflares or active regions. If another definition of a flare is used, the last part includes 10–20% of energy. But in the chromospheric CI line, 80% of the energy belongs to stationary radiation. Later, Saar and Bookbinder (1998a) performed analogous observations of two young dwarfs in the range 1380–1670 Å: G0 star HD 129333 with an axial rotation period of 2.8 days and K2 star LQ Hya with a period of 1.8 days. For a time resolution of 1 s they found low-amplitude flares, which amounted to 8% of the radiation from the transition region of HD 129333 and 11% in LQ Hya, in the summary flux of the CIV+SiIV lines.

Linsky et al. (1994a) studied the red shift of the transition-region lines with respect to photospheric lines for the stars of late spectral types. It is known that on the Sun such shifts are observed above the regions with strong magnetic fields and are caused by the plasma flows descending along the magnetic flux tubes. Linsky et al. considered one of the most active red dwarfs, AU Mic. In its spectra obtained with a resolution of 20 000 in the region of the CIV and SiIV lines, they recorded such a shift of 4.1 ±2.2 km/s. In studying the components of the α Cen AB system, Wood et al. (1997) found that as on the Sun, the red shift of the transition region lines grew with increasing temperature of line formation to $\log T = 5.1$, and then decreased. From the ratio of line intensities OIII], sensitive to the electronic density, at the level of $\log T = 5.14$ they obtained the estimates $\log n = 9.65$ and 9.50 for α Cen A and B, respectively. These estimates agree with the values of electron pressure in the coronae obtained from EUVE data.

Linsky and Wood (1994) observed the flare star AU Mic with 162-s exposures: 17 exposures in the region of the CIV doublet and 22 exposures in the region of the SiIV doublet. The profiles of CIV lines did not show time variations, but their wings up to 200 km/s exceeded by a factor of three the wings of the instrumental profile. The observed profiles of both lines were presented by the sums of two Gaussians with widths of about 30 and 170 km/s with negligible relative shifts of their centers. The widths found noticeably exceed the expected Doppler widths for the temperature of formation of the lines. Narrow components resemble the profiles of the lines observed in solar-active and quiet regions, while the wide components resemble those found in the so-called explosive phenomena on the Sun. They arise on the solar disk with a frequency of several hundreds per second in small regions of several arcseconds and have a lifetime of about a minute. They occupy about 1% of the solar surface, generate about 5% of the radiation in the CIV and SiIV lines, while the broad components in the spectrum of AU Mic yield up to 40% of fluxes in these lines and should occupy about 12% of the stellar surface. The physical essence of the explosive phenomena is not completely clear, probably new magnetic fluxes emerge in old magnetic structures, thus the magnetic fields are annihilated there. The surface fluxes in CIV lines are up to 6×10^5 erg/cm^2 s, which is three times higher than the saturation level estimated by Vilhu (1987). Total radiative losses in the most active regions are estimated as several 10^9 erg/cm^2 s and the appropriate ratio $R_{tot} = 0.23$ is one of the record ratios for the magnetic heating of stellar atmospheres. (The event corresponding to the explosive events on EV Lac in the line CIV λ 1548/51 Å of 6 February 1986 will be described in Sect. 2.4.3.)

Maran et al. (1994) obtained the spectra of AU Mic with a resolution of about 10 000; the observations were performed in the mode of high-speed spectrography with exposures of 0.4 s over 30 min at each of seven subsequent turns about the Earth. They studied the wavelength range of 1345–1375 Å, containing the CI, OI, OV, CII, and FeXXI lines, which are formed at 10^4–10^7 K. Parallel IUE observations proved that during the HST observations the star was quiet. Of particular interest is the line FeXXI λ 1354 Å formed at $T \sim 10^7$ K: it does not display the traces of shifts and asymmetry and its width allows for a nonthermal velocity of not more than 38 km/s. Based on this line and the simultaneous IUE data, the complete distribution of EM in the entire temperature range of the upper atmosphere of the star was first constructed from the observations. Comparison of the intensities of the lines OV λ 1371 Å and λ 1218 Å measured by Woodgate et al. (1992) provided an estimate of the electron density of 5×10^{10} cm^{-3} at $T \sim 250\,000$ K. Comparison of the obtained spectrum of AU Mic with the spectra of the solar flare and the active region proved its similarity to the flare spectrum.

To analyze the profile of the Ly_α line observed from HST in the spectra of cool dwarfs, Gayley (1994) developed the approximate analytical theory of line wings within the concept of radiative transfer with a partial frequency redistribution for a simplified chromospheric model. The theory makes it possible to determine the electron density and its gradient with depth in a certain typical chromospheric layer from the measured surface flux at a distance of 1 Å from the line center and from the spectrophotometric gradient in this point. Applying this theory to the observations of the flare star AU Mic (Woodgate et al., 1992), where the resolution of 0.1 Å at high S/N was achieved, Gayley estimated the range of electron density as 9×10^{10}–3×10^{10} cm^{-3} in the layer forming the wings of the line and expanding with respect to the ionization degree of hydrogen from 0.1 to 0.001. This implies that the profile of Ly_α is the most sensitive to the electron density in the hydrogen ionization region of 0.05–0.10, which corresponds to the temperature plateau in the chromosphere.

Linsky et al. (1995) observed ultraviolet spectra of the components of the binary system Gl 752: the bright component dM3.5 Gl 752 A and the faint component dM8e Gl 752 B (= VB 10). This pair is of particular interest, since one component is earlier and the other is later than the spectral type M5, the boundary separating the stars with convective envelopes and completely convective stars, and the star Gl 752 B is close to the lower limit of the main sequence, where thermonuclear burn of hydrogen ceases and red dwarfs are replaced by brown dwarfs. The observations were performed with a resolution of 0.17 Å in the wavelength range of 1160–1718 Å for Gl 752 Å and 1287–1575 Å for Gl 752 B. In the spectrum of Gl 752 A, one can clearly see the lines CIII λ 1176 Å, SiIII λ 1207 Å, NV λ 1240 Å, OI λ 1304 Å, CII λ 1335 Å, SiIV λ 1400 Å, CIV λ 1550 Å, and HeII λ 1640 Å. But no lines are seen in the spectrum of Gl 752 B obtained by summing 10 exposures with a total duration of 54 min (see Fig. 49).

Saar and Bookbinder (1998b) analyzed the spectra of the eclipse system CM Dra in the MgII region obtained with one-second resolution. They noticed several flares. After subtracting them, on the light curve they found the differences in the wings of entrance and exit point from the eclipse that evidenced the inhomogeneity of the chromosphere, while the differences between eclipses suggested fast evolution of these inhomogeneities.

During HST observations of three G dwarfs in the α Per cluster and in the Pleiades, Ayres (1999) found chaotic variability in the SiIV λ 1393 Å line for times of tens of minutes on all the stars.

Wood and Linsky (1998) thoroughly analyzed the profile of the Ly_α line to study the interaction of stellar wind with interstellar medium, discovered the hot "hydrogen wall" of neutral gas, and estimated the pressure in the stellar wind of several stars, whose parameters were close to the solar ones.

When the high-resolution diffraction spectrograph GHRS/HST was replaced for the echelle spectrograph STIS HST, Pagano et al. (2000) obtained the spectra of AU Mic in the range of 1170–1730 Å with a resolution of 46 000 in four turns of the telescope around the Earth. During 9200 s of the summary exposure of 10 100 s the star was in the quiescent state, 142 emission lines of 28 atoms and ions arising at different levels of stellar atmosphere were identified in its spectra. The spectra were used to study the atmosphere of one of the most active dMe stars. Comparison of the relative intensities of six lines of the multiplet CIII (4) of about 1176 Å in the AU Mic spectrum and in the spectra of solar faculae, in the limb, and the spot umbra showed that the best agreement occurred for facula spectra, which meant that the lines in the AU Mic spectrum were closer to the solar-active regions than to the quiet Sun. Continuing the studies of Linsky and Wood (1994), who found broad and narrow components in the profiles of the CIV and SiIV lines, Pagano et al. found an analogous effect in other transition region lines: narrow and broad components were shifted with respect to the photospheric spectrum for $+2.2 \pm 1.1$ km/s and $+5.4 \pm 2.3$ km/s, respectively. But the earlier suggested change of the shift with the temperature of line formation was not confirmed. The fluxes in narrow and broad components are comparable, FWHM=31 and 103 km/s for narrow and broad components, respectively, and the latter is close to the appropriate value in the mentioned explosive phenomena on the Sun. Analysis of the profile of the coronal line FeXXI λ 1354 Å showed that the nonthermal velocity component of the ions radiating in plasma at $T \sim 10^7$ K was about 70 km/s, almost half their thermal velocity. Thus, these motions are subsonic and the dissipation of shock waves is not significant for the corona heating. The radial velocity of the line does not differ from the velocity of photospheric lines, which suggests that hot plasma is retained by strong magnetic fields. Analysis of the HeII λ 1640 Å emission line suggests that it was excited by collisions (about 40%) and by cascade transitions after the recombination (60%), while the nonthermal velocity component was 48 km/s. The differential emission measure (DEM) curve constructed for AU Mic systematically differed from the solar curve: it reached a maximum at $\log T = 4.7$, this value was maintained till $\log T = 5.4$ or 6.4, and then increased by an order of magnitude near $\log T = 7.0$. The data does not allow one to estimate the temperature at which DEM starts decreasing, while on the Sun this occurs already at $\log T = 6.4$. From the DEM curve found for AU Mic Pagano et al. (2000) estimated the total radiation for the range of $\log T = 4.1$–7.0 as 2×10^8 erg/cm^2 s or 2×10^{-2} $L_{\rm bol}$. From the ratio of components of the intercombinatory multiplet OIV λ 1401/1407 Å they estimated the density at the level of $\log T = 5.25$ as $\log n_e = 10.8$, from other lines at the same level the density was 30 times higher. Obviously this suggests nonuniformity of the medium: low-density structures cover 10–20% of the surface, high-density structures occupy about 0.1%, but yield a comparable contribution to the total emission measure and the summarized profiles of the emission lines.

Pagano et al. (2004) analyzed the echelle spectrum of α Cen A obtained with the resolution of 2.6 km/s at STIS HST. Within the range of 1140–1670 Å they identified and measured 671 emission lines of 37 different ions and molecules of CO and H$_2$. They presented the profiles of the strongest emissions of the transition region SiIII λ 1206 Å, NV λ 1238 Å, SiIV λ 1393 and 1402 Å and CIV λ 1548 and 1502 Å by two Gaussians with average widths of about 43 and 72 km/s and with an average contribution of the wide component of about 45% to the general profile widths. The Gaussians were shifted by several km/s with respect to the photospheric lines to the long-wavelength part, with the shift of narrow components being slightly greater. The widths of the nonthermal components increased from 7.5 to 39 km/s when the emission formation temperature increased from 6000 to 200 000 K. Pagano et al. concluded that the obtained spectrum of α Cen A fit the best the solar spectrum.

Using STIS data, Ayres et al. (2003) studied the forbidden coronal lines in the spectra of late stars, including 12 F7–M5.5 dwarfs. They considered all possible lines of different elements and concluded that only the emissions of FeXII λ 1242 and 1349 Å and of FeXXI λ 1354 Å with formation temperatures of 2×10^6 and 10^7 K, respectively, were promising for the diagnostics of coronal plasma. They found with confidence the emission in Proxima Cen, AD Leo, EV Lac, AU Mic, ξ Boo, κ Cet, χ^1 Ori and ζ Dor with the average FWHM of about 110 km/s but did not reveal it in ε Eri, 70 Oph, τ Cet, and α Cen. The emission of FeXII found in seven stars of the sample had FWHM equal to 40–56 km/s. This means that in both cases the widths of the lines were close to the thermal widths. Comparison of the fluxes in these coronal lines with L_X (0.2–2 keV) revealed that $L_{FeXII}\hat{A}_i\hat{A} \ll L_{1/2X}$, whereas there was a linear correlation between L_{FeXXI} and L_X up to the minimum activity level $L_X/L_{\rm bol}\hat{A}_i\hat{A} \ll 10^{-5}$. The absence of a noticeable Doppler shift of the lines suggests that radiating plasma is localized in magnetic loops rather than in the hot wind.

From the HST STIS data Jordan et al. (2001) obtained slightly different results. In the spectra of $\tilde{A}$ ¥ Eri, 70 Oph A, and $\tilde{A}^a$ Cet they found the forbidden lines of FeXII $\tilde{A} \ll$ 1242 and 1349 $\hat{A}$ ¢ $\hat{A}^a A$. For $\tilde{A}$ ¥ Eri from the measured line parameters they estimated the pressure as $< 7\tilde{A} - 10^{15}$ cm^{-3} K, the column emission measure as $9\tilde{A} - 10^{27}$ cm^{-5}, and the magnetic field as 20 G.

Christian and Mathioudakis (2002) studied the echelle optical spectra of a number of stars of late spectral classes with strong emission in the EUV region and found that the equivalent widths of $H\hat{I}_{\pm}$ emission achieved 8 Å in them, some of them had strong emission in the helium D line, a rotation rate from 5 to 80 km/s, and L_{EUV}/L_{bol} up to 10^{-3}.

Orbited in 1992, the American spacecraft Extreme Ultraviolet Explorer (EUVE) was designed for the investigation of the electromagnetic radiation in the range of 50–760 Å. After the long-term ultraviolet studies with IUE and operation of HST and the X-ray studies at the Einstein Observatory, EXOSAT, and ROSAT (see Sect. 1.4.1), this intermediary range was addressed only by the survey observations with the Wide Field Camera (WFC) and by a few spectral observations of the brightest stars with coronae using the diffraction spectrograph TGS EXOSAT. The working range of EUVE in the temperature range of 2×10^5–3×10^6 K covered plasma radiation from the transition regions to the lower coronae of cool stars. EUVE included 4 grazing-incidence telescopes: 3 coaligned scanning telescopes with 4 photometric bands centered at 100, 180, 400, and 550 Å, and a three-channel telescope-spectrometer mounted perpendicular to them and centered at the ranges of 70–190, 140–380, and 280–760 Å with resolutions of 0.5, 1, and 2 Å, respectively, designed for deep spectral surveys and for simultaneous photometric measurements in the bands centered at 100 and 180 Å. Each channel of the telescope-spectrometer had a field of view of $5°.25$ along dispersion and $2°.1$ in the perpendicular direction. The diffraction gratings with variable pitch were mounted in the convergent beam. Multichannel plates with diameter of 50 mm and a pixel dimension of 29 µm were used as detectors.

By August 1993, the all-sky survey by scanning telescopes was completed. In the course of the project each object was observed for 10–20 s in each of several successive turns of the apparatus around the Earth. As a result, the total exposure for the objects placed on the ecliptic was about 400 s and that of the near-polar objects was about 20 000 s. Thus, by May 1995 about 740 sources of EUV radiation were recorded, among them 270 were F–M stars. Some of them were previously unknown active dwarfs, candidates to the objects of the UV Cet type.

Ball and Bromage (1995) performed photometric monitoring of four such candidates from the results of the EUVE survey and revealed flare activity in all of them. In total, they recorded 14 flares with amplitudes of 0.3–4^m in the U band.

With the help of EUVE it was established that 29 of the 47 flare stars listed by Pettersen (1991) were EUV sources, 25 were dMe stars, and 13 were dM stars, one of the latter was a EUV source. Apparently, the level of saturation of K–M dwarfs in EUV range achieved 10^7 erg/cm^2 s (Vedder et al., 1994).

Mathioudakis et al. (1994) studied 19 stars of late spectral type with ultimately low chromospheric activity in the EUV region and found the flux from the K dwarf Gl 33 in the band 60–180 Å. This result complies with the calculations by Mullan and Cheng (1994), who proved that under acoustic heating coronae with temperatures of about 10^6 K could exist and the surface flux from them could reach 10^5 erg/cm^2 s.

Mathioudakis et al. (1995a) compared the level of extreme ultraviolet (EUV) fluxes from 74 main-sequence stars observed at EUVE with their rotation and found that, similarly to the chromospheric activity in MgII h and k lines, the Rossby number better characterized the level of EUV activity than the stellar rotation period in the sample of stars with different effective temperatures. But for EUV the saturation occurred at lower Rossby numbers than for the chromospheric activity of magnesium.

Observing the flare star DH Leo at EUVE, Stern and Drake (1996) found the modulation of its EUV radiation with a period of 1.05 days, which corresponded to the photometric period of the star.

EUVE recorded quiet EUV radiation from the very-low-mass star VB 8 (Drake et al., 1996). In combination with the X-ray data from the Einstein Observatory and ROSAT, these observations enabled the estimates of the coronal temperature as several million kelvins and only a slight amplitude of EM variations on the interval of about 10 years. This evidences that the activity energy source of the star is a turbulent dynamo rather than a solar-type dynamo.

Tsikoudi and Kellett (1997) studied the EUV radiation of 127 active late stars and recorded 49 of them in one or two (S1 and S2) WFC bands. In addition to 35 flares on 23 stars, on almost half of the objects they found low-amplitude variations at times of 1–2 h to a day that were independent of rotational or orbital modulation; most of them were dKe and dMe stars.

Analyzing EUVE observations of AU Mic in July 1993, Del Zanna et al. (1996) found that the high-temperature coronal component was necessary for representing the emission in the 80–120 Å band if the cosmic abundance of iron was accepted and was not necessary if lower iron abundance was assumed.

The scientific program of the 5-day US–German space experiment ORFEUS carried out in September 1993 from a shuttle included far-ultraviolet studies of active dwarfs. ORFEUS was equipped with a 1-m telescope, its mirror due to special coverage could operate at up to 500 Å, and with a spectrometer with a resolution of 5000 on the variable-pitch diffraction grating. The K–M dwarfs ε Eri, AU Mic, κ Cet, BY Dra, and

α Cen A and B were studied within the program. The emission lines of the transition regions were recorded in their spectra. The electron density was estimated with the spectrum of ε Eri in the region of formation of the lines of 10^9–10^{10} cm^{-3} from the ratio of the intensities of the CIII λ 1175 Å and λ 977 Å lines. The estimates are close to the appropriate solar values (Schmitt et al., 1996a).

In 1999, NASA orbited the Far Ultraviolet Spectroscopy Explorer (FUSE) to study the radiation of space objects in the wavelength range of 900–1200 Å that was inaccessible to IUE and HST spectrographs. FUSE consists of 4 coaligned prime-focus telescopes with off-axis parabolic mirrors of 352 × 387 mm² feeding four Rowland spectrographs with spherical holographic gratings.

Using FUSE, Redfield et al. (2002) observed α Cen A, α Cen B, ε Eri, and AU Mic dwarfs with the resolution of 20 000 and found more than 40 emission lines and blends of HeII, NII, NIII, CII, CIII, SiIII, SiIV, SIII, SIV, SVI, FeIII, OVI, FeXVIII, and FeXIX within the range 912–1180 Å. Most of the lines emerge in the transition regions at 50 000–500 000 K, the chromospheric lines originate at lower temperatures, and the forbidden lines of highly ionized iron appear at the coronal temperature of 10^7 K. The strongest lines of the range, CIII λ 977 Å and λ 1176 Å and OVI λ 1032 Å were stronger by one–two orders of magnitude than other emissions; they were recorded in the ORFEUS experiment as well. The electron density was estimated as $(3-4) \times 10^9$ cm^{-3} from the ratios of the intensities of the CIII lines in the spectra of α Cen A and B. The values are in agreement with the data obtained by Wood et al. (1997) from HST data. Similarly to the HST observations of the CIV and SiIV lines in the spectrum of AU Mic, Redfield et al. (2002) presented the profiles of strong lines by two Gaussians, narrow components with characteristic velocities of 20–30 km/s and broad components with velocities of up to 100 km/s, and confirmed the tendency of an increasing contribution of the broad component with increasing activity of the star and decreasing contribution with growing temperature of formation of the appropriate emission.

Continuing the Ayres et al. (2003) study of forbidden coronal emission from the STIS HST observations, Redfield et al. (2003) used FUSE data. They considered seven F8–M5.5 dwarfs: EK Dra, α Cen A and B, ε Eri, TW Hya, AU Mic, and Proxima Cen. Having considered more than 50 candidates, they concluded that stellar spectra contained forbidden coronal lines of FeXVIII λ 974 Å and FeXIX λ 1118 Å with formation temperatures of 6 and 8 MK, respectively. The line λ 974 Å free of blending was found in the spectra of AU Mic, EK Dra and ε Eri, in all cases the line width was thermal and without a noticeable Doppler shift. In the spectra of these stars they found the emission of λ 1118 Å but noticeably blended by the CI line. A distinct correlation close to linear was found between the luminosities of $L_{FeXVIII}$ and L_X.

Del Zanna et al. (2002) analyzed the upper atmosphere, the transition region, and corona of AU Mic in the quiescent state using the EUVE, STIS HST, and FUSE observational results. They paid attention to the restrictions imposed on such studies by the uncertainties in available atom constants and insufficiently strict selection of lines to be analyzed. According to their calculations, the best agreement for DEM is provided by the model with $P_e = 10^{16}$ cm^{-3} K.

Using FUSE, Simon et al. (2002) studied the regions of subcoronal lines of CIII λ 977 and 1175 Å and OVI λ 1032 and 1037 Å formed at 50 000–300 000 K in the spectra of seven A stars. These lines were found only in the coolest stars with $T < 8200$ K. As a result, Simon et al. (2002) confirmed the validity of the predictions of the modern models on the transition from convective to radiative stellar envelopes and on extension of this transition over the interval of only 100 K.

1.3.2 Models of Stellar Chromospheres

Theoretical consideration of solar and stellar chromospheres in principle reduces to the calculation of the self-consistent problem on the energy balance of the medium heated by a flux of nonradiative energy and completely cooled by radiation from the whole volume. The independent variable in this problem is the power of the input energy flux, the parameters are the conditions at the interface with the photosphere: gravity, density, temperature, and chemical composition of matter. The solution of this problem involves cumbersome calculations of radiative losses of the medium that has a complex chemical composition and considerable gradients of temperature, density, and ionization state, as well as an appreciable optical thickness in frequencies of many lines. But depending on the specific form of the nonradiative energy flux arriving from below, different hydrodynamic and hydromagnetic disturbances can dissipate and transform one into another in the chromosphere, and the calculation of such a stationary system of disturbances involves considerable difficulties. Therefore the observations can be used to solve the inverse problem: the power and the form of the arriving flux of nonradiative energy are estimated from physical parameters of observed chromospheres. Parameters of stellar chromospheres were first determined in the 1960s, and above we enlisted the estimates obtained for the electron density, electron temperature, and the radiation intensities in different lines.

In the first chromospheric models of K5–M3 dwarfs constructed before the quantitative values of the above parameters were obtained, Kandel (1967) considered the stability of such structures on the "mass–temperature" plane, where the temperature parameter was the value at the hydrogen temperature plateau and the mass parameter was the column density above the plateau. (Under thermodynamic analysis of the solar chromosphere Athay and Thomas (1956) found that there are at least two plateaux in the temperature distribution with height as a consequence of the presence of such "cooling" temperature stabilizers as neutral atoms of hydrogen and helium ions in the medium.) On the plane, Kandel calculated the net of hydrostatically equilibrium plane-parallel structures with microturbulent velocity and calculated for each model the expected profiles of the CaII H and K emission lines with partial allowance for the non-LTE effects. Having compared the resulting values with the observations, he restricted the region where stellar chromospheres could exist, but excluded too hot and too dense models, in which hydrogen emission appeared: within the accepted scheme the correct calculation of the model was impossible. This fact reduced the significance of his study as an insight into UV Cet-type stars.

1.3.2.1 Semiempirical Homogeneous Models

The lack of a strict chromospheric theory favored the wide spread of semiempirical models based almost exclusively on the observed characteristics of continuous and line stellar spectra. Semiempirical models are calculated by selecting such a temperature distribution with height at which the calculated continuum and line profiles best match those observed. The distribution of the turbulent velocity with height is selected from the widths of the lines formed at different heights, while the change of density is taken in compliance with the hydrostatic equilibrium condition. Since it is assumed that the chromosphere parameters are independent of the considered region of the stellar surface, in this sense the models can be called homogeneous or one-component models. Since the mechanism for heating of chromospheres is not completely clear, the calculations of semiempirical chromospheric models disregard the energy balance.

For the first time, such calculations for the solar atmosphere were performed using the results of extensive spectral observations from the far ultraviolet to the microwave range (Vernazza et al., 1973). For the extensive set of spectral data they managed to find a section of continuous spectrum or a spectral line that are formed at each height in the photosphere and the chromosphere. As a result, they determined the temperature change throughout the atmosphere height with sufficient reliability and found a plateau with a slow increase of the temperature from 6000 to 7000 K predicted by Athay and Thomas (1956) in the range of heights from 1000 to 2000 km.

Somewhat later, a similar method for the calculation of semiempirical models was applied to the chromospheres of active stars, based on the progress achieved in the understanding of physical distinctions between the quiescent solar chromosphere and floccules. First, it was suggested that the increased luminosity of the chromosphere in active regions is directly stipulated by a steeper temperature gradient, then the idea of such luminosity due to the lower level of the chromosphere in active regions was discussed. In both schemes, chromospheric temperature should be accompanied by higher electron density. But the primary physical distinction of active regions from the quiet Sun is increased magnetic fields, whose closed flux tubes result both in an increase of the temperature gradient and a lowering of the middle level of the chromosphere. Since flare UV Cet stars are magnetic red dwarfs, it is natural that chromospheric models for red dwarfs are constructed reasoning from the models of active regions of the Sun.

Thus far, many researchers have calculated the chromospheric models of cool stars. Of course, due to the absence of sufficient number of high-quality observational data the reliability of stellar models is significantly lower than for solar models. But the calculation scheme used within the concept of radiative transfer in frequencies of optically thick lines is the same. First, one selects or calculates a proper photospheric model corresponding to the spectral type and luminosity of a star under consideration, postulates a certain temperature distribution with height in the atmosphere conjugate to the upper limit of the photosphere, and solves the non-LTE equations of state of hydrogen atoms governing the ionization state of the medium. Then the profiles of emitted fluxes in lines are calculated and compared with the observational results, which allows one to correct the temperature distribution in the course of further iterations. In this case, the absolute density can vary because pressure at the upper boundary of the chromosphere or the mass of matter above the boundary is a free parameter of the problem.

Ayres et al. (1976) were among those who started this kind of study. They calculated plane-parallel one-component hydrostatic models of upper photospheres and lower chromospheres of two dwarfs similar to the Sun – α Cen A (G2) and α Cen B (K1) – from the profiles of the CaII K lines. Both stars in the upper photosphere demonstrate deviations from the radiative equilibrium compared to analogous phenomena on the Sun. For both stars, the column density at the level of 8000 K, at the bottom of the transition region, is the same as on the Sun. Then, Kelch (1978) calculated the photospheric and chromospheric models of two active stars – 70 Oph A and ε Eri – from the measured profiles of the K CaII lines and the fluxes in the MgII h and k lines. His calculations proved that the agreement between the calculations and the observations can be improved, if instead of the earlier assumed constancy of the temperature gradient $dT/d\log m$, where m is the column density, throughout the height of the chromosphere one assumes the same temperature change only in the lower chromosphere and isothermicity in its upper layers, thus approximating the temperature plateau known from the solar chromosphere. In Kelch's models, the temperature gradient is higher by a factor of 1.5–2 than in the chromospheric models of the Sun and nonactive stars. Later, using the data on CII, MgII, SiII, and SiIII ultraviolet lines recorded from the IUE satellite in the spectrum of ε Eri, Simon et al. (1980) performed more detailed calculations of the upper chromosphere of the dwarf. They obtained an identical temperature change below the temperature plateau, while the plateau was somewhat cooler and, in general, their model was closer to the model of bright chromospheric points on the Sun.

From the high-dispersion profiles of the CaII K lines Kelch et al. (1979) constructed the models of photospheres and lower chromospheres of eight F0–M0 main sequence stars, including the pairs of stars of identical spectral types, but with a different level of chromospheric activity. The models proved that active stars displayed nonradiative heating already in the upper layers of the photosphere, which resembles the situation in solar faculae. Such stars have a greater temperature gradient in the lower chromosphere and a hotter and deeper temperature minimum as compared to nonactive stars of the same spectral types. Thus, the best temperature gradient $dT/d\log m$ for the lower chromosphere model of the flare star EQ Vir is 1800 K, while on the quiet Sun this value is 900 K; the model for the nonactive star 61 Cyg B yielded a close value, whereas in the active solar regions it is 1600 K. As nonradiative heating increases, the distance between the CaII K emission peaks – $\Delta\lambda_{K2}$ – decreases and the emission width in the base – $\Delta\lambda_{K1}$ – increases. According to the observations, the models yield the H_α absorption line in the spectrum of the dM0 star 61 Cyg B and the H_α emission line in the spectrum of the flare star EQ Vir.

From the CaII K line profiles Giampapa et al. (1982a) constructed the chromospheric models of emission dwarfs EQ Vir, Gl 616.2, and YZ CMi and nonemission M stars Gl 393 and Gl 411. The models were calculated for a plane-parallel configuration with hydrostatic equilibrium and constant temperature gradients $dT/d\log m$ from the temperature minimum to 6000 K and from 6000 K to 9000 K at turbulent velocities from 1 to 2 km/s. Within the calculated "calcium" chromospheric models that represented well the CaII K line they managed to obtain the H_α emission line for the dMe dwarf YZ CMi and H_α absorption line for the dM star Gl 411 that belonged to the halo population, but failed to match the measured fluxes in the MgII h and k lines. To eliminate the contradiction, they assumed that there were different filling factors of chromospheres for calcium and magnesium floccules. Since they are formed at somewhat different heights, their expected relation to the divergent magnetic field flux tubes can produce this effect. From the comparison of absolute fluxes in the CaII and MgII lines of the stars with a different level of chromospheric activity they concluded that the average filling factor in calcium lines was 0.13, while in magnesium lines it was 0.31.

According to Cram and Mullan (1979), in the calculations of stellar chromospheres the temperature change with height was set by the "joint" scheme: two sections of constant values of the gradients $dT/d\log m$. This method of specification of the chromospheric model artificially reduced the number of free parameters. But the ideology of the modeling method was the same: the model was chosen based on the criterion of the agreement between calculated and observed spectral characteristics.

To avoid laborious calculations within the models of stellar chromospheres through the approach of Ayres et al. (1976), Cram and Giampapa (1987) elaborated a simplified theory of the formation of hydrogen and CaII lines in the atmospheres of cool dwarfs. The simplifications consist in the replacement of the temperature-inhomogeneous chromosphere by an isothermal one and in the appropriate reduction of the calculations of the emitted fluxes in optically thick lines. Using this model, they calculated the expected values of the F_{HK} fluxes and H_α equivalent widths for different values of column density. The nonlinear dependence of W_{H_α} and the monotonic dependence of F_{HK} on column density led to a U-like curve on the plane (F_{HK}, W_{H_α}) for the broad range of column-density values, the curve being dependent on the effective temperature of a star and the temperature of its chromosphere. The H_α equivalent width measured by Stauffer and Hartmann (1986) did not contradict these theoretical expectations, but the comparison of W_{H_α} and F_K in a larger sample of M stars revealed considerable dispersion of the points, which can be associated with different levels of formation of the lines in the chromosphere or with different surface filling factors of the emission regions in these lines (Giampapa et al., 1989). Thus, the strong H_α absorption in the spectra of dM stars is a chromospheric indicator that is probably even more sensitive than CaII emission, and M dwarfs without a chromosphere should have weak H_α absorption and no CaII emission. To find such objects, Fleming and Giampapa (1989) performed spectral observations with a resolution of 0.11 Å of the group of late M stars to $M_V = 17\overset{m}{.}8$ at the multimirror 4.5-m telescope. For VB 8, the weakest emission star in the sample, the ratio R_K is equal to 4×10^{-6}, which is lower than for typical dMe stars and even for the quiet Sun and is equal to the values for M dwarfs with H_α absorption. In the star LHS 2, which is absolutely brighter by $2\overset{m}{.}4$, the ratio R_K is three times lower and the H_α line is not found. Discovery of stars with weak H_α absorption and without calcium emission would be important for the whole concept of magnetism and activity of M dwarfs.

Jordan et al. (1987) proposed a principally new calculation method for the models of stellar atmospheres, which was tested before on the Sun, and applied it to the ultraviolet observations of five G0–K2 dwarfs described by Ayres et al. (1983a). As opposed to the above models based on the calculations within the theory of transfer of the profiles of optically thick lines, this method is based on the analysis of optically thin lines excited by electron collisions. The column emission measures are calculated from the surface fluxes in the lines with different formation temperatures – MgII, SiII, CII, SiIII], SiIV, CIV, and NV. The EM change is constructed within 10^4–3×10^5 K. The application of X-ray data allows this change to be expanded to coronal temperatures. Then, from the density-sensitive ratios of the intensities of emission lines one estimates the absolute electron density on the level of formation of the lines and constructs a complete hydrostatically equilibrium model of stellar atmosphere. Jordan et al. constructed such models covering the upper chromospheres and coronae for the five dwarfs: they encompassed the range of models from the quiet Sun to well-developed active solar regions. The estimates of the column density and pressure in the chromospheres were much higher than those in the earlier chromospheric models of these stars based on the CaII K line profiles.

Mathioudakis et al. (1991) performed ultraviolet observations of the star Gl 182 (=V 1005 Ori) with an abnormally high abundance of lithium. From the ratio of the intensities of the CIII λ 1176 Å and λ 1908 Å and SiIV λ 1396 Å lines that are sensitive to the electron density of radiating matter they estimated density in the stellar atmosphere at the level of $\log T = 4.8$ as 7×10^9 cm^{-3}. By analyzing differential emission measures they estimated the total radiative losses of the quiescent atmosphere of the star within the range of $4.3 < \log T < 5.4$ as 3×10^6 erg/cm^2 s. Then, Byrne and Lanzafame (1994) studied the He I line λ 10830 Å in the spectrum of this star with a resolution of 7000. To interpret the measured equivalent width of 0.17 Å, they constructed the grid of models with non-LTE radiative transfer for plasma from hydrogen, helium, and silicon. The dependence of $W_{\lambda 10830}$ on pressure resembles the situation with the H_λ line: first, as pressure increases, so does the equivalent width of the helium absorption line, which then turns into an emission line. Later, these calculations were supported by the observations of Byrne et al. (1998). In the range of pressures acceptable for the observed fluxes in the Gl 182 H_α and SiII lines they determined the optimal pressure of 0.22 dyn/cm^2 and the temperature plateau as 8450 K. The calculated model agrees with the results of Mathioudakis et al. (1991) for CIII] lines.

From IUE observations of ultraviolet lines in the spectra of three nonactive G–K dwarfs τ Cet, δ Pav, and 61 Cyg A Fernandez-Figueroa et al. (1983) calculated the emission-measure distribution with temperature and constructed models of the transition regions assuming their uniform distribution over the stellar surface.

The total radiative losses in the atmosphere of Gl 380 estimated by Byrne and Doyle (1990)) within the temperature range of $4.3 < \log T < 5.4$ using the emission-measure distribution vs. temperature practically coincide with the appropriate value for the quiet Sun. From the ratio of the intensity of the intercombination lines CIII] λ 1908 Å and SiIII] λ 1892 Å in the spectrum of the star they estimated its electron density as $\log n_e \sim 9.9$ at a level of $\log T = 4.7$, which is close to the solar value and is much lower than on dMe stars.

Thatcher et al. (1991) formulated the problem of constructing the models of stellar atmospheres in the context of the solar atmosphere model by Vernazza et al. (1981), that is, using a set of different lines formed at different levels of the atmosphere, as opposed to the pure "calcium" models of Kelch (1978), Kelch et al. (1979), Linsky et al. (1979b), and Giampapa et al. (1982a) and pure "hydrogen" models by Cram and Mullan (1979, 1985). Thatcher et al. constructed the model of the lower chromosphere of the K2 dwarf ε Eri from the K line profiles and two components of CaII IR triplet, Na D lines, and H_α and H_β. The model was constructed using the solution of non-LTE equations of statistical equilibrium and equations of radiative transfer under hydrostatic equilibrium. Within the calculated model Na D lines were an important diagnostic for the upper photosphere, while the CaII IR triplet provided the localization of the temperature minimum. The depth of H_α and H_β appeared to be sensitive to the gradients of the transition region and to the pressure in its base. On the whole, the chromospheric model of ε Eri developed by Thatcher et al. yields a lower temperature gradient $dT/d\log m$ and thicker temperature plateau than the models constructed by Kelch (1978) and Simon et al. (1980). However, the temperature at the plateau in the former is intermediate between the values provided by the latter models. The model of Thatcher et al. yields a slightly increased level of the temperature minimum and the base of the transition region localized at the level of $\log m = -4.54$. To obtain a more complete solution of the problem, Thatcher and Robinson (1993) observed the spectra of early K stars with a resolution of 55 000. Each star was spectrographed in one or two successive nights in the regions of the Na D lines, CaI λ 4227 Å, green triplet and MgI λ 4571 Å, CaII H, K, and IR triplet lines, H_α and H_β. As indicators of the chromospheric activity at different levels of stellar atmospheres they considered integral fluxes in the calcium and in H_α lines calculated as differences between the profiles of these lines for active and nonactive stars, equivalent widths of H_α and H_β lines, and the depths of their cores. The analysis of the obtained data proved that CaI and NaI lines were suitable for modeling the atmospheres of nonactive and moderately active stars. The MgI λ 4571 Å line is important for the region of temperature minimum. The model should present UV and IR lines of ionized calcium in parallel. They should comply with the depths and fluxes in the Balmer lines, whose central cores are flooded due to the high pressure in the transition region, while the emission wings are due to the high-temperature plateau. The Balmer decrement can serve as an activity indicator. But the equivalent widths of the Balmer lines are poor activity indicators.

For extremely weak chromospheric calcium emission Doyle et al. (1994) constructed the models of late dwarf atmospheres without H_α emission. Their calculations proved that a very weak flux, of about 10^3 erg/cm² s, could leave the atmosphere with a temperature minimum of about 2600 K. Later, Doyle et al. (1998), for the M dwarf Gl 105 B with very low surface flux $F_{HK} \sim 6 \times 10^3$ erg/cm² s and low luminosity $\log L_X < 26.1$, calculated the chromospheric model with a steep temperature rise after the temperature minimum at 2650 K to the thin chromosphere and with total chromospheric radiative losses at the level of 10^5 erg/cm² s.

Houdebine and Doyle (1994a) thoroughly analyzed the effect of different parameters of stellar chromospheres of active M dwarfs on the hydrogen emission spectrum. They found that to reproduce the observed ratios of Ly_α/H_α fluxes, H_α and H_β line profiles with the self-absorption, and the widths of these lines in hydrostatic models with constant temperature gradient $dT/d \log m$, this chromospheric gradient should be rather high and expand up to 8200 K, whereas the transition region should be very thin but of considerable column density, have high pressure and be placed at a level of $\log m \sim -3$. The break in the temperature change separates the formation regions of the Lyman and other hydrogen series. The turbulence and rotation have a slight influence on the line profiles, though the rotation can flood the double-peaked structure of H_α. The profiles depend weakly on the effective temperature. These general results were applied to the detailed analysis of the hydrogen emission of one of the most active dMe stars, AU Mic. Houdebine and Doyle proved that, in addition to the observed H_α and H_β profiles and the Ly_α/H_α ratio, the model reproduced the Balmer decrement, the Balmer jump, and the width of the Ly_α line as well, but overestimated the equivalent H_α line equivalent width by a factor of 3. This contradiction can be removed by the assumption on the inhomogeneity of the chromosphere with the filling factor of 0.3. Analogous considerations of helium emission lead to the idea of inhomogeneity of the transition region. The resulting structure of the stellar chromosphere approaches the existing models of solar flares but requires a strong and continuous flux of nonradiative energy. In this case, the Lyman radiation ensures radiative losses of the transition region, and the Balmer radiation, those of the chromosphere, with the Lyman series heating the upper chromosphere and the Balmer one the lower chromosphere and the region of the temperature minimum.

Continuing the studies of the atmospheres of M dwarfs with minimum and maximum chromospheres, Houdebine et al. (1995) performed detailed calculations of a number of intermediary models to determine the expected spectral features between the extremities over the whole activity range. In other words, they analyzed the general scheme of variation of the H_α line profile in the spectra of M dwarfs with the chromospheres of varying activity outlined by Cram and Mullan (1979) for the following typical situations: high activity with strong emission and weak self-absorption; intermediate activity with emission wings and an absorption core; intermediate activity with strong and wide absorption; low activity with weak and narrow absorption; zero activity with indistinguishable H_α line – zero H_α stars.

The models were calculated on the following assumptions: identical photosphere up to the temperature minimum corresponding to the M4 dwarf; temperature minima of 3000 and 2660 K, hydrostatically equilibrium atmosphere, constant temperature gradient $dT/d \log m$ from the temperature minimum to 8200 K and a transition region from this temperature to 3×10^5 K. The column density above the chromosphere varied from $\log m_o = -3$ to the level of a zero H_α star. The calculations proved that the lines of all hydrogen series in the most active stars should demonstrate emission. As the activity level decreases, the Brackett lines are the first to convert into absorption, then this occurs with the Paschen and Balmer lines. After these lines disappear, Lyman absorption lines should remain in the spectrum. The transition from the Balmer emission to absorption should occur when m_0 decreases by only one order of magnitude. The values of m_0, at which qualitative changes occur in the hydrogen spectrum, strongly depend on the value of the temperature minimum. Having compared the calculated line profiles and the equivalent width obtained within the above atmospheric models with the observations, Houdebine et al. (1995) charted the objects with different chromospheres on the "$\log m_0$–electron density" plane (see Fig. 13).

Then, Houdebine et al. (1996) considered the continuous radiation of stellar chromospheres in the broad range of chromospheric activity from the basal to a very high level in M dwarfs with the same temperature minima. The calculations showed that this continuous radiation depended first on pressure in the transition region and the value of the temperature minimum, and therefore could be used for the diagnostics of the stellar chromosphere throughout its depth. Houdebine et al. found that the ultraviolet continuum was the main channel of chromospheric radiative losses, in particular, the losses for H_α were, as a rule, only a small portion of total losses for the hydrogen lines and continua. Thus, the earlier observed saturation in MgII and some other lines does not mean saturation of magnetic heating of the stellar atmosphere. In the region $\lambda < 915$ Å the Lyman continuum dominates, while in the range of 915 Å $< \lambda < 3650$ Å so does the radiation of the transition region and the

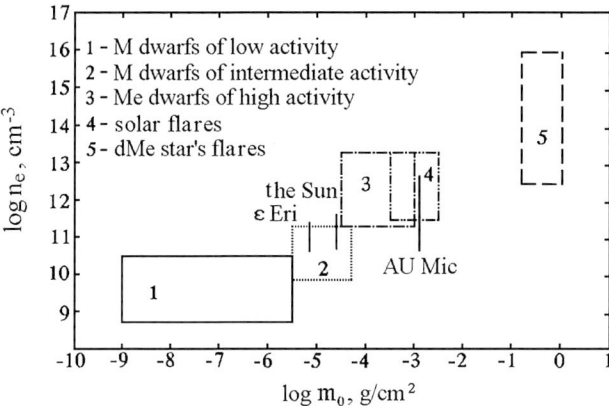

Fig. 13. Localization of chromospheric radiation sources and flares on the electron density–$\log m_0$ plane (Houdebine et al., 1995)

temperature minimum, and in the range $3650\,\text{Å} < \lambda < 40\,\mu\text{m}$, the temperature minimum dominates. The fluxes calculated by Houdebine et al. in the $UBVRIJKL$ bands show the expected excess in the U band and probably in the B band. The stars with strong H_α emission should have an excess of $U - B$ color rather than in $B - V$, which corresponds to the observations. Finally, the stars with H_α emission should be systematically brighter than the stars of the same spectral types with H_α absorption, which Houdebine et al. considered as evidence of the youth of dMe stars and their relation to the T Tau-type stars.

Mauas and Falchi (1994) developed a semiempirical atmospheric model of the quiescent flare star AD Leo, selecting the model parameters to present simultaneously a large number of observed values: stellar continuum from $UBVR$ to $25\,\mu\text{m}$, the Balmer line profiles, Na D doublet, Mg b triplet, λ 4227 Å CaI, CaII K line, and the component of the IR triplet of this ion λ 8499 Å, fluxes in Ly_α and MgII h and k. Eight elements were considered in the calculations under the assumption of solar abundance and turbulent velocities of 1–2 km/s. In the resulting model, the flux in Ly_α was three times higher than the observed one, the observed and calculated fluxes in the magnesium doublet coincided, the flux in the CaII K line was 5 times lower than the expected value, fluxes in the Balmer lines differed from the measured fluxes by $-34, +13, +17$, and $+27\%$ for H_α, H_β, H_γ, and H_δ, respectively. Some of these discrepancies could be attributed to different level of stellar activity at the moments of nonsimultaneous observations in different parts of the spectrum.

Then, Mauas et al. (1997) carried out calculations of semiempirical chromospheric models of two extremely low activity stars – Gl 588 and Gl 628 – using broadband photometry, the hydrogen, calcium, and Na line profiles, and fluxes in the MgII lines. Their models satisfactorily presented the continuum, profiles, and equivalent widths of the CaII K lines, H_α, H_β, and Na

D, as well as fluxes in the magnesium doublet. The models of these M stars with a H_α absorption line were compared with the earlier constructed chromospheric model of AD Leo with ten-times stronger fluxes in MgII and CaII and a hundred-times stronger X-ray flux. The comparison proved that the chromospheric temperature was higher by 2000 K at a height of $\log m_0 = -4$ in the upper chromosphere of AD Leo, the electron density was higher by an order of magnitude, and the transition region started at a lower height.

Jevremovic et al. (2000) considered the dependence of the chromospheric models of M dwarfs on microturbulent velocity at a fixed temperature profile. They developed 12 different models with fixed velocities throughout the chromosphere, with velocities equal to a certain portion of the local speed of sound, and with the velocity profiles composed of three linear sections. The main conclusion is that with the growth of the microturbulent velocity the electron density of the medium decreases, the level of the central minimum decreases in the profile of the H_α emission line, while H_β turns into absorption, and the NaI D line profiles narrow. They estimated the column density for the transition region and for the temperature minimum of CR Dra star, $\log m_0 = -4.3$ and -1, respectively. Then, Jevremovic et al. (2001) constructed the grid of hydrostatic chromospheric models for AD Leo varying the depth of the temperature minimum and the bases of the transition region given $dT/d\log m = $ const. The observations of the H_α line profile in the spectrum of a quiescent star were best presented at the temperature minimum of $\log m_0 = -1$ and in the base of the transition region of $\log m = -4$.

1.3.2.2 Surface Inhomogeneities at the Chromospheric Level

The chromospheres of active stars, as the solar chromosphere, should have considerable surface inhomogeneities, which is confirmed by observations.

In the late 1960s, Krzeminski and Kraft (1967) and Krzeminski (1969) published the results of photometric observations of nine red dwarfs. Spottedness was found only on dMe but not on dM stars. Based on this result and subsequent observations, Martins (1975) advanced the concept of "activity centers" on stars by analogy with active regions or larger activity complexes on the Sun. The conclusion on the spottedness of dMe stars was repeatedly confirmed in later observations (Bopp and Espenak, 1977; Bopp et al., 1981). All the observations suggested the inhomogeneity of stellar chromospheres, but parallel monitoring provided direct evidence.

In 1965, when the spottedness of BY Dra was the greatest, the CaII emission equivalent widths were several times greater than during the following years (Gershberg and Shakhovskaya, 1974).

During spectral monitoring of seven flare stars by the Struve telescope (McDonald Observatory) Bopp (1974b,c) revealed considerable weakening of the Balmer and calcium emissions over an interval of about a day. Since the parallel photometric monitoring revealed no flares and the axial rotation period of the stars was several days, Bopp associated these emission fluctuations with the inhomogeneities of the stellar chromospheres. From the observations of the eclipse YY Gem system Bopp (1974d) and Ferland and Bopp (1976) estimated the longitudinal extension of the emission regions emitting Balmer lines as 20–60°. Later, based on the spectra of YY Gem obtained out of flares, Kodaira and Ichimura (1980) assumed that there was a 2-sector structure of the H_β radiation regions on the main component of the system and a 4-sector structure on the secondary component, for the latter the contrast of active regions at active longitudes being much greater than for the main component. During the following years the sectorial structure of the main component demonstrated changes on the time scale of several months (Kodaira, 1986).

The anticorrelation of the equivalent width of the H_α emission and the brightness of CC Eri were found by Busko et al. (1977), who naturally linked it with the chromospheric active region near the dark spot.

The filling factor of the chromosphere by the hydrogen emission was first directly estimated by Grinin (1979) in analyzing the physical conditions in active regions of flare stars: 6% on AD Leo and 14% on EQ Peg A.

Specific examples of the activity centers were found by Dorren and Guinan (1982b) during photometric studies of five single dwarfs, for which the variations in the CaII emission with time scales of about a week and several years were found in the course of long-term spectral monitoring. On three stars they found low-amplitude brightness variations in the band close to B, and on two of them (12 Oph and 61 Cyg A) brightness and H_α and CaII emission were in anticorrelation, dark photosphere spots and active chromospheric regions typical of the Sun. Similar results were obtained by Chugainov (1983) for ξ Boo AB and HD 1835, Reglero et al. (1986) for the G0V star HD 206860, Barden et al. (1986) for DH Leo, and Strassmeier et al. (1989) for HD 80715.

Radick et al. (1983) carried out photometric observations of 11 F3–K4 dwarfs of those spectrally monitored by Wilson's team and found the anticorrelation of brightness and calcium emission for two or three of them. Later, they studied five fast rotators F8–G2 in the Hyades and found the same effect but with very low variability amplitudes: $0\overset{m}{.}01$ in the total optical brightness and $0\overset{m}{.}03$ in the CaII line flux (Radick et al., 1987).

Kodaira and Ichimura (1982) revealed periodic variations of the intensity of H_β emission in the spectrum of the YY Gem system, whose maximum coincided with the brightness minimum due to spottedness. Andersen et al. (1986) found a distinct anticorrelation of chromospheric ultraviolet lines and HeII line λ 1640 Å with the brightness of EV Lac with simultaneous absence of rotational modulation of the intensities of transition-region lines. One possible explanation is the difference in the filling factors at different heights of the stellar atmosphere. Gershberg et al. (1991b) discovered the anticorrelation of the H_α intensity and brightness of this star in one season and the absence of such dependence during the next year, one year later a similar correlation was found by McMillan and Herbst (1991). The same situation for V 775 Her, VY Ari, and LQ Hya was detected by Alekseev and Kozlova (2000, 2001, 2002, 2003a,b). Byrne (1986) traced a distinct rotational modulation of the

H_α intensity in the dK5e star HDE 19139, on which one of the strongest hydrogen emissions is observed.

Analysis of fast variations of polarization and calcium emission found at almost simultaneous observations of ξ Boo A and HD 206860 made it possible to suggest at least three active regions on each star (Huovelin et al., 1988).

Hallam and Wolff (1981) carried out independent periodogram analysis of the intensities of one of the three emission lines Ly_α, SiII, and MgII in the ultraviolet spectra of six F8-K7 dwarfs: 111 Tau, α Cen A, α Cen B, ε Eri, 61 Cyg A, and 61 Cyg B. For each star up to a dozen spectra were obtained. The periods of close duration and phases found from different lines were identified with axial rotation periods. Thus-found periods from 12 to 47 days are in good agreement with the appropriate values estimated from the calcium emission. On some stars during the period one intensity maximum of one emission and two maxima of the other emission (at the opposite longitudes) were recorded. The deepest modulation was found in the Ly_α line, while the weakest, in MgII. Later, Hallam et al. (1991) applied an improved method of search for the periodicity in nonuniformly distributed data and found the period: 2.8 days for ε Eri and 29 days for α Cen A.

After three four-day monitoring series of the young G0 star χ^1 Ori Boesgaard and Simon (1984) found a distinct modulation of the intensity of CIV lines that preserved the phase for about a year, with an axial rotation period of 5.1 days, which was determined from the variations of calcium emission. The behavior of the HeII λ 1640 Å emission was analogous, and both emissions retained the phase for about a year. The rotational modulation was found in the lines of all ions, except for MgII. An asymmetric CIV light curve can be interpreted as a result of luminosity of a large active region, which after a 1/4 revolution is followed by a second region, and the total high level of this emission flux may be due to the bright chromospheric network in addition to the active regions. Independent estimates of the spottedness of the chromosphere in different lines yielded filling factors from 20 to 50% of the stellar surface.

The intensity of the high-temperature resonance CIV doublet in the spectrum of the G0 dwarf χ^1 Ori over one day, which is equal to 0.22 of an axial rotation period, reduced by a factor of three, while the intensity of the CII line λ 1355 Å, by 50%, then the lines returned to the initial level (Simon, 1986).

In the spectrum of the single very fast rotating star Gl 890, Byrne and McKay (1990) found the rotational modulation of the intensity of MgII ultraviolet lines, while Young et al. (1990) found successive shifts of the centroid of the H_α line to the blue and red sides in the phase with weakening of the MgII line and positive correlation of the intensities of H_α line with the brightness of the star. But in later observations, Byrne and Mathioudakis (1993) recorded an asymmetric light curve, which was presented by the model with two spots of close size spaced in the longitude for 90°. Within this interpretation of the light curve no positive correlation should be observed between the H_α emission and the spots, and Young et al. (1990) actually found the H_α intensity minimum in the vicinity of the broad brightness minimum in the V band.

Newmark et al. (1990) revealed a clear anticorrelation of equivalent emission widths of H_α, H_β, and H CaII and the brightness of DH Leo. As brightness weakened in the R band by 10% the equivalent widths increased by 30%, which meant that there was a real enhancement of emission near the starspots. But between the emission extremes of CaII H and K and CaII IR triplet there was a phase shift for a half-period.

Byrne et al. (1992a) found a distinct anticorrelation between the brightness of the CC Eri star and the intensity of MgII lines: the amplitude of variations of the brightness of these lines achieved 40% for the amplitude of the stellar brightness in the V band of 5%. The variations of the intensities of other high-temperature lines SiII, CII, and CIV were much smaller, they hardly exceeded random scatter.

Bopp and Ferland (1977) did not find rotational modulation of calcium emission in the BY Dra spectrum. Butler (1996) compared the spottedness and the rotational modulation of the emission lines of floccules on BY Dra and AU Mic from the observations carried out the 1980s (Butler et al., 1983, 1987) and established that there was a correlation between the spottedness and the high-excitation emissions on BY Dra, while the frequent flares on AU Mic prevented detection of the correlation. As to the MgII emission, there was no correlation probably because the emission covered almost the whole star, although during the one-week observations in October 1981 the correlation was beginning to show (Butler et al., 1984). From the UBV observations in 1974–1980 Contadakis (1997) discovered alternate periods of strong and weak rotational modulation of stellar brightness with the duration of one–two months on BY Dra, in the periods of increased modulations there occurred a higher frequency of flares and enhanced H_α emission.

Stern and Drake (1996) studied three bright BY Dra-type systems in the extreme ultraviolet. On the light curve of FK Aqr over a quarter of a period, which is about a day, they recorded twice as high radiation that could be due to the active region on one of the system components. From the light curve of DH Leo encompassing approximately eight periods they suspected rotational modulation of brightness. For BF Lyn they noticed brightness variations up to 50%, but observations of the system lasted for less than a period, so it was not clear whether the variations were due to a flare, rotational modulation, or the evolution of the active region.

In 1986, an extensive campaign of comprehensive study of the active G8 star ξ Boo A was undertaken. Magnetometric observations of the line λ 6173 Å and spectral observations in the region of the D_3 helium line were carried out at the American National Solar Observatory McMath. Multicolor broadband polarimetric observations were performed at the five-channel photopolarimeter in the Crimea. Spectral observations in the ultraviolet and in the CaII H and K lines were carried by the IUE satellite and at the Mount Wilson Observatory, respectively (Saar et al., 1988). Analysis of the obtained data revealed clear synchronous changes in the magnetic flux, the intensity of the UV emission of CIV and CII lines, and CaII emission, which was the first evidence of the relation between the magnetic flux of the photospheric field and the emission of the outer stellar atmosphere. The absorption maximum in the helium D_3 line and the maximum of flooding of the cores of Na D lines were also near the magnetic flux maximum. On the basis of comparison of magnetometric and polarimetric data an hypothesis on the existence of four longitudinal sectors with increased magnetic fluxes on the star was advanced.

By analogy with the Sun one can expect that monitoring of stars with an asymmetric distribution of active regions on the disk will show the growth and decline of the ultraviolet emission, X-ray radiation, and magnetic flux at certain phases of the rotation period. This prediction was confirmed in the observations of ε Eri (Saar et al., 1986b) and ξ Boo A (Saar et al., 1986b, 1988).

Frasca et al. (2001) compared photometric and spectral observations of the G0 dwarf HD 206860 and found a distinct anticorrelation of the brightness with the fluxes in the CaII and H_α lines at coincident axial rotation period of 4.74 days. López-Santiago et al. (2003) obtained similar results in studying many lines of the echelle spectra of PW And.

The list of cases when observations suggested surface inhomogeneity of the stellar chromosphere can be expanded. But long-term and homogenous observations of the fluxes in the CaII H and K lines in the spectra of a large number of stars started by O. Wilson (see Fig. 8) are more representative. Analysis of such observations yielded the estimates of the characteristic lifetime of active regions on the stars as 50 days and the characteristic lifetime of the active complex as about 200 days (Donahue et al., 1997). Application of gapped wavelet analysis to long-term observation series for the active stars HD 1835, HD 82885, HD 149661, and HD 190007 made it possible to discover local long-lived active regions that had a tendency, at least three of four stars, to resume in the narrow latitudinal region (Soon et al., 1999).

In the above one-component models, the luminosities of stellar chromospheres were eventually determined from the total mass of matter at the chromospheric temperature. An alternative approach implies the variation of the surface filling factor for active regions (Giampapa, 1980). Above, we mentioned the estimates by Giampapa et al. (1982a) for the average filling factors of chromospheres in the CaII and MgII lines as 0.13 and 0.31, respectively. But analyzing the H_α absorption line in the spectra of dM stars Giampapa (1985) concluded that it was not sufficient to vary only this value, since the lower limit of the filling factor of dM star surfaces was rather high, at least 0.3. Thus, he advanced the hypothesis: the differences in dM and dMe stars were determined by the structure of magnetic fields rather than by the filling factor. On dMe stars the local fields form closed loops, whereas on dM stars these are open structures of the type of solar coronal holes. Almost at the same time, independently of Giampapa's conclusions Byrne et al. (1985) formulated a similar idea: magnetic structures uniformly distributed over the surface were formed on the stars with axial rotation periods of more than 10 days, on faster rotating stars there was a considerable concentration of local magnetic fields, which resulted in more explicit manifestations of activity.

Byrne and Doyle (1990) compared ultraviolet spectra of two dM stars, Gl 380 and Gl 411, and the Gl 900 star with very low emission. The latter is a "marginal dMe star", according to Young et al. (1984), but Doyle and Byrne proposed to replace this term by the notation dM(e). They found that the surface flux in the CIV λ 1548/51 Å line of Gl 900 was higher by an order of magnitude than that of dM stars but was lower by a factor of 3–5 than that of dMe dwarfs. Identical intermediary fluxes were recorded in the Ly_α, CII λ 1335 Å, and SiII λ 1817 Å lines. This complies with the considerations of the dM$\rightarrow$ dMe transition as an increase in the surface filling for solar-type active regions, and the weak H_α line of the Gl 900 star is a result of flooding of the absorption line by emission. However, Turner et al. (1991) made another conclusion on the basis of comparing the intensities of the H_α, CaII H, K, and the IR triplet lines of the nonactive star Gl 1, Gl 735 active dwarf, and the intermediate activity star Gl 887. Assuming that on the intermediate-activity star active regions have the same properties as on an active star, the other regions are as on a nonactive star, they estimated the relative area of different regions on the intermediate activity star from one of the above chromospheric lines. But the solution does not suit the other two chromospheric lines. Therefore they concluded that the observed activity range of M dwarfs was due to the changes in the function of chromospheric heating on the whole star rather than by the variations of the filling factor for active regions. Later, Giampapa (1992) made a similar conclusion. However, the substantial contribution of microflares to the emission of the Balmer lines of the "quiescent" chromosphere found by Alekseev et al. (2003) bring us to new discussions.

Cerruti-Sola et al. (1992) compared absolute profiles of the MgII h and k lines in the spectra of about 40 F6–K5 dwarfs and in the spectra of 22 different regions on the Sun – quiet regions, the bases of coronal holes, faculae, and flares – obtained with comparable spectral resolution. Local fluxes in these lines on the Sun overlapped the range of fluxes averaged over disks on the stars under consideration and the spectra of stars of different activity levels were in good correspondence with different solar regions. This suggests that different MgII emission levels in the spectra of stars of close spectral types are caused by different surface filling factors for magnetic fields and related MgII emission regions. Cerruti-Sola et al. also found that the data for different solar regions fitted the same relations of fluxes in different lines constructed for stars. In principle, the obtained results enable calculations of the surface filling factor for active regions, but the fact of immense luminosity dispersion in MgII lines in different solar structures makes this procedure rather formal.

Panagi and Mathioudakis (1993) collected and analyzed spectral observations in the h_α, CaII H and K, and MgII h and k regions of about 600 K–M dwarfs. They noted systematic growth of $W_{H_\alpha}^{\rm emission}$ toward late spectral types and the growth of $W_{H_\alpha}^{\rm absorption}$ toward earlier types, a systematic decrease of the luminosities L_{CaII} and L_{MgII} toward later dwarfs and similar by a less explicit change of the surface fluxes of F_{CaII} and F_{MgII}. The close correlation of the values of L_{CaII} and L_{MgII} evidences the formation of these emissions in the overlapping atmospheric regions with common surface inhomogeneities. Such a close correlation of luminosities in the Ly_α line and in the X-ray region suggests the closeness of the structures of surface inhomogeneities from the upper chromosphere to corona, whereas the comparison of $W_{H_\alpha}^{emission}$ with L_{CaII} reveals great dispersion, which Panagi and Mathioudakis associated with the essentially different structures of surface inhomogeneities at the appropriate levels of stellar atmospheres. In comparing the values of W_{H_α} and fluxes in CaII, they found the U-like dependence $W_{H_\alpha}(\log F_{CaII})$ theoretically precalculated by Cram and Giampapa (1987).

Andretta and Giampapa (1995) developed the method of estimating the filling factor for chromospheric active regions of F–G stars based on the observations of the helium lines D_3 and λ 10830 Å and calculations of their expected intensities in stellar active regions. These lines appear in absorption in active solar regions, but are invisible or very weak in the quiescent photosphere and in sunspots. Therefore there should be magnetic regions beyond starspots. To calculate possible maximum equivalent widths of the helium absorption lines they used the method applied earlier by Cram and Mullan (1979) to calculate the H_α equivalent widths on active dwarfs. They took the temperature structure of the solar quiescent atmosphere, following Vernazza et al. (1981), that was then shifted to deeper and denser atmosphere layers, for each shift they calculated the plane-parallel hydrostatic chromospheric model with regard to the non-LTE ionization of hydrogen, the contribution to the electron density of metals and molecules and helium ionization by the isothermal corona with exponentially decreasing pressure, the corona being in hydrostatic equilibrium with the chromosphere. The temperature structure was combined with the photospheric models of F and K stars. The calculations showed that the helium triplet lines arose under the scattering of the photospheric radiation at two levels: in the upper chromosphere at $T_e \sim 7000 - 8000$ K and in the region of the temperature plateau at $T_e \sim 20\,000$ K. In the upper chromosphere, the population of the lower levels of these lines occurs due to recombinations and a cascade transition after the ionization by coronal ionization. At the temperature plateau, this occurs due to collisional excitations, so that the role of this plateau starts dominating as the chromosphere moves deep inside to higher densities. The prevailing collisional excitation, as in the case with the H_α line, results in the appearance of the emission line λ 10830 Å. Calculations showed that for F–G stars the maximum equivalent width of this absorption line was close to 0.45 Å. For the D_3 line such a situation can occur at the much higher density typical of flares. The comparison of the H_α and D_3 lines in solar structures of different activity level confirmed the validity of the method. Thus-found filling factors from the D_3 lines are within 0.03–0.32 for the F–G stars under consideration, from the line λ 10830 Å within 0.02–0.62.

For the five G dwarfs, for which the estimates of the photosphere filling factor for strong magnetic fields are available, the chromosphere filling factors from the D_3 line are systematically much lower than the "magnetic" filling factors, but the latter are rather close to the chromosphere filling factor from the line λ 10830 Å.

Let us briefly sum up the results of studying stellar chromospheres. The concept of heating of the atmospheres of medium- and low-mass stars by nonradiative energy fluxes originally developed for the Sun did not contain a concrete definition of the physical nature of these fluxes, but it explained the main features of stellar chromospheres: strong emission of Balmer, calcium, magnesium, and other numerous UV lines, a broad range of absolute luminosities of the lines, extremely weak and the strongest stellar chromospheres, the correlation of their luminosities with masses and stellar rotation, systematic changes of the content of chromospheric emission along the spectral sequence. Accumulated observations evidence the essential surface inhomogeneity of these components of stellar atmospheres. The known structure of the solar chromosphere provides heuristic considerations on the possible character of such inhomogeneities of stellar chromospheres, but no satisfactory solution of this inverse problem, an estimate of the parameters of surface inhomogeneities from the observed integral fluxes, has been obtained yet. Consideration of different starting assumptions – the model of a plane mosaic of active and nonactive elements at different chromospheric levels at the expense of the divergence of magnetic flux tubes, the considerations on the broad brightness ranges of active regions and the surface filling factors on them – yielded no result. Probably, the reason is the lack of insight into the structure of the magnetic field that eventually governs the structure of the stellar chromosphere. Available homogeneous models of stellar chromospheres provide only the first approximation of these structures, and it is obvious that the parameters of the radiating medium will differ in the homogeneous and inhomogeneous models. This statement is illustrated by Fig. 14, which presents some results of the construction of semiempirical models of quiescent chromospheres. The solid curves show the models developed by Baranovskii et al. (2001a) based on the profiles and equivalent widths of the H_α, H_β, and H_γ lines in the spectrum of EV Lac observed with the Shajn telescope in the Crimea. The calculations were carried out assuming that the chromosphere was either homogeneous or the active regions radiating the emission lines occupied 1/2 and 1/3 of the stellar surface, the three variants are marked by 1, 2, and 3, respectively. The second chart shows the temperature structures of the models, while the bottom chart illustrates the change of the electron density. In the calculations, the "joint" scheme of the temperature structure was not used, and the model of a homogeneous chromosphere has a distinct temperature plateau, where the main part of radiation in the Balmer lines is formed. Comparison of curves 1–3 in Fig. 14 shows that the structures of the calculated models of

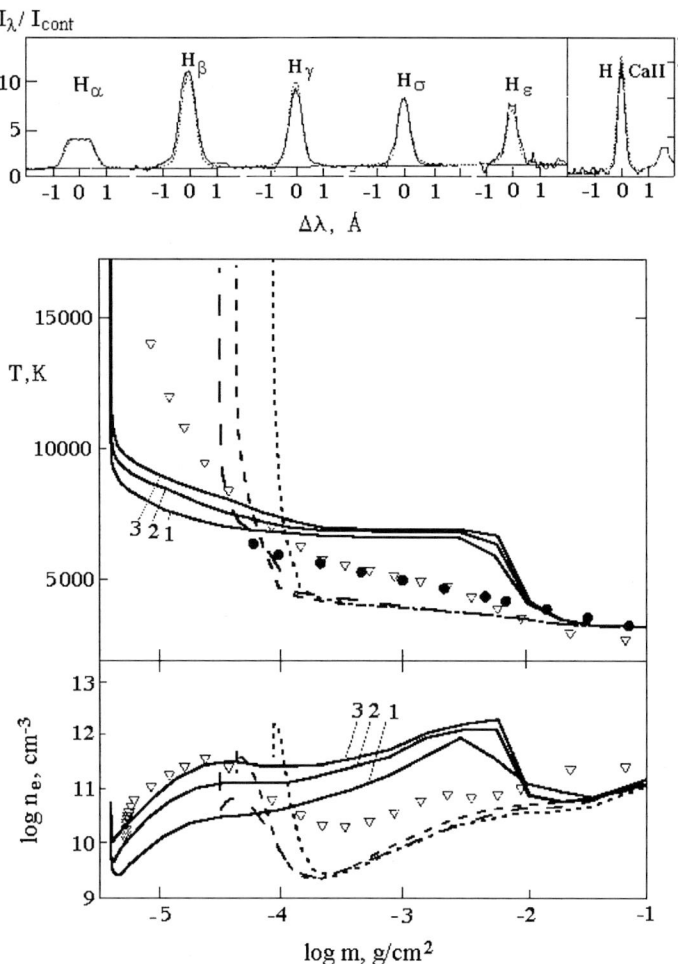

Fig. 14. Semiempirical models of active regions of the chromosphere of EV Lac are shown by *solid*, *dotted*, and *broken* curves. *Triangles* illustrate the model of the quiescent chromosphere of AD Leo by Mauas and Falchi (1994), *circles* correspond to the model of the lower chromosphere of YZ CMi based on the calculations by Giampapa et al. (1982a). At the top, the observed emission line profiles in the spectrum of EV Lac (*solid lines*) are represented by the total radiation of active regions and three microflares (*dotted line*) (Alekseev et al., 2003)

active regions differ from the homogeneous chromosphere model by the higher temperature of matter on the plateau, less extended isothermal region, and an earlier smooth rise to the high-temperature region. The H_α and H_β emissions start forming practically at the same depths as in the model of a homogeneous chromosphere, but in the active region model they go higher and their optical thicknesses are 3–4 times greater, electron densities are increased in them by a factor of 2 to 4.

Soon, these results were essentially supplemented by Alekseev et al. (2003) who analyzed the radiation from the chromosphere of EV Lac in 1994 using much better observational data – high-resolution echelle spectrograms from the Nordic Optical Telescope. The analysis showed that to present simultaneously recorded profiles for the lines from H_α to H_ε and H CaII, one should take into account not only the spatial filling factor of the chromosphere but, which is more important, the physical inhomogeneity of the radiation source. The discovered broad wings of emission lines show that in a quiescent star a considerable portion of the line chromospheric emission is due to microflares whose amplitudes in broad photometric bands are lower than the registration threshold of individual bursts. Consideration of two 15-min intervals before and after the fast flare of 30 August 1994 (23:19 UT) with $\Delta U = 0\overset{m}{.}85$ showed that at each interval the recorded radiation could be presented as a sum of radiations of active regions and three different microflares, while the latter provided the main contribution to the radiation in the hydrogen lines. A presentation of the EV Lac spectrum out of a flare is shown in the top chart of Fig. 14. In the diagrams of temperature and density structures broken lines with long dashes show the models of active regions constructed after subtracting the two above observed spectra of the appropriate contributions of microflares. The closeness of these curves to each other evidences sufficient correctness of the procedure of accounting of microflares. The dashed curves in Fig. 14 present the model of the active region on EV Lac during the flare of 30 August 1994. On the whole, the models of "pure" active regions, as compared to the models 1–3, display noticeably lower electron temperatures of active regions and the regions themselves go down to deeper atmospheric layers.

As follows from Fig. 14, on the planes (T_e, $\log m$) and (n_e, $\log m$) the models of Mauas and Falchi (1994) and Giampapa et al. (1982a) in the temperature regions below 10 000 K are localized inside the band occupied by the Crimean models. A noticeable similarity of the homogeneous model of the chromosphere of AD Leo to the models of active regions of EV Lac can be associated with the observed lower contribution of microflares to the spectrum of AD Leo, since the frequency of flares on this star is two–three times lower than on EV Lac.

1.4 Coronae and Stellar Winds

The solar corona has been known since antiquity: during total solar eclipses a silver glow of the corona flares up for several minutes around the solar disk covered by the Moon, then it disappears without a trace when the smallest bright crescent of the Sun appears. Solar eclipses showing corona have always been the most impressive performances in the celestial theater.

Emission lines in the solar corona spectrum were long a mystery. Identification of the lines with multiply ionized metals by B. Edlen in 1942 became a scientific sensation. This discovery unambiguously led to the estimate of the temperature of coronal plasma as a million degrees. (G.A. Shajn was one of the first who formulated the hypothesis on very hot solar corona: during the eclipse of 1936 he did not find the absorption lines of H and K CaII in the photospheric spectrum of the Sun scattered by the corona and explained this fact by the scattering on fast electrons of very hot plasma.) Today, the concept of a high-temperature corona is a fundamental of heliophysics. The coronal temperature is determined by several independent methods: from the vertical gradient of coronal density, from its blackbody radiation at meter radio waves, from high-temperature lines in the region of extreme ultraviolet, from their existence and widths. Such direct and independent observations are used in constructing the models of solar corona, because here, as for the chromosphere, physical bases of high temperature are known, but since there is no complete certainty about the heating mechanism, in constructing coronal models the energy considerations are usually disregarded.

The structure of the solar corona is determined by frozen-in magnetic fields and, as a rule, is far from spherical symmetry. The global structure of the corona is subject to slow changes correlated with the 11-year activity cycle. The local magnetic fields of active regions are associated with small-scale formations, coronal condensations, which are well seen in the X-ray photographs of the Sun and in which matter density and temperature are somewhat higher than beyond the structures. In the regions adjacent to spots, the temperature of the coronal plasma achieves 10^7 K. In observations with high spatial resolution, the coronal condensations are seen as a system of loops and arches (see Fig. 15). The characteristic length of the condensations is 10^9–10^{10} cm or 0.01–0.1 $R_\odot$, magnetic fields in them achieve several hundred gauss, and at matter density of about 10^9 cm^{-3} the Alfven velocity is close to 3×10^9 cm/s. The regions of corona free of loops form the so-called coronal holes, in which magnetic fields are practically radial and the plasma temperature is slightly reduced. In other words, the solar corona is not a homogeneous hot atmosphere, but, as in the chromosphere, an ensemble of plasma structures of different scales and configurations governed by the local magnetic fields. According to Schmitt (1996), 90% of the X-ray flux from the solar corona comes from several active regions that in total occupy only 1% of its disk.

Nowadays, the solar corona is observed from the Earth and from spacecrafts over the whole range of electromagnetic radiation. The X-ray radiation of the Sun as a star in the bands of 0.5–3 Å, 1–8 Å, 8–20 Å, and 44–60 Å was regularly recorded already for more than three solar cycles. The solar wind, the outmost layers of the expanding corona that, according to Parker (1963), cannot be in static equilibrium with the interstellar medium, is directly sounded from the spacecrafts.

Kahn (1969) used Parker's hydrodynamic theory of the solar wind (1963) as an initial point in constructing the first semiempirical model of corona

Fig. 15. The loop structure of the solar corona photographed on 6 November 1999 in the 171 Å line from TRACE (NASA Lockheed Martin Solar and Astrophysics Laboratory)

and stellar wind of a flare star. He used the results of Lovell (1969), who recorded the flare on YZ CMi on 19 January 1969 in the radio-frequency region and found a 4–5-min delay of the radio burst at 240 MHz with respect to the burst at 408 MHz. Lovell interpreted the radio-range radiation as plasma oscillations in the stellar corona and associated the delay with the propagation of a shock wave in the medium with decreasing density. Considering these data and setting reliable values for the temperature of the stellar corona and the shock-wave propagation velocity, Kahn constructed the complete density and velocity profile of the expanding stellar corona and estimated its secular mass loss as $3 \times 10^{-12} M_\odot$/year.

Zirin (1976) proposed an optical method of studying stellar coronae: owing to hard radiation a noticeable population of the neutral helium metastable level $2s^3 S$ appears in stellar atmospheres, which results in the formation of the absorption IR line λ 10830 Å. Zirin discovered this variable intensity line in the spectra of ε Eri, 70 Oph A, and 61 Cyg A. Later, Zarro and Zirin (1986) compared the measured equivalent widths of this He line in the spectra of about 70 F0-K7 stars with the appropriate ratios $L_X/L_{\rm bol}$ and found a linear correlation between $W_{\lambda 10830}$ and $\log(L_X/L_{\rm bol})$ for F7 and later stars. To interpret the relation, they used the considerations on the spottedness of stellar coronae. However, Lanzafame and Byrne (1995) analyzed the absorption profile of this He line in the spectrum of M dwarf V 1005 Ori and showed that it could be reproduced within the semiempirical model of the chromosphere and the transition region disregarding additional ionization by particle beams or X-ray radiation from the corona.

Elegant prominences of cold gas floating in the hot corona are one of the most beautiful scenes on the Sun, such structures cannot be seen on other stars. But there are indirect evidences of the existence of stellar prominences. Above we described in detail the case of the fast rotator Gl 890, and BD+22°4409 and RE 1816 + 541. Later the prominences in the atmospheres of rapidly rotating G dwarfs in the α Per cluster were suspected by Collier Cameron and Woods (1992) and Barnes et al. (1998), and in the binary systems OU Gem and BF Lyn by Montes et al. (2000). Collier Cameron (1999) thoroughly analyzed the structures, considering the problems of their experimental detection and the theory of their retention at considerable distances above the stellar surface.

1.4.1 Soft X-ray Emissions: X-ray Photometry and Colorimetry

Information on stellar coronae can be directly obtained only from a few channels. The main source of information is the thermal X-ray radiation of the coronal plasma. The discovery of the radiation was not a surprise or accident: since the late 1950s the coronae of stars of different effective temperatures and luminosities have been calculated within the theoretical considerations on the generation of the flux of nonradiative energy in the subphotospheric layers and its dissipation in the atmosphere (see the review by de Jager, 1976). It was only a matter of time for space technologies to achieve the level sufficient for the detection of the expected weak point sources. Thus, when the X-ray radiation of red dwarfs was found, long-term monitoring or repeated observations were needed to confirm that this was weak radiation of quiescent coronae rather than that of strong sporadic flares.

Using the Astronomical Netherlands Satellite (ANS) with an aperture diameter of 10 cm, Mewe et al. (1975) recorded the X-ray radiation from Capella and Sirius within the range of 0.2–0.284 keV and established only the upper limit of the X-ray radiation of the K dwarf ε Eri equal to 5×10^{27} erg/s.

The first article that reported the direct detection of the X-ray radiation of the corona of the active dwarf was published by Nugent and Garmire (1978): on 19–25 August 1977 they recorded the radiation of the α Cen system during the A-2 experiment on the American satellite HEAO-1. The experiment was run in the scanning mode in the energy ranges of 0.18–0.44 and 0.44–2.8 keV. On scanning, the width of the FWHM band was about 1°.5 along the scanning direction and about 3° in the perpendicular direction. Over 3 days they obtained about 50 scans of the region of α Cen. In the summary scan one could distinctly see the signal of the system, which in the band of 0.18–0.44 keV corresponded to a luminosity of $L_X \sim 3 \times 10^{27}$ erg/s. Using the same scans, Cash et al. (1979) studied the region of the 40 Eri triple system, composed of a K1 star, a white dwarf, and a dMe star. They concluded that the X-ray source found was most probably the K1 star. Later, Walter et al. (1980) used the HEAO-1 data to study four G0–K2 dwarfs – χ^1 Ori, ξ Boo, 70 Oph, and HR 6806 – with the same authenticity of 3σ. They suspected that the stars were coronal radiation sources. Using the HEAO-1 data, Ayres et al. (1979) studied the signals of 30 G-M stars, including 10 dMe objects, and, in addition to 40 Eri C, found the X-ray radiation from BY Dra and AD Leo of $L_X/L_{\rm bol} = 8 \times 10^{-4}$ and 1×10^{-3}, respectively. During the A-1 experiment at HEAO-1, when the sky was scanned within the energy range of 0.5–20 keV with the rectangular field of view $1° \times 4°$, Ambruster et al. (1994b) recorded quiet X-ray radiation of EV Lac at the level of 4×10^{28} erg/s between flares. Tsikoudi (1982) considered the vicinities of 70 known flare stars using the A-2 experiment observations, and in 13 cases found nonflare signals exceeding 3σ. In the range of 2–20 keV these signals corresponded to the luminosity L_X varying from 5×10^{28} to 6×10^{29} erg/s and the ratios $L_X/L_{\rm bol}$ varying within 10^{-4}–10^{-2}.

A revolutionary advance in the X-ray studies was made by the second American High-Energy Astrophysics Observatory (HEAO-2), Einstein Observatory with a 56-cm aperture diameter, launched in 1978. Among the devices installed on the apparatus, the most efficient was the position-sensitive imaging proportional counter (IPC) with a field of view of $1° \times 1°$ and an angular resolution of about $40''$. The counter was a thousand times more sensitive than previous devices and was able to separate 32 bands in the energy range from 0.15 to 4.0 keV. Another device, the high-resolution imager (HRI), made it possible to reach a spatial resolution of $4''$, but with a somewhat lower sensitivity and without spectral resolution. From December 1978 to August 1979 the Einstein Observatory carried out the Stellar survey, during which the X-ray radiation of 143 stars was recorded, including 32 G-M dwarfs (Vaiana and 15 coauthors, 1981). The detectors of the Einstein Observatory made it possible to reveal surface fluxes of soft X-ray radiation from stars within the range of 10^3–10^8 erg/cm^2 s, while the lower limit of this range was more than an order of magnitude lower than the flux from solar coronal holes and the upper limit was more than 10 times greater than the flux from solar-active regions. The analysis of these data formed a basis for the modern stellar X-ray astronomy: it was established that the X-ray radiation was intrinsic not only in previously known binary systems with accretion disks and the RS CVn-type systems, but also in the stars of all spectral types and luminosities, except for late giants and supergiants. In each spectral type and luminosity class the range of X-ray luminosity was 2–3 orders of magnitude, but average X-ray luminosities of G–K dwarfs amounted to 10^{27}–10^{28} erg/s (see Fig. 16), whereas bolometric luminosities within this spectral interval decreased considerably.

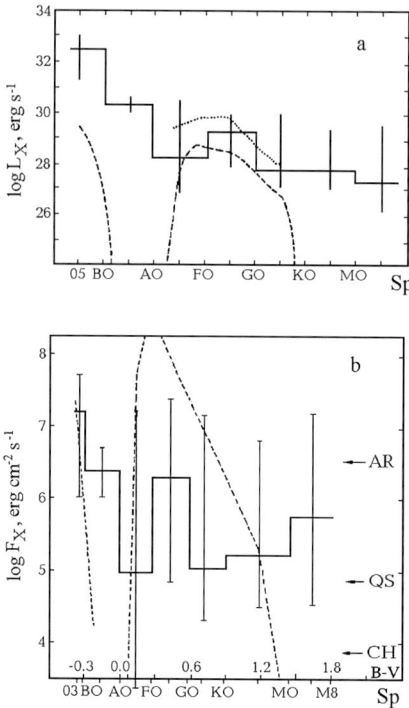

Fig. 16. Average luminosities (**a**) and average surface fluxes (**b**) in the soft X-ray range detected by the Stellar survey from the Einstein Observatory. AR is the level of a solar active region, QS is the level of the quiet Sun, and CH is the level of coronal holes; *dashed* and *dotted curves* show expected values based on the results of theoretical calculations of acoustic heating of stellar coronae (Vaiana, 1980)

Such a high X-ray luminosity of medium- and low-mass dwarfs weakly depending on the effective temperature contradicts the traditional theoretical calculations and thus proves the inconsistency of the underlying conception of acoustic heating of stellar coronae. On the other hand, the large range of L_X in every spectral type showed that classical stellar parameters, effective temperature and luminosity, were insufficient to explain X-ray radiation of the stars. Thus, the magnetic field related to rotation started playing the most important role in understanding of the solar-type stellar activity.

Observations within the Stellar survey and some other programs of the Einstein Observatory enabled important conclusions on the properties of stellar X-ray radiation to be made, supplementing the above results of Vaiana and 15 coauthors.

Schmitt et al. (1985) constructed histograms of the distribution of luminosities L_X of more than 120 A5–F7 stars and found that coronal emission started abruptly at about F0, where the external convective zone appeared in stars. In the sample under consideration, there is no correlation between

L_X and $v\sin i$, which also evidences the decisive role of convection in the appearance of coronal emission. On the other hand, Golub (1983) and then Bookbinder (1985) found a noticeable decrease of L_X in stars later than M5, on which convection involved the whole star. Earlier, the correlation of X-ray luminosity and the rotational velocity was found from different stellar samples within the above limits: $L_X/L_V \sim v_{\rm rot}^{2.8}$ (Katsova, 1981a), $L_X \sim (v\sin i)^2$ (Pallavicini et al., 1981), $L_X \sim \Omega$ (Walter, 1981). This provided an impetus for including the dynamo theory in the concept of stellar X-ray activity.

To find a parameter governing X-ray luminosity, Johnson (1981) considered photometric and kinematic characteristics of 23 closest stars (up to 6.5 pc to the Sun), from which the Einstein Observatory recorded X-ray radiation. The parameter was not found, but he suspected the correlations of X-ray with optical luminosities and of X-ray luminosity with the intensity of the CaII emission. Later, using the data observed by the Einstein Observatory, Johnson (1986) selected about 200 stars close to the Sun (up to 25 pc) that were not program objects and constructed the function of X-ray luminosity up to 3×10^{26} erg/s for this sample independent of L_X. He revealed for the brightest young disk stars $L_X = 1 \times 10^{29}$ erg/s, while for the brightest old-disk stars the luminosity was only 2×10^{28} erg/s.

Walter (1982) showed that the exponential dependence $L_X/L_{\rm bol}$ on the period and the power law with a break of about 12 days equally well described the observational data. Marilli and Catalano (1984) compared the luminosities of L_K, L_{CIV}, and L_X of late F–M stars with their rotational periods and found exponential dependences of the luminosity on the period, which yielded the following relations independent of the spectrum

$$L_{CIV} \sim L_K^{1.2} \quad \text{and} \quad L_X \sim L_K^{2.6} \ . \tag{8}$$

The latter relation is close to that obtained by Ayres et al. (1981a)

$$F_X/F_{\rm bol} \sim (F_{MgII}/F_{\rm bol})^3 \tag{9}$$

and the relation found by Pallavicini et al. (1982)

$$F_X \sim F_{CaII}^{2-3} \ . \tag{10}$$

For a spatially limited sample of K dwarfs Neff et al. (1987) found the relation

$$F_X/F_{\rm bol} \sim (F_{MgII}/F_{\rm bol})^{2.7} \ . \tag{11}$$

Mewe et al. (1981) considered the sample of 20 cool stars with CaII emission and found a correlation of the surface fluxes F_X and F_{HK}, which was amplified up to $r = 0.93$, when F_{HK} was substituted for ΔF_{HK}, the excess of this flux above the basal level of CaII emission. Schrijver (1983) completed multifactor analysis of 66 F–G–K stars including flux densities in soft X-rays F_X, flux densities F_{HK}, and general stellar parameters – radius, mass, and luminosity – and found the close relation between F_X and ΔF_{HK}

$$\langle F_X \rangle = 3.4 \times 10^5 \langle \Delta F_{HK} \rangle^{1.67} \,. \tag{12}$$

The relation occurs in the range of 4–5 orders of magnitudes of F_X and does not depend on the spectral type, luminosity class, and the multiplicity of a star. Considering only close in time observations and using EXOSAT observations, Schrijver et al. (1992) determined more precisely the exponent of ΔF_{HK} up to 1.50 ±0.20. The nonlinearity of the relation between the characteristics of the chromosphere and corona radiation suggests that this relation is stipulated not only by the filling factor: the averaged structure of radiative coronal regions changes with the averaged activity level over the stellar surface. (Later, this conclusion was specified within the two-temperature models, the relation between the components of which systematically changed with L_X.) But Rutten et al. (1989) found that for M dwarfs the values of ΔF_{HK} were lower at least by an order of magnitude than those obtained from (12) for F–G–K stars. They associated this effect with heating by the flux of mechanical energy from deeper atmospheric layers up to the temperature minimum.

Teplitskaya and Skochilov (1990) suggested the existence of a basal level of X-ray radiation of F–K main-sequence stars, that is, as for the quiet Sun in the minimum, due to the emission of bright X-ray points. They constructed the curve of the basal level that resembled a similar curve for the emission from the transition region and found a rather close relation between ΔF_{HK} and ΔF_X:

$$\log \Delta F_X = -6.03 + (1.86 \pm 0.11) \log \Delta F_{HK} \quad (r = 0.92) \,. \tag{13}$$

The X-ray radiation of emission K–M dwarfs is consistent with the general tendency found by Linsky et al. (1982): toward more active stars the contribution of high-temperature components in the radiation of their atmospheres increased (see Fig. 12). Indeed, both the surface fluxes in the transition-region lines and the soft X-ray radiation of dMe stars increase considerably as compared to the flux from a quiet Sun, and the luminosities L_X of such stars substantially exceed the luminosity of the emission lines of the chromosphere and the transition region, whereas the solar corona is much weaker than the chromosphere. The X-ray luminosity of dMe stars exceeds that of the Sun by a factor of 100–1000, while L_X/L_{bol} is approximately 10^{-2}–10^{-3} for the stars and 10^{-6} for the Sun (Giampapa, 1987). For 13 M dwarfs Byrne and Doyle (1989) found a practically linear correlation of the coronal radiation and the emission of the upper chromosphere in the lines HeII λ 1640 Å and CIV λ 1550 Å:

$$\log F_X = 0.87 + 1.04 \log F_{HeII} \quad (r = 0.96) \,,$$
$$\log F_X = -1.97 + 1.13 \log F_{CIV} \quad (r = 0.97) \,, \tag{14}$$

which is valid over three orders of magnitude.

Mathioudakis and Doyle (1989a) compared the surface fluxes F_X and F_{MgII} in the group of dMe, dKe, dK, and dM stars and found that for the same values of F_{MgII} the F_X fluxes of emission dwarfs were greater by approximately an order of magnitude than in nonemission K–M stars. The relation is close to $F_X \sim F_{MgII}$, which differs significantly from the relation obtained by Ayres et al. (1981b) for F–G–K stars, which can be due to the saturation of the considered fluxes in the emission K–M dwarfs.

Young et al. (1984) proposed to estimate the emission of H_α as the difference between the observed equivalent width and the maximum equivalent width of the line in absorption in the stars of the same spectral type. They compared the values of L_{H_α} with L_X and found that the correlation of these values was close to linear: $L_{H_\alpha} \sim 0.2 L_X$. Commenting on this result, Young et al. (1989) noticed that in addition to the Cram hypothesis on the heating of the chromosphere by coronal radiation, a similar correlation occurs if the whole stellar atmosphere is heated by the same source of nonradiative energy. Following the method of Young et al. (1984), Doyle (1989b) calculated the equivalent widths of the H_α emission in the spectra of about 50 M dwarfs, determined the surface fluxes F_{H_α} and luminosities L_{H_α}, and compared these values with the X-ray radiation. He found the correlation

$$\log L_X = 1.11 \log L_{H_\alpha} - 2.56 \quad (r = 0.93) \,. \tag{15}$$

According to Cram, though this correlation is rather close to $L_{H_\alpha} \sim L_X$ and suggests noticeable heating of the chromosphere of M dwarfs by their hot coronae, direct energy estimates for the active star YZ CMi showed that the heating of the corona accounted for not more than 50% of the radiative losses of the chromosphere.

In low-resolution Einstein observations, the temperatures of stellar coronae were determined using the theoretical models of radiation of optically thin very hot low-density plasma. The lines of highly ionized atoms excited by an electron collision dominated in this radiation at temperatures up to 10^7 K, while the contribution of continuous, free–free, recombination, and two-photon radiations was relatively small. At higher temperatures, a free–free radiation of hydrogen and helium started prevailing. Such calculations for the plasma of the given chemical content and temperature were convolved with the response curves of X-ray detectors. Comparison of the calculation results with the observations enables the estimate of coronal temperatures. (Initially the calculations involved the results for normal chemical content calculated by Kato (1976) and Raymond et al. (1976). In the 1970s–1980s, the RS model based on the results of Raymond and Smith (1977) and Raymond (1988) was applied. Today, the most frequently used is the MEKA model based on the results of Mewe et al. (1985) and Kaastra (1992) and its modifications.)

Within the isothermal model, Schrijver et al. (1984) determined the temperatures and column emission measures of stellar coronae, $\zeta = EM/4\pi R_*^2$, and found that for dwarf stars these temperatures were, on average, equal to 2 MK and decreased for later spectral classes. Having considered the X-ray radiation within the model of static magnetic loops, Schrijver et al. concluded that the size of the structures varied considerably. Multifactor analysis revealed a close relation between the column emission measure, the temperature of coronae, and the rotational period

$$\zeta = 10^{28.6\pm0.2} T^{1.51\pm0.16} P_{\rm rot}^{-0.88\pm0.14} . \qquad (16)$$

The relation covers all the sample stars and overlaps three orders of magnitudes of ζ and periods from one day to three months. Ayres et al. (1981b) found that the region occupied by dwarfs – X-ray sources in the Hertzsprung–Russell diagram coincided with the region of the emission sources in the line CIV λ 1548/51 Å, but there were no X-ray sources – the stars with a strong cold wind.

Helfand and Caillault (1982) analyzed 270 fields, about 240 square degrees recorded by the Einstein Observatory to identify the additional X-ray sources with 1720 stars brighter than 10^m found in these fields. They identified 70 sources and most of them were F–G–K type stars.

Summing up the X-ray observations of M dwarfs made by mid-1982, Golub (1983) listed 35 objects and noted that their luminosities, as in F–G dwarfs, differed by up to 3 orders of magnitudes, but there was a systematic decrease of L_X on redder stars with $B - V > 1^m\!\!.7$. All the determined coronal temperatures fell into a narrow range from 2 to 5 MK. There was a tendency of growing temperature with increasing L_X and decreasing axial rotation period.

Fleming et al. (1988) studied the complete sample of early M dwarfs found as X-ray sources in the fields of program stars. They found that 42% luminosity of such dwarfs was $L_X > 6 \times 10^{27}$ erg/s, which corresponded to the sample of such stars based on optical properties, and confirmed the relation of L_X with stellar rotation and mass. Comparison of F_X, F_{H_α}, and F_K revealed the correlation of these values, but chromospheric emission was less dependent on rotation than the X-ray one, and in energy could be maintained by this hard radiation.

Mangeney and Praderie (1984) compared the X-ray radiation of 44 main sequence stars measured by the Einstein Observatory with their Rossby numbers. They used the effective Rossby numbers $\mathrm{Ro}^* = v_m/2\Omega L_C$, where v_m is the maximum velocity of convective flows and L_C is the depth of the convective zone calculated for such zero-age stars, and obtained

$$L_X \sim (\mathrm{Ro}^*)^{-1.2\pm0.1} . \qquad (17)$$

The relation covers stars of 0.5–20 solar masses.

Maggio et al. (1987) analyzed the observations of a spatially limited sample of about sixty F–G dwarfs by the Einstein Observatory and found that the stars involved in the binary systems had the same X-ray luminosities as single stars of the appropriate spectral types.

Fleming et al. (1989) compiled a sample of 128 F–M stars selected according to the value of F_X among nonprogram X-ray sources. The optical study of the sample yielded spectral types, luminosities, distances, L_X, $v \sin i$, and $v_{\rm rot}$. From this most complete sample of X-ray sources limited by the flux they found the correlation $L_X \sim (v \sin i)^{1.05 \pm 0.08}$ for single stars. Since L_X did not correlate with $\Omega \sin i$, they concluded that the correlation was due to the radii of the stars involved in both values. This result differed from the earlier conclusions based on the analysis of the samples of X-ray sources selected on the basis of optical properties. Apparently, the samples selected on the basis of F_X were enriched at the expense of the objects with saturated X-ray luminosities, while in the samples based on optical properties unsaturated objects dominated. Later, similar results were obtained by Schachter et al. (1996), who considered optical and X-ray characteristics of the sources detected by the Einstein Observatory during the Slew Survey when repointing the apparatus from one program object to another. The sample of 809 thus-recorded sources included objects of rather high luminosity. Single F7–K5 stars selected from the sample demonstrated a linear correlation of L_X with $v \sin i$ and stellar size and a correlation with the Rossby number as $L_X \sim {\rm Ro}^{-0.4}$, but there was no correlation of L_X with the angular rotation rate of the stars.

Barbera et al. (1993) analyzed statistically the spatially limited sample of K–M dwarfs in which 257 stars observed with IPC were identified as X-ray sources. They found that these sources formed a representative subsample of the initial sample of about 1700 late stars from three optical catalogs of near stars. The statistical significance of the decline in F_X after the spectral type M5 and a decrease in L_X with age was proved: for K–M dwarfs inhabiting the young disk population the average luminosity was three times higher than in analogous old-disk stars. However, later Fleming et al. (1995) concluded that kinematic classes should not be used as indicators of the age of the nearest stars.

Vilhu et al. (1986) estimated the upper limit of the coronal radiation in soft X-rays: for G0 stars and later spectral types $F_X/F_{\rm bol} < 10^{-3}$. The solar-active region was far below the saturation level and the filling factor on the Sun was well below unity.

Favata et al. (1995) compared the levels of X-ray radiation L_X/L_V, the content of lithium, and rotation in three stellar samples of G–K stars: from the Einstein sky survey, from the Pleiades cluster, and for active binary systems of the RS CVn type. They found that in the Pleiades all the three parameters were in good correlation with each other, whereas for the Einstein survey stars the content of lithium and rotation were in good correlation, but they did not correlate with L_X/L_V. In the sample of binary systems, no correlation between these parameters was found.

The following results were obtained by the Einstein Observatory for specific red dwarfs.

In March 1979, in IPC observations of Proxima Cen Haisch and Linsky (1980) first recorded the X-ray radiation of a quiescent corona of a flare star, and against this background the radiation of strong flares was found. A thorough analysis of these observations by Haisch et al. (1981) yielded the estimates of the X-ray luminosity of the star $L_X = 1.5 \times 10^{27}$ erg/s, the temperature of the coronal plasma of 4×10^6 K and the ratio $L_X/L_{\text{bol}} = 2.2 \times 10^{-4}$. It should be noted that for the Sun this ratio is 1.3×10^{-6}, whereas for the Sun completely covered by active regions it would amount to 5×10^{-5}; but the surface flux F_X for Proxima Cen is equal to only 0.22 of the appropriate value in the solar-active region. The volume emission measure of the Proxima Cen corona is estimated as 2×10^{49} cm^{-3}, whereas for the Sun it is 5×10^{49} cm^{-3}. The measured flux from the star was presented by Rosner et al. (1978) within the plasma-loop model developed for the solar corona. Hence, it follows that magnetic flux tubes link sunspots of different polarity and form stable discrete arches, plasma loops, in which plasma can move along but not across the axes. Such an arch structure spreading up to the heights of the order of the solar radius determines to a considerable extent the properties of the quiescent corona. Theoretical analysis of the arch-shaped structures in the solar atmosphere (Rosner et al., 1978) showed that the basic equation of plasma loops was the equation of stationary thermal balance. For not too large loops, whose size is less than the height scale, and consequently, the gas pressure along the arch can be considered constant, the following similarity law is valid:

$$T_{\max} = 1400(ph)^{1/3} . \tag{18}$$

Here, T is the temperature at the top of the arch, p is the pressure in it, and h is the height of the loop. The numerical coefficient depends weakly on temperature. Haisch et al. (1981) concluded that the X-ray radiation recorded from Proxima Cen could be caused by the corona composed of plasma loops of length over 10^{10} cm and covering the whole star. During the repeated parallel Einstein Observatory and IUE observations of Proxima Cen in August 1980 Haisch et al. (1983) noticed the X-ray radiation of the quiescent stellar corona that corresponded to the same temperature as a year ago and a 4-times lower luminosity. Analysis of the data led to the conclusion that during this season the coronal loops occupied 6% of the stellar surface, while the necessary surface flux achieved 10^7 erg/cm^2 s.

Golub et al. (1982) found that both components of the system α Cen were X-ray sources: $L_X(G2) = 1.2 \times 10^{27}$ and $L_X(K1) = 2.8 \times 10^{27}$ erg/s. Although the K1 star was weaker by $\Delta V = 1^{m}.3$ than the G2 component, it had a stronger corona. The observations were analyzed using the plasma loop model developed by Rosner et al. (1978) and modified by Serio et al. (1981). The modification made it possible to avoid the restriction on the size of loops; for A and B components of α Cen the model yielded loops with the filling factor of 0.05 and 0.15 and the surface fluxes $F_X \sim 3 \times 10^5$ and 4×10^5 erg/cm^2 s.

Using the HRI Einstein Observatory, Cash et al. (1980) repeated the observations of the 40 Eri system and found that the sources of X-ray radiation are K1 and dMe stars: $L_X(K1) = 5 \times 10^{27}$ erg/s and $L_X/L_{\rm bol} = 4 \times 10^{-6}$, $L_X(dMe) = 2 \times 10^{28}$ erg/s and $L_X/L_{\rm bol} = 1.5 \times 10^{-3}$, that is, the optically weakest component of the system was the brightest in X-rays. Since the activity cycle of 40 Eri is estimated as 10–13 years, the star is close to the Sun in all parameters.

Using the Einstein Observatory, Caillault (1982) studied five spotted stars: AU Mic, HD 218738, YY Gem, CC Eri, and HD 216803. As X-ray sources, these dKe-dMe stars were weaker by one–two orders of magnitude than the RS CVn-type systems, but three of these dwarfs complied with the relation ($L_X/L_{\rm bol}$, rotational period) found for RS CVn variables, which suggested that the high luminosity in X-rays was due to fast rotation rather than binarity.

Agrawal et al. (1986) observed seven K–M dwarfs with the Einstein Observatory. The average X-ray luminosities of the stars out of flares were close to 2×10^{28} erg/s. Multicolor analysis of three brightest late dMe stars – Gl 729, Gl 735, and Gl 791.2 – yielded the estimate of the coronal temperature as $\sim 3 \times 10^6$ K and the emission measure as 1.2–2.9×10^{50} cm^{-3}. They analyzed all observations of flare stars performed using this equipment and found that red dwarfs and the RS CVn-type systems satisfied the correlation $L_X = 10^{-3.23 \pm 0.22} L_{\rm bol}$ with a correlation coefficient of 0.94, but nonflare dM stars noticeably deviated from it. For the sample of late dwarfs, Agrawal et al. revealed neither the correlation between $L_X/L_{\rm bol}$ and the axial rotation period, nor the correlation of L_X with $v_{\rm rot}$. These results differ significantly from the results obtained before for earlier dwarfs. They got a new explanation within the above concept of the saturation of the X-ray luminosity for the most active dMe stars completely covered by active regions. This concept made it possible to understand the results obtained by Fleming et al. (1989) and Schachter et al. (1996) for much greater stellar samples.

The system Wolf 630 AB was studied by Swank and Johnson (1982) using the solid state spectrometer (SSS) onboard the Einstein Observatory. SSS had the field of view of 6', sensitivity to plasma radiation of $T > 2 \times 10^6$ K, and a spectral resolution of 5–10. The obtained energy distribution within 0.5–4.0 keV could not be presented as the radiation of isothermal plasma, thus a satisfactory presentation was achieved for two-component plasma with temperatures of 6.5×10^6 and 4×10^7 K, and on the assumption of reduced content of iron to 60% of the solar content. This assumption is concerned with the absence of highly ionized iron lines in the range of 0.7–1.0 keV seen in the spectrum of RS CVn-type systems. In the X-ray radiation of Wolf 630, $EM = 1.8 \times 10^{51}$ cm^{-3} and $L_X = 3.0 \times 10^{28}$ erg/s were estimated for the low-temperature component, while for the high-temperature component, $EM = 0.7 \times 10^{51}$ and $L_X = 0.9 \times 10^{28}$, respectively.

In analyzing the flare star Gl 867 A over an interval of three days, Agrawal (1988) found a decrease in the level of its X-ray radiation by 40% and assumed that it could be due to the surface inhomogeneity of the corona. The multicolor X-ray photometry of the star was presented within the two-temperature model: $T_1 = (1.4-2.4) \times 10^6$ K and $T_2 = (1.2-1.7) \times 10^7$ K, $L_{X1} = (3-8) \times 10^{28}$ and $L_{X2} = (1-2) \times 10^{29}$ erg/s.

Giampapa et al. (1985) considered jointly the X-ray and ultraviolet observations of eight solar-type stars to determine the applicability of the hydrostatic loop model of the solar corona to them. For the Sun and α Cen B, they obtained a reasonable correspondence of the one-component loop model and the positions of the coronae on the "pressure-filling factor" plane. In particular, the Sun at the activity minimum has long loops ($h \sim 7 \times 10^{10}$ cm) with low pressure $p \sim 0.1$ dyn/cm^2 and a filling factor close to unity, whereas at the activity maximum the Sun has compact loops ($h \sim 7 \times 10^9$ cm) with high pressure $p \sim 1.5$ dyn/cm^2 and a filling factor of about 0.03. But in a general case the use of nonsynchronous observations of stellar transition regions and coronae did not provide a confident localization of stellar coronae on the mentioned plane. Probably, it would be more expedient to consider a multitemperature corona and add low-temperature loops contributing to the ultraviolet emission rather than to X-rays. As Golub et al. (1982) proved earlier, from the position of a corona on the "pressure-filling factor" plane one can find the filling factor, if all coronal loops are considered identical and their heights are equal to the scale of pressure heights.

Stern et al. (1986) analyzed the observations of several brightest active cool stars in the Hyades. They concluded that the isothermal model could not represent the observations. The ensemble of loops with a certain temperature distribution and equal maximum temperature or the two-temperature model would be more suitable for this purpose, though less physical. The models with significant variations of the transverse section of loops would not do either. The most acceptable model contains an ensemble of small loops ($h < 10^{10}$ cm) with high pressure (above 400 dyn/cm^2) with a maximum temperature of 10–15 MK and a filling factor of up to 0.1–0.2; on the Sun such loops are typical of flares.

Ambruster et al. (1987) studied the observations of 19 active late dwarfs carried out with the Einstein Observatory to detect the variations of their X-ray emission at time scales from minutes to hours. They used the χ^2 method modified for the detection of nonperiodic variability of weak sources. Such a variability was revealed on 16 stars, for 40 Eri C its characteristic time was close to 150 s, while for the others it was more than 1000 s and the amplitude was up to 30%. The events of this variability did not fit the general energy spectrum of flares constructed using individually recorded flares, and the observation gaps due to the operation mode of the satellite did not allow them to determine the nature of this variability.

Wide-angle systems of the Einstein Observatory made it possible to study not only the X-ray radiation of individual stars but also that of stellar clusters.

Below, we consider the results for the nearest clusters, in which F and later dwarfs are available.

Stern et al. (1981) studied the X-ray radiation of stars in 27 fields with the size of a square degree in the center of the Hyades. It was found that for a half of the 85 considered members of the cluster $L_X > 4 \times 10^{28}$ erg/s, more than 80% F and G dwarfs were X-ray sources with an average luminosity of $\langle L_X \rangle \sim 10^{29}$ erg/s and this value varied over an order of magnitude. The strongest X-ray source was the F0V star 71 Tau with $v \sin i \sim 200$ km/s, and on the whole one can suggest the correlation $L_X \sim v_{\rm rot}^2$. G0–G8 dwarfs found in the Hyades in X-rays on average are 30 times as bright as the active Sun, while the G1 dwarf HD 27836 is brighter by a factor of 300. Combining the results of these X-ray observations with the ultraviolet observations at IUE, Zolcinski et al. (1982) constructed the differential emission measure curves for temperatures within 3×10^4–10^7 K for the model of the transition region and static coronal loops. The resulting curves were similar to each other and to the appropriate solar curve. They enabled estimates of the parameters of the stellar coronae of four F5–G1 dwarfs: temperatures within 1–46 MK, the filling factors of the stellar surface within 0.002–1, the density at the base of the coronae within 2×10^9–3×10^{10} cm^{-3}, and heights within 4×10^9–3×10^{11} cm. Micela et al. (1988) carried out a more detailed analysis of the Hyades: they used an improved algorithm for the detection of sources, considered the complete series of IPC observations of the cluster, 63 fields. Some 66 of 121 cluster members in the studied fields were identified as X-ray sources. The comparison of the resulting luminosity function for the Hyades with those of other stellar groups of the same age led to the conclusion that the dependence of L_X on the stellar age was a function of the spectral type: L_X of young solar-type stars in the Hyades was higher than in the field stars, L_X of K–M stars in the Hyades corresponded to the young-disk stars, whereas L_X of K–M old-disk stars was much lower.

At 4 square degrees, Caillault and Helfand (1985) found 61 X-ray sources in the Pleiades, of which 44 were cluster members. They found that X-ray luminosities of F stars in the Pleiades were close to the L_X of the stars of this spectral type in the Hyades and the field stars, which evidenced slow evolution of L_X of these stars. From the uniform sample of about 30 F stars the authors did not confirm that the dependence of L_X on rotation started abruptly from F6. For G stars of the cluster the average luminosity is $\langle L_X \rangle = 4 \times 10^{29}$ erg/s, which exceeded by more than by two orders of magnitudes that of the Sun and was higher by only 60% than for the solar-type stars in the Hyades. The Sun completely covered by active regions would have ten times lower L_X than the brightest G star in the Pleiades Hz II 253 with $L_X = 2 \times 10^{30}$ erg/s. Therefore, Caillault and Helfand (1985) assumed that the structure of coronae of young stars differed from the solar corona by a greater number of loops. The absence of the rotational modulation evidenced a large number of uniformly distributed active regions.

Micela et al. (1990) analyzed all 14 IPC images of the Pleiades region and, using an updated algorithm for data processing, found that for slowly rotating stars $L_X \sim (v \sin i)^2$, while fast-rotating K dwarfs did not fit this relationship, but they are brighter and thus are preferable for detection. In the Pleiades, K dwarfs are brighter by an order of magnitude in soft X-rays than similar stars in the Hyades and brighter by two orders of magnitude than K field dwarfs.

Gagné and Caillault (1994) analyzed the observations of the region of the stellar cluster of the Orion nebula, during which 65 of 157 F6–M5 cluster members were identified as X-ray sources. The comparison of $v \sin i$ of 29 stars and the rotation periods of 8 stars with L_X/L_{bol} showed that in these samples fast rotators had lower ratios. This means that for the youngest stars the rotation is not the only factor that governs the activity level and suggests that the accretion disk can be such an independent factor.

Concluding the survey of the Einstein Observatory results it is worth noting that IPC observations covered only 10% of the sky, which enabled measuring of X-ray fluxes from more than 35 000 stars in about 4000 fields. The methods of data analysis were brought to a state of near perfection by the team of the Harvard-Smithsonian Center for Astrophysics and the Palermo Astronomical Observatory. The data obtained and processing results are available on CD-ROM. This experience was then widely used in space experiments.

EXOSAT Observatory. Two years after the Einstein Observatory completed its operation, in May 1983 the European X-ray satellite EXOSAT with an aperture diameter of 28 cm was launched. Unlike the Einstein Observatory, the satellite was put on a highly eccentric orbit, which made it possible to observe selected objects continuously for up to three days. EXOSAT was equipped with instruments designed for the investigation for soft (LE) and medium (ME) X-ray emission. Soft X-rays were recorded by two telescopes of grazing incidence optics with four changeable filters, which to a certain extent narrowed the band of 0.05–2 keV; multichannel plates were used as detectors. At a field of view of 2 arcmin the angular resolution was 20 arcsec. The instrument for medium X-rays included two blocks, each composed of 4 proportional counters designed for the range of 1–20 keV; the axes of the blocks were slightly misaligned for simultaneous measurement of a source and an adjacent background. This instrument did not produce images; the angular resolution was determined by FWHM = 45′; spectral resolution in the region of 3 keV was equal to 3. Although EXOSAT was less sensitive than the Einstein Observatory, its advantage was a wider spectral band.

During the investigation of different characteristics of stellar activity of the UV Cet-type stars carried out in 1983 in the Crimea a satisfactory linear correlation between X-ray luminosities L_X of such stars measured by that time and time-averaged luminosities of their optical flares $\langle L_{opt}^{fl} \rangle$ was established (Gershberg and Shakhovskaya, 1983). This result was practically unnoticed, but then between October 1984 and January 1985 three independent studies with identical results were published by Whitehouse (1985) and Skumanich (1985), each for 9 flare stars, and by Doyle and Butler (1985) for 18 flare stars. The latter authors found a close correlation between L_X and time-averaged flaring luminosity in the U band. Combining the Crimean data and those of Doyle and Butler, Shakhovskaya (1989) obtained the following relation from 23 stars

$$L_X = (4.4 \pm 1.9) \langle L_{\text{opt}}^{\text{fl}} \rangle \ . \tag{19}$$

Apparently, the linear relation of the optical radiation of sporadic flares and X-ray emission of quiescent coronae can be caused either by a common energy source, or direct heating of coronae by flares. This issue will be considered in more detail in the following chapter.

Smale et al. (1986) studied the dM4.5e star 1E0419.2+1908 found occasionally in the course of EXOSAT X-ray observations of the region near T Tau and then found in the archive of the Einstein Observatory. Analysis of the data from both observatories made it possible to estimate the parameters of the quiescent corona: the temperature of 4×10^6 K, $L_X(0.2$–4.0 keV$) = 6 \times 10^{27}$ erg/s (IPC), $L_X(0.02$–2.5 keV$) = 1.5 \times 10^{28}$ erg/s (LE), $EM = 5 \times 10^{51}$ cm^{-3}. Within the spherically symmetrical homogeneous model these values yielded the height scale of the isothermal corona of 4.5×10^9 cm and mean-square density of 1.3×10^{10} cm^{-3}, which does not differ much from the parameters of the solar corona, whereas the application of the correlations of Rosner et al. (1978) to the loop structure resulted in nonrealistic estimates of characteristic coronal loops.

Jordan et al. (1987) constructed spherically symmetrical models of hydrostatically equilibrium upper atmospheres of five G0–K2 stars on the basis of joint consideration of their ultraviolet spectra and X-ray luminosities. The models covered the range of structures from those similar to the quiescent solar atmosphere to the structure corresponding to a well-developed solar-active region.

Within the spherically symmetrical models of stellar coronae, Katsova et al. (1987) developed the method for estimating n_0, electron density, on the basis of the models. Using Einstein Observatory and EXOSAT X-ray observations, they estimated n_0 for 42 F8–M6 dwarfs (see Fig. 17). For this purpose, they took the coronal emission measures and temperatures, so that the resulting values of n_0 were averaged over the whole stellar surface. The upper envelope curve of the points corresponds to maximum values of n_0 for dwarf stars that belong to different spectral types. As to the found range of n_0, it should be noted that for the region of large polar coronal holes on the Sun $n_0 \sim 6 \times 10^7$ cm^{-3}, for quiet and active solar regions it was 3×10^8 cm^{-3} and up to 10^9 cm^{-3}, respectively; in the most dense stationary condensations above sunspots $n_0 \sim 10^{10}$ cm^{-3} and up to 10^{11} cm^{-3} in the flare loops. Thus, within the developed formalism, the objects placed in the upper part of Fig. 17 should have coronae with mean characteristics close to the densest

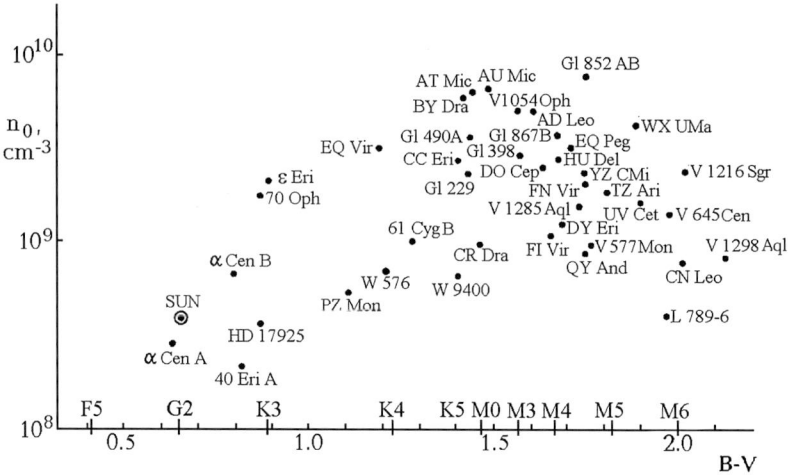

Fig. 17. Electron densities n_0 in the bases of red dwarf coronae calculated from X-ray emission within spherically symmetrical models (Katsova et al., 1987)

condensations above sunspots. The upper envelope curve of the points grows monotonically for G–K–M3 stars and then has a clear decline. Katsova et al. (1987) explained the character of the envelope with the systematic growth of the filling factor at the ascending branch and the transfer to smaller loops and the mechanism of coronal heating by microflares in late M dwarfs. Later, analyzing the Balmer decrement of red dwarfs in the quiescent state Katsova (1990) suspected that the ratio F_{H8}/F_{H_γ} correlated with L_X: for the stars with higher X-ray luminosity the ratio was higher, i.e., the decrement was flatter.

Observing 7 M dwarfs close to the Sun, Schmitt and Rosso (1988) found that EXOSAT values of L_X were close to the appropriate estimates for such stars based on the Einstein Observatory data and confirmed the systematic decrease of L_X to late spectral types and the presence of structures with temperatures of about 10^7 K.

Analyzing the X-ray spectra of the stars with increased luminosity – Capella and σ^2 CrB – within the range of 10–150 Å, Mewe et al. (1987) proved that the allowance for the expansion of coronal magnetic flux tubes with height considerably improved the presentation of observations. In this case, the tubes with a temperature of about 5 MK expanded by 30–50 times, whereas the tubes with hotter plasma (up to 30 MK), increased only by a factor of 2.5–4. An indirect confirmation of the validity of this idea is the fact that coronal condensations above bipolar magnetic fields on the Sun are greater by an order of magnitude than the underlying photospheric faculae. A more complete theory of static plasma loops in stellar coronae was developed by Ciaravella et al. (1993). They studied the influence of pressure at the loop base and its length on thermal conductivity, which governed the loop energy balance, temperature distribution, density, and EM. In particular, they found that for a given loop length the temperature at its top was a minimum when the pressure at the base was maximum.

Johnson continued the studies of the Wolf 630 system at EXOSAT and found a cool stellar corona on one of the weakest flare stars: the coronal temperature of VB 8 was close to 6×10^5 K. For VB 8 Tagliaferri et al. (1990) estimated L_X as 6×10^{26} erg/s within the range of 0.1–3.5 keV and found that the observed ratio $L_X/L_{\rm bol}$ corresponded to that on earlier M dwarfs with convective envelopes, whereas VB 8, whose mass is $0.12 M\odot$, should be completely convective.

Based on EXOSAT LE and ME X-ray observations of one of the fastest rotators Gl 890, Rao and Singh (1990) found that the estimated temperature and luminosity were the same as for slowly rotating stars, which could be due to the saturation of the dynamo mechanism on Gl 890.

Haisch et al. (1990a) considered ultraviolet and X-ray observations of the YY Gem system and concluded that each of the components was completely covered by active regions, where MgII emission was formed, since the latter was rather stable. Transition-region lines demonstrated a certain rotational modulation, while the X-ray light curve had a deep minimum (to 50%) due to low-lying coronal loops or the X-ray emission concentrated at the eclipsed component.

Rao et al. (1990) studied the red dwarf BD+48°1958 A by EXOSAT LE and ME instruments and estimated the parameters of its quiescent corona as $T = 18 \times 10^6$ K, $L_X(0.1-4.0\,{\rm keV}) = 3 \times 10^{29}$ erg/s and $EM = 1.4 \times 10^{52}\,{\rm cm}^{-3}$.

Using EXOSAT observations, Pollock et al. (1991) revised the observational data of the Einstein Observatory for the Gl 867 AB system. They concluded that the components of the system contributed to the total radiation as 3:1 and that the change of the X-ray emission of the corona found by Agrawal (1988) from the Einstein Observatory was due to flares occurring on both components rather than to the spottedness of the corona. It is worth noting that between 1980 and 1984 X-ray luminosity of one of the components changed by a factor of three.

Above, we mentioned the principle of estimating the coronal temperature through theoretical calculations of the radiation of optically thin hot plasma of the given chemical composition. However, the limited range of examined wavelengths and low spectral resolution of the X-ray Einstein IPC usually allowed only a one-temperature model of the corona to be elaborated. Nevertheless, the higher-resolution SSS observations by Swank and Johnson (1982) at the Einstein Observatory yielded the two-temperature model of the Wolf 630 corona, in which one component had a temperature of several million kelvins and the other, an order of magnitude higher value. Then, Majer et al. (1986) noticed that the same temperatures were obtained for different stars observed by the same instrument, while the observations of the same star by different instruments systematically yielded different temperatures. They suggested that the estimated temperatures were governed not so much by the corona physics as by the different sensitivity curves of the detectors, since the shape of the curves was rather complicated. Further studies confirmed these suggestions (Schmitt et al., 1987). Schrijver and Mewe (1986) undertook the first attempt to simulate EXOSAT data within the concept of the differential emission measure (DEM) and compared them with the calculated spectra of the loop structure. Jordan et al. (1987) constructed DEM from separate well-determined spectral lines, whereas Schrijver and Mewe presented the whole EUV spectrum at once. Pallavicini et al. (1988) examined the possibility of wide-band EXOSAT observations to estimate temperatures, emission measures, and luminosities of stellar coronae. They established that there was a continuous temperature distribution up to 10^7 K in the coronae of late stars in the quiescent state, but the measurements of EXOSAT filters prohibited construction of the DEM distribution and the selection of a multitemperature solution among different versions. Schmitt et al. (1990) studied the validity of two-temperature models of stellar coronae obtained from wide-band IPC/Einstein data. To this end, they considered the observations with rather high S/N of 130 stars later than A0 and found that this ratio greatly affected the resulting parameters of the models and the observational data could also be well presented within a physically more reliable model on a continuous EM distribution with temperature. The low-flux objects can be presented only by the isothermal model, while the objects with higher fluxes can be presented within two-temperature models, with the relations of the obtained components being different for different stellar groups: in M dwarfs the hot component ($T > 10^7$ K) was found in combination with the component with $T \sim 3 \times 10^6$ K and EM of the hot component exceeded that of the cooler one, whereas in F and G dwarfs, as well as on the Sun, the high-temperature component was very weak or absent. Probably, the two-temperature models can account for the real situation when the radiative losses of the coronal plasma were predominant cooling mechanisms and over certain temperature intervals, as these losses grow with increasing temperature, the loops become stable to temperature disturbances. On the other hand, the two-peak DEM model suggests that the corona consists of two different ensembles of quasi-static magnetic loops with different maximum temperatures.

EXOSAT was operational from May 1983 until April 1986. Over this period, during 45 sessions with a total duration of about 300 h 25 UV Cet-type stars were examined. The results of these studies were summed up by Pallavicini et al. (1990a). The following results were obtained for quiescent coronae. In spite of the relatively narrow spectral range of considered objects, the X-ray luminosities of quiescent coronae L_X (0.05–2.0 keV) calculated from the observations varied within a wide range from 1.4×10^{27} erg/s (Proxima Cen) to 6.7×10^{29} erg/s (YY Gem). These values did not display any correlation with axial rotation periods, while the correlation coefficient between L_X and $v_{\rm rot}$ is rather low, 0.37, but the dependence $L_X = 10^{27} v_{\rm rot}^2$ found by Pallavicini et al. (1981) from the observations at the Einstein Observatory matched well the points of this comparison. A reliable dependence between

L_X and L_{bol} with the correlation coefficient $r = 0.95$ was found by Pallavicini et al. (1990a) from EXOSAT data

$$\log L_X = -9.83 + 1.21 \log L_{\text{bol}} . \tag{20}$$

The correlation is close to $L_X \sim L_{\text{bol}}$ and is probably due to the saturation of stellar corona by active regions, since such a correlation is not valid for dM stars. During the long-term monitoring of red dwarfs slow intensity variations were observed: at times of the order of tens of minutes and hours. When different observational sessions of the same objects separated by several months were compared, the distinctions in the level of their intensity were usually equal to 10–20% and never exceeded 100%. To test the hypothesis of Butler and Rodonò (1985) suggesting that quiet radiation of stellar coronae is a superposition of numerous microflares, all EXOSAT data were analyzed using the dispersion analysis, the autocorrelation function, and the power spectrum. For this purpose, 26 runs of observations were divided into intervals of 2 and 5 min, and the dispersion of the X-ray intensity was equal to 10 to 20% in most cases and to about 40% only in some cases. In the resulting autocorrelation functions, the e-fold decay time was 30–60 min and never shorter than 10 min. Thus, though EXOSAT detected independently the variability of dMe stars at all time intervals from several minutes to hours, none of the stars provided definite evidence of the variability at times shorter than several hundred seconds, which could be associated with microflares. Pallavicini et al. (1990a) mentioned that this conclusion complied with the rigorous data analysis by Butler and Rodonò (1985) and by Ambruster et al. (1987). Earlier, conclusions practically coinciding with the results of Pallavicini et al. (1990a) were obtained by Collura et al. (1988), who analyzed the EXOSAT observations of 13 flare stars using the technique developed by Ambruster et al. (1987) for analyzing the Einstein observations of active red dwarfs.

The research results for time variations of the X-ray emission of active stars noticeably differ from the solar data. On the whole, the Sun as an X-ray source is to a greater extent more variable than red dwarf stars. The long-term observations for F_X fluxes within the ranges of 0.5–4 Å and 1–8 Å with a time resolution of 5 min revealed the variations from fractions of a per cent to 2–3 orders of magnitude. The greatest of them are due to flares, but the variations of lower amplitudes result from the development of the 11-year cycle, birth of new and death of old active regions, rotation of the Sun nonuniformly covered by such regions, the processes occurring in active regions, the fluctuations of coronal heating, and the exit of new magnetic fluxes. For the Sun, $L_X(1-8\,\text{Å})/L_{\text{bol}} = 10^{-9}$, therefore the variations of F_X are so significant. The amplitudes of the variations due to the solar rotation are up to 10, while the changes between the extremes of the solar cycle are 100–200. The amplitudes of fluxes from quiescent stellar coronae do not exceed 2–3 times because the sample includes the most active stars with the surface filling factors for active regions that are higher by orders of magnitudes than the appropriate magnitudes on the Sun (Pallavicini, 1993).

The ROSAT Observatory was orbited in June 1990 and was operational for about 9 years. The Observatory carried a German Wolter-type I X-ray telescope with an aperture diameter of 83 cm and the above-described English Wide-Field Camera (WFC). Two position-sensitive proportional counters (PSPC) and High Resolution Imager (HRI) very similar to the Einstein Observatory imager were mounted on a carousel within the focal plane turret. The PSPC efficiently recorded the radiation in two bands: 0.1–0.28 and 0.5–1.5 keV. The sensitivity of this system was much higher than in the previous experiments, and the main noise in the instrument was due to the diffuse X-ray background. The field of view of the device was about 2°, and the resolution was 15–25 arcsec on the axis and about 160 arcsec at a distance of 50 arcmin from the axis.

The first PSRC ROSAT all-sky survey, the RASS program, was undertaken between August 1990 and January 1991. In sky scanning with the orientation of the satellite to the Sun each object was accessible for at least two days and was measured for 20 s each 96 min. The RASS program made it possible to discover about 60 000 X-ray sources, 1/3 of them being the stars with hot coronae. The important advantage of the RASS program was that it enabled direct statistical studies of X-ray sources based on their X-ray characteristics and not from earlier known manifestations of the activity in other wavelength ranges. Upon completion of RASS the studies were run in the mode of individual pointings of the apparatus with exposures of up to 40 min.

In considering the ROSAT observations of red dwarfs later than M5, Fleming et al. (1993) found that the upper border of the $L_X(M_V)$ curve continued the saturation curve outlined from earlier stars, while the L_X/L_{bol} ratio remained constant, at about 2×10^{-3}, with M_V ranging from 8^m to 20^m. They concluded that there were no grounds to believe that M dwarfs were less efficient in creating coronae than hotter stars.

The RASS program and individual pointings of ROSAT made it possible to observe an almost complete spatially limited sample of 114 K and M dwarfs in the vicinity of up to 7 pc from the Sun. F_X fluxes were measured for 87% of sample stars, the upper limits of the fluxes were determined for 13% (Schmitt et al., 1995; Fleming et al., 1995). From the calculated values of L_X a luminosity function of such stars was constructed, which covered almost 4 orders of magnitude. Comparison of L_X and the hardness of X-ray emission suggested that the stars that were brighter in X-rays had hotter coronae. Comparison of ROSAT and Einstein Observatory data did not reveal the cases when L_X varied more than by a factor of two. The analysis of the sample did not provide any evidence of a decrease in the L_X/L_{bol} ratio for lowest-mass stars that were completely convective but suggested that the coronae of dM stars were systematically cooler and weaker than those of dMe dwarfs. The correlation of L_X/L_{bol} and the radiation hardness with metallicity, i.e., with age, was established, but no correlation of these values with kinematic classes, which required cautious consideration of these classes as indicators of the age for close stars, was revealed.

Marino et al. (1999) studied the variability of M dwarfs in the X-ray range of 0.1–2.4 keV from the PSPC data. They considered all M dwarfs from the CNS3 catalogs found in the course of individual pointings of ROSAT and recorded not further than 48 arcmin from the center of the fields. The total number of such stars was 55 and they were observed in 86 sessions. The duration of exposures was shorter than in EXOSAT observations, but the sensitivity was much higher. For 29 of 32 sources, for which more than 1000 pulses were recorded, the variability was found at the level of significance of over 99%. This result suggests that all M dwarfs are variable in X-rays and the detection of this variability is limited only by the statistics of recorded quanta. For the stars with $M_V \sim 13^m$, which is typical of completely convective red dwarfs, there was no noticeable decrease in the portion of variable sources and the variability was independent of the average level of X-ray luminosity. Comparison of normalized distributions of L_X of the considered M stars yielded results similar to those for young variable stars in the molecular cloud of ρ Oph and in solar flares, which suggests that stellar flares are responsible for the discovered variability of M dwarfs.

Then, in a similar way, Marino et al. (2002) studied the X-ray variability of F7–K2 dwarfs. Analyzing the statistics of photons from 40 objects in 70 sessions, they found 8 variable sources in 10 sessions and suspected long-term changes, presumably of a cyclic character, for 9 stars observed in the beginning and at the end of ROSAT operation. The amplitudes of the brightness variations were lower on F–K stars than on M dwarfs.

Ambruster et al. (1994a, 1998) compared the radiation of the chromospheres (MgII), the transition regions (CIV), and the coronae (ROSAT PSPC) of 6 near K dwarfs of close age, whose rotational periods were within the range from 8 h to 6.6 days. The comparison of their surface fluxes with the rotational periods showed that F_{CIV} saturated at $P_{\text{rot}} < 3$ days, and the saturation of F_X occurred at even shorter periods, while saturation of F_{MgII} required periods of about 5 or even 11 days.

Later, Micela et al. (1997) analyzed the observations of 12 close K4–M6 halo and old-disk stars at PSPC and WFC ROSAT and found that the half of considered stars displayed variations of L_X to 50% on time scales from hours to days, for GJ 191 such changes occurred over six months and, as compared to the data of Schmitt et al. (1995), such variations were found at even greater time intervals. For three brightest stars of the sample – GJ 845, GJ 866, and GJ 887 – two-temperature models were constructed, which demonstrated rather low temperatures of the components, 1.7×10^6 and 5.8×10^6 K, and as opposed to more active dwarfs, the EM_2 of high-temperature components was lower than EM_1 of the low-temperature component. In comparing the models obtained from nonsimultaneous data, the temperatures of the components were found to be constant, while L_X varied with the change of emission measures, and there was a close correlation of these values: $EM_2 \sim (EM_1)^3$.

Pan and Jordan (1995) found that EM, temperature, and electron pressure in the source of quiescent X-ray emission of CC Eri corresponded to the well-developed active region on the Sun.

Pan et al. (1995) considered the IPC (Einstein Observatory) and PSPC (ROSAT) observations of the flare star EQ 1839.6+8002 discovered during a strong flare by the Ginga apparatus. They presented its X-ray emission in the quiescent state by the two-temperature model: $T_1 = 1.5 \times 10^6$ and $T_2 = 7.6 \times 10^7$ K, $EM_1 = 4 \times 10^{50}$ and $EM_2 = 6 \times 10^{50}$ cm^{-3}.

Giampapa et al. (1996) carried out long individual observations of 11 M dwarfs at ROSAT PSPC. Their program included dMe stars with $L_X/L_{bol} \sim 10^{-4}$ and dM objects with $L_X/L_{bol} \sim 10^{-7} - 10^{-6}$ (see Fig. 18). In particular, they recorded the X-ray emission from Gl 411 and Gl 754, which

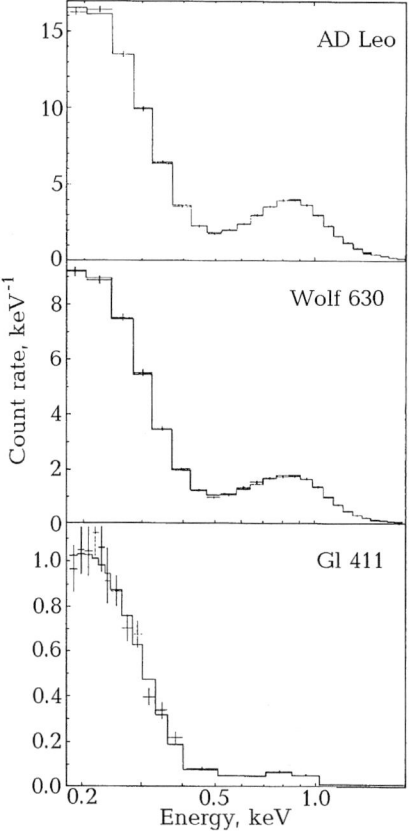

Fig. 18. Spectra of three M dwarfs recorded by ROSAT PSPC (Giampapa et al., 1996)

earlier were considered as stars without coronae. On the basis of the analysis of the obtained data they concluded that the coronae of such stars, probably except for Barnard's star corona only, contained two different components of the X-ray emission: the soft component with a temperature of 2–4 MK and the hard component with $T \sim 10$ MK. The temperature estimates are more reliable in calculations with reduced content of heavy elements (up to an order of magnitude) in the coronae as compared to that on the Sun. The hard coronal component on dMe stars makes a systematically greater contribution to the total X-ray emission than on dM stars and is responsible for the major part of variations observed in X-rays. Giampapa et al. (1996) modeled hydrostatic coronal loops for such components and found that the soft component could be presented by small-size loops ($h < R_*$) and high pressure, whereas the hard-component loops required either a small filling factor (< 0.1), large loops ($h > R_*$), and high pressure, or a very small filling factor ($\ll 1$), small loops ($h < R_*$) and very high pressure. From the calculations, they concluded that the soft component of the coronal radiation of dMe stars was formed in the quiet active regions, while the hard component occurred in compact nonstable flare formations. The compact loops of the soft component should give a natural explanation of the observed evolution of angular momentum of late main-sequence stars.

Using ROSAT PSPC observations of solar-type stars in young clusters, Maggio et al. (1997) thoroughly simulated the loop structure of coronae. They used the hydrostatic loop model modified by Serio et al. (1981) for arbitrary size that was described by the system of two equations linking four loop parameters: the temperature at the top, pressure at the base, the height, and the heating of a unit volume. Specifying the distribution of the plasma temperature and density along the loop height, one could calculate its integral optically thin X-ray emission. Then, upon normalizing to the sensitivity curve of a detector, the obtained spectrum could be compared with the observed one. Applying the developed algorithm to seven G–K dwarfs from several young clusters, they considered the models with identical loops and with two kinds of loops. In the resulting models, in all cases a hot component with a maximum temperature of $(0.6-4) \times 10^7$ K was available, which never occurs on the Sun, and three oldest stars from the Hyades had a cool component with $T_{\max} = (1-6) \times 10^6$ K, typical of the quiet Sun. However, in the low-temperature loops, the pressure exceeds typical solar values by two orders of magnitude, and these loops contribute 30–70% to the total X-ray luminosity of these stars. The above algorithm yielded the following characteristics of the corona of the F9 dwarf HR 3625: low-temperature (1–2 MK) short ($< 10^8$ cm) low-pressure ($p_0 > 6$ dyn/cm^2) loops covering a large part of the stellar surface and high-temperature (over 7 MK) loops related to active regions. The characteristics of the latter loops could vary from $h < 10^9$ cm at $p_0 > 10$ dyn/cm^2 and a filling factor less than 0.5% of the surface to $h \sim 10^{10}$ cm at $p_0 > 5-10$ dyn/cm^2 and f ~ 0.1 (Maggio and Peres, 1997). Upon similar analysis of X-ray data for eight G dwarfs in the solar vicinity Ventura et al. (1998) concluded that the most suitable presentation of these data was provided by the models with two types of loops: "cool" loops with $T \sim (1.5 - 5) \times 10^6$ K and p_0 varying from 2 to 100 dyn/cm^2 and "hot" loops with $T \sim (1 - 3) \times 10^7$ K and p_0 varying from 10^2 to 2×10^4 dyn/cm^2. Ciaravella et al. (1997) proved the advantages of the two-temperature modeling using the SIS ASCA data.

The fact that the hottest components of the quiescent stellar coronae have the temperatures close to the plasma temperature in solar flares suggests that the unresolved stellar flares can be responsible for these hot components. Within the Kopp–Poletto model (1992), Güdel et al. (1998) calculated the summary effect of a great number of such flares. Using the observed solar-power distribution of flares with respect to energy and the correlation of the flare duration and its full energy, they found that at simultaneous existence of several hundreds of magnetic loops with flares occurring in them the summary light curve of such loops in soft X-rays should display smooth oscillations, similar to the observed ones, while the numerically obtained bimodal distributions of $EM(T)$ reminded the appropriate distributions recorded on the four solar-type stars.

Additional data on calculated models of some quiescent coronae with flares developed on their background are presented in Sect. 2.4.1.

The Catalog of Chromospherically Active Binaries by Strassmeier et al. (1988) includes about 200 systems, 163 of them were found within the RASS program. Dempsey et al. (1997) studied the radiation of 35 binary stars of the BY Dra type and found that for these dwarf systems the dependence of F_X on the rotational period was much weaker than in RS CVn systems: $F_X \sim P_{\rm rot}^{-0.2 \pm 0.3}$ and $P_{\rm rot}^{-0.6 \pm 0.1}$, respectively. They did not find evidence of a basal level of X-ray emission of these stars. The temperatures of both components in the systems of both types found within two-temperature corona models were practically identical – 2 and 15 MK, but EM in RS CVn systems were higher by almost an order of magnitude and the dependences of temperatures and emission measures on rotation were identical for the systems containing the components of different luminosities.

Using the simplest geometrical model of inhomogeneous stellar corona, Güdel and Schmitt (1996) estimated the electron density of the EK Dra corona as 3×10^{10} cm^{-3} from the measured modulation of its X-ray emission recorded in the RASS experiment. For another star with X-ray rotational modulation, 47 Cas, they failed to construct an unambiguous coronal model and obtained two density estimates: 2×10^{10} and 2×10^{11} cm^{-3}.

Piters et al. (1997) compared the values of ΔF_{HK} for 215 F–G–K stars calculated from the Mount-Wilson data and the values of F_X calculated from RASS data obtained within several days and found the relationship

$$F_X/F_{\rm bol} \sim (\Delta F_{HK}/F_{\rm bol})^2 , \qquad (21)$$

which is independent of the spectral type and luminosity, while the scatter of points near the dependence was explained by the measurement accuracy; the hardness of the X-ray spectrum of dwarf stars grew systematically with F_X.

Using new instruments, Schmitt (1997) repeated the study carried out in 1985: he considered the X-ray emission of a practically complete sample of A–F–G stars in the solar vicinity up to 13 pc and even with the most sensitive observations with long exposures did not find radiation from stars earlier than A7. This can be considered as an argument in favor of the universal character of corona formation on stars with outer convective zones. He found that L_X correlated with the kinematic age of stars and with the hardness of their X-ray spectra, while the range of F_X measured on the stars was close to the appropriate range of different structures on the Sun, from coronal holes to active regions.

Hünsch et al. (1998, 1999) published the catalog of RASS observations of 980 optically bright dwarfs and subgiants of A–K spectral types and the catalog of RASS observations of 1252 nearest stars.

Fuhrmeister and Schmitt (2003) analyzed about 30 000 X-ray light curves obtained within the RASS program and discovered 1207 variable sources. Among the sources identified to optical objects there were about 450 F–M stars. In addition, many flares of the sources invisible in a quiescent state were recorded.

Caillault et al. (1992) performed preliminary observations of the Pleiades and found 47 X-ray sources, of which 36 were cluster members. A half of these sources were not revealed by the Einstein Observatory, on the other hand, a half of the sources discovered by the Einstein Observatory were not detected by ROSAT, which evidence noticeable variability of the sources.

In studying the central nucleus of the Pleiades, Schmitt et al. (1993b) found that in the range of 0.5–1.8 keV the radiation density was 4×10^{29} erg/s pc^3, which was higher by two orders of magnitude than in the solar vicinity.

Using long-exposure images, Stauffer et al. (1994) studied 3 regions in the Pleiades and identified 317 X-ray sources, of which 171 were cluster members. In a radius of 25 arcmin from the centers of considered fields practically all cluster members of G and later types were identified. Thus, for the first time the X-ray luminosity function of G–K–M stars was constructed without statistical estimates for the objects with determined upper limits of luminosity only. Apparently, there were low-mass cluster members among the found sources, which were not identified as such in the optics. The constructed dependence $L_X/L_{\rm bol}(v_{\rm rot})$ revealed the correlation of saturated type for the objects with $B - V > 0\overset{m}{.}6$: in each of the considered $B - V$ intervals there was a fast growth of L_X for the rotation velocities of up to 15 km/s and then saturation for the velocities of about 100 km/s; the dispersion of rotational velocities for low-mass cluster members was a dominating factor that governed the dispersion of L_X. About 35% of B, A, and early F stars in the Pleiades were identified as X-ray sources, but the properties of their coronal radiation supported the hypothesis that invisible low-mass satellites were responsible for this radiation. Continuing the studies of Stauffer et al. (1994), based on PSPC data, Gagné et al. (1994a, 1995) constructed one- and two-temperature models of a dozen of the brightest sources related to the stars of late spectral types. They found that the coronae of K and rapidly rotating G dwarfs were hotter than the coronae of F and slowly rotating G dwarfs. For F–G–K stars they found a dependence of the coronal temperature on L_X/L_{bol}.

Sciortino et al. (1994) studied the variability of the X-ray emission of stars in the Pleiades using the ROSAT PSPC images obtained within six months and found variability due to flares only. The comparison of the Einstein Observatory and ROSAT data showed that over 10 years the variations of the X-ray brightness were only slightly greater than over 6 months. Thus, Sciortino et al. concluded that this fact was stipulated by permanent high surface filling of young stars by active regions.

Barbera et al. (1993) and Fleming et al. (1993) came to different conclusions concerning the difference of stellar X-ray emission of the objects with radiative core and fully convective structures, which could be explained by a different age composition of the stellar samples considered in the solar vicinity. To avoid this, Hodgkin et al. (1995) analyzed the indicators of chromospheric and coronal activity – L_{H_α}/L_{bol} and L_X/L_{bol}, respectively – using a large sample of low and very low mass stars, slightly more or less than 0.3 solar masses, in the Pleiades. The H_α emission was found in all cluster members with $M_I > 7\overset{m}{.}5$, and the ratio L_{H_α}/L_{bol} reached its maximum within 0.3–0.4 solar masses, then rapidly decreased, i.e., the chromospheric activity decreased on fully convective stars. The situation with X-ray activity is more complicated. Since for weak X-ray sources, the members of the Pleiades cluster, ROSAT PSPC measured only the upper limits of L_X, the determined change of the ratio L_X/L_{bol} depends on the specific consideration of these sources: if they are disregarded and only really measured sources are taken into account, L_X/L_{bol} reaches a maximum at somewhat higher masses than L_{H_α}/L_{bol}, then it also decreases, according to Barbera et al. (1993). If we take into account the upper limits of L_X as well, the conclusions of Fleming et al. (1993) are confirmed.

The analysis of long-term observations of the central square degree of the region in the Pleiades allowed Micela et al. (1996) to identify the sources with $L_X > (2-3) \times 10^{28}$ erg/s, to measure the fluxes from 99 of 214 cluster members, and to estimate the upper limits of F_X for the rest. Inside the central 20' region they found all early M dwarfs, confirmed the dependence of L_X on the age for G dwarfs revealed by the Einstein Observatory, and extrapolated it to K and M stars. G dwarfs in the Pleiades displayed an almost linear dependence of L_X on the rotational velocity up to 100 km/s, but dK members of the Pleiades showed a weak dependence on the velocity, which required a certain additional parameter, in addition to rotation and age, to understand the X-ray emission of the stars. As opposed to the Hyades, the X-ray luminosities of dG and dK stars in the binary systems are of the same order of magnitude as the appropriate values of single stars. This means that the evolution of L_X in single and binary stars to the age of the Pleiades is independent of their involvement in binary systems. Up to 15% of the members of the Pleiades displayed a variation of F_X up to twice on an interval of 10 years.

Then, Micela et al. (1999a) analyzed the HRI ROSAT data for the Pleiades. This more sensitive instrument with higher spatial resolution made it possible to supplement the list of X-ray sources of the cluster by 34 objects, restrict the estimates of the upper limits of L_X of weak sources, and continue the dependence $L_X(v_{\rm rot})$ toward slow rotators.

Marino et al. (2003a) analyzed the variability of stars of late spectral types in the Pleiades based on PSPC observations and divided them into two groups: 42 F7–K2 and 61 K3–M stars. Application of the Kolmogorov–Smirnov statistics to the time series of photon fluxes revealed an equal percentage of significant variations over short times – hours – in both groups. The analysis of amplitude-distribution functions showed that over short times K3–M stars were variable to a greater extent than F7–K2 dwarfs. For such stars in the Pleiades and the field the variability over short times depended on X-ray luminosity, whereas for later stars this dependence was not valid. The lowest-mass stars are characterized by fast variations, but not rotational modulation or cycles, whereas variations associated with rotary modulation and cyclic variability are more typical of F7–K2.

Prosser et al. (1995) studied the open stellar cluster NGC 6475, whose age is about 2×10^8 years, and identified 120 early F–M dwarfs – new weak members of the cluster – from X-ray emission with $L_X > 10^{29}$ erg/s. The upper envelope of the points on the plot of L_X (Sp) is analogous to that in the Pleiades, while the scatter of the values of L_X is apparently due to binary systems and fast rotators.

The X-ray properties of the IC 4665 cluster similar to those of the members of the Pleiades, a cluster of close age, were studied by Giampapa et al. (1998).

The central part of the Orion nebula was studied by Gagné et al. (1994b) using HRI ROSAT. They identified 361 sources with $L_X(0.1 - 2.0 \text{ keV}) > 10^{30}$ erg/s and 55 of 87 F6–M5 optical cluster members as X-ray sources. The relation between $L_X/L_{\rm bol}$ of young stars and rotation was not found from this sample.

Pye et al. (1994) observed 11 fields in the region of the Hyades: at a general coverage of 18 square degrees at an area of 3.5 square degree their sensitivity increased by a factor of 5 as compared to the RASS program, and the X-ray emission of 75% of 70 stars of the cluster was found. This made it possible to construct the luminosity function up to $L_X(0.1 - 2.4 \text{ keV}) = 5 \times 10^{27}$ erg/s. They found that K dwarfs involved in the binary systems were systematically brighter than single K stars, but since these were all wide pairs with periods of more than a year, the cause of the higher brightness of the binary system components was not clear. For three dozen stars the two-component models of coronae were constructed, since the isothermal models did not provide a satisfactory presentation of the observations. Stern et al. (1995) studied the Hyades cluster from the RASS data covering the region of $30° \times 30°$. Some 187 of the 460 known cluster members were identified as X-ray sources with

$L_X > (1-2) \times 10^{28}$ erg/s. In comparing the results with the measurements of the Einstein Observatory, they found the brightness changes by no more than a factor of two, i.e., there were no manifestations of cyclicity, when one could expect variations up to an order of magnitude. Apparently, the point is that in these young stars the long-term dynamo mechanism has not been established as yet, as on the Sun, and a small-scale mechanism without long cycles operates there. Stern et al. measured F_X of 90% G dwarfs and from the constructed X-ray luminosity function estimated the average luminosity of G dwarfs in the Hyades as $\langle L_X \rangle = 10^{29}$ erg/s.

Randich and Schmitt (1995) obtained 42 images of the Praesepe cluster that covered the area of $4° \times 4°$. On the composite map of the cluster they achieved the limit of $L_X \sim 2 \times 10^{28}$ erg/s. Only 40 of 255 G–K–M members of the cluster were identified as X-ray sources, i.e., a percentage much lower than in the Hyades. This distinction is particularly explicit for G dwarfs: in the Hyades they all are identified as X-ray sources, whereas in the Praesepe – only 28%. This contradiction was removed by Franciosini et al. (2003), who observed the Praesepe with the EPIC mounted on XMM-Newton and detected all F–G stars of the cluster, about 90% of K dwarfs and more than 70% of M dwarfs. As a result, the X-ray luminosity function of solar-type stars appeared to be the same, as in the Hyades.

Prosser et al. (1996) studied X-ray emission of the members of the α Per cluster of the age of about 50 million years. About 80 of the 222 detected sources were identified with the known cluster members, whereas in the central part of the cluster all K and the majority of M dwarfs were identified as X-ray sources. In the sources with $v \sin i < 15$ km/s L_X grows together with the velocity in the range $-4.3 < \log L_X/L_{\text{bol}} < -2.9$ and at about 50 km/s reaches saturation. A noticeable decrease of this ratio for G–K at $v > 50$ km/s found in this and some other young clusters was called supersaturation by Randich (1997). Probably this effect is associated with the change of the structure of coronal loops on the fastest rotators and with a slight shift of the emission beyond the sensitivity limits of the detector (Stern, 1999; Jeffries, 1999). F and G dwarfs of the α Per cluster have greater L_X than similar stars in the Pleiades, for K dwarfs the distinctions are less and for M dwarfs of both clusters the distribution functions of the X-ray luminosities are very close.

Using ROSAT observations, Randich et al. (1996b) studied the Coma cluster and found that X-ray properties of the cluster members were similar to those in the Hyades and differed noticeably from those in the Praesepe, although all the three clusters are of almost the same age (see Fig. 67): in the Coma, as in the Hyades, almost all late F and G stars were recorded, whereas in the Praesepe such dwarfs had lower luminosity. The Coma cluster is distinguished by the deficit of K and later stars. The X-ray observations revealed 12 new members of the cluster.

Jeffries et al. (1997) carried out a PSPC survey of the field containing the young open cluster NGC 2516, whose age is 1.1×10^8 years. Only 6 of 159 sources found in the range of 0.5–2.0 keV were identified within 0.1–0.4 keV.

As many as 65 sources were reliably or hypothetically identified with the known members of the cluster, whereas most of the others were apparently low-mass cluster members. The maximum of the ratio $L_X/L_{\rm bol} \sim 10^{-3}$ falls on the late G and early K dwarfs. The reduced metallicity of the stars of this cluster results in qualitative distinctions in the X-ray luminosity distribution functions from other clusters. Then, Micela et al. (2000) studied this cluster from the ROSAT HRI observations. They found 12 new X-ray sources and identified 12 sources as K and M dwarfs, probable cluster members. Comparing the X-ray luminosities of red dwarfs of this cluster and the Pleiades, they concluded that the distinctions in the metallicity, did not influence within a factor of two, the saturation level of X-ray emission.

Jeffries and Tolley (1998) identified most of the 102 X-ray sources with weak stars, members of the young but relatively far open cluster NGC 2547 and mentioned that X-ray observations had certain advantages for detecting weak cluster members.

Using ROSAT PSPC data, Hünsch et al. (2003) investigated the open clusters NGC 2451 A and B and in each of them identified tens of active stars.

Micela et al. (1999b) used the ROSAT HRI data to study the X-ray emission of stars in the region of the open cluster Blanco 1 placed high above the Galaxy plane. One could expect that it differed from the other stellar clusters. From the number of X-ray sources identified with the cluster members they determined that the cluster was rather young, and from the L_X of G–K dwarfs its age seemed to be closer to that of the Pleiades rather than to the α Per cluster. Many low-mass cluster members were found among the X-ray sources as well. Pillitteri et al. (2003) continued this study up to M dwarfs. They specified the affiliation of X-ray sources to this cluster, constructed X-ray luminosity functions, compared them with the appropriate functions of the Pleiades, NGC 2516, and α Per cluster, and proved the increased metallicity of Blanco 1 resulting in the increased X-ray luminosity, probably at the expense of coronal emission lines.

Franciosini et al. (2000) used ROSAT HRI to study the open cluster NGC 3532, whose age is intermediate between the ages of the Pleiades and the Hyades. They found about 50 X-ray sources with luminosity above 4×10^{28} erg/s, 15 of them were identified optically.

From the ROSAT PSPC/HRI data Barbera et al. (2002) investigated the open cluster NGC 2422, whose age is close to that of the Pleiades. They found 78 sources, 62 of them were identified with optical objects, of which 80% were late stars. The X-ray luminosity function for F–K stars of this cluster was indistinguishable from that of the Pleiades.

Fleming (1998) noted that X-ray observations could reveal unknown nearest stars. As a rule, the lists of the nearest stars were compiled on the basis of observations of proper motions; but young dwarfs have low velocities, thus, these objects are underrepresented in the lists. From the RASS data Fleming identified 54 M dwarfs within 25 pc. Earlier, near the north pole of the Galaxy

238 X-ray sources were found within the RASS program. Using wide-angle optical images, Richter et al. (1995) discovered five flare stars.

Using ROSAT PSPC data, Pizzolato et al. (2002) revised the "activity–rotation" relation established from the Einstein Observatory measurements. They considered 115 field stars and 136 stars in the Pleiades, the Hyades, IC 2602 and α Per clusters in the range of $0.^{m}5 < B - V < 2.^{m}0$ with the known rotational periods and confirmed the existence of two X-ray radiation regimes: in one of them X-ray luminosity $L_X \sim P_{\mathrm{rot}}^{-2}$, in the other the saturation regime is achieved when $L_X/L_{\mathrm{bol}} \sim 10^{-3}$ and L_X varies from 5×10^{30} for $B - V = 0.^{m}65$ to 10^{29} erg/s for $B - V = 1.^{m}5$.

Golub (1983) extrapolated the relationships between the X-ray emission intensity of individual loops and the strength of the appropriate magnetic field established for the Sun to M dwarfs. Already from the first X-ray observations of red dwarfs he concluded that the observed X-ray emission of such stars should correspond to magnetic fields with the strength analogous to that in active regions of the Sun, and covering almost the whole stellar surface, or to the stronger fields covering a smaller area of the stellar surface. On the basis of the subsequent observations Saar and Schrijver (1987) found a close correlation between the average magnetic fluxes $\langle fB \rangle$ from G–K–M dwarfs and soft X-ray fluxes from them. The Sun satisfied this correlation. With allowance for Saar's refinement (2001), this correlation is close to linear:

$$\langle F_X \rangle = 6100 \langle fB \rangle^{0.95 \pm 0.05} . \tag{22}$$

Combining 22 and 12 refined by Schrijver et al. (1992), we obtain

$$\langle \Delta F_{HK} \rangle = 0.06 \langle fB \rangle^{0.62 \pm 0.14} , \tag{23}$$

which is valid up to $\langle fB \rangle \sim 300$ G.

1.4.2 EUV and X-ray Spectroscopy

The EUVE apparatus enabled direct images and the first spectra in the intermediate region between UV and X-rays to be obtained and gave the results of EUVE multicolor photometry of the objects of the considered activity type. According to Mathioudakis et al. (1995a), radiative losses in EUV are comparable with the losses in the X-ray region, while a considerable part of the radiative losses of low-activity dwarfs falls exactly in the EUV range. Below, we present the results of EUVE spectral observations of such objects: the observation program included the objects discovered in the course of the earlier WFC and EUVE sky surveys in EUV. The Atlas by Craig et al. (1997) contains the EUV spectra of four G dwarfs (κ Cet, χ^1 Ori, α Cen AB, and ξ Boo), six K dwarfs (GJ 117, ε Eri, BF Lyn, LQ Hya, GJ 702 AB, and VW Cep), and ten M dwarfs (YY Gem, YZ CMi, AD Leo, Prox Cen, GJ 644, AT Mic, AU Mic, FK Aqr, EV Lac, and EQ Peg).

The system α Cen (G2V+K1V) was one of the first observed in all four photometric EUVE bands, these measurements led to the conclusion on the radiation of the two components: $T_1 = 8.5 \times 10^5$ K, $EM_1 = 1.5 \times 10^{50}$ cm^{-3}, and $T_2 = 10^5$ K, $EM_2 = 5 \times 10^{49}$ cm (Vedder et al., 1993). However, the spectral lines in the EUV range give more reliable estimates of the temperature of radiating plasma, and Mewe et al. (1995a) carried out spectral observations of the system by EUVE. The recorded spectrum of α Cen was presented by a linear combination of isothermal plasma structures: the hot component with a temperature of about 3 MK, less hot with a temperature of 0.1 MK, and probably a very hot – with a temperature of several tens of megakelvins. The spectral lines of the FeX, XII, XIII, and XIV ions sensitive to the electron density yielded estimates of $(2-20) \times 10^8$ cm^{-3} for a temperature of 1–2 MK. Later, Drake et al. (1997) found that the DEM of this system had a minimum at $\log T = 5.5$ and a maximum at $\log T = 6.3$. Analyzing the abundances of the elements in the corona of α Cen AB, they established that the chemical composition of the corona differed from that of the photosphere, and, as on the Sun, the elements with low FIP were enriched by almost a half as compared to the elements with high FIP.

Monsignori Fossi and Landini developed a numerical method for analyzing differential emission measure (DEM). Using the method, they presented EXOSAT and IUE observations of AU Mic. Using the DEM model, they calculated the expected spectra of the star for the wide range of short-wavelength (70–190 Å) and medium-wavelength (140–380 Å) EUVE spectrographs and compared them with the observations. The lines of high-ionized ions of iron FeXIII–FeXXVI, OIV and HeII Ly_α dominated in these spectra. Using EUVE observations, they revised the DEM model and obtained the following parameters of the AU Mic corona in the quiescent state: $L_{EUV} = 4 \times 10^{29}$ erg/s, $EM = 6 \times 10^{51}$ cm^{-3}, and from two pairs of FeXXI and FeXXII lines sensitive to the electron density, the upper limit of the value was estimated as $n_e < 10^{12}$ cm^{-3} (Monsignori Fossi and Landini, 1994; 1996; Monsignori Fossi et al. 1996). Monsignori Fossi et al. (1995a) similarly analyzed the flare star AT Mic using EUVE and IUE data. In the quiescent state, the lines of high-ionized iron ions prevailed, while the ions of FeXVIII and lower ionization stages were presented more poorly. The DEM function had a minimum near 10^6 K and a maximum near 10^7 K. Similar results were obtained in analyzing another binary flare star EQ Peg: 13 iron ions lines up to FeXXIV were identified in the wavelength range of 94–360 Å and the same extremes of the DEM function were found (Monsignori et al., 1995b).

Schmitt et al. (1996b) analyzed the EUV spectrum of the K dwarf ε Eri and found the emission lines of iron from FeIX to FeXXI with a maximum intensity in FeXV–FeXVI. From these iron lines they constructed DEM, which displayed a wide maximum; such a DEM was presented within the coronal model with two kinds of magnetic loops with maximum temperatures about 3 and 7 MK. From the FeXIII and FeXIV lines they estimated the electron densities, which were similar to those in the active regions of the Sun. With

154   1 Flare Stars in a Quiescent State

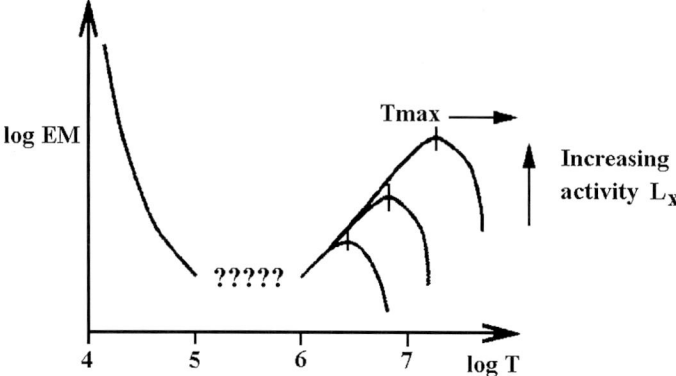

Fig. 19. Schematic dependence of the differential emission measure of stellar coronae upon the temperature and activity level (Drake, 1996)

account of the found densities and emission measures they estimated the size of coronal loops as 3×10^9 cm and the filling factor as $f \sim 0.9$.

Landi et al. (1997) analyzed the EUV spectrum of κ Cet, identified the iron lines from FeIX to FeXVIII, and, incorporating the IUE data, constructed DEM, whose maximum was about 3 MK, as in solar-active regions.

Analyzing the restrictions on the stellar corona parameters observed using EUVE, van den Oord et al. (1997) found that the minimum expansion of magnetic loops from the base to the top was from 2 to 5, the coronae of α Cen should be composed of the loops of two different kinds, whereas for χ^1 Ori loops of one kind were sufficient.

Summing up the results of three years of EUVE operation, Drake (1996) noted that the studies from this apparatus yielded the following conclusions on the general character of DEM: within $T \sim 10^4$–10^5 K there was a decline of the function, from 10^5 to 10^6 K there was a poorly determined minimum, then a rise started, which achieved a maximum in the range of 2×10^6–10^7 K depending on the general activity level of the star (see Fig. 19). Thus, the two-temperature corona model is the approximation of the structure with a continuous but not monotonic DEM(T) distribution. Further, of principal significance were direct spectroscopic estimates of electron density in stellar coronae: the lines from FeXII to FeXXII in the wavelength range from 91 to 327 Å allowed an estimate of n_e as 10^9–10^{15} cm^{-3} at 2×10^6–10^7 K (Brown, 1994). The estimates of n_e provided a considerable advance in the analysis of the inner structure of coronae, since in combination with the measured EM they allowed for the estimates of the characteristic volume of luminous matter, which, if a certain plausible geometry is specified, makes it possible to estimate the filling factor by the loop structure and the characteristic strength of magnetic fields. Thus, if from the analysis of FeXIX–XXII lines one assumes $n_e \sim 10^{13}$ cm^{-3}, compact stationary loops with $h \sim 100$–1000 km and B $\sim$ 1 kG appear at $T \sim 10^7$ K, which substantially differ from the solar situation, where $n_e \sim 10^9-10^{11}$ cm^{-3}, $T \sim 2 \times 10^6$ K, $h \sim 20\,000$ km, and B ~ 10 G. For intermediate activity stars, ε Eri and ξ Boo A, n_e estimates from FeXIII and XIV lines yielded 10^{10} cm^{-3}. For the even less active pair α Cen AB from the FeIX–XIV lines $n_e \sim 10^9$ cm^{-3}, as in active solar regions. Probably the highest density estimates correspond to filaments inside the coronal loops. Thus, the EUVE results suggested that the growth of the X-ray flux for more active stars was due to the increase of the characteristic density rather than the volume of the luminous matter.

The first studies of the abundances of heavy elements in stellar coronae were started at EUVE to find an analogy with the solar so-called FIP effect. The effect consists in the increased content of the elements with first ionization potential (FIP) below 10 eV with respect to that of the elements with higher first ionization potentials. Laming et al. (1996) suspected this effect for ε Eri; but although the star is noticeably more active than the Sun, the value of the FIP effect in it definitely did not exceed the solar value. In the coronae of α Cen AB system, Drake et al. (1997) for the first time reliably detected an analog of the solar FIP effect with a twice higher content of the elements with low FIP than the elements with high FIP. Later, a distinct solar-type FIP effect was found by Laming and Drake (1999) and Drake and Kashyap (2001) in the corona of ξ Boo A, whose activity level is higher by an order of magnitude than that of the Sun.

Drake et al. (2000) compared EUVE spectral observations of ε Eri (K2V) and ξ Boo A (G8V) with Yohkoh solar observations near the maximum of its cycle on 6 January 1992 to find the solar coronal structures with $EM(T)$ corresponding to the coronae of these intermediate-activity stars. The intermediate status is illustrated by the following figures: solar L_X is 2×10^{26} erg/s at the activity minimum and 5×10^{29} erg/s at the maximum, for the most active solar-type dwarfs $L_X > 10^{30}$ erg/s, whereas $L_X(\varepsilon$ Eri$) = 3\times 10^{28}$ and $L_X(\xi$ Boo A$) = 9\times 10^{28}$ erg/s, respectively. Drake et al. constructed the $EM(T)$ dependences for the whole solar disk and for the sections limited by isophotes 1/100 and 1/10 from the maximum brightness on the disk and found that these curves differed by maximum EM values, the positions of the maxima on the temperature scale, and the slopes of the curves beyond the maxima: if one approximates $EM \sim T^\beta$, then $\beta = 1.1$, 1.6, and 3.8 for the listed $EM(T)$ dependences. The $EM(T)$ curve of the brightest solar regions is the most similar to the curves of the considered stars, but its maximum is approximately 70 times lower. Since the electron density in the coronae of these stars and the brightest solar-active regions was estimated by close values, Drake et al. concluded that to achieve the activity level corresponding to that of ε Eri and ξ Boo A, a filling factor for solar-active regions approaching unity should be sufficient. But an even higher activity level requires either an increase in the density of the coronal plasma, or an increase of its radial extension, i.e., the formation of structures differing from solar-active regions. The known increase in the coronal temperature with increasing L_X argues in favor of such qualitative changes. Probably at f ~ 1 a new mechanism of flaring activity is launched due to the interaction of neighboring active regions, which engages stronger and higher-temperature processes than the flare activity in separate active regions on the Sun.

For complete quantitative analysis of the EUVE data, Drake (1999) performed detailed computations of the sensitivity of all systems of the apparatus to the radiation of optically thin astrophysical plasma within the temperature range of 10^4–10^8 K.

The Japanese–American X-ray satellite ASCA (Advanced Satellite for Cosmology and Astrophysics) was orbited in February 1993 with a period of 96 min. The satellite carried 4 identical X-ray grazing-incidence telescopes designed to operate within the range of 0.5–10 keV and had an angular resolution of $1'$–$3'$; at the focus of the telescopes 2 solid-state spectrometers (SIS) were installed with a spectral resolution of 50 and a field of view $20' \times 20'$ and two gas imaging spectrometers (GIS) with a spectral resolution of 13 and a diameter of the field of view of $50'$. The main advantage of ASCA over its predecessors was the possibility of studying the region of higher energies and higher effective area.

Drake et al. (1994b) observed the active G1.5 dwarf π^1 UMa with SIS ASCA and found stable radiation at the level of $\log L_X = 28.7$. The analysis of the obtained spectra showed that isothermal RS and MEKA models could not represent the observations even with varying chemical composition but the two-temperature MEKA model with $T_1 = 4.0$ and $T_2 = 7.4$ MK and $EM_1 = EM_2 = 2 \times 10^{51}$ cm^{-3} with noticeably reduced content of N, O, Ne, Si, S, and Fe well represented the observations.

Gotthelf et al. (1994) observed the α Gem system using SIS ASCA. The developed algorithm of image restoration made it possible to separate the radiation of YY Gem from that of Castor AB, then the quiet radiation of YY Gem was presented as a four-temperature coronal model; the two-temperature model could also yield a good presentation of the observations, but only if the abundances of heavy elements were decreased by a factor of 5–10 as compared to their content in the solar photosphere.

The studies of solar-type stars based on ASCA data by Güdel et al. (1997a,b) will be described in detail in Sect. 3.2.2.

The ASCA X-ray spectrum of α Cen AB was analyzed by Mewe et al. (1998a) within the concept of multitemperature corona and using DEM. It appeared that the corona was close to isothermal with $T \sim 3.5 \times 10^6$ K. The content of neon, silicon, and iron in it was close to that in the solar atmosphere, but there was a deficit of oxygen and a several-fold excess of magnesium. When the system was observed by the BeppoSAX/LECS apparatus, which enabled simultaneous investigation of the range from 0.1 to 10 keV with rather high spectral resolution, Mewe et al. (1998b) found a two-temperature corona, whose cold component was recorded before by EUVE and ROSAT, and the hot one by the Einstein Observatory, EUVE, and ASCA. The established content of iron corresponded to the solar photosphere, but differed from the α Cen photosphere enriched by metals. However, there is no confidence that the hot component was not due to flare activity.

Using ASCA, Singh et al. (1999) observed two very fast rotators, HD 197890 and Gl 890. Analysis of these data and the observations of YY Gem showed that at least two temperature components should be available at each of these sources. In the YY Gem coronae the content of metals was definitely reduced, in two others it was preferential. Considering ROSAT and ASCA results in the general sample of 17 G–K–M dwarfs, the authors confirmed the clear correlation of $L_X/L_{\rm bol}$ with the rotational period and the Rossby number. In the subsample of 10 stars observed from ASCA, they found that the temperature of the hot component correlated with the ratio. For all active dwarfs with $\log(L_X/L_{\rm bol}) > -3.7$ the content of iron was noticeably lower than the solar value, whereas for less-active stars it differed by no more than a factor of two.

Using EUVE and ASCA, Gagné et al. (1999) studied the active corona of the F8 dwarf HD 35850. In the short-wavelength and medium-wavelength ranges of EUVE they identified 28 iron lines from FeIX to FeXXIV. From FeXXI lines they estimated the upper limit of the electron density $\log n_{\rm e} < 11.6$. In the short-wavelength range they confidently discovered a continuum. On the obtained DEM curve one can see two distinct maxima at $\log T = 6.8$ and 7.4. During one-week observations no strong and long flares were recorded on the star, but there were traces of weak flares and probably rotational modulation. In comparing the nonsimultaneous EUVE and SIS ASCA spectra they found a similarity of the obtained DEM curves, but the ASCA spectra required stellar corona noticeably depleted in heavy elements. The surface X-ray fluxes from HD 35850 are comparable with those from EK Dra, and at an age of about 100 million years, $v \sin i \sim 50\,{\rm km/s}$ and high L_X this star can be considered as an extremely active main-sequence F dwarf.

Favata et al. (2000a) analyzed X-ray observations of the flare star AD Leo with IPC Einstein, PSPC ROSAT, and SIS ASCA, and concluded that in six strong X-ray flares recorded by these instruments magnetic loops, where the flares were developed, had similar parameters. Hence, they accepted that such loops were typical of the quiescent corona as well. This suggests that the parameters of these magnetic loops corresponded to the compact solar loops with 10-fold increase of pressure, 3-times higher loop diameter-to-height ratio, and low surface filling factor. This conclusion is an alternative to the suggestion that an increase in L_X of active stars occurs at the expense of the increase of the filling factor. However, in this case one can expect an increase in L_X only by 1.5–2 orders of magnitude, whereas for the most active dwarfs it should be increased by 3–4 orders. Considering the above results of Ventura et al. (1998) on two different magnetic-field structures, one can assume that the growth of the filling factor has a decisive role for the stars with low and medium activity, whereas for the most active stars the most important is the growth of the loop density. This conclusion coincides with that by Drake et al. (2000) on the structure of coronae of ε Eri and ξ Boo A obtained from EUVE data. According to the estimates of Favata et al. (2000a), thick-section magnetic loops of the AD Leo corona are characterized by a density of $5 \times 10^{10}\,\text{cm}^{-3}$, a pressure of 70 dyn/cm^2, and a height $h \sim 0.3 R_*$, and at f ~ 0.06 they yield $EM \sim 2 \times 10^{51}\,\text{cm}^{-3}$. From the spectra obtained by SIS/ASCA Favata et al. constructed two-temperature models of the corona of AD Leo both for the solar content of the elements and for varying compositions; in the second case $T_1 = 5.6 \times 10^6$ and $T_2 = 10^7$ K, $EM_1 = 1.6 \times 10^{51}$ and $EM_2 = 2.3 \times 10^{51}\,\text{cm}^{-3}$ at the deficit of [Fe/H] = -0.7 and [O/H] = [Si/H] = 0.3.

Covino et al. (2001) analyzed the corona of the young K2 star LQ Hya from the ROSAT PSPC (November 1992) and ASCA (May 1993) observations. The former were presented by the 1T MEKAL model with $T \sim 8$ MK, $EM \sim 1.2 \times 10^{53}\,\text{cm}^{-3}$, and $Z \sim 0.09$. The latter were presented by the 2T MEKAL model with $T = 9$ and 15 MK, $EM = 4.4 \times 10^{52}\,\text{cm}^{-3}$ and $3.2 \times 10^{52}\,\text{cm}^{-3}$, and $Z \sim 0.13$. Slow brightness variations with amplitudes of up to 2–3 were associated with EM variations at practically constant temperature.

As to the above studies dealing with chemical anomalies of stellar coronae, it should be noted that Drake (1998) assumed that the observed effects could be stipulated by the enrichment of coronae by helium. However, in 2001, in the beginning of the Chandra and XMM-Newton era, while surveying the chemical composition of stellar coronae, Drake (2002) did not mention this concept.

In April 1996, the Italian–Dutch X-ray Astronomy Satellite BeppoSAX was orbited. The satellite had a uniquely wide spectral coverage ranging from 0.1 to 300 keV and the wide-field cameras designed for studying the variability at long time intervals and detecting nonstationary phenomena. There were four X-ray concentrator spectrometers with apertures of 124 cm^2 each and proportional counters in the focal planes: a low-energy concentrator spectrometer (LECS) designed for the range of 0.1–10 keV and three medium-energy concentrator spectrometers (MECS) for the range of 1.3–10 keV, a high-pressure gas scintillation proportional counter for the range of 4–120 keV, and a proswich detection system (PDS) for the range of 15–300 keV that contained four scintillators with a total area of about 800 cm^2. The wide-field cameras were designed for the range of 1.8–28 keV, had a fields of view of 20°, angular resolution of 5' and spectral resolution of 5 near 6 keV.

LECS and MECS instruments installed on the satellite recorded the spectra of the active binary system VY Ari and the flare star AD Leo within the range from 0.1 to 10 keV (Favata, 1998). The spectrum of VY Ari was presented by the two-temperature model with $T_1 = 1 \times 10^7$ and $T_2 = 2.4 \times 10^7$ K, the ratio $EM_1/EM_2 = 0.44$, and a content of iron lower by a factor of 2.5 than that on the Sun. The preliminary analysis of the spectra of AD Leo yielded the two-temperature model of the quiescent stellar corona that is quite close to the model constructed using ASCA.

Sciortino et al. (1999) observed AD Leo and EV Lac by BeppoSAX and compared the results with the ROSAT PSPC data. The ROSAT spectrum of EV Lac and AD Leo were presented by the one-component isothermal MEKAL model and by the two-component MEKAL model, respectively, with the metallicity of both stars lower by an order of magnitude than the solar value. For the BeppoSAX data, the two-component model was insufficient for any metallicity value, but the three-component model provided a satisfactory presentation of the observations of AD Leo at noticeably reduced metallicity and the observations of EV Lac for a slightly reduced content of metals. The models with uniform magnetic loops were unsuitable, and the best correspondence was achieved in the model of two loop types with dominating hundreds of compact loops with relatively low maximum temperature, heights of about $0.1 R_*$ and a filling factor of ~ 0.01 and tens of heavily elongated loops with very small filling factor that were responsible for the high-temperature radiation.

During 5 days in November 1998 Tagliaferri et al. (2001) observed the YY Gem system within the range of 0.1–10 keV. At two intervals out of flares they obtained the spectra of the quiescent state that were presented within the two-temperature MEKAL models with $T_1 = 7.7$ and 3.8 and $T_2 = 23$ and 16 MK at metallicities $Z = 0.2$ and 0.4, respectively. In addition, they suspected radiation within the range of 20–30 keV.

On the border of the 20th and 21st centuries NASA and ESA orbited new-generation high-sensitivity and high spectral resolution X-ray telescopes Chandra and XMM-Newton that opened a new era in X-ray astronomy.

Chandra contains four concentric grazing-incidence X-ray telescopes with Wolter-type I mirrors of outer diameters up to 123 cm. This optics feeds advance CCD imaging spectrometer (ACIS), low- and high-energy transmission grating spectrometers (LETGS and HETGS), and high-resolution cameras. ACIS ensures high-resolution images and moderate resolution spectra can be obtained. HETGS and LETGS operate in 0.4–10 keV and 0.08–0.2 keV ranges, respectively. Both have spectral resolving powers of about 1000. HRC is a descendant of HRIs that were installed at Einstein and ROSAT, its field of view is $30' \times 30'$. Chandra has 2^m higher sensitivity as compared to ROSAT and ASCA and its unprecedented spatial resolution is $0''5$. LETGS can record simultaneously OVII triplets about 22 Å, NVI about 29 Å and CV about 41 Å, which are widely used in the diagnostics of coronal plasma, in the range of > 100 Å, FeXXII lines can be used for this purpose.

The X-ray Multi-Mirror Mission (XMM-Newton) is designed for observations within the range of 0.2–15 keV. The observatory consists of three modules, each includes an X-ray telescope composed of 58 coaligned gilded mirrors with an outermost diameter of 70 cm, with a resolution of about $4''$ in the field center. After two mirror modules, identical reflection grating spectrometers (RGS) are mounted that intercept about a half of the converging beams and construct spectra within the range of 0.3–2.1 keV with a resolution of 100–500. The other half of the beams passing the spectrometer gratings continue on to the EPIC MOS, European Photon Imaging Camera, with mosaics of 7 MOS CCD, which cover the field with a diameter of about 0°5. The third module contains the telescope with a fast-reading pn-CCD $6 \times 6\,\text{cm}^2$ in size with a sensitivity of up to 15 keV and a resolution of about $4''$ in the focus. In addition to X-ray telescopes, XMM-Newton carries a 30-cm Ritchey–Chretien system intended for simultaneous observation in the range of 1700–6500 Å in the central 17 square minutes of the field of view of the X-ray instruments. The XMM-OM system is equipped with a high-speed photometer with broadband filters and a grism for obtaining low-resolution spectra.

Based on Chandra LETGS observations, Audard et al. (2003) divided the components of the L 726-8 system in the X-ray range and in the HRC-S camera obtained a zero-order image of the system that was in excellent agreement with optical ephemeris.

Fleming et al. (2003) observed M8 dwarf VB10 using Chandra ACIS and found a quiescent corona with an X-ray luminosity of 2.4×10^{25} erg/s and $\log L_X/L_{\text{bol}} = -4.9$. These values comply with the fact that the star was discovered in X-rays only during the flare. The obtained luminosity is lower by two orders of magnitude than that of earlier M dwarfs. Nevertheless, the existence of corona on VB 10 contradicts the hypothesis on the absence of permanent coronae on such cool stars.

Ness et al. (2002) analyzed Chandra LETGS observations of 10 cool stars, whose spectra displayed clear qualitative distinctions: there is a strong continuum at the active ε Eri and AD Leo in the range of 10–20 Å, which is absent on nonactive α Cyg A and B. On the latter stars, the intensity of OVII lines (21.6 Å) exceeded that of OVIII lines (18.97 Å), whereas for ε Eri and AD Leo the opposite relations were observed. Applying the theory of relative intensities to Ly_α and He-like triplets for various highly ionized atoms, Ness et al. carried out systematic analysis of all proper emissions. The following results were obtained:

- OVII lines: ε Eri $n_e \sim 10^9$–$10^{10}\,\text{cm}^{-3}$, $T = 2.2$–$3.0\,\text{MK}$; AD Leo $n_e = 8 \times 10^9\,\text{cm}^{-3}$, $T = 2.2$–$3.5\,\text{MK}$; YY Gem $n_e = 2 \times 10^{10}\,\text{cm}^{-3}$, $T = 2.2$–$3.6\,\text{MK}$;
- NVI lines: ε Eri $n_e = 2 \times 10^{10}\,\text{cm}^{-3}$, $T = 2.0$–$2.5\,\text{MK}$; AD Leo $n_e = 2 \times 10^{10}\,\text{cm}^{-3}$, $T = 2.0$–$2.9\,\text{MK}$;
- CV lines: ε Eri and AD Leo $T = 1.4\,\text{MK}$;
- SiXIII lines: ε Eri $n_e < 10^{14}\,\text{cm}^{-3}$, $T = 8\,\text{MK}$; AD Leo $n_e = 4 \times 10^{13}\,\text{cm}^{-3}$, $T = 9$–$11\,\text{MK}$, YY Gem $T = 12.5\,\text{MK}$;
- MgXI lines: ε Eri $n_e = 4 \times 10^{12}\,\text{cm}^{-3}$, $T = 6$–$7\,\text{MK}$; YY Gem $T = 8.7\,\text{MK}$;
- NeIX lines: ε Eri $n_e = 10^{11}\,\text{cm}^{-3}$, $T = 3.3$–$4.4\,\text{MK}$; AD Leo $n_e < 4 \times 10^{10}\,\text{cm}^{-3}$, $T = 4.5$–$5.3\,\text{MK}$; YY Gem $n_e < 2 \times 10^{11}\,\text{cm}^{-3}$, $T = 4.7$–$5.6\,\text{MK}$.

Apparently, the parameters for different ions of the same star were scattered, mostly due to various temperatures of formation of the appropriate ions and coronal heterogeneity. Comparison of the ratios of intensities of OVIII/OVII, NVII/NVI, and NeX/NeIX resonant lines with X-ray luminosity revealed a positive correlation: L_X grows with an increase in these ratios.

Using Chandra ACIS, Tsuboi et al. (2002) discovered X-ray emission of brown medium-age dwarf TWA 5B, a companion of a T Tau-type star, at the level of 4×10^{27} erg/s within 0.1–10 keV. They concluded that up to the age of 10^7 years the saturation of X-ray emission occurred at the level of $L_X/L_{\mathrm{bol}} \sim 10^{-3} - 10^{-4}$, then this ratio decreased to 10^{-5}, and the coronal temperature started decreasing even earlier. The MEKAL model of this dwarf yielded $T = 3.5$ MK, $EM = 4 \times 10^{50}$ cm^{-3}, and $Z = 0.3$.

With Chandra LETGS, Maggio et al. (2002) observed AD Leo in the range of 6–180 Å during 15-h observations out of strong flares. They identified most of the 110 detected emissions, including iron lines from XVI up to XXIII, C, N, O, Ne, Mg, Si, S, and Ni with formation temperatures varying from 0.6 to 15 MK. They constructed DEM overlapping this temperature range and from He-like triplets determined the electron density: 2×10^{10} cm^{-3} for $T = 1.5$ MK in OVII, $< 3 \times 10^{11}$ cm^{-3} for $T = 4$ MK in NeIX and 1×10^{14} cm^{-3} for $T = 6$ MK in SiXIII.

Using Chandra ACIS, Preibisch and Zinnecker (2002) investigated the very young cluster IC 318, found 215 X-ray sources with masses varying from 0.15 to 2 solar masses, and identified them with 80% of the cluster members.

By means of Chandra HETGS and VLA, Brown et al. (2002) observed on 29/30 March 2001 active coronae in the short-period binary system ER Vul, composed of G0 and G5 dwarfs. The X-ray monitoring was run within 1.8–40 Å, radio monitoring was performed at 3.6 and 20 cm. The observations revealed a wide range of coronal temperatures from 2 (OVII) up to 30 (FeXXIV) MK with smooth variations of the level up to $A_X \sim 2$.

Raassen et al. (2003) were the first to obtain and analyze the spectrograms of each of the components of the α Cen system recorded by Chandra LETGS within the range 10–180 Å. From the ratios of the intensities of spectral lines they found that the corona of the K1 component was slightly hotter than that of the G2 dwarf. Then, using the 2T MEKAL program, they constructed the two-component models of the coronae from DEM of each component of the system and obtained the following parameters: for K1 star $T_1 = 1.2$ and $T_2 = 2.2$ MK, $EM_1 = 1.0 \times 10^{49}$ cm^{-3} and $EM_2 = 1.9 \times 10^{49}$ cm^{-3} and for G2 star $T_1 = 1.1$ and $T_2 = 2.0$ MK and $EM_1 = EM_2 = 1.1 \times 10^{49}$ cm^{-3}. On both components of the system, which is somewhat older than the Sun, they found the FIP effect, which apparently was slightly stronger on the K1 star.

Drake and Sarma (2003) observed the V 471 Tau system composed of white and red dwarfs by Chandra LETGS. In the range $\lambda < 50$ Å the radiation of the corona of the red dwarf prevailed, and from the carbon and nitrogen resonance lines they found $[C/N] = -0.38$. This value is intermediate between the nonevolutionary stellar matter and that characteristic of red giant. In the opinion of Drake and Sarma, this is stipulated by the contamination of the atmosphere of the red dwarf when its companion was in the stage of a red giant that preceded the appearance of the white dwarf.

Sanz-Forcada et al. (2004) considered four F–K stars with different luminosity and activity levels, including K dwarf ε Eri to study the FIP effect. Using Chandra LETGS-HRC they determined the content of 11 elements in the corona from the emission lines with the formation temperature varying from 50 000 up to 300 000 K, found increased content of calcium and nickel, and suspected the tendency to a reduction of the content of the elements with an increase of atomic weight.

One of the first objects studied with XMM-Newton was the Castor ABC system (Güdel et al., 2001b). During the observations, the components A and B were resolved in X-rays for the first time. A somewhat stronger luminosity of the A component was measured. Using EPIC data and the VMEKAL algorithm three-component models of coronal plasma were constructed for YY Gem and Castor AB from low-dispersion spectra: for YY Gem $T_1 = 3.5$, $T_2 = 7.8$ and $T_3 = 16.5$ MK, $EM_1 = 0.9 \times 10^{52}$ cm^{-3}, $EM_2 = 1.3 \times 10^{52}$ cm^{-3}, and $EM_3 = 0.5 \times 10^{52}$ cm^{-3}, for Castor AB $T_1 = 2.0$, $T_2 = 7.8$ and $T_3 = 20.4$ MK, $EM_1 = 0.6 \times 10^{51}$ cm^{-3}, $EM_2 = 5.6 \times 10^{51}$ cm^{-3}, and $EM_3 = 1.0 \times 10^{51}$ cm^{-3} with a distinct FIP effect. From OVII lines the coronal density of cool components of YY Gem and Castor AB coronae were estimated as a few 10^{10} cm^{-3}. The recorded light curve of YY Gem with three eclipses enabled the construction of atmospheric models of the components to find the change of density with height and the prevailing localization of active regions at middle latitudes that is consistent with the Doppler mapping (Hatzes, 1995).

During simultaneous observations of YY Gem on 29/30 September 2000 by Chandra and XMM-Newton, Stelzer et al. (2002) found a good agreement between the calibrations of these instruments for the wavelengths and intensities. The high-resolution spectra obtained by the spacecrafts and averaged over the whole observation session encompassing about 75% of the orbital period duration demonstrated in the range of 5–133 Å a rich emission spectrum of highly ionized atoms: SiXII-XIV, NiIX, FeXVII-XXII, OVII and VIII, CVI, NeIX and X, NVII, MgXII, SX, and XII. The strong triplet OVII containing resonant, intercombination, and forbidden lines ensured estimation of the parameters of the "cold" component of the coronal plasma from their relative intensities: temperature of 2–3 MK and $n_e < 2 \times 10^{10}$ cm^{-3}. Even in the case of the strongest emission no self-absorption effect was revealed. From the low-resolution spectrum recorded by EPIC XMM-Newton in the beginning of the session, when the YY Gem system was the quietest, using the 3T VMEKAL algorithm, the following parameters of the coronal plasma were found: $T_1 = 2.4$, $T_2 = 7.4$, and $T_3 = 20.8$ MK, $EM_1 = 2.2 \times 10^{51}$ cm^{-3}, $EM_2 = 14 \times 10^{51}$ cm^{-3}, and $EM_3 = 2.9 \times 10^{51}$ cm^{-3} with an appreciable deficiency of iron and some other elements. EM_2 and EM_3 decreased during the eclipse in the system.

Stelzer and Burwitz (2003) carried out simultaneous spectral and photometric observations of the Castor system using Chandra and higher-sensitivity XMM-Newton instruments. The summary medium-resolution spectrum of Castor AB recorded by XMM-Newton EPIC was presented by the 3T VMEKAL model with temperatures of 3.1, 9.0, and 20 MK and close emission measures $(1.7–3.1) \times 10^{51}$ cm^{-3}; the high-resolution spectrum obtained on XMM-Newton RGS enabled the estimate of the ratio of the emission intensities as $T \sim 2$ MK and $n_\mathrm{e} \sim (0.5–1.0) \times 10^{10}$ cm^{-3}. The high-resolution spectra of each component, A and B, obtained using Chandra LETGS displayed the domination of OVIII, OVII, FeXVII, and NeIX emissions.

Heterogeneity of the active corona was found in XMM-Newton RGS observation of the eclipse in the α CrB system: according to Güdel et al. (2003a), the corona of the G component of this system is essentially asymmetrical and there are areas with a density of 10^9–3×10^{10} cm^3.

Krishnamurthi et al. (2001b) observed the central area of the Pleiades by Chandra within a range of 0.1–10 keV and discovered 57 sources, some of them could be low-mass stars.

Harnden et al. (2001) studied the central part of this cluster using Chandra and found more than 150 new sources. Probably due to low metallicity the X-ray luminosities of G and K stars in NGC 2516 were lower than in the Pleiades.

Using XMM-Newton RGS and Chandra LETGS and HETGS, Ness et al (2003) investigated the ratios of the intensities of FeXVII, OVII, and NeIX lines but failed to find the effect of optical thickness.

1.4.3 Microwave and Shortwave Emissions

In addition to thermal X-ray emission and optical emission of the forbidden lines of highly ionized ions of iron, calcium, and some other metals excited by electron collisions, the solar corona radiates electromagnetic waves arising at free–free transitions, as a result of plasma oscillations and gyroresonant and gyrosynchrotron interactions of fast electrons and local magnetic fields. The coherent emission induced in the cyclotron maser and under plasma oscillations has a many orders of magnitude higher brightness temperature and a high degree of polarization from the incoherent emission. In principle, by analyzing such an emission one can estimate the density of the emitting plasma and the characteristics of the relevant magnetic fields. But the significant diversity of the mechanisms of radio-frequency emission and the large number of independent parameters governing these emissions often makes ambiguous not only parameter estimates but even the identification of the specific emission mechanism.

According to Dulk (1985), at wavelengths of 3 cm and shorter the Sun looks like a homogeneous disk with a brightness temperature of about 15 000 K that has bright active regions with a complex polarization pattern. By 30 cm the brightness of the disk slowly increases to $T_\mathrm{b} \sim 50\,000$ K and that of active regions to $(1-2) \times 10^6$ K. Dark coronal holes with $T_{\rm b} \sim 30\,000$ K emerge there. At meter wavelengths the temperature of the whole disk achieves $T_{\rm b} \sim 10^6$ K, the active regions become indistinct, and the size of the solar disk becomes 1.5 times larger than in the optical range. But large magnetic loops containing the main mass of the luminous matter at great heights are usually localized at low latitudes, which results in an asymmetry: the radio-Sun is greater in the equatorial dimension.

As in studying the X-ray emission of red dwarfs, initially microwave flares simultaneous with the optical flares were found, and only launching of the Very Large Antenna (VLA) in New Mexico allowed the microwave emission of quiescent flare stars to be detected.

After a number of failures in the detection of such a quiet radiation, during which the estimates of its upper limit subsequently decreased from 50 to 1–2 mJy (see the survey by Bastian, 1990) the quiet microwave emission of red dwarfs was first successfully revealed by Gary and Linsky (1981). With 25-m VLA antennas they observed 6 stars that were not included in close pairs and had high X-ray luminosity. At 6 cm they found the emission of χ^1 Ori (G0) at the level of 0.6 mJy and UV Cet (dM5.5e) at the level of 1.6 mJy. For π^1 UMa, ξ Boo, 70 Oph, and ε Eri they found only the upper emission limits. Each program star was monitored for several hours, which made it possible to consider the recorded emission as the quiet corona emission rather than that of stellar flares. The flux measured from UV Cet corresponded to the brightness temperature of the radio-frequency emission source $T_{\rm b} = 10^8 (R_*/r_{\rm s})^2$ K, where $r_{\rm s}$ is the characteristic size of the source. Due to the small optical thickness of the stellar corona at 6 cm that was responsible for the observed X-ray emission, the recorded microwave emission could not be explained by the free–free emission. Thus, Gary and Linsky used the mechanism of gyroresonant emission of thermal electrons that provided a greater optical thickness in relatively strong magnetic fields. They concluded that the mechanism occurred at 6 or lower harmonic in the magnetic fields of at least 300 G. Then, Topka and Marsh (1982), using all 27 acting 25-m VLA antennas, over 20 minutes of observations found the microwave emission of each component of the EQ Peg system (dM3.5e+dM5.5e): at 6 cm the density of recorded fluxes was 0.7 and 0.4 mJy (see Fig. 20). Simultaneous detection of the effect at both components justifies the conclusion that these short-term observations dealt with quiet coronae rather than flares. Topka and Marsh also decided that the discovered emission could not be stipulated by the thermal emission of the X-ray corona, but could be presented within the gyroresonant emission on the condition that the size of the radiating region was several times greater than the optical size of the star. Fisher and Gibson (1982) confirmed the results of Gary and Linsky (1981) and Topka and Marsh (1982) and determined quiet radiation of the flare star YZ CMi at 6 cm as 0.5 mJy and the radiation of UV Cet at 21 cm at the level of 1.1 mJy.

To continue the program, Linsky and Gary (1983) observed 14 other late stars. At 6 cm in the course of 2-to-4-h sessions they found the varying

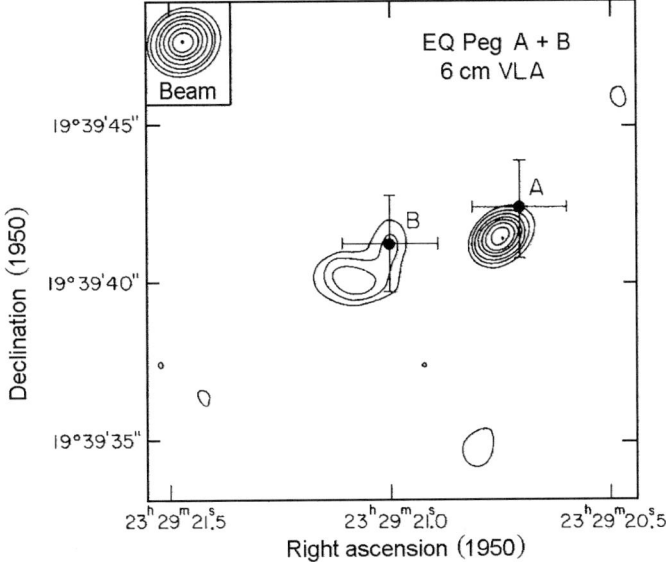

Fig. 20. VLA map of microwave emission of the EQ Peg region; *top left corner*: VLA beam (Topka and Marsh, 1982)

radiation of YY Gem and Wolf 630 AB binary systems, while the upper limit of F–K dwarfs was estimated as two orders of magnitude lower than L_R of dMe stars. The YY Gem system was recorded in five of six sessions, the Wolf 630 AB system in four of five sessions, and all the dMe stars found by that time in the microwave range had a luminosity $L_R \sim (1\text{--}50) \times 10^{13}$ erg/s Hz. The measured fluxes from YY Gem and Wolf 630 AB could also be presented within the gyroresonant emission of thermal electrons. However, the required source size exceeding the stellar size by 3–4 times and the magnetic fields of hundreds of gauss that should correspond to photospheric fields of $10^3 - 10^4$ G raised doubts as to the validity of the model. Thus, Linsky and Gary (1983) concluded that the fluxes from UV Cet and YY Gem measured at 6 cm could be better explained by gyrosynchrotron emission of fast electrons whose required density was lower by 3 orders of magnitude than that of thermal plasma electrons responsible for the X-ray emission of the stars. The advantage of the gyrosynchrotron (GS) model is that with the growing electron energy the efficiency of their microwave emission increases rapidly, and consequently the number of particles necessary for the presentation of the observed flux decreases. Therefore, the model requires smaller sources of lower density than the gyroresonant (GR) model. Later, Gary (1986) concluded that both the power distribution of nonthermal electrons and the small addition of plasma with $T \sim 4 \times 10^8$ K to plasma with $T \sim 10^7$ K were suitable to form the recorded microwave emission. This small addition should be considered as an independent component of the stellar corona. In the course of six observational sessions Linsky and Gary (1983) suspected the correlation of the radio flux with spottedness of YY Gem, which can be considered as direct evidence of the relation between the microwave emission and magnetic fields of the star.

To explain the polarization features of the quiet microwave emission of active red dwarfs, whose surfaces are usually spotted to a much greater extent than the Sun, Gibson (1983) proposed the model of a pair of spots: at equal magnetic fluxes the size of a leading sunspot is much less than that of an appropriate tail spot, so that the magnetic-field strength above the leading spot is much higher and consequently the emission above the leading spot is formed at a certain fixed frequency at a much higher level than above the tail spot. This asymmetry can result in noticeable polarization of the summary microwave emission. However, the emission can be polarized both in the sources and along the propagation trajectory in the stellar atmosphere.

During five, three-hour intervals at different phases of the axial rotation period Cox and Gibson (1984) observed the AU Mic star by VLA. They found rotational modulation at all wavelengths. At 2 cm, the modulation amplitude was minimum, probably because the filling factor was close to unity. Apparently, at 2 cm the emission was thermal, at two other wavelengths it was nonthermal. Gary (1985) performed 14-h observations of the YY Gem system with the orbital period of 20 h and found the eclipse at all three wavelengths.

Kundu and Shevgaonkar (1985) observed UV Cet and YZ CMi at 6 and 20 cm. They interpreted the data obtained as gyrosynchrotron emission of nonthermal electrons with a power distribution with respect to energy occurring in the sources whose size was equal to 2–3 R_* for UV Cet and 4–6 R_* for YZ CMi and L 726–8 A. The proposed nonthermal electrons can fill stellar coronae as a result of flares or events similar to solar noise storms, so that the microwave emission should be considered as the time-averaged response of the quiescent corona to the sporadic occurrences of energy particles in it rather than its permanent emission. The magnetic field on the photospheric level is estimated as several kilogauss from the lifetime of nonthermal particles of the order of an hour.

Using VLA, Pallavicini et al. (1985) observed 5 known flare stars and about a dozen chromospherically active G–K dwarfs and found the radiation of UV Cet, EQ Peg and YZ CMi at 6 cm: at about 1 mJy with low circular polarization for the first two stars and rather unstable radiation at the level of 2–8 mJy with the circular polarization of 50–90% for YZ CMi. For YY Gem and EQ Vir and all observed G–K dwarfs, including χ^1 Ori, they obtained only the upper limits of microwave emission. The upper limit of the flux from χ^1 Ori appeared to be 3 times lower than the value estimated by Gary and Linsky (1981). To explain this contradiction they suggested that during the observations of Gary and Linsky the flare of a recently found satellite M dwarf was recorded (Gary, 1985).

Lang and Willson (1986a) continued the studies of the decimeter radiation of YZ CMi at two wavelengths close to 21 cm. They confirmed the variability of the radiation on an hour scale, but they also found its narrow-band feature:

at $\Delta\nu/\nu \sim 0.1$ the variations of intensities occurred at each frequency band independently. Considering the measured level of X-ray emission of the star, one could expect in the decimeter region a thermal radiation flux two orders of magnitude lower than the recorded one. Within the gyroresonant model the size of the source should be $\sim 200 R_*$ and the magnetic field in this scale should achieve hundreds of G, which would require absolutely unrealistic photospheric fields. Thus, one might expect that the solution could be found in the gyrosynchrotron model. But neither the model of thermal free–free emission, nor those of gyroresonant emission of thermal electrons or gyrosynchrotron emission of nonthermal electrons suit the narrow-band emission of the whole stars. However, long-term narrow-band emission with high brightness temperature and strong circular polarization, which have no close analogs on the Sun, can be interpreted within the model of quasicontinuous bursts of coherent maser emission or plasma oscillations. It should be noted that Lang and Willson (1986a) estimated that in both cases the electron density should be about $10^{10}\,\mathrm{cm}^{-3}$, and the magnetic field strength at this level should be of several hundred gauss.

During the observations on the Molonglo interferometer in 1985, Vaughan and Large (1986a,b) found nonflare radiation at 35 cm on the CC Eri and AT Mic flare stars. During one session, the flux from CC Eri was at the level of 1 mJy and disappeared in a day, while the level of the flux from AT Mic was within 4–8 mJy during the four sessions within six weeks and in one of them meaningful variations at an interval of about an hour could be suspected.

Observing UV Cet in 1985, Kundu et al. (1987) recorded a flux of 2.8 mJy at 6 cm, which was twice that observed previously at the same frequency by VLA (Gary and Linsky, 1981; Fisher and Gibson, 1982; Kundu and Shevgaonkar, 1985). Thus, either there was a very slow weak flare or this was due to the flux variations on the time scale of years. The radiation of UV Cet at 6 cm was not polarized; the radiation at 20 cm was noticeably more variable. Usually, $F_6/F_{20} > 1$, which corresponds to an optically thick source. But when $F_6/F_{20} < 1$, there probably was a flare that is more clearly seen at 20 cm. In the observations of AU Mic a flux of 0.8 mJy without considerable polarization was recorded at 6 cm, and a flux of 4 mJy with a polarization of 70–90% was recorded at 20 cm. It is noteworthy that shortly before a flare occurred at 20 cm that was invisible at 6 cm. Developing the model of optically thick gyrosynchrotron radiation of nonthermal electrons for UV Cet quiet radiation at 6 cm, Kundu et al. (1987) estimated the necessary density of such particles as a few $10^7\,\mathrm{cm}^{-3}$ in magnetic fields of hundreds of gauss. Such a small addition of electrons should not be noticeable in the total X-ray emission of the star. Further, they showed that, as opposed to the solar corona where the height scale with respect to pressure is some hundredths of the solar radius, in the gyroresonant (GR) and gyrosynchrotron (GS) models of stellar coronae, where the height scale is comparable with R_*, the process of precipitation of fast particles down from the corona must play an important role: it should be the main mechanism of loss of particles determining their lifetime. In the

GR model these losses are much higher than in the GS model, which imparts certain advantages to the latter. Jackson et al. (1989) observed several other flare stars. Each star was observed for several 5-min intervals with 10-s time resolution at 6 and 20 cm. Near each wavelength the radiation at two frequencies separated by 50 MHz was measured. During 2 h of monitoring at 6 cm UV Cet displayed smooth changes of nonpolarized radiation from 2.5 to 4.0 mJy, whereas L 726-8 A showed much greater variability and at 20 cm both components were more variable than at 6 cm. No evidence of narrow-band emission was found. A weakly varying flux from Wolf 630 was recorded at 6 cm, it contained the constant part at the level of 0.7 mJy and two 10–15-min flares with the amplitudes of $0\overset{m}{.}6$ and $0\overset{m}{.}8$. A weakly varying nonpolarized flux from YZ CMi was recorded at the level of 0.45 mJy. On the whole, the quiet radiation of dMe stars, as a rule, smoothly changing over hours and days, in some respects was closer to solar flares than to quiet radiation of the Sun. If one takes $r_s \sim R_*$, then for flare stars at 6 cm $T_b = 10^8$–10^{10} K and at 20 cm, 10^9–10^{11} K, which is much higher than for the quiet Sun and even for nonthermal solar flares. Apparently, in both cases the mechanism of gyrosynchrotron emission of nonthermal electrons is realized in magnetic fields above active regions, but it is not clear why the electrons live so long or are permanently generated in stellar atmospheres. Thus, the Sun, as a much weaker source of radio emission than active dwarfs, cannot always be a model for interpreting the radio-frequency emission of stars.

After several years of observations of flare stars in the microwave range and the discussion on the nature of the radiation recorded in them Dulk (1987) compiled the table of main characteristics of different radiation mechanisms, which was further elaborated by Linsky (1988), see Table 9.

Table 9. Characteristics of radio-emission mechanisms

Mechanism	Source Size	Effective or Brightness Temperature	Circular Polarization	Time Variability
Thermal	$\gg R_*$	$\sim 10^4$ K	about zero	years
Gyroresonance	$\geq R_*$	$\sim 10^7$ K	low	low?
Gyrosynchrotron and synchrotron	$\leq R_*$	10^8–10^{10} K	$\leq 30\%$	from minutes to hours
Cyclotron maser ($\lambda \geq 30$ cm)	$\ll R_*$	to 10^{20} K	to 100%	from milliseconds to hours
Plasma oscillations ($\lambda \leq 30$ cm)	$\ll R_*$	up to 10^{17} K	10–90%	from milliseconds to days

In 1985, Kundu et al. (1988a) conducted the first simultaneous observations of four flare stars (UV Cet, EQ Peg, YZ CMi, and AD Leo) by EXOSAT and VLA. Radio observations were run continuously for 7–10 h at two close frequencies near 6 and 20 cm with a time resolution of about 7 s. X-ray monitoring was executed in the range of 0.04–2 keV (LE) with a time resolution from 60 to 600 s. All the stars were found in the quiescent state in both wavelength ranges, at 6 cm the flux from UV Cet was about 1 mJy; from EQ Peg, 7–13 mJy; from YZ CMi and AD Leo, about 0.3 mJy. These values of the fluxes are 40, 230, 38, and 17 times higher than the expected values of free–free coronal radiation estimated from their X-ray emission measures at $T \sim 10^7$ K, which can be explained either by 2–3 orders of magnitude higher brightness temperature or by the source sizes exceeding R_* by 2–20 times. This directly points to the difference of the physical mechanisms of X-ray and radio emissions. This conclusion is supported by the absence of any correlation between the time variations of the fluxes.

In 1986, Jackson et al. (1987b) observed dMe stars (that were not observed before) with VLA and for nine of them detected emission at 20 cm at the level of 0.4–5.2 mJy and for 7 stars at 6 cm within 0.2–12.8 mJy. Each object was observed for about an hour. An extended observational program during the second epoch did not reveal new radio-emitting flare stars (White et al., 1989a). According to their statistics for 83 objects, the microwave emission was found from 42% flare stars, 42% dKe-dMe stars, 65% BY Dra-type stars and from the only of 22 nonemission stars, Barnard's star. They considered various global characteristics of these stars and found that starting from dM5.5 the number of radio-frequency emission sources diminished noticeably, the mean portion of radiating objects grew with increasing L_X of red dwarfs, and according to their kinematic properties such objects mainly belonged to the young disk.

Willson et al. (1988) observed 16 K–M dwarfs at 6 cm by VLA. One-hour monitoring revealed a flux at the level of 0.45 mJy only from the dM2e star Gl 735, the strongest X-ray source in the sample. Similarly to the estimates of the mechanism of microwave radio-frequency emission for the other dwarf stars, the GR model in this case requires a size of the source of 20–30 R_* with a magnetic field strengths of hundreds of gauss, whereas for the GS model the density of nonthermal electrons of 10^7 cm^{-3} is sufficient, the linear size of the source of 10^8 cm and the magnetic fields are weaker by an order of magnitude. Then, Lang and Willson (1988) observed YZ CMi by VLA near 20 cm in 15 narrow bands of 3.125 MHz with a time resolution of 10 s. The responses of the left- and right-circular polarization were recorded in turn at time intervals of 10 min. They recorded a slowly changing radiation with a characteristic change time of minutes and with 100% left-circular polarization. During several 10-min intervals in the band with a total width of 50 MHz there was a noticeable change of F_ν, i.e., the radiation was narrow-band with $\Delta\nu/\nu \sim 0.02$, at other time intervals F_ν did not change noticeably. Lang and Willson interpreted the polarized emission as the coherent emission of an electron maser in a magnetic field of about 300 G.

During two seasons, Caillault et al. (1988) observed seven rapidly rotating BY Dra-type stars with known rotational periods by VLA at 6 cm and did not discover a constant flux. In 1985, a flux of 0.6–1.0 mJy was recorded from CC Eri, HD 218738, BY Dra, and EV Lac. During 3 nights in 1986 the radiation was found only from BY Dra and the measured flux varied from 0.8 to 1.5 mJy. For 22 dMe stars no correlation was found for the luminosities L_X and L_6. For 14 stars no correlation between L_6 and L_6/L_{bol} and the rotational period was revealed. Thus, Caillault et al. concluded that separate active events rather than quiescent corona were responsible for the recorded microwave emission. Then 16 K–M fast rotators were added to the research program and each program star was observed from two to four times over the interval from several days to a month (Drake and Caillault, 1991). Eight sample stars were identified reliably and four stars were detected supposedly as radio-emission sources. F_6 of four stars displayed variations at least by a factor of three, the flux from Gl 867 B varied within 14–0.2 mJy on a monthly interval. Drake and Caillault found that the radio-frequency emission of M dwarfs was independent of whether the star was single or a member of a close binary system, while for M dwarfs the following relation was valid

$$L_6 \sim L_X^{0.65 \pm 0.1} . \tag{24}$$

Later, Güdel (1992) formulated the following correlation for K dwarfs with widely ranging rotational period

$$L_R \sim P_{\text{rot}}^{-1.9} . \tag{25}$$

Slee et al. (1988) ran simultaneous observations of 24 chromospherically active stars, including six active dwarfs, at 6, 3.6, and 2 cm by VLA and the 64-m telescope of the Parkes Observatory. A flux from YZ CMi was found with $F_6 = 0.5$ mJy, from AT Mic $F_6 = 3.2$ and $F_2 = 1.6$ mJy, from V 1054 Oph $F_2 = 1.4$ mJy with the left-circular polarization of 28%, from Gl 890 $F_2 = 1.4$ mJy. No microwave emission was recorded from AU Mic and Gl 182. The mean brightness temperature of measured emissions was 1.2×10^8 K. Apparently, all measured fluxes corresponded to the quiet radiation of coronae or weak flares, since stronger fluxes from these objects were recorded before.

VLA observations of 12 M dwarfs in a wide range of L_X/L_{bol} ratios by Kundu et al. (1988b) made it possible to detect noticeable microwave emission from four stars. Three of them had a maximum L_X/L_{bol} ratio, and on the fourth star, Gl 461, no flares and H_α emission were recorded before.

Güdel and Benz (1989) observed UV Cet by VLA at 90, 20, 6, 3.6, 2, and 1.3 cm. During two 5-h sessions no flares exceeding 3σ of the noise were found, and the order of magnitude of the fluxes at 20 and 6 cm was close to the earlier estimates, at higher frequencies they were obtained for the first time. No meaningful polarization was recorded at any frequency. During the first session the spectrum of microwave emission was measured within the range from 20 to 2 cm, in the second, from 6 to 1.3 cm. The obtained spectra revealed the statistically reliable U-like shape with a minimum at 3.6 cm and with spectral indices of -0.3 and $+0.4$ for low- and high-frequency branches, respectively. In the second session, for the main component of the L 726-8 A system at 6, 3.6, and 2 cm the fluxes of 0.8, 0.4, and 0.55 mJy recorded, i.e., the minimum was also at 3.6 cm. The brightness temperature of UV Cet at 1.3 cm was close to the temperature of the X-ray emission for the microwave source $(1-2)R_*$ in size. Güdel and Benz presented the observed spectrum as a sum of two power spectra and obtained the spectral indices of -0.42 for $\nu < 10$ MHz and $+2$ for $\nu > 10$ MHz. They concluded that the high-frequency branch could not be due to the free–free plasma radiation responsible for the X-ray emission of the corona, but could be explained by an optically thick cyclotron emission at 4–6 harmonics of the plasma at a magnetic field strength at the photosphere of 600–2100 G and a density of $3 \times 10^{9-10}$ cm^{-3}. This mechanism is not suitable for interpreting the low-frequency branch of the recorded spectrum, in which gyrosynchrotron or synchrotron radiation should be taken into account.

In 1992, Rucinski (1994) observed YZ CMi with VLA. At wavelengths of 3.6, 2, and 1.6 cm in two sessions he recorded the fluxes of 4–8, 2–7, and 3 mJy, respectively. The first observational session, during which maximum fluxes were found, lasted only 11 and 13 min; thus one cannot conclude with confidence that no flares occurred over that time.

White et al. (1989b) proposed the model for the generation of microwave radiation on the stars with separate active regions and high-strength photospheric magnetic fields. The model suggests that the nonthermal gyrosynchrotron radiation is emitted from these regions and the expected radiation spectrum depends on the vertical gradients of the magnetic-field strength and the density of matter. In particular, the negative spectral index for optically thick radiation is achieved at low values of these gradients and for low harmonics of 5–10. If the model is valid, constant sources of electrons with energies of 20–200 keV should exist on a star and there should be some evidence of precipitation of these electrons into the chromosphere. White et al. (1994) considered the restrictions imposed on the models of the dMe coronae by the radio data collected by that time. According to the estimates, if the hot element of the corona with a temperature of about 2×10^7 K, which is usually found on such stars, was in the lower corona, where the strength of magnetic fields due to the proximity of the photosphere should be more than 1 kG, then due to higher optical thickness at 15 GHz it would be revealed by radio observations. Since observations for the considerable sample of dMe stars did not justify the expectations, White et al. concluded that the hot coronal component was localized at a height of the order of R_* and was the cooling flare plasma.

Fomalont and Sanders (1989) surveyed all stars with $\delta > -28°$ at a distance of not more than 5 pc from the Sun with VLA at 6 cm. Among 241 objects they found microwave emission at the level of $F_6 > 1$ mJy only for five known flare stars: UV Cet, YZ CMi, AD Leo, Wolf 630, and EQ Peg.

In 1988, Caillault and Drake (1991) ran 21 VLA observational sessions of BY Dra at 6 cm to reveal the rotational modulation of the quiet microwave emission, but failed to reveal it, in their opinion, because of the small angle between the rotational axis of the star and the line of sight, about 30°, or because the magnetic loops, in which radio-frequency emission occurred, were very long.

In 1988, Spencer et al. (1993) observed YZ CMi at 6 cm by the wide-band interferometer at Jodrell Bank composed of the 76-m Lovell antenna and the 25-m telescope Mark II. They revealed stellar radiation varying at hourly intervals with an average level of 1.4 mJy.

The next step in the development of radio-astronomical techniques after VLA was the intercontinental Very Long Base Interferometer (VLBI). The first successful observations of dMe stars at VLBI were carried out by Phillips et al. (1989) within the framework of an astrometric program. Their VLBI consisted of two 70-m antennas (Madrid and Goldstone), 100-m Effelsberg antenna, and VLA. In observations at 3.6 cm, three of six program stars were found: YZ CMi, Wolf 630, and EV Lac. The recorded flux was at the level of 2–5 mJy and the size of sources was at least R_*. Using VLBI composed of VLA and the largest radio telescopes in Arecibo (Puerto Rico), Green Bank (West Virginia) and Effelsberg (Germany), Benz and Alef (1991) ran four 13-min observation sessions of YZ CMi and recorded a smoothly varying correlated flux with 80% circular polarization at the level of 1.3 mJy. The upper limit of the source size was equal to 3.4 times the stellar diameter and $T_b > 4 \times 10^8$ K, which exceeded the temperature of slowly varying radio emission of coronal condensations on the Sun by orders of magnitude and required the nonthermal emission mechanism. It is still unclear how the necessary energy particles are retained for hours in the stellar atmosphere. Then, Benz et al. (1995) at VLBI composed of the above four telescopes, the 76-m Lovell telescope and the 40-m Owens Valley dish (California) observed AD Leo and the EQ Peg binary system at 18 cm. In one of the two sessions they estimated the properties of quiet radiation of AD Leo: the flux of 0.75 mJy, a source size of less than 3.7 stellar diameters, and $T_b > 2 \times 10^9$ K, which was higher by an order of magnitude than in the strongest solar flares and much higher than the appropriate temperature of YZ CMi. They suggested that the quiet radiation of active dwarfs was physically related to IV-type solar stationary bursts, which continue for many hours after strong flares, or to noise storms, in which these bursts transform and then last for days at characteristic $T_b \sim 10^8$–10^{10} K. If the suggestion is correct, the radio-emissive volume of dMe stars should be determined by the sizes of closed coronal magnetic loops. Using the VLBI composed of the above six radio telescopes, Alef et al. (1997) studied the binary YY Gem system at 18 cm during the phase of main eclipse. They estimated the upper limit of the size of the radio source as 2.1 photospheric diameters of the star and $T_b = 1.1 \times 10^9$ K. The high brightness temperature and low circular polarization corresponded to gyrosynchrotron emission. The rather symmetrical radio image of the star made it possible to claim that the magnetic coronal loops trapping radio-emitting relativistic electrons achieved heights comparable with the stellar radius and were distributed rather isotropically. Benz et al. (1998) observed the system L 726-8 AB at 3.6 cm and found both components, while the UV Cet component, brighter in the radio range, was resolved at least into two components remote by $(4-5)R_*$ and with the noticeable variation of relative brightness during a 6-h session. Within the gyrosynchrotron model the recorded radiation requires coronal loops of several stellar radii and field strength between 20 and 130 Gs. In VLBI observations at 3.6 cm, Pestallozzi et al. (2000) found a level of the flux from YZ CMi of about 3 mJy and the left-circularly polarization up to 60%, for AD Leo, 0.5 mJy without noticeable polarization. They concluded that the size of the radio image of YZ CMi was 1.7 ±0.3 stellar diameters, that of AD Leo was less than 1.8 stellar diameter, the brightness temperature of YZ CMi and AD Leo was 7×10^7 K and more than 5×10^7 K, respectively.

Güdel et al. (1993) observed simultaneously 12 M dwarfs by ROSAT (0.1–2.4 keV) and VLA (3.6 and 6 cm) and found a close correlation between the X-ray and microwave emission

$$\log L_X = (1.06 \pm 0.10) \log L_R + (14.5 \pm 1.3) \quad (r = 0.95) \ . \tag{26}$$

This correlation became slightly weaker when nonsimultaneous observations were considered as well. The revealed linear correlation valid for the interval of three orders of magnitude continued with a slight change of slope toward stronger stellar coronae for the objects of the type of RS CVn, FK Com, and Algol. Then, Güdel (1992) studied 12 active K0-K8 dwarfs, among them 5 single stars, 5 components of spectrally binary systems, and 2 apparent binary objects using VLA. All the program objects were known as strong X-ray sources with fast rotation and/or strong calcium emission. For 7 of them nonflare radiation was found at 3.6 cm, while their luminosities were comparable with L_R of M dwarfs. L_R of studied K dwarfs was in good correspondence with the correlation $L_R \sim L_X$ previously found for M dwarfs. Faster-rotating K dwarfs were found to be stronger radio sources. The Sun, as a too-weak microwave source, does not satisfy the correlation $L_R \sim L_X$ that is common for active stars of different type and expands to 5 orders of magnitude. However, the solar flares satisfy the correlation, in particular in the maximum phase (Güdel and Benz, 1993; Benz and Güdel, 1994). This correlation, common for active stars, suggests that either the coronal heating and particle acceleration are of common origin or continuously accelerated particles are thermalized and heat the corona.

During observations at 6 and 3 cm by the Australian Telescope Compact Array (ATCA) Lim (1993) found a flux from the dM4e star Rossiter 137B close to that recorded from UV Cet. Since the former is ten times further from the Sun than the latter, the radio luminosity of Rossiter 137B is higher by 2 orders of magnitude than that of UV Cet. Rossiter 137B is very young, less than 10^8 years, a fast rotator with $P_{\rm rot}$ not exceeding 9 h; and has apparently not reached the main sequence and is in the post-T Tau stage.

Reasoning from the $L_R \sim L_X$ correlation for coronally active stars, Güdel et al. (1994a,b) selected 15 nearby solar-type stars with a high level of X-ray emission from the Catalog of Bright Stars and from the ROSAT data and observed them at 3.6 cm by VLA. For Gl 97, Gl 755, EK Dra, and HD 225239 during 45-min sessions quiet microwave emission at the level of 0.18–0.33 mJy was found, which differs from the values expected from the correlation (L_R, L_X) for later active dwarfs by no more than 60%. The age and rotational rate of Gl 97 are close to those of the Sun. EK Dra is a "young Sun", its age is $\sim 7 \times 10^7$ years, the emission lines of its chromosphere and transition region are very strong, the X-ray flux well satisfies the correlation with the axial rotation period. HD 225239, which based on the metal deficiency and the spatial velocity is related to the old-disk population, unexpectedly displayed radio luminosity that was several times higher than that of the strongest solar flares. The cause of the strong radio luminosity of these active stars is unclear.

Güdel (1994) summarized the results of the first decade of studying the microwave emission of the stars of late spectral types, for which the discovery of the slowly changing component was a surprise. This component, whose intensity exceeds by orders of magnitude the free–free radiation of the transition region of the Sun and slowly changing thermal gyroresonant radiation of the regions with strong magnetic fields cannot be presented within the stationary corona. A concept of continuous or recurrent filling of a corona by the high-energy electrons, whose emergence could be related to flare phenomena, should be introduced. The characteristic times of variation of the radiation are hours and days, and there is no evidence of such variations over minutes. The low level of circular polarization of the radiation could be stipulated by relativistic electrons at a not very hard energy distribution in strong magnetic fields and small optical thickness of the source. On the other hand, mutual compensation of the polarization in the total radiation of many separate sources is possible. Thus, the investigation of the microwave emission of stars detected magnetic nature of the activity, close physical connections of the quiet radiation with flare processes, and the hot component of the corona with the acceleration of particles to relativistic energies.

Using ACTA, Robinson et al. (1994) discovered at 6 cm the emission of 2 mJy from the young rapidly rotating K0V star HD 197890 with strong non-stationary EUV radiation; however, there is no confidence that the star already reached the main sequence.

Pallavicini et al. (1990b) analyzed the Einstein Observatory and EXOSAT observations of Castor (α Gem) and concluded that the quiet X-ray emission of this system was due to the invisible red dwarf, a component of one of the A stars forming the system. HRI and PSPC ROSAT observations confirmed the X-ray emission of the Castor A+B system, but the microwave studies performed at the Very Large Antenna at 3.6, 6, and 20 cm with an order of magnitude higher spatial resolution showed that such a radiation was emitted only from the A component (Schmitt et al., 1994). Apparently, the X-ray emission of the system is also emitted by this component. The radiation at 20 cm changed from 0.3 to 1.2 mJy on one day and from 0.3 to 0.7 mJy on another with a measurement error of 0.1 mJy. X-ray and microwave studies led to the conclusion that the optically invisible low-mass component was responsible for the observed activity of α Gem, but its parameters have not been estimated unambiguously as yet.

Güdel et al. (1995b) discovered strong corona at the F0V star 47 Cas: strong X-ray emission corresponding to $L_X \sim 3 \times 10^{30}$ erg/s and strong microwave emission at a level of 10^{15} erg/s Hz were recorded. The X-ray emission was presented by the two-temperature corona model with temperatures of 2 and 10 MK and emission measures of 4×10^{52} cm^{-3} and 10×10^{52} cm^{-3}, with the "cool" component displaying rotational modulation, while the X-ray flares were recorded only in the "hot" component. The microwave emission of the star at 3.6 cm was recorded by VLA at a level of 0.63 mJy, it corresponded to $T_b = 8 \times 10^7$ K and was in good agreement with the correlation (L_R, L_X) for later active dwarfs. Apparently, 47 Cas is the youngest F0 single star with an extremely high level of magnetic activity, which agrees with fast rotation, $v \sin i = 95$ km/s, and an age not exceeding that of the Pleiades.

Güdel et al. (1995c) studied the sample of 24 brightest, according to RASS data, F and G stars at VLA at 3.6 cm. They looked for possible duplicity among the sample stars and found that at least for the F0V star HD 12230, one of the nine sources of microwave emission found, except for EK Dra, was definitely the star itself rather than the weak invisible component. The star belongs to the Pleiades moving cluster, and its age is 50–70 million years. Its X-ray emission is modulated with a period of about a day, with the cool component of the corona modulated to a greater extent than the hot one, as in EK Dra.

Güdel and Benz (1996) collected 33 VLA measurements for 20 dKe-dMe stars in the range from 20 to 2 cm and found that in most cases the intensity of radio-frequency emission decreased with increasing frequency, as one would expect in the gyrosynchrotron model, but in several measurements a more complicated change of F_ν suggested a contribution of thermal gyroresonant emission occurring at the plasma temperature of 10 MK and magnetic fields of about a kG.

To check the existence of a minimum in the radio spectrum at 3.5 cm found by Güdel and Benz (1989) and to relate the increase of the radio-frequency emission at higher frequencies to known excesses of active stars at millimeter wavelengths and in the IR range (Mathioudakis and Doyle, 1991b, 1993), Leto et al. (2000) observed five flare stars in the wavelength range of 35–7 mm in 1996 by VLA. But they measured the fluxes from UV Cet, V 1054 Oph, and EV Lac only at 3.5 cm, in the other cases only the upper limits were estimated.

Lim and White (1995) observed four fast-rotating G–K dwarfs in the Pleiades at 3.6 cm by VLA. For three stars microwave emission was detected. On the fastest G dwarf rotator HII 1136, a flare was recorded and after its decay a nonflare radiation was determined at a level of 0.16 mJy. From the fastest K dwarf rotator HII 1883, the quiet radiation was recorded in two sessions separated by three months at the level of 0.10 and 0.05 mJy. The K0 dwarf HII 625 displayed slowly changing emission at a level of 0.16 mJy. Such fluxes correspond to the luminosity $L_R \sim (1–3) \times 10^{15}$ erg/s Hz. The discovery of microwave emission of the open cluster members showed that solar-type stars that recently reached the main sequence were equally strong radio sources, as the nearby field stars with equally fast rotation. The radio fluxes of the stars averaged over the surface are comparable with the appropriate values of the T Tau-type stars, which suggests that the saturation mechanism is realized in both cases.

Between 1990 and 1994, Lim et al. (1996) ran four ACTA observational sessions of Proxima Cen at 3.5, 6, 13, and 20 cm, but did not record quiet radio-frequency emission. However, using the least estimates of the upper limits of the fluxes at 3.5 and 6 cm, 0.12 and 0.11 mJy, respectively, they obtained for the two temperatures of X-ray plasma, 3 and 20 MK, the lower estimates of the magnetic-field strength in the coronal loops from 400 to 1040 G and the upper estimates of the filling factor. These observations fit into two different models of radio-frequency emission sources: the filling factor is low at B > 400 G, or it is high at low B. Using the data of X-ray observations and (24), Lim et al. estimated the parameters of the photospheric magnetic field of the star: $B \sim 1\,\mathrm{kG}$ and $f \sim 0.1 - 0.2$.

Güdel and Zucker (2001) presented the microwave emission of about a dozen stellar coronae recorded during 20 observational sessions within the framework of the gyrosynchrotron model. They found that the index of the power distribution of the energy of radio-radiating electrons was mainly within 2–3.5. Such a hard distribution of energy particles on the Sun is observed only in strong flares. In their survey of cool stars in the centimeter wavelength range, Güdel and Audard (2001) stated that the X-ray emission of almost all stars found in this range was close to $L_X/L_{\mathrm{bol}} \sim 10^{-3}$, while the level of radio-emission saturation was $L_R/L_{\mathrm{bol}} \sim 10^{-18}$ Hz^{-1}.

Using VLA, at 3.6 cm Krishnamurthi et al. (1999) sought for microwave emission from four M7–M9 dwarfs and two brown dwarfs: Gl 229 B, LHS 2065, Kelu 1, Gl 569 B, VB 8, and VB 10. The accumulation time varied from 2.5 to 3.5 h, but no emission was recorded in either case. If the objects were removed at the distance of UV Cet, the upper limits would vary from 0.2 to 0.8 of the flux from UV Cet.

At the time when the gyrosynchrotron model seemed to become generally accepted for the interpretation of nonthermal microwave emission of stars, Stepanov et al. (1999) and Stepanov (2001) proved that plasma radiation near the frequency of the second harmonics excited by the instability of beams of energy electrons trapped in magnetic flux tubes is a serious alternative to the GS radiation of stellar coronae in the quiescent state. To excite the radiation, a density of energy electrons lower by 6–7 orders of magnitude than that of the thermal plasma is sufficient, the brightness temperature of this emission can reach 10^{14-16} K, it is characterized by noticeable polarization of a variable spectrum, and in the case of high-temperature stellar coronae this radiation is excited much more efficiently than on the Sun.

1.4.4 Stellar Winds and Far-IR Emission

Parker (1958) discovered the solar wind by means of computations, as Leverrier and Adams found the planet Neptune "at a tip of a pencil". However, Parker had less-known predecessors whose results were reviewed by Ponomarev and Rubo (1965). The main conclusions of the Parker theory were that the solar corona could not be a hydrostatic structure continuously transforming into the interstellar medium but that it could be described by a particular solution of the nonlinear differential hydrodynamics equation for the continuously expanding atmosphere. The theory became widely accepted almost immediately because of the elegant derivation and because the solution provided a simple and natural explanation for a number of observed facts concerned with the structure of comets, the brightness of the zodiacal light, the variation of cosmic rays, and various geomagnetic and geophysical phenomena. Further, Parker's model was revised on the basis of direct measurements of physical parameters of interplanetary medium at different distances from the Sun and with the transition from a spherically symmetrical purely hydrodynamics scheme to the concept of solar magnetosphere containing open and closed magnetic flux tubes.

Higher densities and temperatures of the coronae and higher strengths of photospheric magnetic fields of active dwarfs suggest that they should have solar-wind analogs, which are stronger than the prototype. As stated above, the first quantitative model of the stellar wind of an M dwarf was constructed by Kahn (1969). He estimated the secular loss of stellar matter at the expense of expanding corona as $3 \times 10^{-12} M_\odot$/year, which exceeded the appropriate solar value by a factor of 300.

Mullan et al. (1989a) were the first to find the stellar wind from a late dwarf based on optical data: during the eclipse phase of the eclipse system V 471 Tau composed of red and white dwarfs they found absorption lines of neutral and ionized metals in the ultraviolet spectrum, i.e., the spectral details occurring when the white dwarf was examined through the atmosphere of the red one. The discovered absorption components were shifted blueward by 700–800 km/s, which definitely exceeded the parabolic velocity of the K2 dwarf and their common character was preserved for at least 4 months. The multiplicity of the absorption components evidenced the nonstationarity and/or nonsphericity of the stellar wind. From Mg lines, Mullan et al. estimated the secular mass loss as $3 \times 10^{-11} M_\odot$/year, which exceeds the appropriate solar value by three orders of magnitude.

Then, winds were sought in the far-IR and millimeter wavelength regions primarily using the observational results from the Infrared Astronomical Satellite (IRAS). The spacecraft carried a telescope with an aperture of 57 cm and 62 infrared detectors. Between January and November 1983 IRAS scanned 98% of the sky in the regions of 12, 25, 60, and 100 μm, while 72% of the sky was scanned more than three times and 24% twice. IRAS recorded about

280 000 point sources with a characteristic accuracy of coordinates $30'' \times 7''$ and a photometric accuracy of 5–15%.

Initially IRAS data were analyzed irrespective of stellar winds. Mullan et al. (1989b) compared the IRAS data for the samples of 78 flare red dwarfs and 38 dM stars. For the stars with identical stellar magnitudes in the K band they found that the brightness of dMe stars at 12 µm was systematically and noticeably higher, by 70%, than that of dM stars. At 12 µm they recorded the overwhelming majority of the stars of both samples, but only 15 dMe stars and probably 2 dM stars were identified as IR sources at 100 µm; 9 dMe and no dM stars were recorded in all IRAS bands. Thus, dMe stars are definitely stronger IR sources than dM stars. The fact that the $V - K$ color indices of dMe stars also differ systematically from analogous values of dM stars suggests that IR excesses of dMe stars spread to 2.2 µm. For those flare stars, from which fluxes were recorded at 100 and 60 µm, a noticeable increase in intensity toward the long-wavelength range was observed. The authors suggested that since the total duration of IRAS observations of each object did not exceed 100 s, which was a sum of tens of independent measurements, it was due to stationary emission of dMe stars, rather than flares on them. They concluded that the IR excesses could not be due to circumstellar dust. If the reason was synchrotron radiation, one should expect noticeable fluxes of about 1 mm. Based on the IRAS data for the BY Dra-type stars, Chugainov and Lovkaya (1989) suspected that dust circumstellar structures were responsible for the discovered IR excesses from 11 of 22 considered stars. Tsikoudi (1989, 1990) compared IRAS data for 36 active F–K dwarfs and 30 nonactive stars of the same spectral types and found that 27 active stars were recorded at 12 µm, half of them at 25 µm and only 2 stars at 60 and 100 µm. However, no systematic distinctions between these two samples were revealed: with the correlation coefficient of 0.96 for active stars she obtained the relation

$$L_{12} = 10^{-2.6 \pm 0.3} L_{\text{bol}}$$

and

$$L_{12} = 10^{-2.7 \pm 0.2} L_{\text{bol}}$$

for nonactive stars. Only for the active stars Gl 113.1 and Gl 735 were noticeable excesses at 12 µm found, for ε Eri and 70 Oph probable excesses were found at 60 and 100 µm. Then, Mathioudakis and Doyle (1991b) considered the radiation of seven active M dwarfs in the range from the ultraviolet to the IR region, including IRAS data. All objects were recorded at 12 µm, 5 at 25 µm, 2 at 60 and one at 100 µm. Clear IR excesses were discovered at Gl 65 AB, Gl 644 AB, and Gl 873. Mathioudakis and Doyle noted that these excesses could result from the radiation of relict dust or synchrotron emission of relativistic particles emerging in microflares, but did not select one of the mechanisms.

By that time, the concept of gaseous shells expanding around hot stars had been developed and confirmed by observations: at certain frequencies they become opaque for the free–free radiation, and the fluxes going from them at these frequencies can be used for estimating their sizes and kinematics. Within this concept Mullan et al. (1992a) calculated the expected radiation spectrum of stellar wind from cool stars and determined that the frequency range within which it should be opaque and radiate as an absolutely blackbody was within tens of micrometers, the "photospheric radius" of the wind should vary as $\nu^{0.6-1.2}$ and the flux expected from YZ CMi was estimated as 14 mJy. Since, within the Rayleigh–Jeans approximation the blackbody radiation is proportional to ν^2, the optically thick wind should provide a certain addition to the radiation of the stars in the far IR region. Such an excessive radiation of flare stars had indeed been discovered already. To check these calculations, Mullan et al. (1992a) performed special observations at 0.8 and 1.1 mm. Only for six of 13 program stars were rather unreliable results obtained. However, combination with the IRAS and VLA data for YZ CMi, Wolf 630, and EV Lac yielded fluxes that were in rather good correspondence with the estimates. Paradoxically, the microwave emission successfully interpreted within the gyrosynchrotron emission of stellar coronae matched the single power spectrum with the expected index with the IRAS data for the far-IR range and the measurements by Mullan et al. in the millimeter range.

To find stellar winds of G dwarfs younger than 250 million years, Vikram Singh (1995) observed HD 39587, HD 72905, HD 115383, and HD 206860 by VLA at 6, 3.6, and 2 cm. At 6 cm meaningful signals were recorded from the first two stars, at 3.6 cm only from HD 39587. Within the thermal radiation of spherically symmetrical expanding shells the obtained estimates correspond to the secular mass losses of $(2-3) \times 10^{-10} M_\odot$/year for the two first stars and of $< 10^{-11} M_\odot$/year for two others.

Another attempt at interpreting the excessive radiation of active late dwarfs in the far-IR region was associated with direct confrontation of this radiation with the manifestations of magnetic activity of these stars. First, for 14 K–M dwarfs, for which the measurements of magnetic fields and IR fluxes were available, Katsova and Tsikoudi (1992) found a correlation of magnetic fluxes fB with the radiation in the 12-μm band that exceeded the values expected for blackbody radiation at the temperature estimated from the $B-V$ color index. They also suspected the correlation between the excessive IR radiation and the rotational rate. Then they compared the excessive IR radiation of G–K–M dwarfs with the fluxes in the soft X-ray range and found that in the region of G–K dwarfs, where $\log L_X/L_{bol} = -5.5-\!-3$, there was a correlation of these radiations, while at high $\log L_X/L_{bol}$, typical of flare M dwarfs, there was a saturation in X-rays (Katsova and Tsikoudi, 1993). Considering the found dependences Katsova and Tsikoudi justified why IR excesses could not be caused by interstellar dust or synchrotron radiation and advanced the hypothesis on the responsibility of stellar spots covering a considerable part of the stellar surface for the generation of the radiation. However, later they accepted the model of cool $(T \sim 10^4 \text{ K})$ stellar wind (Katsova et al., 1993).

Probably, the solution to this complicated problem was found by van den Oord and Doyle (1997), who noticed that the optical thickness of the wind for cool stars was much lower than for the hot ones, and on cool stars the wind was hotter than the photosphere, whereas for hot stars this relation was the reverse. These differences introduce significant changes in the theory of expanding shells as radio sources, and, according to the estimates of van den Oord and Doyle, the mass loss by dwarf stars at the expense of stellar wind did not exceed $10^{-12} M_\odot$/year. But on the other hand, several years earlier, Badalyan and Livshits (1992) considered possible stellar analogs of solar streamers, coronal beams in which the effective outflow of solar coronal plasma occurred. They found that with allowance for the X-ray observations of the densities at the base of stellar coronae (Katsova et al., 1987) at five considered F8–M0 dwarfs one could expect mass loss at the expense of these structured stellar winds at a level of $(1–2) \times 10^{-11} M_\odot$/year.

Solar studies and gasdynamic calculations showed that expanding stellar wind separated from the colliding incident interstellar gas by two shock waves isolated by a "hydrogen wall" of heated and compressed neutral gas. Such a "hydrogen wall" was initially found by Linsky and Wood (1996) in HST observations of the Ly_α line in the spectra of α Cen A and α Cen B. Then, they (Wood and Linsky, 1998) found analogous structures in 61 Cyg A, 40 Eri A, ε Ind, and in the giant λ And. From the parameters of the "walls" found they estimated the pressure in stellar winds p_w and found its correlation with the surface X-ray flux

$$p_\mathrm{w} \sim F_X^{-1/2} . \tag{27}$$

This correlation is valid for the Sun at different phases of the activity cycle. Hence, it follows that as stellar activity attenuates, one should expect strengthening of the wind, and consequently an increase of the secular mass loss. Further UV observations of several more solar-type stars allowed Wood et al. (2001) to estimate the rates of mass loss for stellar wind as $2\dot{M}_\odot$ for α Cen and less than $0.2\dot{M}_\odot$ for Proxima Cen, to find a correlation between the mass-loss rate and the X-ray emission level as $\dot{M} \propto F_X^{1.15 \pm 0.20}$ and to suspect evolutionary weakening of the wind as $\dot{M} \propto t^{-2.00 \pm 0.52}$ (Wood et al., 2002).

Independent considerations on the generation of strong stellar winds as a result of considerable coronal ejections of matter were developed by Houdebine et al. (1990) in analyzing the flare on AD Leo of 28 March 1984, the spectrum of which contained a short-term blue wing of the H_γ emission line expanded up to 80 Å.

The major result of studying stellar coronae is the ascertained universality of these structures practically for all F and later stars: stars without coronae are absent or are extremely rare. The investigation of stellar coronae is one of the rapidly developing branches of solar–stellar physics, which, as did many other branches, grew conceptually and instrumentally from the solar physics. The features of this branch are determined by the amazing research object, stellar coronae, whose structure is practically "detached" from visible stellar surfaces, but depends on such global parameters as rotation, the depth of the convective zone, and the intensity of mixing of matter in it. This situation, though paradoxical at first sight, when the outermost atmospheric layers are determined by the internal structure of the star, can be easily explained: the physical state and spatial structure of stellar coronae are determined by magnetic fields, which are generated in the subphotospheric layers. This concept, common for the Sun and stars with convective shells, first made it possible to understand the phenomena, for which it sufficed to simply scale solar processes to stronger and hotter stellar coronae. Further extrapolation enabled qualitative interpretation of the phenomena nonexistent on the Sun: very hot component of the coronae and strong microwave emission.

1.5 Heating Mechanisms for Stellar Atmospheres

In the previous sections we presented experimental data on different atmospheric layers of active stars as thoroughly as possible. Below, we summarize the data analysis results that yielded the conclusions on possible heating mechanisms only as qualitative descriptions, because thus far no unambiguous and generally accepted solution has been found for the problem of heating of stellar atmospheres, though there are numerous hypotheses.

The idea of heating of stellar atmospheres by nonradiative energy fluxes released from subphotospheric layers, which explains the existence of hot chromospheres and coronae, has been under development for many decades. The fluxes can be associated with different hydrodynamic and magnetohydrodynamic processes observed in the solar atmosphere and expected in stellar atmospheres, while the considerable diversity of the processes is due to the mobility and appreciable three-dimensional stratification of all physical parameters of the medium. First, the heating of the solar chromosphere was explained by acoustic waves generated by convection, then, by the field of velocities of 5-min oscillations, Alfven and other magnetohydrodynamic waves, the Joule heating by electric currents, the quasistationary annihilation of magnetic fields, and fast annihilation of the fields both in the nonstationary low-energy processes, microflares, and in high-energy flares. But for the most active dMe stars the radiation of quiescent chromospheres and coronae is about 10^{-3} of the bolometric luminosity, i.e., higher by orders of magnitude than on the Sun, and it was not clear if the solar mechanisms were sufficient for the heating of such strong stellar atmospheres.

Fosbury (1973) concluded that nonthermal components in the widths of the H_α, CaII H and K and MgII h and k lines in the stellar spectra were due to acoustic waves. However, Blanco et al. (1974) from the measurements of absolute luminosity of the CaII K emission line concluded that the acoustic waves were sufficient to maintain the chromosphere only on the stars with an effective temperature of above 5000 K, whereas for cooler stars this flux calculated from the Lighthill–Proudman theory was insufficient: for the coolest stars it is lower by two orders of magnitude than the required value. When Linsky and Ayres (1978) showed that the losses for the radiation of ultraviolet MgII h and k lines exceeded approximately by a factor of three those of the violet calcium lines, the insufficiency of the heating mechanism by acoustic waves became even more convincing.

The fact that the solar chromosphere, transition region, and corona radiate mostly above the magnetic regions and there are clear correlations between the appropriate radiation fluxes and magnetic fluxes of active regions resulted in the conclusion on the domination of magnetic heating mechanisms. These mechanisms can be reduced to immediate dissipation of magnetohydrodynamic waves or to the generation of the beams of fast particles in the course of development of different instabilities and subsequent thermalization of the particles.

As stated above, Mullan (1975a) proposed the concept of dMe stars as magnetic red dwarfs, in which strong magnetic fields enabled the heating of high-density chromospheres by magnetohydrodynamic waves. According to his estimate, such a heating can provide the radiative losses of chromospheres estimated by Blanco et al. (1974).

The estimates of radiative losses with regard to ultraviolet lines led Basri and Linsky (1979) to the conclusion of the existence of chromospheres with minimum nonradiative heating due to acoustic waves, which is compensated by radiative losses in the emission lines and at H^-. In the solar floccules the chromospheric losses for the radiation of resonance doublets MgII and CaII are 10 times higher, and since the intensity of the CaII K line correlates with the magnetic-field strength, the necessity of involving the magnetic heating mechanism became obvious. The insufficiency of the acoustic heating of the atmospheres of active red dwarfs was demonstrated by Haisch and Linsky (1980), who established that the losses of the corona of the dK5e star EQ Vir for radiation, heat conductivity, and wind were higher by two orders of magnitude than the heating expected from this mechanism.

The most ponderable argument against the concept of purely acoustic heating of stellar atmospheres was the detection of X-ray emission of the coronae of the stars of different spectral types and the weak dependence of radiation on the effective temperature (see Fig. 16).

Stein (1981) considered the dependence of the flux of acoustic and various hydromagnetic waves leaving the subphotospheric convective layers toward the atmosphere on the general stellar parameters and showed that the observed very weak dependence of radiative losses of stellar chromospheres and coronae on the effective temperature corresponded to Alfven or slow magnetohydrodynamic waves.

Comparing the absolute fluxes of ultraviolet emission lines formed in the chromospheres and in the transition regions and the X-ray emissions of G–K dwarfs, Ayres et al. (1981a) found that the ratios R_{hk}, R_{lines} and R_X

correlated, but between R_{hk} and R (chromospheric lines) the correlations were linear, whereas between R_{hk} and R (transition-region lines) and R_X they were nonlinear. The difference between the dependences of ratios R_{hk} and R_{CIV} and (5) on the effective stellar temperatures noted by Linsky et al. (1982) evidence certain differences in the heating mechanisms of the chromospheres, transition regions, and coronae. Later, Schrijver (1990) found that the exponent κ in the relation $F_{CIV} \sim B^\kappa$ was between the exponents in the analogous expressions relating F_{HK} and F_X with B. This fact made it possible to admit the existence of two heating mechanisms of the transition region: one analogous to the heating of the chromosphere and the heat conductivity from the corona. Along with this, as stated above, the widths of emission lines measured by Ayres et al. (1983b) led to the conclusion on the commonness of the heating mechanisms of the transition regions in the stars of different luminosity from the solar-active regions to M dwarfs.

Cram (1982) calculated the model of such a dMe star chromosphere, which provided the relations of the equivalent widths of Balmer lines corresponding to the observations, and showed that in this case the coronal X-ray emission was sufficient to heat a stellar chromosphere. This hypothesis was supported by the closeness of the total losses for radiation of ultraviolet lines of the chromosphere and transition-region and for the X-ray emission of the corona found by Giampapa et al. (1981b). Further, upon calculating the atmospheric temperature distribution of the emission measure from the observations of ultraviolet lines in the spectra of three nonactive G–K dwarfs τ Cet, δ Pav, and 61 Cyg A and constructing the transition-region models, Fernandez-Figueroa et al. (1983) obtained an ambiguous result: the thermal conductivity of τ Cet was insufficient to compensate for the radiative losses, whereas that of δ Pav and 61 Cyg A could compensate for these losses and the models of the transition regions of these two stars corresponded to the X-ray fluxes measured on them.

Marcy (1983) found that for G–K stars the dependence of the calcium emission on the magnetic field and the effective stellar temperature corresponded to the heating of chromospheres by magnetohydrodynamic waves, while the correlation of the field and soft X-ray emission from coronae evidenced that coronae were heated by Alfven waves.

An important advance in studying the heating mechanisms of stellar atmospheres was achieved owing to the concept of differential emission measure (DEM) that enabled the first calculations of the whole energy in the atmospheric layers, where optically thin radiators prevailed. The models elaborated within this concept by Jordan et al. (1987) made it possible to calculate radiative losses at different levels, the heat conductivity from corona to underlying layers, required strength and height distribution of heating sources. It was shown that the distribution of the emission measure between 2×10^4 and 10^5 K could be explained if the nonthermal broadening of the lines was due to the energy release caused by the passage of Alfven or slow magnetohydrodynamic waves equilibrating the local radiative losses. But the outflow of matter, the diffusion and geometry of the corona make the situation somewhat ambiguous.

Continuing the studies of Basri and Linsky (1979) and Mewe et al. (1981), Schrijver (1987) found the minimum level of surface fluxes in the main emission lines of the chromosphere and the transition region that was essentially dependent on the effective stellar temperature, the so-called basal level of chromospheric activity that has no correlation with the X-ray luminosity and is observed in old and slowly rotating stars and in the centers of solar supergranules far from noticeable magnetic fields. This level was associated with the acoustic heating whose presence determined the weak chromospheric emission in low-activity stars.

To check if the acoustic waves were sufficient to maintain the basal chromospheric activity level of M dwarfs, Mullan and Cheng (1993) thoroughly studied the plane-parallel atmospheric model with effective temperatures of 3000 and 4000 K. They found that the equilibrium state of stellar atmospheres could indeed be realized at equal dissipation of weak shock waves emerging under the upward propagation of acoustic waves emitted by subphotospheric convection, and the radiative losses of the atmosphere in the MgII and Ly_α lines. Later, Mullan et al. (1995) showed that in the sample of over 80 nearby red dwarfs practically all dM stars had surface X-ray fluxes of less than $10^{5.2}$ erg/cm^2 s, and practically all dMe stars had greater fluxes, while the maximum fluxes on acoustic heating of the coronae, according to the calculations of Mullan and Cheng (1994), were $10^{5.0-5.1}$ erg/cm^2 s.

Mathioudakis and Doyle (1992) considered the surface fluxes in the MgII lines of about 160 K–M dwarfs and found that the strong dependence of the basal level F_{MgII}^{min} on the effective stellar temperature confirmed its acoustic nature.

Using the detailed calculations within the model of acoustic heating, Buchholz et al. (1998) presented the basal levels of calcium and magnesium emission of F0–M0 dwarfs within a range of two orders of magnitude to the accuracy of the factor of 2.

The excessive emission above the basal level correlates with rotation and is of magnetic origin, but it is not clear if the additional heating occurs under the dissipation of turbulence or magnetohydrodynamic waves, or the fast annihilation of magnetic fields generates the fluxes of accelerated particles, which cause the further heating of the ambient plasma.

Musielak et al. (1990) calculated the energy fluxes generated as transverse waves of magnetic tubes coming out of the convective zone and found that the fluxes were sufficient to maintain the observed X-ray emission of F–G–K dwarfs.

Schrijver and Aschwanden (2002) considered the heating of solar and stellar coronae by an ensemble of a great number of coronal loops. They found an expression for the energy flux spent for corona heating as a function of the magnetic-field strength at the loop base, its extension, and the velocity of gas motion at the base, and satisfactorily presented the observed X-ray emission of active stars with rotational periods of more than 5 days.

For the most active fast rotators, the very young stars and the components of close binary systems, Vilhu and Rucinski (1983) found the upper limit of surface fluxes in the transition region lines, the saturation level. Then Vilhu et al. (1986) and Vilhu (1987) extrapolated the saturation concept to the maximum surface fluxes in the MgII and CIV lines and in soft X-rays. They advanced an hypothesis that maximum fluxes corresponded to the saturation due to the fact that the filling factor of magnetic structures on the stellar surface reached values close to unity. In this case, the reverse relation of the magnetic field to convection and differential rotation is switched on, which restrains further growth of nonradiative energy and thus prevents further nonradiative heating of the stellar atmosphere. On the whole, the saturated fluxes in the X-ray and in the spectral lines formed in the chromosphere and the transition region are only 10 times lower than the estimates of the mechanical energy flux in the subphotospheric convective zone. Thus, the saturated fluxes can indicate the maximum efficiency of the conversion of the mechanical energy into the heating of outer layers of stellar atmospheres. Since the heating is maximum where the magnetic fields are the strongest, the saturation is realized when the star is completely covered by the strongest magnetic fields, which occurs when the magnetic and gas pressures in the photosphere are equal. But there are other points of view on the saturation nature. Thus, Doyle (1996b) concluded that this effect in the ultraviolet could be caused by the disregard of the radiative losses by hydrogen, which should be great on rapidly rotating dwarfs. Jardine and Unruh (1999) suggested that the saturation of X-ray emission of fast rotators was due to the effect of centrifugal forces, which at the level of corotation essentially disturbed the corona and initiated the above-mentioned large prominences retained by the magnetic tension at loop tops.

Summing up the results of the Conference on Mechanisms of Chromospheric and Coronal Heating, Linsky (1991) noted that the problem of heating of quiescent stellar chromospheres could be directly related to the conclusion of solar studies: the quiet solar chromosphere can be heated by the dissipation of the energy of magnetohydrodynamic shock waves in the so-called bright points related to the elements of the chromospheric network. Such points are observed only in the places of localization of magnetic structures, their lifetime is 100–200 s, magnetic fields in them are of 10–20 G, and they repeatedly occur in the same magnetic elements. Only two mechanisms of Linsky's extensive survey on heating mechanisms for stellar atmospheres can be applied to the objects considered in this book: short-period acoustic waves that guarantee a weak basal level of the atmospheres of late stars and slow magnetohydrodynamic waves propagating from the convective zone to the chromosphere, heating the lower and middle chromosphere under shock-wave dissipation, and easily excited under the interaction of other waves.

Over a number of years, the insufficiency of the acoustic heating of the atmospheres of late stars was emphasized, and M stars occupied the region in the Hertzsprung–Russell diagram, where the insufficiency was the most evident. (After the revision by Bohn (1984) the theory of acoustic heating became much more efficient, but he apparently disregarded the weakening of the acoustic wave flux in the photosphere. Thus, it is not clear which portion of the large flux of 10^6–10^7 erg/cm^2 s ascertained by him can be spent for the heating of the chromosphere and corona.) However, it should be noted that the predictions of the acoustic theory were always compared with the most active dMe stars. But the acoustic heating can also play an important role in the atmospheres of dM stars, where the flux from the chromosphere is only 10^4–10^5 erg/cm^2 s (Giampapa et al., 1989).

Apparently, the consideration of acoustic and magnetic heating as mutually exclusive mechanisms was, to a certain extent, exhausted by Cuntz et al. (1999), who calculated the chromospheric models of single K0–K3 dwarfs with rotational periods from 10 to 40 days, in which the surface magnetic component and the nonmagnetic component were heated by the waves in the longitudinal flux tubes and by acoustic waves, respectively. They found that the heating and maximum chromospheric emission were largest in rapidly rotating stars, while the stars with very slow rotation had the basal level of chromospheric emission. This can be illustrated by the example of the G8 V star τ Cet (Rammacher and Cuntz, 2003).

Fawzy et al. (2002) considered the Ca and Mg emission in F5-M0 dwarfs in a wide range of activity levels and concluded that at the chromosphere bases acoustic heating prevailed, while in the middle and upper chromospheres, heating by magnetohydrodynamic waves dominated, and in the upper chromosphere and the overlying transition region and corona, nonwave heating, probably reconnection of magnetic flux tubes, dominated.

Thus, chromospheric emission on the whole can contain three different components: the basal level due to the acoustic heating and the emission caused by the magnetic heating owing to large-scale and/or turbulent magnetic fields. The third component is effective in low-mass stars, without regular fields of the solar type (Durney et al., 1993).

Different variants of atmospheric heating of atmospheres by flares were discussed repeatedly, those by an ensemble of individually recorded events, and numerous microflares inaccessible for such a registration with characteristic energies of 10^{30} erg (Butler et al., 1986) and nanoflares with energies of the order of 10^{24} erg (Parker, 1988). Of course, heating by flares is not an alternative to magnetic heating, but one of its realizations.

As stated above, the concept of coronal heating by flares was advanced after the discovery of the linear correlation of X-ray luminosity of flare stars with the time-averaged optical luminosity of flares (see (19)). At the same time, numerous short-life bursts of hard X-rays were found on the Sun and the possibility of heating of the solar corona by the bursts was actively discussed. Therefore, when Butler and Rodonò (1985) found the fine structure in the EXOSAT records of soft X-ray emission of UV Cet, EQ Peg, and Proxima Cen out of bursts with characteristic times of about 20 s and $L_X \sim 2 \times 10^{30}$ erg, and when temporal correlation of X-ray bursts and bursts in the H_γ line was found during parallel optical and X-ray observations of UV Cet, they concluded that these were microflares and that in X-rays they observed were not stellar coronae heated by microflares but microflares themselves. (It should be noted that since the energy of bursts is about 2×10^{30} erg, they most probably correspond to compact solar flares rather than solar microflares with a typical energy of 10^{27} erg or Parker's nanoflares with an energy of 10^{24} erg.) Then, Butler et al. (1988) found the linear correlation of total energies E_{H_α} and E_X, common for stellar and solar flares and covering a range of four orders of magnitude. Butler (1992) found that solar flares recorded in the range of 8–12 Å also satisfied this correlation, while the flare on the RS CVn-type star II Peg continued it for two orders of magnitude toward higher energies. Later, the arguments both in favor and against the substantial role of microflares in the energy of stellar atmospheres were obtained. Thus, Butler et al. (1986) believed that the established variations of F_X at times of 100–1000 s in newly processed observations of dKe and dMe obtained at the Einstein Observatory supported the concept. Mathioudakis and Doyle (1990) continued the correlation of E_X and E_{H_α} of flares to the quiescent state of active M dwarfs. Considering the results of studying the quiescent-state spectrum of EV Lac that proved a considerable contribution of microflares to the radiation of Balmer lines out of flares (Alekseev et al., 2003), the conclusion of Mathioudakis and Doyle seems to be a natural continuation of the conclusion of Butler et al. (1988). Young et al. (1989) found a rather close correlation between the luminosity of the H_α emission and quiet X-ray luminosity of the coronae: $L_{H_\alpha} \sim 0.2 L_X$, which may suggest the heating of the entire upper atmosphere from a single source or the coronal heating of the chromosphere. But, Haisch (1989) noticed that the X-ray emission in a very broad range of intensities was recorded in solar flares of the same H_α importance. Tsikoudi and Kellett (1997) detected slow and small-amplitude variations of the brightness of active stars in the EUV range and concluded that the processes, which they called milliflares, were important for coronal heating. But the detailed analysis of the EXOSAT X-ray observations carried out by Collura et al. (1988) and then more thoroughly by Pallavicini et al. (1990a) did not confirm the hypothesis on the decisive contribution of microflares. Besides, Schmitt (1993) did not find the expected effects of microflares in ROSAT observations.

When the power energy spectra of flares are extrapolated to the region of supposed microflares with the same spectral index, which is determined from individually recorded optical and X-ray flares and is usually equal to 0.7–0.9, the contribution of microflares to the total flare radiation appears to be low. However, even for the Sun there are observational and theoretical evidences that in the microflare range the spectrum of frequency energy distribution should be softer than within the range of stronger events. Therefore, even on the Sun the contribution of microflares should be higher than in the estimate with a constant spectral index for the whole range of flare energies. Finally, recently data confirming the significant role of microflares on active late stars were obtained. Based on the EUVE observations of 28 flares on active solar-type stars, for 47 Cas and EK Dra $\beta \sim 1.2 \pm 0.2$ within the energy range of 3×10^{33}–6×10^{34} erg (Audard et al., 1999). Considering EUVE data for F–M stars, Audard et al. (2000) concluded that for flares on F–G dwarfs $\beta > 1$, as for K–M stars it is more probable that $\beta < 1$. But Kashyap et al. (2002) analyzed the EUVE/DS observations of FK Aqr, V 1054 Oph, and AD Leo and found that for these late dwarfs $\beta > 1$. At $\beta > 1$ $E_{\min} = 10^{29}$–10^{31} erg is sufficient for the time-averaged luminosity of flares to be comparable with the total luminosity of a quiescent corona, and such $E_{\min}$ are typical of flares on the Sun.

This conclusion was later confirmed by Güdel et al. (2003b). From 37-day EUVE monitoring of AD Leo and overlapping BeppoSAX monitoring they found that $1.0 < \beta < 1.4$ for EUV and X-ray flares of this star. These results were obtained by two independent statistical methods and under independent consideration of EUVE, LECS, and MECS BeppoSAX data. Simultaneously, they developed the calculation method for the expected DEM for the stochastic flare ensemble, whose success provided another argument in favor of the entire statistical model of flare activity. Later, Arzner and Güdel (2004) developed a refined analytical approach to the analysis of light curves based on flare superposition that could be responsible for coronal heating.

In some solar-physics studies, flares, microflares, and nanoflares are considered as the processes of reconnection of magnetic flux tubes in the current sheets, and discussions are mainly focused on whether these processes can be responsible for the quiet radiation of the solar corona (Parker, 1988; Oreshina and Somov, 1997). However, it is noteworthy that if the calculations of energy release in such processes seem to be rather plausible, the mechanism of formation of the structures in turbulent plasma seems rather problematic because their extension exceeds their thickness by several orders of magnitude. One may think that at this point a transition is needed from an idealized continuous medium of classical magnetohydrodinamics to the fractal concept.

Studying the active corona of the F8 dwarf HD 35850 by EUVE and ASCA, Gagné et al. (1999) concluded that the variability of its EUV radiation could be presented by the model of continuous flares, having integral energy spectrum with the spectral index 0.8 and capable of heating the corona.

If corona conductivity contributes to the heating of a quiescent chromosphere, during flares coronal heating occurs through the evaporation of the chromosphere initiated by energy particle beams from the corona. This conclusion was made by Güdel et al. (1996) on the basis of simultaneous X-ray and microwave observations of UV Cet and confirmed by the analysis of the flare on Proxima Cen of 12 August 2001 cited below (Güdel et al., 2002).

1.5 Heating Mechanisms for Stellar Atmospheres

Discussing permanent microwave emission of stellar coronae, Kundu et al. (1987) noticed that fast particles emerging in flare processes were responsible for the radiation and should efficiently precipitate, and this could be an important mechanism of coronal heating for the chromosphere.

Güdel (1997) considered the $EM(T)$ distributions in the solar-type stars and showed that the observed two-peak distribution could be due to the sum of two components: the hot component stipulated by the cooling plasma of strong flares in large loops and the cold component that is due to the large number of microflares.

One of the explanations of the revealed close correlation of the thermal X-ray and gyrosynchrotron microwave emissions of stellar coronae and solar flares (Benz and Güdel, 1994) consists in the common origin of the corona heating and particle acceleration. Already in 1985, Holman (1986) formulated a hypothesis that nonthermal electrons necessary for the gyrosynchrotron emission could appear in the current sheets, in which the Joule losses are the heating mechanism of stellar coronae. Airapetian and Holman (1998) developed this idea: they considered two mechanisms related to electrical currents, within which the Joule heating of plasma to millions of K and the acceleration of electrons to subrelativistic energies occurred simultaneously. In the first model, electrons are accelerated by electrical currents within classical current sheets in stellar coronae. The second model deals with MHD turbulence: in small-scale regions in the presence of current sheets it excites ion sound waves, which intensify the heating and accelerate particles in the transition regions of stellar atmospheres. In the first model, one should expect a linear correlation of L_X and L_R, in the second, a power correlation with the exponent of 0.6–0.8. Recently, Podlazov and Osokin (2002) concluded that the coronal heating and generation of fast particles occurred simultaneously within the "avalanche" concept.

Let us consider the results of the studies dealing with particular aspects of the general problem of heating of stellar atmospheres.

The analysis of several hundreds of ultraviolet spectra in the region of the CII line 1335 Å showed that strong emission of the high-temperature line was common for early F dwarfs and its intensity was comparable with the value recorded in the most active solar-type stars (Simon and Landsman, 1991). The emission maximum is achieved near F0V and decreases toward late A dwarfs. For stars earlier than F5, the CII emission does not correlate with the rotational rate, which confirms that the chromospheres of A–F stars are heated under shock dissipation of sound waves.

Garcia López et al. (1993) studied the He D_3 line in the spectra of 145 F–G stars and found that near the spectral class F0 the depth of the convective shell was sufficient to maintain the chromosphere by acoustic heating; the lack of a dependence on rotation and the Rossby number in such stars evidences the absence of magnetic heating in them, which appear only in F5 stars.

Wood et al. (1997) analyzed the SiIV and CIV line profiles obtained from HST in the spectra of stars of late spectral types and presented these profiles by the sum of two Gaussians with substantially different widths. They made the following conclusions on the heating mechanisms of the transition region. The widths of narrow components correspond to subsonic thermal velocities and can be caused by the dissipation of turbulent motions or the Alfven waves, whereas the wide components corresponding to supersonic motions can arise at microflares. The wide component of the profile of the line SiIV λ 1394 Å with FWHM = 109 km/s was observed in the spectrum of

α Cen A, which is practically identical to the Sun and in which analogous wide components are seen in the spectra of the so-called explosive phenomena. But on the star this wide component contains up to 25% of the total energy of the line, whereas on the Sun, it contains not more than 5%. The connection of wide components of the profiles and microflares is supported by high velocities of plasma eruptions from the regions of annihilation of magnetic fields, as well as the established correlation of the portion of energy in this component of the energy of the whole line profile with the level of activity of the star and increased temperatures of coronae of the most active stars.

Saar et al. (1994c) found that in the spectrum of AD Leo from 10 to 40% of the emission in the SiIV lines were caused by low-amplitude flares, while Saar and Bookbinder (1998a) found that for fast rotators G0V EK Dra and K2V LQ Hya the radiation of the CIV and SiIV emissions was approximately for 10% determined by such flares. In the appropriate relationships, according to Saar and Bookbinder, the transition region is heated by microflares and permanent magnetohydrodynamic waves. Ayres (1999) obtained similar results from the observations of three G dwarfs. But the width of the coronal line FeXXI λ 1354 Å in the spectrum of AD Leo was rather narrow, which points to the low contribution of shock waves to the corona heating (Pagano et al., 2000).

Studying the ultraviolet spectra of ε Eri obtained from STIS HST and the energy balance in the upper atmosphere, Sim and Jordan (2002) concluded that the surface filling factor was $\sim$0.2 in the transition region and approached unity in the corona. Then, from STIS HST and FUSE data they considered the heating of the upper atmosphere of ε Eri (Sim and Jordan, 2003). They measured the widths of more than 50 emission lines in the range of 1030–2300 Å, calculated for each of them the width of the nonthermal component and found that the value grew from 7.5 km/s at a level of the formation temperature of 5000 K to 21.3 km/s at a level of 30 000 K and then remained practically constant at this level up to 14 MK. The profiles of strong lines CIV, SiIII, SiIV, NV, and OIV were presented by two Gaussians. It appeared that with increasing temperature the contribution of the wide component systematically decreased. Sim and Jordan concluded that the narrow component was stipulated by Alfven waves reaching the corona, but tending toward the idea of microflares by Wood et al. (1997) they left the question of the nature of the wide component open. An essential contribution of wide components to Balmer line profiles was found for EV Lac by Alekseev et al. (2003).

2
Flares

Short-term flares are the most easily accessible to observational manifestations of the activity of red dwarfs. Being the last link in the chain of magnetohydrodynamic processes occurring on the variable stars under consideration, flares are a challenge for those who examine their mechanism to reconstruct the whole chain of magnetohydrodynamic processes.

The general problem of studying flare activity of UV Cet-type stars can be split into two particular problems: elaboration of a physical model of an individual flare and presentation of an ensemble of flares within a certain statistical model. The first problem should be solved by analyzing the relatively few flares, for which the most detailed observations are available. The second problem can be solved using as long and uniform an observational series as possible. This approach determined the structure of this chapter. In the first section, we describe several strong flares to provide a general idea of the phenomena of stellar nonstationarity. In the two next sections, we consider the principal statistical properties of flare activity, temporal characteristics of flares and their energetics. Then we survey and analyze the observational data underlying the physical modeling of flares and finally present simulation results.

Before proceeding to the description of stellar flares let us recall the principal properties of solar flares.

Solar flares are one of the most impressive phenomena of the solar activity. They were discovered in the mid-19th century but have been studied intensely only since the 1940s. In visual observations, a solar flare is a rapid increase in the brightness of small regions, a few tenths of a per cent of the solar disk, near groups of solar spots. If observations are carried out through a specially selected color filter picking out the radiation of the red H_α hydrogen line, the contrast of flares against the background of the quiet solar surface multiplies and the frequency of recorded flares consequently increases. The fast burning is usually followed by a phase of maximum brightness of comparable duration, which gradually turns into the decay phase, which is several times longer.

Most solar flares develop near spots, but for about 10% of flares the relation to spots is not evident, and all flares occur in the facular areas.

Visual observations and filming reveal the extraordinary diversity of solar flares: visible patterns of the development of such flares are unique and their temporal and energetic characteristics are very diverse. Thus, a typical flare lasts one hour, but the fastest manage to burst and die out in several minutes, while the strongest flares last for many hours. The total energy of optical emission of the weakest and strongest flares recorded on the Sun is 10^{26}–10^{27} erg and 10^{30}–10^{31} erg, respectively.

During solar flares, the heated middle layers of the atmosphere, the chromosphere, intensely radiates at optical wavelengths. Therefore, as long as flares were studied only within the optical range, they were called chromospheric flares. But radio-astronomical and X-ray observations showed that optical flares accounted only for the secondary effects of this complicated phenomenon. It was established that solar flares are caused by magnetohydrodynamic and plasma processes related to strong magnetic fields of solar spots. Spots of opposite polarity are coupled in the solar atmosphere by magnetic flux tubes, and the beginning of a flare is usually recorded as an appearance of a large number of charged high-energy particles in the bases of the arches formed by flux tubes. The particles are found from the strong burst of nonthermal X-ray emission. Moving along the flux tubes, these particles heat plasma confined in the tubes to high temperatures and induce a longer thermal glow of flares in the X-ray range. The excitation of deeper and denser layers by these energetic particles leads to optical glow of the lower chromosphere. In some cases, the particles in the flares have such a high energy that nuclear reactions occur, which is evidenced by the recorded gamma quantum spectral line corresponding to the annihilation of electron–positron pairs.

The high-energy particles arising at the beginning of a solar flare also induce the distortions that propagate upward in the solar corona. These distortions are manifested as nonthermal radiation in a wide wavelength range from centimeters and decimeters to tens of meters. It is noteworthy that there are several types of radiation that differ in the generation mechanism, place, time, polarization and frequency characteristics, and their time changes.

Different motions of matter occur during solar flares. Ten minutes before the beginning of many flares, prominences come into motion high above the active region; these are odd filament structures of relatively cold plasma "suspended" in the hot corona. These motions of the prominences are one of the most reliable precursors of flares. Movies of flares made in the light of the red hydrogen line show rapid motions of gas in the lower chromosphere. But the most important are the eruptions of matter apparently associated with the flares, they can be traced from the photospheric level to interplanetary space.

In the strongest solar flares, the so-called white light flares, when the flare continuum is seen in the optical range, the distortions expand to the middle photosphere. Thus, during solar flares the short-term processes involve the vast space from the visible solar surface to the orbit of the nearest planets. The contribution of radio emission to the total flare energy is minimum, the total radiation energies in the optical and X-ray ranges are of comparable order of magnitude, but the kinetic energy of matter involved in flares exceeds the electromagnetic radiative energy by an order of magnitude.

Solar physics in general, and the physics of solar flares in particular, are the major impetus for the development of cosmic electrodynamics and magnetohydrodynamics. Two types of stable magnetic topological structures with a limited energy reserve in a limited volume were discovered experimentally and theoretically on the Sun: magnetic flux tubes and arcades. Solar flares of two types – compact and two-ribbon – are associated with these structures. Compact flares occur in one or two loops with fast energy release at which the plasma temperature achieves 10^7 K and nonthermal particles emerge. Two-ribbon flares occur in active regions with dark filaments, which evidence a definite structure of the magnetic field. It is believed that in this case a neutral layer is formed, and the motion of the filament provokes the reconnection of the magnetic flux tubes. Strong electric fields are induced and the arcades of loops are formed, whose behavior is analogous to the behavior of loops in compact flares. The energy of two-ribbon flares exceeds that of compact flares by a factor of 10–1000.

The dominant point of view on the immediate energy sources of solar flares supposes that the sources are nonpotential magnetic fields arranged in loop structures, whose bases are in the photospheric regions of different polarity and whose tops are in the corona.

Zirin and Ferland (1980) thoroughly compared various characteristics of solar and stellar flares. Two aspects should be considered in comparing these structures.

First, we cannot see stellar disks and, consequently, individual structures on the surface of any star, as on the Sun. Therefore there are only few indirect data on the correlation of stellar flares with active regions. Thus, Busko and Torres (1978) noted a weak correlation of the visibility of spots with the flare frequency, but from the observations of EV Lac and BY Dra in the B band Mavridis and Avgoloupis (1993) suspected that the season of high flare activity is followed by the high-spottedness season, and Contadakis (1995) found the alternation of 1–2-month intervals during which either the flare activity and the rotational modulation of the brightness of EV Lac intensified in turn, or both manifestations of stellar activity disappeared or substantially weakened. Later Contadakis (1997) suspected the situation on BY Dra and CE Boo was similar. Leto et al. (1997) analyzed the flare activity of EV Lac in 1967–77 and found that in 1970 flares were concentrated at a certain longitude, in the following year this longitude was the place of maximum spottedness. Doyle (1987) found that in 1973–76 there were definitely more flares on one hemisphere of EV Lac than on the other. Mavridis et al. (1995) found that flares on BY Dra occurred more often within a certain range of axial rotational phases. Doyle and Mathioudakis (1990) and then Butler et al. (1994) concluded that in the YY Gem system the best visibility of spots was in the quadrature phase and flares were drawn to this phase.

Secondly, even on Proxima Cen, the star nearest to the Sun, the number of optical quanta per unit time is 10^{15} times lower than the appropriate number of solar quanta. This strongly restricts all experimental stellar studies.

These two circumstances make practically impossible the application of the same technique and the same tools to the studies of the Sun and flare stars. Therefore, the determination of the substantial effects of the observational selection should be taken into account in comparative studies.

2.1 General Description of Stellar Flares

During a strong flare, the optical luminosity of a star increases by several fold over tens of seconds. The character of stellar radiation changes markedly. Continuous emission floods the absorption details of the short-wavelength region of the optical spectrum. The intensity of emission lines of the chromospheres and the transition region from the chromosphere to the corona observed in the optical and ultraviolet parts of the spectrum sharply increases. Many emission lines observed in the optical range broaden noticeably. The intensity of thermal coronal radiation increases, so does the temperature of the coronal plasma. Over a wide frequency range, nonthermal radiation emerges and propagates in the corona. Flares on red dwarfs are distinguished by extreme diversity: the light curves of the flares are unique, even those occurring on the same star successively during one night (see Fig. 21). The burning time of a flare varies from fractions of a second to several minutes, the time of decay varies from seconds to many hours. Absolute luminosities in the optical range of recorded flares range from 10^{26} to 10^{32} erg/s, as the total energy of the optical emission of flares cover at least the range of 10^{28}–10^{34} erg. Flares differ substantially in the relative intensities of optical, radio and X-ray emissions, in the relative contribution to the optical emission of continuous and line emission, in the broadening and symmetricity of emission lines, in the temperature and emission measure of hot coronal plasma, in the temporal sequence of the occurrence of flares in the optical and radio ranges, and in the frequency and polarization characteristics of radiation. Such a diversity of recorded flares, the relatively short time of investigation of the phenomena, which can be insufficient to detect the strongest and rarest events, and the inaccessibility of the weakest flares for observations due to the permanent radiation of flare stars – all this makes describing a typical flare a rather ambiguous task. Therefore, instead of artificial construction of a certain average flare let us consider several real most studied flares.

The flare on AD Leo of 18 May 1965 was recorded by Gershberg and Chugainov (1966) in the Crimea during simultaneous photometric and spectral observations at the 64-cm meniscus telescope and at the 2.6-m Shajn

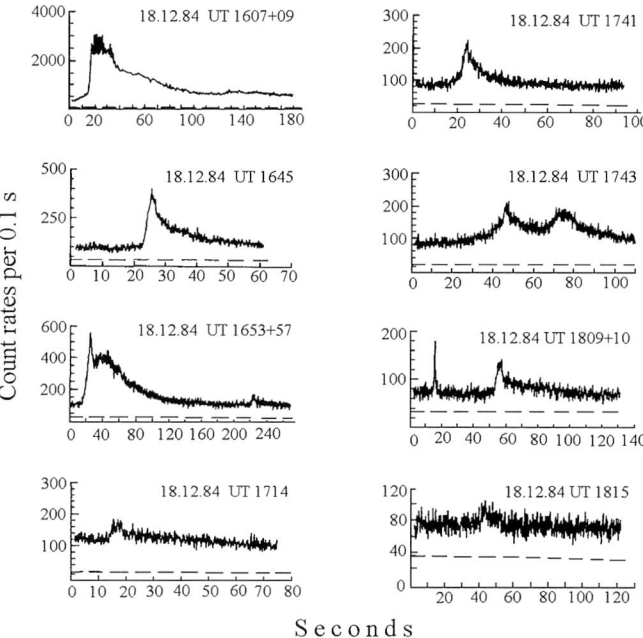

Fig. 21. Light curves of the UV Cet flares recorded by the 6-m telescope of the Special Astrophysical Observatory, Caucasus on 18 December 1984 during a 2.5-h photoelectric monitoring in the U band (Beskin et al., 1988)

reflector, respectively. Figure 22 shows the light curve of the flare and indicates the time intervals of spectral exposures. At the phase of flaring up, the brightness of the star in the blue band increased by more than 5 times, approximately over a minute. At the maximum brightness of about 20 s the luminosity of the flare in the B band was about 3.3×10^{30} erg/s. Several minutes after the main maximum a fast decrease was followed by a slower decline. Then 12 min after the main flare maximum a secondary, smoother and lower-amplitude brightness peak occurred. Figure 22 shows the changes of the spectrum of AD Leo during the flare. The first spectrogram of the flare was obtained in the photographic region of 18 s during a single run of the star along the spectrograph slit. One can see the moment of the flare maximum brightness in the spectrum. On the whole, the spectrum of AD Leo in the blue region at this time changed beyond recognition: the absorption of CaI λ 4227 Å became indistinguishable from the fluctuations of the continuous spectrum intensity, all hydrogen emission lines strengthened abruptly, the half-widths of the H_γ and H_δ emission lines grew from 5–6 Å to 10–11 Å, the emission line of neutral helium λ 4471 Å appeared, which is invisible in the spectra of quiet stars. In the second spectrogram, further strengthening

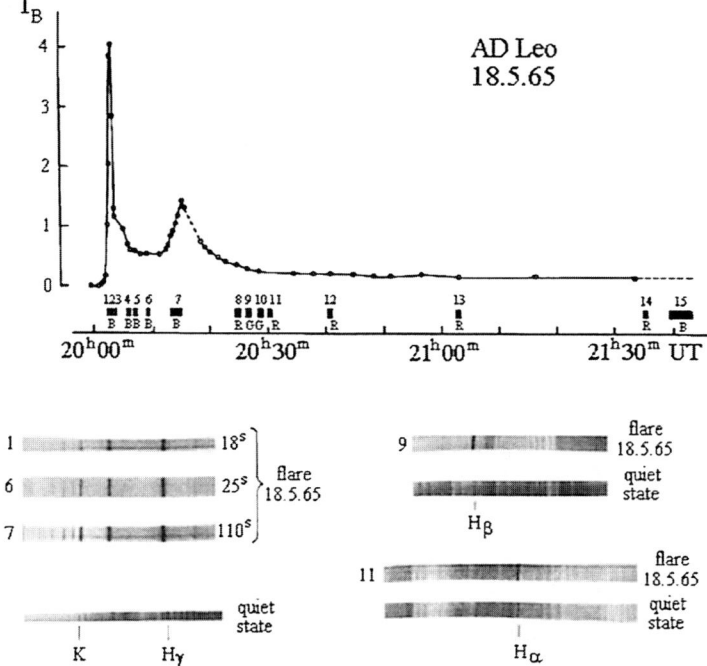

Fig. 22. Flare on AD Leo of 18 May 1965. *Top*: the light curve in the B band. Numbered rectangles mark the time intervals when the spectrograms were obtained in the blue (B), red (R), and green (G) regions of the spectrum. *Bottom*: spectra of the flare, numbers on the left of the spectra correspond to spectrogram number (Gershberg and Chugainov, 1966)

of the hydrogen and calcium lines was found and quite measurable helium lines of λ 4026 Å and λ 4471 Å were seen; the absorption of CaI λ 4227 Å was also rather low but reliably recordable. In the third spectrum, the hydrogen lines strengthened, the absorption of CaI λ 4227 Å increased, the helium line λ 4026 Å disappeared, and the line λ 4471 Å was difficult to distinguish. The fourth and fifth spectrograms were rather similar, the equivalent widths of the hydrogen lines achieved the maximum values of $W_{H_\gamma} = 65$ Å and $W_{H_\delta} = 50$ Å, the widths of the lines slightly decreased as compared to the first image. If we neglect the continued narrowing of hydrogen lines, the sixth spectrum would repeat the third one in the intensity of hydrogen and calcium lines, in the absorption of CaI λ 4227 Å, in the difficult-to-measure traces of the λ 4471 Å line. The seventh image was obtained in the blue region of the spectrum during a four-fold run of the star along the spectrograph slit and fell on the rise of the brightness to the secondary flare maximum. Here, the H_γ line was overexposed, though the emissions to H_{11} were measurable, the absorption depth of CaI λ 4227 Å slightly decreased and the λ 4471 Å line became more distinct. The equivalent width of the H_α line measured in spectra 8 and

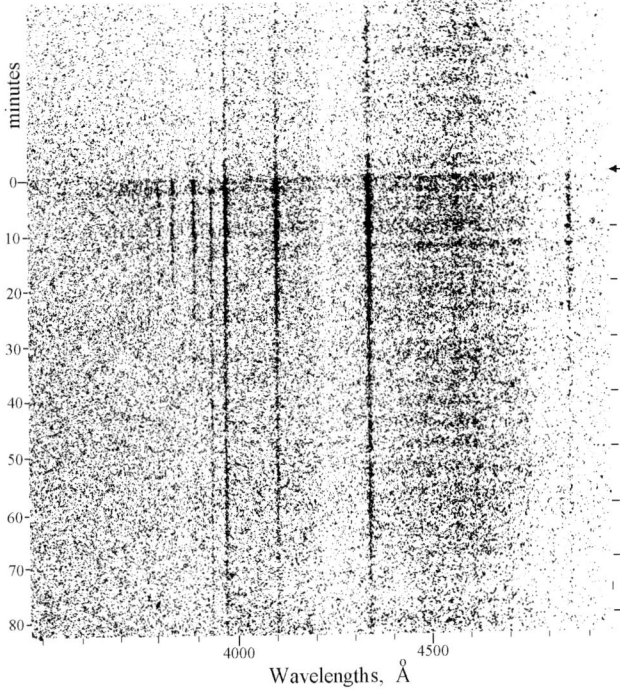

Fig. 23. Spectrogram of the flare on EV Lac of 11 December 1965, figures on the left show the time since the flare beginning (Kunkel, 1967)

11–14 varied from 20 to 14 Å at 8 Å in the quiet star, while W_{H_β} in spectra 9–10 was about 50 Å as compared to 7 Å out of the flare; the HeII λ 4686 Å line was not seen in images 9–10. The last spectrum 15 obtained in the blue region shows that even one and half hours after the brightness maximum the equivalent width of H_γ line noticeably exceeded the normal value.

The flare on EV Lac of 11 December 1965 (Kunkel, 1967) was observed by the 90-cm reflector of the McDonald Observatory in Texas. The moving cassette of the spectrograph made it possible to obtain a clear picture of the flare spectrum (see Fig. 23). From the very beginning of the flare the Balmer lines were sharply strengthened, while the continuous emission, well seen only during the first three minutes after the maximum brightness, almost completely flooded the absorption line of CaI λ 4227 Å and noticeably weakened the absorption details of the normal spectrum in the region of larger wavelengths. The Balmer lines were traced in the spectrogram to H_{11}, while from λ 3750 Å, where the emission lines merged due to low resolution, to λ 3500 Å, which was the sensitivity threshold of the equipment, and the level of continuous radiation was practically constant. The increased radiation was observed in the lines even one hour after the maximum brightness, the decay in the CaII K line was the slowest. From the spectrum of the

198   2 Flares

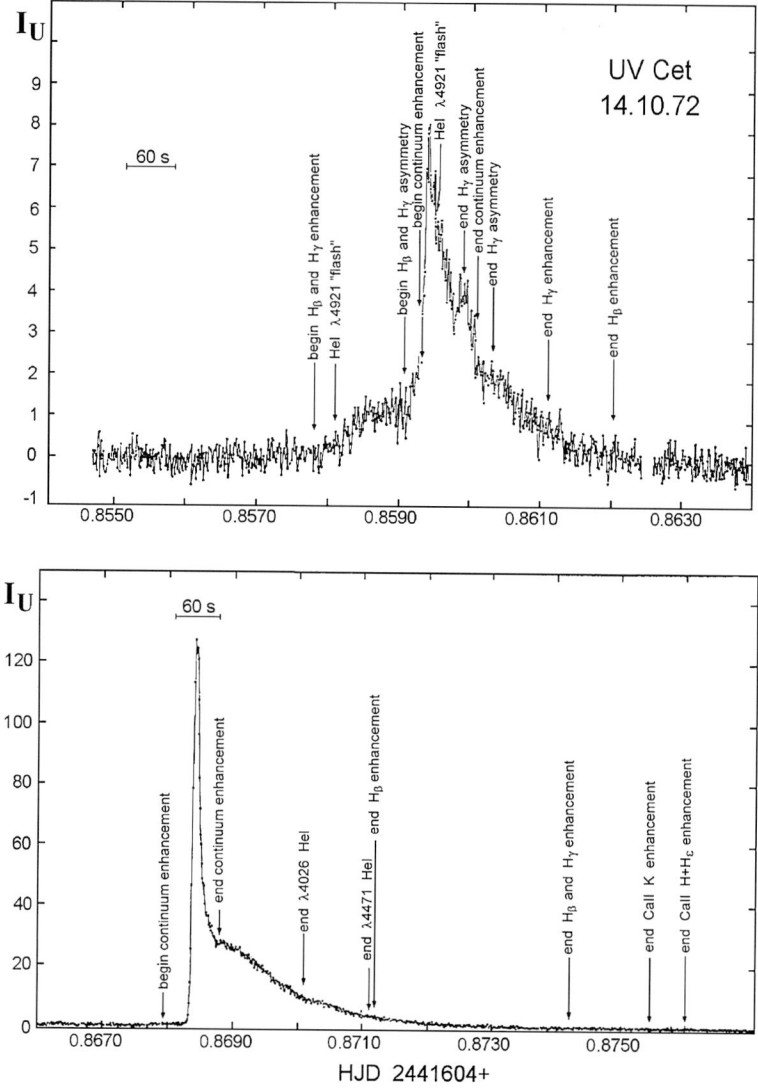

Fig. 24. Flares on UV Cet of 14 October 1972. The sequences of the development of spectral features of flares are specified on the light curves recorded in the U band (Bopp and Moffett, 1973)

flare, Kunkel obtained one of the most reliable photographic estimates of the emission Balmer jump in the intrinsic radiation of the flare and the Balmer decrement at different stages of the flare.

Two contiguous flares on UV Cet (Bopp and Moffett, 1973) were recorded on 14 October 1972 during simultaneous photometric and spectral observations with the 76-cm reflector and the 2.1-m Struve telescope at the

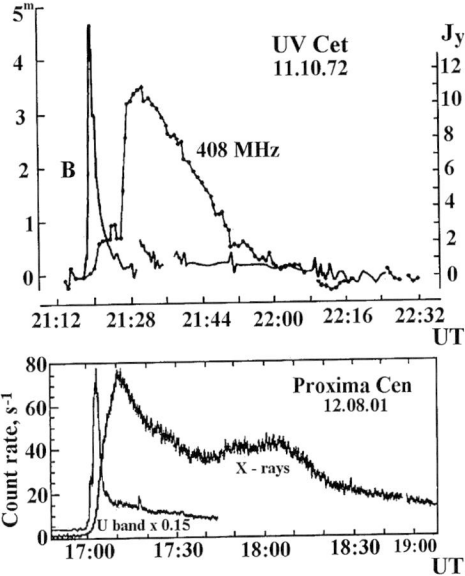

Fig. 25. Responses of stellar coronae to the disturbances resulting in impulsive optical flares: (a) flare on UV Cet of 11 October 1972 recorded in the B band and in the radio range (Lovell et al., 1974); (b) flare on Proxima Cen of 12 August 2001 recorded in the U band and X-rays (Güdel et al., 2002)

McDonald Observatory in Texas. As one can see from Fig. 24, the initial slow burning of the first flare was due to the strengthening of the line emission, then the continuum had a fast rise to the sharp brightness maximum, during which the luminosity of the flare in the U band achieved 2×10^{28} erg/s. The second flare was 15 times stronger and started from an abrupt rise of continuum. During the flares, detailed data on the time of appearance and vanishing of the flare spectral features were obtained. In the spectrum of the first flare the effect of asymmetry of the emission line profiles was localized on the light curve with the highest time resolution: the red wings of hydrogen lines were approximately twice as long as the appropriate blue wings; less explicit asymmetry was recorded also in the CaII K line. No asymmetry was observed in the spectrum of the second flare.

The flare on UV Cet of 11 October 1972 (Lovell et al., 1974) was observed simultaneously at 408 MHz by the Jodrell Bank radio telescope (England) and in the B band by the 76-cm reflector of the Stephanion Observatory at Peloponnese (Greece). Figure 25 shows the light curves of the flare in the different wavelength ranges. The radio flare started practically simultaneously with the optical flare, but the abrupt peak of radio emission occurred 8 min after the optical maximum, when the optical flare decayed. The radio flare faded almost 5 times slower but the total energy of optical emission exceeded the total energy of radio emission of the flare by several orders of magnitude.

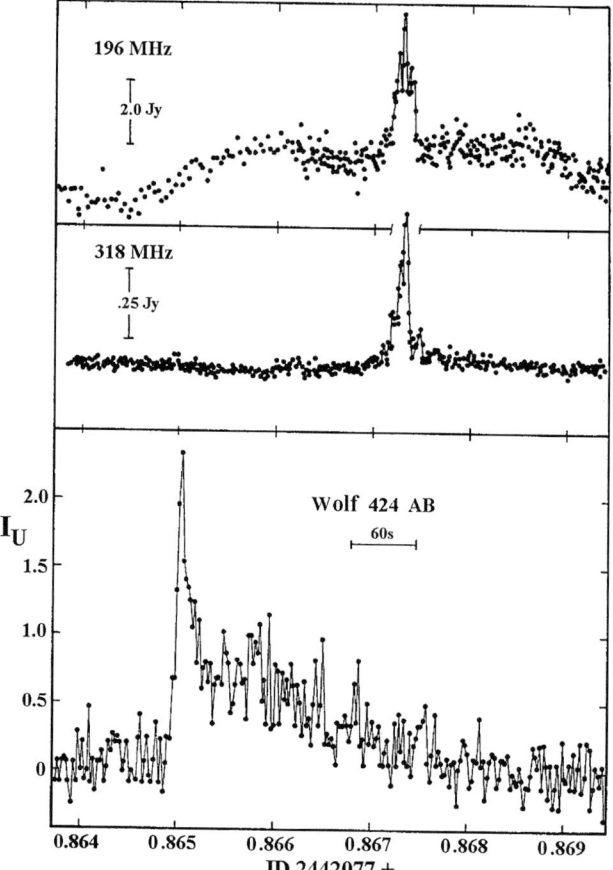

Fig. 26. Flare on Wolf 424 of 30 January 1974 recorded at two radio frequencies and in the U band (Spangler and Moffett, 1976)

The flare on Proxima Cen of 12 August 2001 (Güdel et al., 2002) was detected by XMM-Newton simultaneously in X-rays and the optical range. Figure 25 illustrates the similarity of the coronal response to the optical burst or to its cause in the radio and X-ray ranges.

The flare on Wolf 424 of 30 January 1974 (Spangler and Moffett, 1976) was recorded at 196 and 319 MHz by the radio telescope in Arecibo (Puerto Rico) and in the U band by the McDonald Observatory. Figure 26 shows the light curves of the flare for the three wavelength regions. One can see that practically simultaneously with the abrupt peak of the optical radiation, which at maximum achieved 1×10^{28} erg/s, a smooth rise of radio emission started at a frequency of 196 MHz, two minutes later a fast and short radiation peak occurred at both radio frequencies.

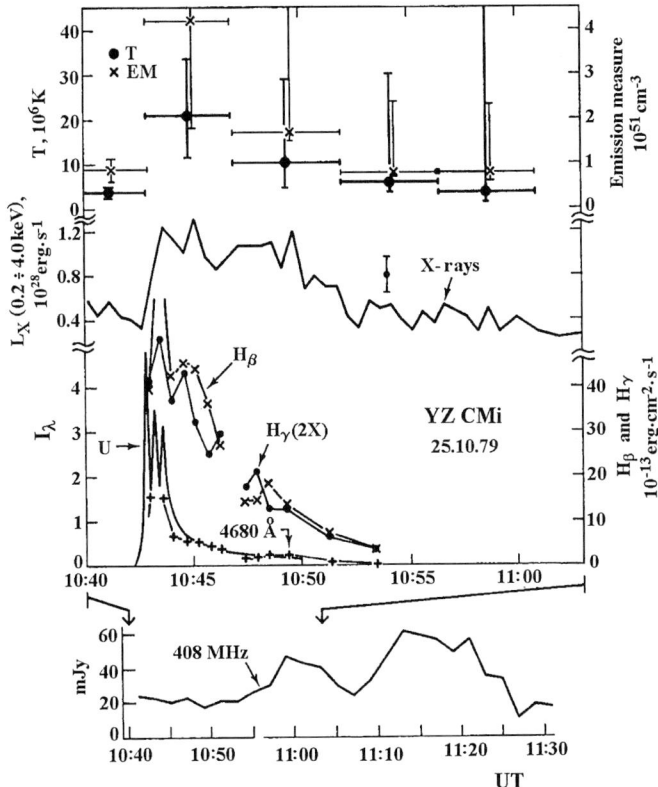

Fig. 27. Flare on YZ CMi of 25 October 1979. *Bottom*: light curves in the radio region, in optical continuum near λ 4680 Å, in the U band, in Balmer lines H_γ and H_β, and in soft X-rays; *Top*: the time changes of temperature and emission measure of coronal plasma (Kahler et al., 1982)

The flare on YZ CMi of 25 October 1979 (Kahler et al., 1982) was recorded in the course of cooperative program that involved the X-ray Einstein Observatory, seven optical telescopes with diameters of 61–230 cm, and seven radio telescopes with diameters ranging from 26 to 300 m. Figure 27 shows the light curves of the flare in different wavelength ranges and the temperatures and emission measures of coronal plasma calculated from the X-ray emission. Over 35 s the luminosity of the flare in the U band achieved a maximum of 2×10^{29} erg/s and after two additional short peaks faded relatively rapidly; a similar behavior was observed in the radiation of the continuum near λ 4680 Å. Radiation in H_β and H_γ lines and the X-ray emission achieved the maximum a little later than the optical continuous emission and weakened much more slowly. The maximum flare luminosity in soft X-rays (0.2–4 keV) was 8×10^{28} erg/s, in this case the temperature of the coronal plasma achieved 20 MK, and the emission measure was 4×10^{51} cm^{-3}. The total flare radiation in the soft X-ray region was close to the total radiation in the optical range. The radio emission at a frequency of 408 MHz was recorded only 17 min after the optical maximum. At higher radio frequencies the flare was not reliably detected.

The flare on Proxima Cen of 20 August 1980 (Haisch et al., 1983) was recorded in the course of coordinated extraterrestrial observations. The upper plot of Fig. 28 shows the light curve of the flare in the soft X-ray region obtained at the Einstein Observatory, and the time intervals of the spectrography in the ultraviolet region performed from the IUE satellite. In the lower plot the records of four spectra in the range of 1100–2000 Å are presented. The analysis of the data showed that the maximum luminosity of the flare in X-rays achieved 2×10^{28} erg/s; the total radiation in this range was about 4×10^{31} erg, the radiative losses for the line ultraviolet emission – to the Ly_α line and the transition region lines from the chromosphere to the corona – were lower by an order of magnitude than the X-ray emission of the flare. The temperature of the coronal plasma in the flare was 6–7 times higher than in quiet corona and reached a maximum value of 27 MK immediately before the maximum of X-ray luminosity of the flare.

The flare on AD Leo of 28 March 1984 (Rodonò et al., 1984) was observed within the cooperative program including four telescopes of the European Southern Observatory in Chile, the 91-cm reflector of the McDonald Observatory, the Very Large Antenna in New Mexico, and the IUE satellite. Figure 29 shows the light curves of the flare in different wavelength ranges. The optical light curve of the flare in many respects resembles that of 18 May 1965 on the same star (Fig. 22): both had fast main maxima, lower-amplitude secondary peaks, and the same characteristic times of all details of the light curves. The luminosity of the flare of 28 March 1984 in the U band achieved the maximum value of 1×10^{30} erg/s 30 s after the burning, the fluxes in the Balmer lines $H_\gamma - H_{11}$ increased approximately two-fold, fluxes in the CaII K line, by a factor of 1.4. The emission at 2 and 6 cm was recorded in the radio range, and the weakening of stellar brightness in the K band (2.2 μm) approximately by 3% was found between the main and secondary maxima of the optical flare. 15–25 min after the main flare maximum the ultraviolet observations showed that the MgII λ 2795 Å line was still enhanced, almost 2-fold, the blend of FeII λ 2600 Å, by 9-fold, and the continuum near λ 2500 Å, was 10 times stronger than the appropriate emissions of the quiet star.

The flares on AU Mic of 15 and 16 July 1992 (Cully et al., 1993) were discovered during four-day monitoring with the Extreme Ultraviolet Explorer (EUVE): in the 65–190 Å band with a time resolution of 100 s (Fig. 30).

Burning of the strong flare on 15 July 1992 lasted for an hour and a half, then the maximum luminosity in this band remained at the level of 10^{30} erg/s for about 2 h, the process of decay lasted for more than a day: first there was an exponential decay with a characteristic time of 1.3 h, then a slower decrease of brightness, and the whole decay phase lasted about one and a half days. The total radiative energy of the flare was 3×10^{34} erg and the volume emission

2.1 General Description of Stellar Flares

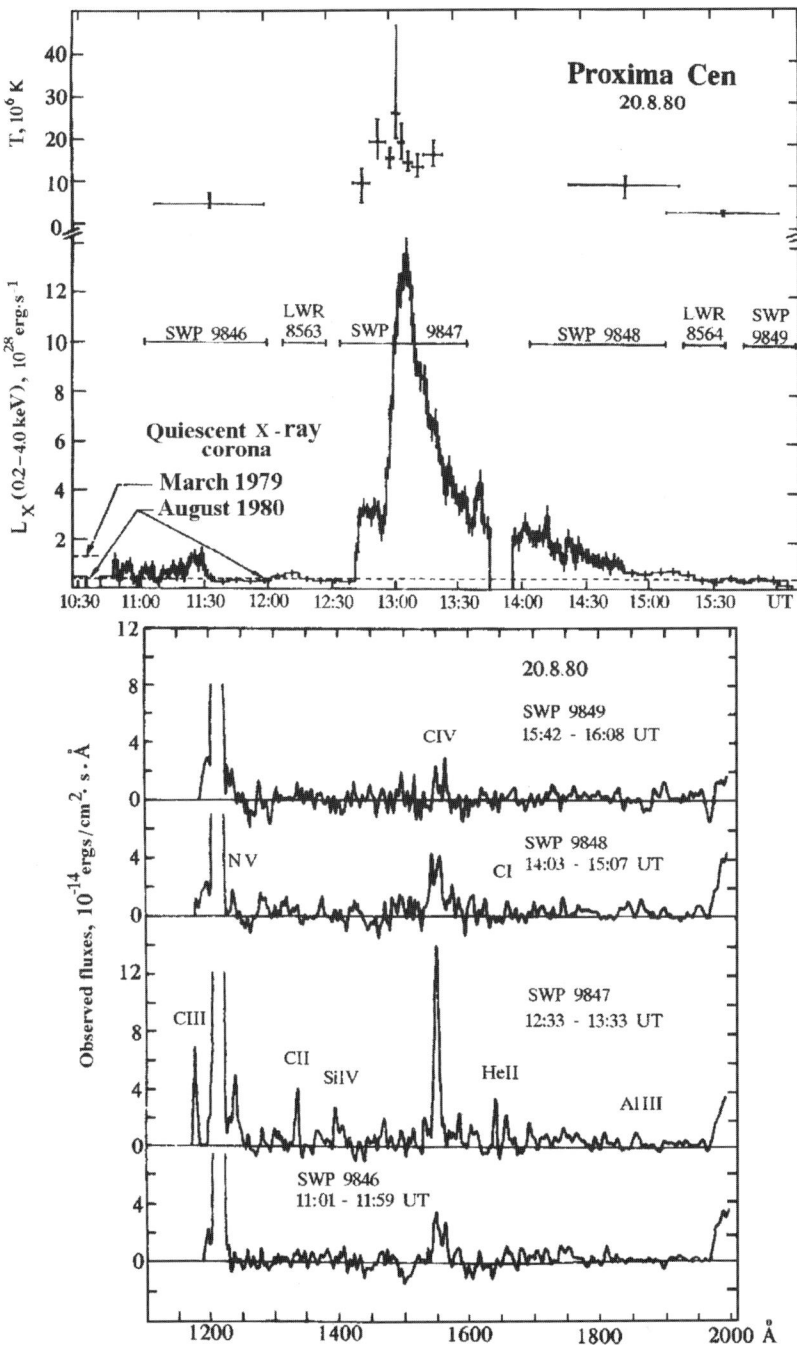

Fig. 28. Flare on Proxima Cen of 20 August 1980 (Haisch et al., 1983)

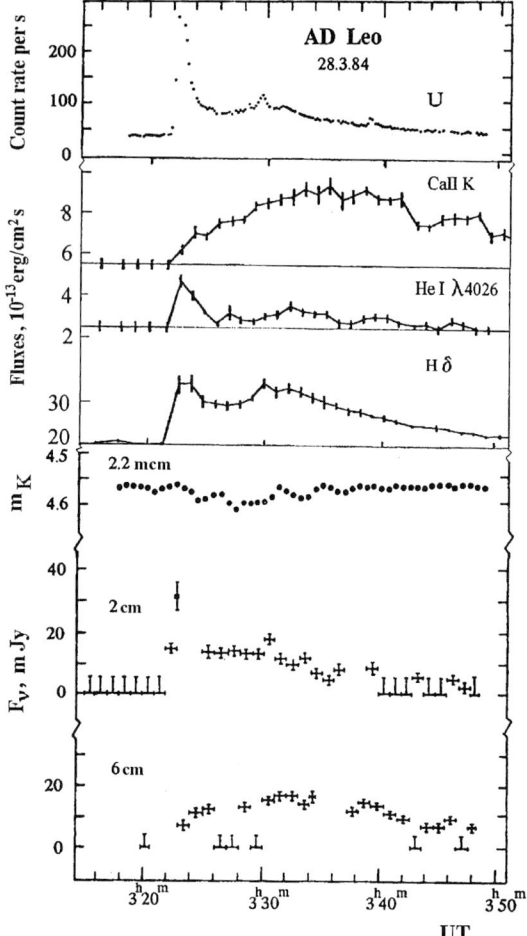

Fig. 29. Light curves of the flare on AD Leo of 28 March 1984 in the U band, in calcium, helium and hydrogen emission lines, in the K band and at 2 and 6 cm in the radio range (Rodonò et al., 1989)

measure was estimated as 6×10^{53} cm^{-3}, assuming that the temperature of the radiating plasma was 30 MK. A day after on the descending branch of the strong flare, there was a slightly weaker flare with a maximum luminosity of 6×10^{29} erg/s, a duration of about 3 h, a total energy of EUV radiation of 2×10^{33} erg, and a volume emission measure of 3×10^{53} cm^{-3}. As to the energy, these EUV flares belong to the strongest events recorded on the UV Cet-type stars. In addition to the strong flares, weaker peaks are seen in Fig. 30.

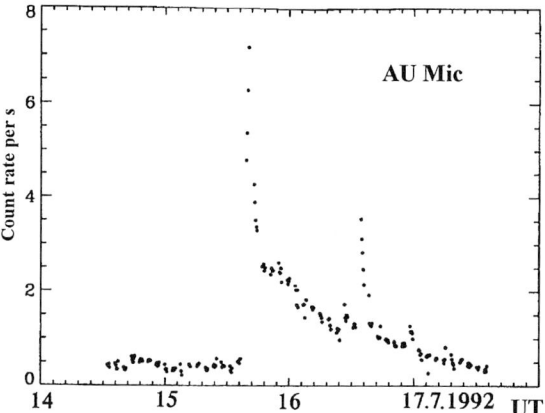

Fig. 30. The light curve of AU Mic in the far ultraviolet (65–190 Å) recorded by EUVE on 14–18 July, 1992 (Cully et al., 1993)

The above flares belong to the most thoroughly studied flares on the UV Cet-type stars and are the strongest and longest processes of the considered type. The vast majority of recorded flares on red dwarfs are shorter and weaker. However, there are no grounds to think that the weaker phenomena are more uniform: a certain uniqueness and diversity of solar flares, whose energies approximate to weak stellar flares, suggests that weak flares are very diverse as well.

2.2 Temporal Characteristics of Flares

We will consider two of the most common time characteristics of stellar flares: duration and time distribution. The character of the development of flares, i.e., internal temporary structures at different wavelengths, will be considered in Sect. 2.4.

2.2.1 Time Scales of Flares

Available data on the duration of stellar flares are determined both by the intrinsic properties of flares and the capacities of recording equipment. During the first visual observations brightness peaks lasting from fractions of a minute to tens of minutes were recorded. When the visual monitoring was replaced by photographic observations, the characteristic duration of the exposures limited the possibilities of detection of fast flares, and the recorded durations of many hundreds of flashes revealed mostly in several nearest stellar clusters varied from 5–10 min to many hours.

Continuous photoelectric monitoring provides the most complete data on the duration of optical flares. The data for several series of long-term observations are presented in Table 10.

Table 10. Distribution of durations of optical flares (min)

	< 1/3	1/3–1	1–3	3–9	9–27	27–81	>81
EV Lac	1	23	42	69	71	13	3
CN Leo	5	16	35	38	17		
UV Cet	5	21	38	33	11		
YZ CMi		5	16	31	8	2	
AD Leo	10	32	45	5	1		
8 stars	8	24	28	12	1		

The first row of Table 10 shows the results for the EV Lac flare star monitored in the Crimea: in 1986–95 the star was regularly examined within international cooperative campaigns. 227 flares were recorded over 307 h of photoelectric $UBVRI$ monitoring with a resolution of 12 s (Alekseev and Gershberg, 1997a). The table presents the distribution of duration of 222 flares. The data show that almost 2/3 of flares on this star lasted from 3 to 27 min. Usually, these observations were performed throughout the night from evening to morning, thus the decrease in the number of events longer than half an hour is a real fact and not the result of observational selection.

The next three rows of the table show the data on the flares on CN Leo, UV Cet, and YZ CMi obtained from the data of Moffett (1974), who in 1971–72 monitored the stars with a resolution of 1–4 s by four telescopes at the McDonald Observatory (USA). The duration of continuous monitoring of one star rarely exceeded 1.5–2 h, thus the small number of long-term flares recorded by Moffett on these stars is largely due to the observational selection. According to the statistics of flares on these three stars, 2/3 and more flares lasted from 1 to 9 min.

The fifth row of the table lists the data on the flares on AD Leo based on the observations of Pettersen et al. (1984a) by two telescopes at the McDonald Observatory with a time resolution varying from 1 to 10 s (1974–1979). In the first four rows complete durations of flares are presented, but for AD Leo Pettersen et al. cited the durations of the descending branches of flares, from brightness maxima to half-brightness. The full time of flare decay is usually several times longer. Thus, to compare the data on AD Leo with previous rows, one should shift them to the right by one column at least. Then, it will appear that the distribution of duration of flares on AD Leo is practically the same as for the three stars observed by Moffett.

Finally, the sixth row of the table shows the observational results for 8 flare stars – YZ CMi, HU Del, GQ And, V 577 Mon, EQ Peg, Wolf 424, UV Cet, and CN Leo – obtained in 1982–85 at the 6-m telescope of the

Special Astrophysical Observatory, Caucasus, using the measuring complex MANIA (Beskin et al., 1988). These observations were performed with a time resolution of 3×10^{-7} s during relatively short time intervals. Over 35 h of monitoring 118 stellar flares were recorded and for 73 of them the light curves with a resolution of 0.1 s were constructed to estimate the duration of flares. Table 10 shows that more than 2/3 of flares of this sample were 1/3 to 3 min long. The small number of long-duration flares is due to the short observational time for each star, and the increase in the portion of fast events is due to the higher resolution of MANIA as compared to standard photometers and the higher efficiency of recording of weak bursts by the larger telescope.

Having analyzed the first observations of flare stars in clusters and in the solar vicinity, Haro and Chavira (1955) suspected that flares on brighter stars were longer, on average. This conclusion was supported by the photoelectric observation of five stars by Kunkel (1969b), Gershberg and Chugainov (1969), and Chugainov (1974), the number of flares on stars in the solar vicinity was greater. Later, using the duration of flares at the level of a half-maximum brightness as the characteristics of their duration, Kunkel (1974, 1975a) considered about 600 flares he had detected on the three stars and concluded that statistically these values depended on the absolute luminosity of the star

$$\langle \log T_{0.5} \rangle = \text{const} - 0.12 M_V . \qquad (28)$$

Then, the relation was confirmed by the observation of about fifty flares on four stars by Busko and Torres (1978). On average, solar flares develop slower than flares on dKe–dMe stars, which qualitatively supports the conclusion. Table 10 also contains an indirect confirmation of the Kunkel conclusion. However, it should be noted that the range of the durations of flares on each star substantially exceeds possible systematic shifts of their average values from one star to another at the expense of different luminosity of the stars. Thus, Kunkel (1973) found that in the sample of 140 flares the duration of the longest flare was 10 times longer than the average duration of flares in the sample.

There are two points of view on the weak dependence of the average duration of flares on the luminosity of a flare star. On the one hand, for higher luminosity stars the density of matter in the atmosphere is systematically lower, which can stipulate slower development of flares. On the other hand, weak flares, as a rule, are shorter, and the contrast with the background of a quiet photosphere results in rather bright and consequently longer flares recorded on higher-luminosity stars. Certainly, one cannot exclude joint action of the effects, but the influence of the observational selection apparently prevails (Pettersen, 1989a).

To understand the physical sense of these processes, one should know not only the characteristic durations of stellar flares, but also the ultimate values of the durations, i.e., the durations of the shortest and longest stellar flares.

Figure 31 shows the light curves of the shortest flares recorded by the 6-m telescope of the Special Astrophysical Observatory by MANIA (Beskin et al.,

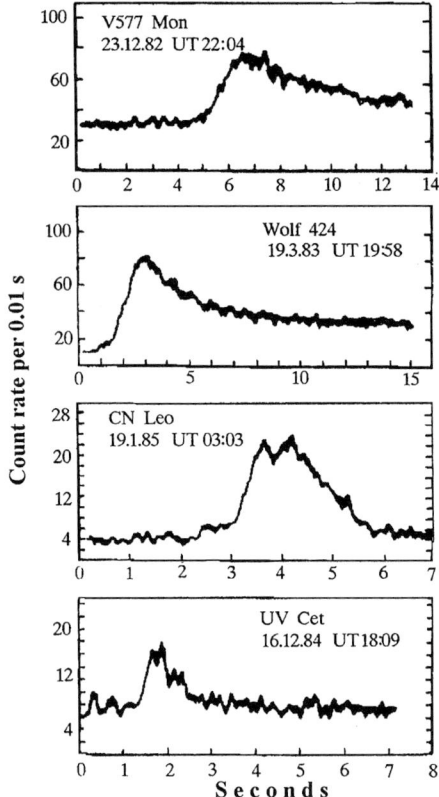

Fig. 31. Fast flares on four UV Cet-type stars (Beskin et al., 1988)

1988). It leaves no doubts about the reality of a few seconds long stellar flares. With high reliability an optical flare on EV Lac of 2.4 s was recorded on 24 February 1984 by the space station Astron (Gershberg and Petrov, 1986). There were doubts about the stellar origin of the flares of one second and shorter recorded by standard one-channel photometers. Some confidence was gained when monitoring with high-speed two-channel photometers produced in the early 1990s in Kiev and Byurakan was started, which enabled simultaneous recording of stellar brightness in two spectral bands. The flare on EV Lac of 26 August 1990 recorded in Terskol with a time resolution of 0.05 s (Zhilyaev and Verlyuk, 1995) and two flares on EV Lac of 7 August 1994 23:43 and 23:59 UT (see Fig. 32) and the flare on V 577 Mon of 10 January 1997 06:12 UT recorded with a resolution of 0.1 s in Byurakan and Pueblo, respectively (Tovmassian et al., 1997), detected stellar flares with a total duration of not more than 0.3 s. To find ultimately weak stellar flares with HST, Robinson et al. (1995) tried to observe one of the weakest flare stars CN Leo in the range of 2400 Å, where the maximum contrast of the flare and the stellar photosphere

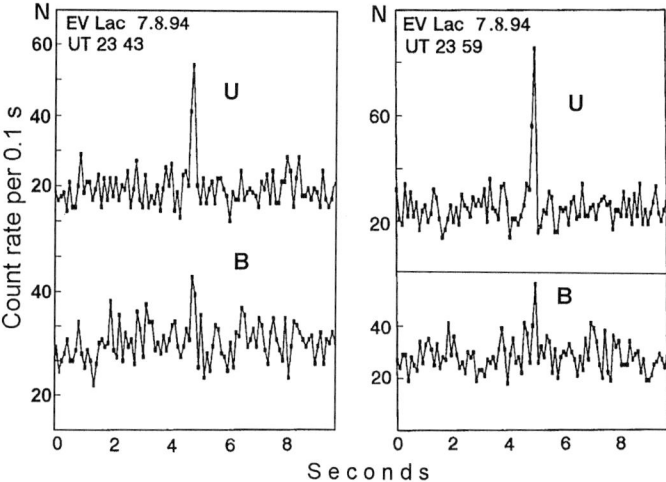

Fig. 32. Two very fast flares on EV Lac recorded with a time resolution of 0.1 s using the two-channel photometer of the Byurakan Astrophysical Observatory (Tovmassian et al., 1997)

was expected. Using the method of statistical photometry, they picked out 32 several-seconds long flares with details to 0.1 s from the two-hour monitoring of stellar brightness. The energy spectrum of the flares (see below) was in agreement with that of the flare spectrum of the star established on the basis of observations in the optical range.

The data on the duration of the longest stellar flares are rather fragmentary. The longest flare (of the above 227 flares) of 29 August 1990 on EV Lac lasted for more than 1.5 h. On this star, Rojzman and Shevchenko (1982) and Roizman (1983) recorded flares of 4.5 and 5 h, respectively. A very strong X-ray flare on this star was more than 4 h long (Favata, 1998). The beginning of the flare on YZ CMi of 19 January 1969 was recorded in Armagh (Northern Ireland), and its end, in Chile, the total duration of the event exceeded 4 h (Kunkel, 1969a). The flare on EQ Peg recorded by ROSAT on 6 August 1985 lasted More than 2.5 h (Poletto, 1989). The flare on BY Dra of 1 October 1990 was recorded during the all-sky survey in the far ultraviolet using the ROSAT WFC. It was recorded during three successive passages of the star in the camera's field of view, so that the full duration of the flare was about 4.5 h (Barstow et al., 1991). On 19 October 1990, ROSAT recorded a flare on AZ Cnc in soft X-rays during six revolutions, i.e., it lasted at least eight hours (Fleming et al., 1993). The flare on the T177 star in the Orion Nebula cluster recorded photographically in Tonantzintla (Haro and Parsamian, 1969) lasted about 20 h, while the flare on T48 star in the same cluster recorded during spectral monitoring lasted for several hours (Carter et al., 1988). The longest known flare on the UV Cet-type stars that lasted more than one and a half days was the flare on AU Mic 15 July 1992 (Cully et al., 1993) (see Fig. 30).

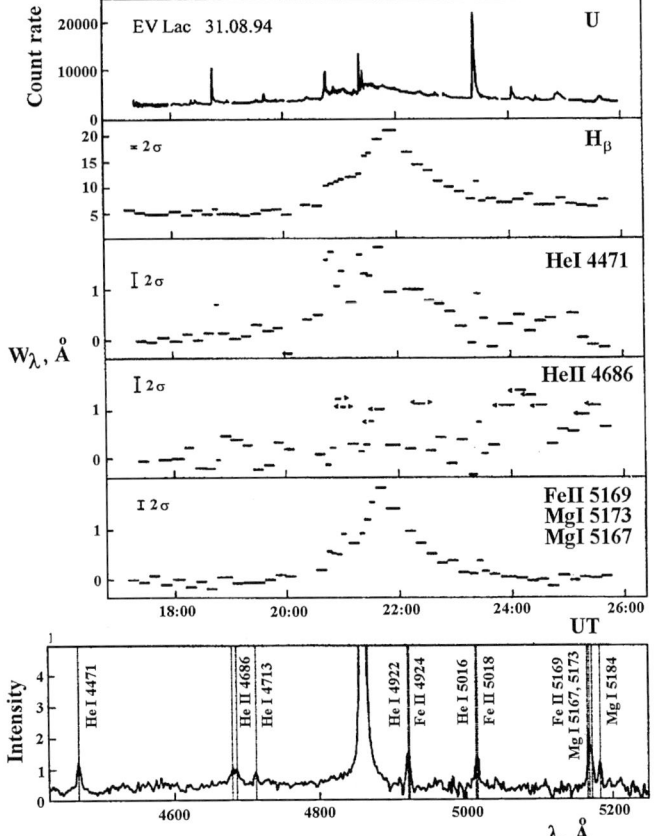

Fig. 33. Results of photometric and spectral monitoring of the flare on the red dwarf EV Lac on the night between 31 August and 1 September 1994 obtained in the Crimea: the light curve in the U band, the equivalent widths of the emission lines H_β, He λ 4471 Å, HeII λ 4686 Å, the blends of FeII λ 5169–5173 Å, and the summary spectrum of active states of the star (Abranin et al., 1998b)

Thus, the durations of stellar flares cover a wide range from fractions of a second to tens of hours, i.e., more than 5 orders of magnitude.

All the above long-duration events had the characteristic light curves with abrupt brightness increase and long decay. In addition, many-hour substantial increases of brightness with almost symmetrical ascending and descending branches of the light curves are sometimes recorded on red dwarfs (see EV Lac light curve in Fig. 33 (Abranin et al., 1998b)). Probably, some slow flares recorded photographically in stellar clusters (see, for example, Fig. 1 in Parsamian, 1980) are just the same events. Gurzadian (1986) developed the theory of the light curves of slow flares assuming that not only direct but also scattered glow of flares occurring behind the stellar limb was observed in these cases.

2.2.2 Time Distribution of Flares

The time distribution of flares is one of the major characteristics of the flare activity. Prior to astrophysical analysis, the consideration of the time distribution makes it possible to distinguish the type of stellar variability, which presumably can determine the physical nature of such activity. Thus, if the distribution is strictly periodical with rather short periods, the activity of red dwarfs could be associated with global phenomena of the type of stellar pulsations. A strict Poisson distribution would evidence the physical independence of individual flares in the observed sequence of such events, the local character of the processes, and apparently the dependence of the flare-triggering mechanism and the course of its development on several independent physical parameters. Intermediate situations can be stipulated by different factors: physical relation of contiguous flares, the modulating effect of stellar rotation on the visibility conditions of a low number of active regions on its surface, and the randomizing effect of the weather on the fulfillment of ground-based observations.

The first researchers of flares on UV Cet-type stars, who spent tens of hours to record one flare, unanimously marked random character of their time distribution. But, with the beginning of regular photoelectrical observations of red dwarf stars that increased the frequency of recorded flares by an order of magnitude, some quasiperiodicity of the distribution was called into question (Andrews, 1966a,b). The character of the time distribution of flares became one of the most topical issues. To solve the problem, a working group was set up in 1967 by the Commission on Variable Stars of the International Astronomical Union. Over a number of years, the Group had been organizing the campaigns of round-the-clock observations of such objects. Annually, the Working Group selected four to seven UV Cet-type stars and recommended certain 10–20 days for monitoring. For ease of comparison of the data obtained by various observers, a standard format for publishing the observational results was agreed (Andrews et al., 1969). By the end of 1975, the Working Group arranged 36 cooperative campaigns that involved astronomers from more than 20 observatories from Australia, Great Britain, Hungary, Greece, India, Italy, New Zealand, Norway, Poland, Southern Africa, the USSR, the USA, Yugoslavia, and Japan. The summary results of the cooperative campaigns, the analysis of uniformity and completeness of the obtained data were published by Gershberg (1972a) and then supplemented by Shakhovskaya (1979).

The analysis of the results obtained during the first cooperative observations again led to the conclusion about the existence of quasiperiods in the time distribution of flares (Osawa et al., 1968; Higgins et al., 1968; Chugainov, 1969b; Jarrett and Eksteen, 1972; see also Gershberg, 1969 and Jarrett and Grabner, 1976). However, it was questionable whether all suspected characteristic intervals between flares – from an hour to tens of hours – on all considered flare stars (YZ CMi, V 1216 Sgr, and UV Cet) was either divisible by a day, or kept an integer number of times within an interval divisible by a day. Apparently, the small number of considered flares combined with the daily variation of the observation schedule led to false quasiperiods. This assumption is rather plausible in the context of Evans's communication (1975) about the detection of the period of 29.53 days in flare distribution, which suggests an obvious effect of Moon phases in the observation schedule of flare stars. The subsequent autocorrelation analysis of flare activity of UV Cet, YZ CMi, and Wolf 630 (Kunkel, 1971, 1973; Lukatskaya, 1976) did not reveal any periodicity in the sequence of flares on the stars. However, it should be noted that there are various opinions on the technique and efficiency of such an analysis as applied to the set of discrete brightness bursts (Oskanian and Terebizh, 1971a; Kunkel, 1975a; Lukatskaya, 1976).

Oskanian and Terebizh (1971bc) were the first to perform statistically strict and full consideration of the time distribution of flares based on cooperative observations: they considered more than 300 flares on UV Cet and about 100 flares on YZ CMi recorded during 1967–70. The authors analyzed the distributions of small identical time intervals along a number of recorded flares within them, the distributions of time intervals since the start of observations till the first flare, between consecutive flares, and in continuous observations, with respect to the number of flares recorded within the intervals. As a result, they concluded that all these distributions in general did not contradict the assumption about the Poisson distribution of the time sequence of flares, though the number of very close flares exceeded their expected number for such a distribution. (This conclusion was confirmed by Haupt and Schlosser (1974) in observations of 94 flares on UV Cet.) Since the excess of close flares was concerned with the events separated from each other by tens of seconds, Oskanian and Terebizh assumed that this excess was caused by flares with several fast bursts near the phase of maximum, i.e., the flares whose light curves were similar to those shown in Fig. 27.

Later, similar statistical studies were carried out using the data of more photometrically uniform observations. Thus, Melikian and Grandpierre (1984) in studying the distribution of small time intervals with respect to the number of flares – about a hundred bursts on UV Cet recorded during 1978–1982 at Maidanak – found a similar excess of the number of close flares over the number expected from the Poisson distribution. Using the result of Moffett (1974), Lacy et al. (1976) analyzed the distributions of intervals between flares during continuous monitoring and concluded that these data did not contradict the Poisson distribution of flares. Pettersen et al. (1984a) came to a similar conclusion upon analyzing 115 flares on AD Leo recorded at the McDonald Observatory: as in the previous case, the criterion χ^2 did not disprove the random time distribution of flares at the 0.05 significance level. This conclusion was valid for the joint consideration of observations of AD Leo from the McDonald Observatory and the Bulgarian National Astronomical Observatory (Pettersen et al., 1986a).

Different results were obtained by Pazzani and Rodonò (1981) who considered the time distribution of 424 flares on UV Cet, 123 flares on EQ Peg, and 80 flares on YZ CMi recorded in Catania, Sicily. They analyzed the distributions of time intervals from the beginning of observations till the first flare, the time intervals between successive flares, the number of flares on small intervals of specified duration and the ratios of recorded and expected (for the Poisson distribution) number of flares on the intervals of continuous monitoring and concluded that the share of contiguous flares essentially exceeded the expected value. Thus, a purely random distribution of flares was of small probability. This conclusion was drawn on the basis of independent consideration of flares as a whole and of individual peaks on light curves, therefore the excess of close bursts was definitely not provided by the secondary peaks. The conclusion on the nonrandom distribution of flares on UV Cet and EQ Peg is the more unexpected, since photometric observations record the total brightness of two components, both are flare red dwarfs, which definitely should randomize the observed total time distributions of flares.

It is still not clear why Texas and Sicily observations with identical photometric uniformity and distribution of monitoring over the day yielded different conclusions, whereas Byurakan and Texas observations with differing photometric uniformity and distribution of monitoring yielded identical conclusions. Nevertheless, one can conclude that the observed time distribution of flares on UV Cet-type stars definitely differs from strictly periodic and is random to some extent. The principal cause of deviations of the observed distribution from the Poisson one toward the increased frequencies of contiguous flares is not clear yet. Probably, it is due to weak precursor flares that provoke subsequent stronger flares (Moffett, 1974). Other possible explanations are that the per cent of sympathetic flares induced by the general cause at close time moments but at different sites of the stellar surface is very high or the effect of the type of active longitudes leading to nonrandom distribution of flares on the stellar surface is very strong (Pazzani and Rodonò, 1981).

Andrews (1982) considered the season of 1975, when many flares were recorded on EV Lac: 50 flares were detected over 208 h of monitoring at six observatories. He found an alternation of 5–6 day intervals with high and low level of flare activity and attributed this effect of flare grouping to the coexistence of 4 or 5 active regions. Probably, this also explains the longitudinal asymmetry of flares on EV Lac found in 1970 by Leto et al. (1997).

Examining the time variations of flare frequencies from photographic monitoring, Szecsnyi-Nagy (1990) considered 1414 flares on 448 stars recorded in the Pleiades by seven observatories over 3300 observation hours. He selected 17 stars with not less than 10 flares detected on each and found that flares were distributed randomly on five stars, whereas there were time variations of flare frequencies occurred on the other 12: hyperactivity periods were 4–20 times shorter than low-activity periods, but flares during the former were 5–25 times more frequent than during the latter.

The high number of binary systems among flare stars repeatedly raised the question as to the effect of this factor on the level of flare activity. Kunkel (1975b) suspected the modulation of U_O (see Sect. 2.3.1.1) in the statistical data for flares in the system V 1054 Oph. But Lacy et al. (1978) did not find regular variations of flare level in the system L 726-8 AB, though due to its large orbit eccentricity the distance between components varied by a factor of 4.2. Rodonò (1978) made a similar conclusion in studying the system EQ Peg AB.

Finally, it should be noted that the trains of four flares recorded over 12 min in the system Wolf 424 (Moffett, 1973) and over 2.5 h in the system YY Gem (Doyle et al., 1990c) hardly match any of the considered statistical models.

2.3 Flare Energy

The energy of flares is another major quantitative characteristic of the flare activity of stars. Strictly speaking, the most general qualitative definition of a flare as "fast release of a noticeable amount of energy that disturbs the steady state of the star on the whole or a part of it" is directly concerned with this parameter. As the time distribution of flares, the characteristic energy gives a hint on the set of physical processes associated with the flares under consideration. Thus, the total radiative energy of supernovae and the kinetic energy of the shell removed under a burst require a thermonuclear explosion that should involve a considerable part of the stellar mass. The energy of flares of novae is lower by several orders of magnitude, and the flares are caused by thermonuclear explosions that involve only a small part of the stellar mass.

Flares on the UV Cet-type stars are accompanied by energy release practically in all ranges of electromagnetic spectrum. By analogy with solar flares, one should expect that strong gas motions should also accompany these stellar flares and generate fluxes of fast particles.

2.3.1 Energy of Optical Flare Emission

All energy parameters of optical emission of flares on the UV Cet-type stars – the absolute luminosity at maximum brightness and during other phases, total radiative energy, the rate of luminosity increase under burning, etc. – are determined directly by photometric observations. For absolute calibration of observational results, stellar luminosity out of the flare is used, which is determined in absolute units from the known stellar magnitude and distance to the star.

As a result of several thousand hours of photoelectric monitoring from observatories placed in different parts of the world, thus far more than 3000 flares have been recorded on the stars of the considered type. The main contribution to the databank was made by the international cooperative programs of 1967–75 and by intensive observations of red dwarfs beyond the framework of these programs carried out at the observatories in Sicily, the Crimea, Chile, Texas, Southern Africa, Greece, Uzbekistan, Armenia, and Japan. The data on many tens of flares were published by Kunkel (1968, 1973), Cristaldi and Rodonò (1970, 1973), Moffett (1974, 1975), Bruevich et al. (1980), Pettersen et al. (1984a). Rather complete lists of publications about recorded flares can be found in statistical studies by Crimean researchers (Gershberg and Chugainov, 1969; Gershberg, 1972a; Shakhovskaya, 1979; Gershberg and Shakhovskaya, 1983) and in reports of the IAU Commission 27 on Variable Stars (Chugainov, 1979; Gershberg, 1982; Gershberg and Shakhovskaya, 1985, 1988, 1991).

Photographic observations of flare stars in clusters were started in Mexico, continued in Italy, and then run intensively in Armenia, Hungary, Georgia, and Bulgaria. Mainly, they resulted in the identification of these objects and determination of their number and spatial distribution. The most complete summary of the observational results is presented in the monograph by Mirzoyan (1981), the Catalog of the Pleiades Flare Stars (Haro et al., 1982), and the reviews by Mirzoyan (1986, 1990). The total number of thus-recorded flare stars exceeds 1200 with more than 4000 flares detected on them. Careful analysis of these observations allows one to obtain certain parameters of flare energy on flare stars in clusters (Krasnobabtsev and Gershberg, 1975; Kunkel, 1975a; Korotin and Krasnobabtsev, 1985).

As stated above, under observations in the standard photoelectric system UBV, the intrinsic flare radiation is rather "blue", while flare stars are the reddest objects. The contrast leads to a fast increase in the amplitude of flares when passing to observations in a band with shorter effective wavelength. The overwhelming majority of flares were recorded in U and B bands and in the appropriate photographic systems. However, this does not mean that U-band observations are always more informative. Because flare stars are red, photometric measurements in violet rays conflict with the quantum noise of the recorded radiation flux earlier than those in the range of longer wavelengths. For weaker objects, the greater amplitude of the useful signal, flare, does not compensate for the increasing noise in the record of the photometric standard, stars out of flares. Therefore, for each telescope equipped with a photometer with a rather low level of intrinsic noise, there is such a critical visual stellar magnitude determined by the telescope size, radiation colors of stars and flares, and viewing conditions that photoelectric observations aimed at recording of flares on stars brighter than this critical magnitude should be carried out in the U band, whereas on weaker stars, the B band should be used.

Considerable diversity of flare light curves complicates comparison of the energy of various events of this kind. The problem of quantitative determination of some average flare activity required to compare the activity of a star at different periods or the activity of various flare stars is even more ambiguous. Originally, such values as average frequencies, average amplitudes, average flare energies and total flare energy attributed to full duration of monitoring were used for this purpose (Gershberg and Chugainov, 1969; Gershberg, 1969; Gurzadian, 1971a; Sinvhal and Sanwal, 1977). However, these average values are heavily influenced by observational selection.

On the basis of consideration of photoelectric observations of about 90 light curves of flares on the UV Cet-type stars recorded before the beginning of cooperative observations, Gershberg and Chugainov (1969) estimated absolute flare energy, burning and decay rates, and the duration of flares. They made an attempt to find a correlation among these values and between them and absolute luminosities of flare stars. This paper considerably affected the ideology of subsequent cooperative campaigns.

The first strict statistical research of the energy of flare activity of red dwarfs on the basis of cooperative observations of four of the most active stars – YZ CMi, AD Leo, EV Lac and UV Cet – was carried out by Oskanian and Terebizh (1971a). They estimated the luminosity range and the total energy of optical flares, distributions of their amplitudes and energy, and constructed the frequency functions of these parameters. They noticed that the functions of the first three brighter stars differed essentially from those of the absolutely weaker UV Cet, on which even weaker flares could be observed. They also estimated the possible contribution to total flare emission by the events that were beyond the detection threshold.

The effects of the observational selection can be revealed, if, instead of average flare characteristics, the distributions of these characteristics are considered. Reasoning from this, Kunkel (1968) introduced into examination the distributions of flares over absolute stellar magnitudes of flare radiation at maximum brightness. Chugainov (1972b) proposed to consider distributions of flares over equivalent durations

$$P = \int [(I_{\text{flare}} - I_0)/I_0] \, dt \,, \tag{29}$$

where I_{flare} is the radiation flux from a star recorded during a flare and I_0 is the flux recorded from a quiet star. If, in the distribution of flares with respect to equivalent durations $n(P)$, we proceed from flare number to their average frequency $\nu = n/T$, where T is the time of monitoring, and from equivalent duration to total flare energy $E = PL_*$, where L_* is the stellar luminosity out of the flare in the appropriate photometric band, we obtain the observed energy spectrum of flares $\nu(E)$ (Gershberg, 1972a). Analysis of these two distributions determines the modern concepts of statistical properties of the energy of flare activity. It should be noted that considering 386 flares detected by Moffett (1974) on eight red dwarfs, Lacy et al. (1976) found a practically linear correlation between average flare energies in the U band and stellar luminosity in this band over five orders of magnitude

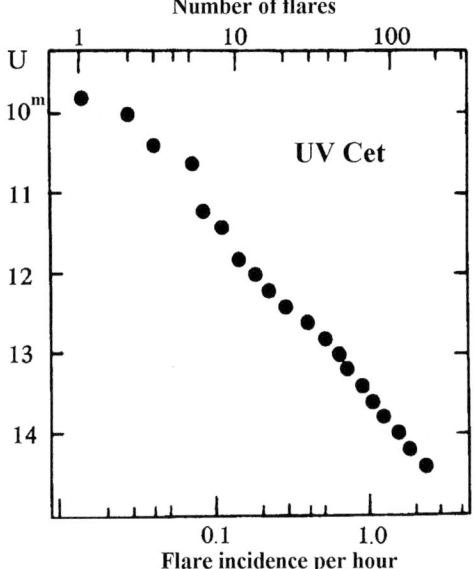

Fig. 34. Distribution of flares on UV Cet over maximum brightness (Kunkel, 1968)

$$\langle \log E_U \rangle = \log L_U^* + 2.0 \pm 0.7 \ . \tag{30}$$

They obtained similar relations for total energy of flares. Later, these relations became widely used, but one should be careful, because in extrapolations to the Sun they yield an error of several orders of magnitude.

2.3.1.1 Spectrum of Maximum Flare Brightness

It is obvious that the flare distribution over the absolute stellar magnitude at maximum brightness in physical terms corresponds to the spectrum of maximum flare radiation. Figure 34, reproduced from the paper by Kunkel (1968), shows the spectrum of the maximum brightness of flares on UV Cet recorded over 78 h of monitoring in 1967 at Cerro Tololo (Chile). The logarithm of the cumulative frequencies $\tilde{\nu}(U < U_0)$, i.e., the logarithm of frequencies of flares, whose brightness in the U band at maximum brightness exceeded the brightness corresponding to the value U_0, is plotted along the x-axis. Experimental points fit well the straight line corresponding to the equation

$$\tilde{\nu}(U - U_0) = \exp[1.04(U - 13.58)] h^{-1} \ . \tag{31}$$

Hence, it follows that flares on UV Cet occur on average every hour, at maximum brightness of the visual magnitude in the U band is at least $13^m\!.58$.

Kunkel (1973) proved that the relation

$$\tilde{\nu}(m) = \exp[a(m - m_0)] \tag{32}$$

Table 11. Parameters of spectra of maximum flare brightness (Kunkel, 1975a)

Star	Observational Epoch	Photometric Region	Number of Flares	a	m_0
Gl 15B = GQ And	1969.3	B	15	0.83	16^m2
Gl 54.1 = YZ Cet	1969.7	U	7	0.9*	17.5
Gl 65 = UV Cet	1966–71	U	802	1.06	14.0
Gl 166 C	1968.0	U	38	0.98	15.0
Gl 206	1967.9	U	3	1.0*	15.3
Gl 229	1970.0	U	2	1.0*	17.2
Gl 234 = V 577 Mon	1969.1	U	35	1.24	14.8
Gl 278C = YY Gem	1971.2	U	10	1.0*	14.3
Gl 285 = YZ CMi	1969–70	U	85	1.10	15.0
Gl 388 = AD Leo	1965.3	U	27	0.91	14.5
Gl 406 = CN Leo	1969.1	U	38	1.19	15.1
Gl 473	1972.2	U	10	1.0*	14.7
Gl 493.1	1969.1	U	3	1.0*	16.1
Gl 494 = DT Vir	1970.3	B	2	1.0*	16.0
Gl 540.2	1969.4	U	3	1.0*	17.6
Gl 551 = V 645 Cen	1969.2	U	28	0.83	14.8
Gl 616.2	1970.4	B	5	1.0*	16.7
Gl 644 = V 1054 Oph	1968–70	U	125	1.14	14.5
Gl 719 = BY Dra	1970.5	U	9	1.0*	15.1
Gl 729 = V 1216 Sgr	1970.5	B	10	1.0*	16.6
Gl 735 = V 1285 Aql	1970.6	B	5	1.0*	16.8
Gl 799 = AT Mic	1967–70	U	80	0.76	14.5
Gl 803 = AU Mic	1970.6	U	31	1.03	13.8
Gl 815	1969	B	7	1.0*	16.4
Gl 860 B = DO Cep	1969.7	U	10	1.0*	17.2
Gl 866	1971.7	U	7	1.0*	15.9
Gl 873 = EV Lac	1970.7	U	67	1.00	14.9

* Postulated values

represented all the spectra of maximum flare brightness that he had constructed. From observations of 12 flare stars he calculated a and m_0 and found that all the calculated values of a were within a rather narrow range 0.76–1.24. The average value of a appeared to be very close to unity, and differences of individual values of this parameter from unity, as a rule, were not greater than the determination error of these values (Kunkel, 1975a). Therefore in considering stars with a few detected flares one could postulate $a = 1.0$ (or 0.9) and determine the only free parameter m_0 from observations. The results of the analysis of flare activity of 27 flare stars are summarized in Table 11 reproduced from Kunkel's review (1975a).

The parameter m_0 evidently represents the observed stellar flare activity and allows one to predict the expected number of flares, which should be recorded over the set duration of observations with the equipment that is able to detect flares with certain minimum amplitude $\Delta m_{\min}$. It is obvious

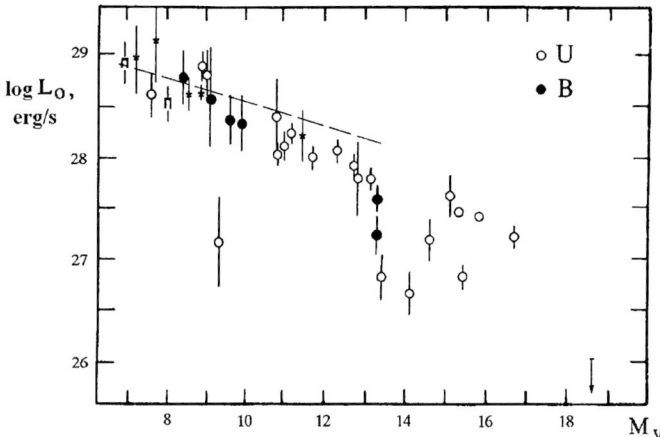

Fig. 35. Absolute luminosities at maximum brightness of flares with a frequency $\nu = 1h^{-1}$ on flare stars of different luminosity (Kunkel, 1975a); *asterisks* mark the data by Busko and Torres (1978); *broken line* shows 1% of bolometric luminosity

that in (32) the difference of apparent stellar magnitude can be replaced by the difference of absolute stellar magnitudes, but the absolute magnitude M_0 corresponding to m_0 characterizes the physical properties of flare activity rather than the observed pattern. Figure 35, based on Kunkel's data (1975a), shows the flare luminosity at maximum brightness obtained by transition from the absolute stellar magnitudes M_0 to luminosity energy units L_0. In addition to flare stars in the solar vicinity listed in Table 11, there are two signs Π corresponding to flare stars in the Pleiades. These data were obtained in analyzing the observations of flare stars in this cluster: there are two groups of flare stars of similar brightness on which flares with a duration close to the characteristic time of photographic exposures were recorded. Then, the flare frequencies on these stars were estimated on the assumption that the samples of flare stars satisfied the ergodicity principle: a set of flares recorded over a time T on n stars was close enough to the set of flares recorded on each of them over the time nT. It follows from Fig. 35 that such an estimate of L_0 for flare stars in the Pleiades is in good agreement with individual values determined for the brightest stars in the solar vicinity.

Using the points in Fig. 35 one can plot the envelope from above with confidence. Obviously, it shows the maximum of flaring activity of flare stars, if for the measure of this activity we accept the luminosity of flares occurring with a frequency of $\nu = 1h^{-1}$. Most likely, there are flare stars that fill in the plane of the figure below the envelope; however, observers prefer to examine the most active flare stars, thus the data on low-activity flare stars are scant and accidental.

The main conclusion from Fig. 35 is that on the most active UV Cet-type stars the maximum absolute luminosity of flares occurring with an average frequency of $\nu = 1h^{-1}$ in the U and B bands reaches 10^{29} erg/s for K stars of middle subtypes and systematically decreases for late M stars. On the other hand, according to the calculations by Kurochka and Rossada (1981), when the Sun approached the epoch of maximum activity in 1980, on average each hour there occurred a flare, whose maximum luminosity in all lines and continua of the hydrogen spectrum amounted to a few 10^{25} erg/s, in the B band the radiation of such a flare should be at least an order of magnitude lower. Thus, stellar flare activity similar to the analogous solar activity covers the range of at least 4–5 orders of magnitude of L_0.

2.3.1.2 Spectrum of Flare Energy

Analysis of the dependence of the average frequency of flares on their total energy $\nu(E)$, i.e., the energy spectrum of flares, is today the most widespread method of statistical consideration of the energy of stellar flares (Gershberg, 1972a, 1985; Krasnobabtsev and Gershberg, 1975; Lacy et al., 1976; Shakhovskaya, 1979; Walker, 1981; Byrne, 1983; Gershberg and Shakhovskaya, 1983; Pettersen et al., 1984a; Korotin and Krasnobabtsev, 1985; Mavridis and Avgoloupis, 1987; Ishida et al., 1991).

Figure 36 shows the energy spectra of the flare stars YZ CMi and BY Dra constructed in the double-logarithmic scales from the B band observations (Gershberg and Shakhovskaya, 1983). To reduce the influence of random spread, cumulative frequencies $\tilde{\nu}(E)$ are used in Fig. 36 instead of average frequencies of flares with the energy E, values of $\tilde{\nu}(E)$

$$\tilde{\nu}(E) = \int_E^{E_{\max}} \nu(E)\, dE \ . \tag{33}$$

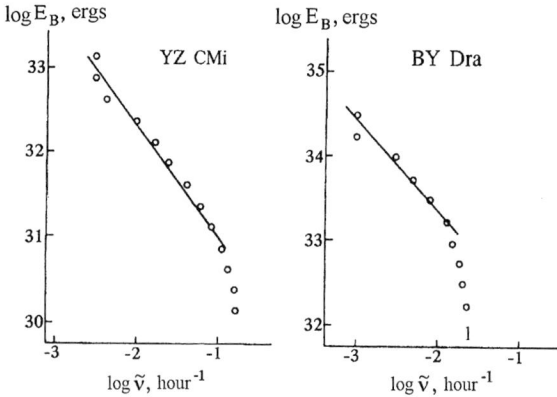

Fig. 36. Distribution of flares on YZ CMi and BY Dra over the total radiative energy in the B band, the cumulative energy spectra of flares (Gershberg and Shakhovskaya, 1983)

The figure proves that for rather strong flares the relation between $\log E$ and $\log \tilde{\nu}$ is close to linear, and weak flares demonstrate a sharp slope.

On each flare star in the solar vicinity over several tens of hours of photoelectric monitoring a number of flares can be recorded, which is sufficient for the construction of the energy spectrum of flares, but the frequency of photographically recorded flares is lower by 1–3 orders of magnitude. This fact prevents construction of individual energy spectra of flares on flare stars in clusters. But, as was mentioned, it seems plausible that flare stars of practically identical age and initial chemical composition, identical brightness and, hence, identical mass in clusters should have similar flare activity, that is, an identical energy spectrum of flares. In this case, flares recorded over the time T in any cluster of stars of similar brightness can be considered as the flares that occurred on the same flare star over the time Tn, where n is the total number of flare stars of similar brightness in this cluster. The energy spectra of flares constructed within the framework of this ergodicity hypothesis are called group spectra.

Figure 37, reproduced from the paper of Korotin and Krasnobabtsev (1985), presents the group energy spectra of bursts on flare stars in the Pleiades. Each group spectrum is constructed for flare stars in the range of apparent brightness $\Delta m = 2^m$, flare stars in the neighboring groups are overlapped by half, and all together they cover all flare stars in the Pleiades. Group spectra are given as dependences $\tilde{N}(L_{\max})$, where $\tilde{N}$ is the cumulative number of flares and $L_{\max}$ is the absolute flare luminosity at the maximum. Since in photographic observations of flare stars in clusters the exposures Δt take many minutes, they are comparable with the duration of flares and are identical for all flares on all stars of a cluster, the values of $L_{\max} \Delta t$ differ from E only by a multiplier of about unity, which can be estimated from observations (Krasnobabtsev and Gershberg, 1975). Hence, the values plotted in Figs. 36

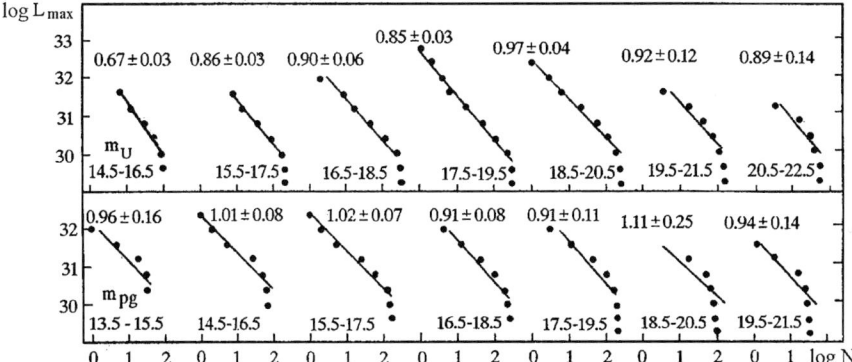

Fig. 37. The group energy spectra of flares on the stars of different luminosity in the Pleiades (Korotin and Krasnobabtsev, 1985). Above each spectrum the value of the spectral index β is specified, below is the range of brightness of the stars included in this spectrum

and 37 differ only in constant shifts. Thus, the two figures can be considered together and one can suggest that the linear relation of $\log E$ and $\log \tilde{\nu}$ for strong flares and the abrupt break of the energy spectrum in the range of weak flares is characteristic of all flare stars. Let us discuss the thus-found general structure of the observed energy spectra of flares and consider some formal consequences of the dependence $\tilde{\nu}(E)$ and physical meaning of the established power character of this dependence.

A. The structure of observed energy spectra of flares. First, one should make sure that the break of the spectrum in the region of weak flares is the result of observational selection. Indeed, the criterion of reality of a flare is usually a certain excess of the amplitude of burst over the width of the noise path. At low E the fastest flares satisfy this criterion, whereas smoother flares with equal total energy E are attributed to noise. The closeness of the break of energy-spectrum to the detection threshold for flares was established in observations (Chugainov, 1972b). (We note that trying to connect the frequency of the energy-spectrum break with intrinsic properties of flare activity Rosner and Vaiana (1978) erroneously took published cumulative frequencies for average frequencies.)

It is obvious that $\tilde{\nu}$ of the last point ahead of the break in the linear part of the spectrum in Figs. 36 and 37 corresponds to the mean frequency of flares that are recorded independently of the detection threshold. In other words, only above the break does the energy spectrum of flares contain physically significant information on the properties of stellar flare activity. On the other hand, ν_{break} should depend on the absolute luminosity, since on brighter stars only absolutely stronger flares can be noticed against the background of permanent radiation of the photosphere. The flare/star contrast depends on the used effective wavelength of photometric band, and on the recording technique, sensitivity, time resolution, intrinsic noise and other properties of the equipment. However, according to the calculations by Shakhovskaya (1979), when using a sample of uniform data the break of the observed energy spectra of flares, as a rule, is rather abrupt, thus the break frequency and the appropriate energy of flares are determined with sufficient confidence.

In the region of strong flares the basic uncertainty of the considered energy spectra of flares is due to the limited scope of the sample: the points corresponding to the strongest flares, are statistically least reliable, since by definition they are based on the minimum number of flares. However, as for flares of any physical nature, there should be a restriction on the maximum energy of an individual flare, in the region of high energy in the axes in Figs. 36 and 37 the cumulative energy spectrum of flares $\tilde{\nu}(E)$ should necessarily turn into a horizontal line. If observations do not definitely show such a transition, maximum-energy flares have not been recorded as yet on the star under consideration.

Results of the most numerous constructs of cumulative energy spectra of flares on the UV Cet-type stars from photoelectric monitoring are presented in the publication by Gershberg and Shakhovskaya (1983). Table 12, reproduced

Table 12. Parameters of cumulative energy spectra of flares (Gershberg and Shakhovskaya, 1983)

Star	M_V	B T*	B N	B α	B β	U T*	U N	U α	U β
1	2	3	4	5	6	7	8	9	10
Gl 15B= GQ And	$13^m.29$	46	16	28.38±0.03	0.98±0.11				
Gl 65(A+B)= UV Cet	15.35+15.89	849	320	24.64±0.07	0.84±0.06	95	205	26.04±0.05	0.89±0.06
Gl 166C= 40 Eri C	12.73					16	30	26.73±0.18	0.90±0.14
Gl 234AB=V577 Mon	13.08					8.8	28	19.96±0.10	0.67±0.12
Gl 268= Ross 986	12.62	37	10	20.15±0.04	0.68±0.08				
Gl 278C= YY Gem	8.26	92	6	14.54±0.07	0.49±0.05	89	6	8.98±0.03	0.32±0.04
Gl 285= YZ CMi	12.29	999	170	22.61±0.03	0.76±0.04	117	89	26.65±0.05	0.89±0.09
Gl 388= AD Leo	10.98	1010	54	13.55±0.08	0.48±0.06	198	53	16.67±0.05	0.57±0.05
Gl 406=CN Leo	16.68	21	59	31.57±0.05	1.10±0.09	42	70	41.12±0.08	1.43±0.11
Gl 473(A+B)= Wolf 424(A+B)	14.98+15.2					2.6	11	27.43±0.09	0.93±0.16
Gl 494= DT Vir	9.4	77	6	13.50±0.05	0.47±0.09				
Gl 616.2=BD+55°1823	8.9	411	20	18.83±0.03	0.64±0.04				
Gl 644(A+B)=V1054 Oph	10.79+10.80					11.4	11	14.20±0.16	0.49±0.19
Gl 719= BY Dra	7.9	1064	35	27.58±0.07	0.89±0.08	234	19	20.38±0.03	0.68±0.04
Gl 729=V1216 Sgr	13.3	227	16	25.24±0.04	0.90±0.06				
Gl 799(A+B)= AT Mic	11.09+11.2					11.5	22	20.44±0.08	0.66±0.09
Gl 815(A+B)	9.8+11.8	69	8	27.21±0.05	0.91±0.07				
Gl 860B= DO Cep	13.3					54	22	20.74±0.08	0.72±0.09
Gl 867B= L 717-22	11.8					6.6	21	42.26	1.4±0.4
Gl 873= EV Lac	11.65	1348	126	29.48±0.05	0.97±0.05	380	117	26.76±0.05	0.89±0.06
Gl 875.1= GT Peg	10.6	76	30	17.30±0.02	0.60±0.04				
Gl 896(A+B)= BD+19°5116	11.33+13.4	485	142	20.56	0.68	165	70	17.6	0.59
Gl 896B= EQ Peg	13.4	58	9	21.01±0.06	0.71±0.06				

* T in h

from the paper, shows the values of initial data: times of monitoring and numbers of recorded flares in the U and B bands separately – and the coefficients of power presentation of the energy spectrum of flares

$$\log \tilde{\nu} = \alpha - \beta \log E \qquad (34)$$

obtained for each of the considered flare stars above the break of an energy spectrum. Korotin and Krasnobabtsev (1985) published the most complete results on the group energy spectra of flares on flare stars in clusters. In both studies, the observations of flares in blue and violet rays were analyzed separately, but the energy spectra obtained were close enough. Figure 38 combines the results of the studies of stellar flares in blue rays: the individual spectra of flares on flare stars in the solar vicinity and group spectra of flares on flare stars in the Pleiades and Orion. The plot shows only the significant parts of the energy spectra located above the breaks of spectra.

It should be noted that to construct the group energy spectrum providing a value of the spectral index β, it suffices to detect twenty to thirty flares in

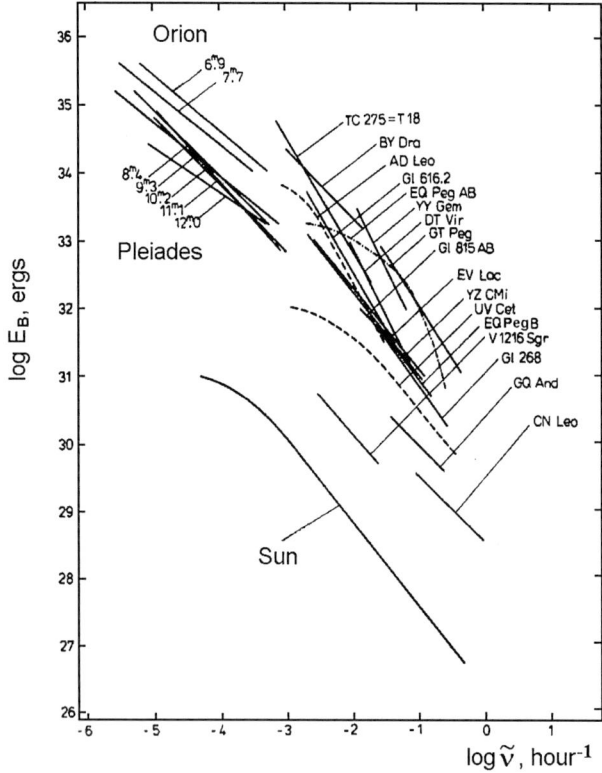

Fig. 38. Energy spectra of flares on stars in the solar vicinity, in the Pleiades and Orion clusters, and solar flares (Gershberg et al., 1987)

the considered group of flare stars. But to localize the spectrum in Fig. 38, to transfer from the cumulative number of recorded flares on all stars of considered brightness to the number of flares on a star, one should know the number n of flare stars of similar brightness in a stellar cluster. This is determined following the Ambartsumian method (1969) assuming that the sequence of flares on a number of similar flare stars can be presented by the Poisson law. If this hypothesis is valid and the average frequency of flares on all flare stars is the same, the number of stars on which k flares occurred over the time t is determined by the relation

$$n_k = n(\nu t)^k \exp(-\nu t)/k! \,. \tag{35}$$

Hence, it follows that the number of flare stars, on which over this time no flares occurred can be calculated using the formula

$$n_0 = n_1^2/2n_2 \,. \tag{36}$$

In other words, from the number of stars on which one or more flares occurred one can determine the full number of flare stars in a cluster as $n = \sum n_k$. However, to get not too coarse a determination, the denominator of (36) should be rather reliable, say, more than 5. Not all considered groups of flare stars satisfy this condition; therefore the number of such spectra in Fig. 38 is rather small.

Figure 38 shows that for flare stars in general the power dependence of the cumulative spectrum of flares $\tilde{\nu}(E)$ can be traced in the energy range of about seven orders of magnitude, but on the spectra of flares on individual stars and groups of stars of identical luminosity the dependence can be traced only in the energy range not exceeding 2 orders of magnitude, because of the effect of observational selection in the region of weak flares and short observational time for recording the strongest flares. It is obvious that the upper limit of the band occupied in Fig. 38 by the energy spectra of stellar flares corresponds to the peak efficiency of optical emission of flares of the considered type. The lower limit of the band, as in Fig. 35, is due to the inclination of observers to study the most active stars.

The energy spectra of flares on three flare stars shown in Fig. 38 demonstrate appreciable deviations from the power dependence $\tilde{\nu}(E)$. The curved lower part of flare spectrum (shown by the chain line) in the binary system EQ Peg is caused apparently by the heterogeneity of the used observational data: as noticed first by Byrne and McFarland (1980), the cumulative energy spectrum of flares constructed on the basis of independent series with different ν_{break} can yield more than one break in the cumulative spectrum. In this case it is difficult to separate the physically significant part of the observed spectrum of flares. Curved upper parts of the energy spectra of flares on EQ Peg, UV Cet, and AD Leo suggest that we are approaching the strongest flares on these stars. Any more resolute statements are premature, because some similar curvatures of energy spectra of flares considered earlier as indications of E_{max} "straightened" later for increasing number of recorded flares. Another distorting effect of the energy spectra of flares was suspected by Doyle and Mathioudakis (1990) in examining the YY Gem system. They found that the energy spectra of flares during the eclipse phase and out of it differed. They explained this by stronger flares between the system components when their interaction was intense.

Figure 39 shows the energy spectrum of solar flares reproduced from the paper of Kurochka (1987), who on the basis of observations of more than 15 500 flares in H_α rays calculated the energy of their complete radiation in all lines and continua of hydrogen series starting from the Lyman series. Three of the above-discussed elements of the energy spectra of stellar flares are seen: the break in the region of weak flares due to observational selection, the linear section corresponding to the power dependence $\tilde\nu(E)$ and the beginning saturation section. The physically significant part of this energy spectrum of solar flares is transferred to Fig. 38. The transition from $\tilde{N}(E)$ in Fig. 39 to $\tilde\nu(E)$ in Fig. 38 is obvious, because the duration of monitoring of the Sun is known, while the transition from the energies of the total hydrogen radiation calculated by Kurochka to the values of E_B is less unequivocal. Using the relations of total hydrogen radiation of solar flares to the radiation in the H_α line provided by Kurochka and Stasyuk (1981) and estimates of the ratio E_B/E_{H_α} ~10–20 for the number of stellar flares (Gershberg and Chugainov, 1966; Gershberg and Shakhovskaya, 1971; Kulapova and Shakhovskaya, 1973; Worden et al., 1984; Rodonò, 1986a), one can accept that, considering the weak continuous emission of solar flares, the total hydrogen radiation of such flares exceeds their radiation in the B band by an order of magnitude. Despite such a rough estimate in which an error of a factor of 3–5 cannot be excluded, the cumulative energy spectrum of solar flares in Fig. 38 matches the general picture of the energy spectra of flares. Similarly to Fig. 35, Fig. 38

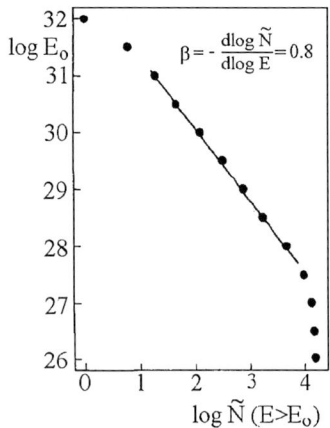

Fig. 39. Cumulative energy spectrum of solar flares (Kurochka, 1987)

evidences that the level of flare activity of the Sun is approximately 4 orders of magnitude lower than of the most active red dwarfs.

Thus, the observations show that the power character of the energy spectrum of flares is a universal feature of stellar flares of the considered type. A number of interesting consequences follow from this experimental fact.

B. Spectral indices of flare energy spectra. Let us assume that the differential energy spectrum of flares $\nu(E)$ in the whole energy range $E_{\min}$–$E_{\max}$ is presented by a uniform power dependence. In this case, the total radiation of all flares over the time T sufficient to realize the whole energy spectrum of flares can be written as

$$\mathcal{E} = T \int_{E_{\min}}^{E_{\max}} E\nu(E)\mathrm{d}E = -T \int_{E_{\min}}^{E_{\max}} E(\mathrm{d}\tilde{\nu}/\mathrm{d}E)\mathrm{d}E$$

$$= \begin{cases} T \times 10^\alpha \beta (E_{\max}^{1-\beta} - E_{\min}^{1-\beta})/(1-\beta), \\ T \times 10^\alpha \ln(E_{\max}/E_{\min}). \end{cases} \qquad (37)$$

The bottom relation is valid for $\beta = 1$, the upper, for the other cases.

Hence, it follows that the value of the spectral index β determines which flares make the major contribution to $\mathcal{E}$: at $\beta < 1$ these are rare but strong flares, at $\beta > 1$, frequent weak flares.

Figure 40 presents the values of spectral indices of energy spectra of flares depending on the absolute luminosity of flare stars. The values of β for stars in the solar vicinity are taken from Table 12, for cluster stars, from the paper by Korotin and Krasnobabtsev (1985). Since the energy spectra of flares recorded in blue and violet rays are rather similar, average values of β are plotted for the stars for which these parameters are available in both photometric systems.

Figure 40 shows that $\beta = 0.4$–1.4. Since the errors in determining β usually do not exceed 0.10–0.15 it is clear that there are flare stars with spectral indices of cumulative energy spectra of flares above and less than unity. However, in Table 12 only 2 of the 23 objects have $\beta > 1$, and for one of them the estimate of this value is obtained with the greatest error.

For flare stars in the solar vicinity between spectral indices β and absolute stellar magnitudes M_V there is a weak positive correlation: $r(\beta, M_V) = 0.60 \pm 0.14$, which causes low confidence in the coefficient of linear regression between the values

$$\beta = (0.047 \pm 0.015)M_V + 0.18 \pm 0.19 . \qquad (38)$$

Finally, Figure 40 suggests the existence of the dependence of β on the age of the flare star. Indeed, the age of each of the next of the three investigated stellar clusters – Orion, the Pleiades, and Praesepe – exceeds the age of the previous cluster by an order of magnitude. In Fig. 40 spectral indices of group energy spectra of flares in these clusters form sequences distinctly spaced for β: the older the cluster, the less is the value of β. Further, among flare stars in the

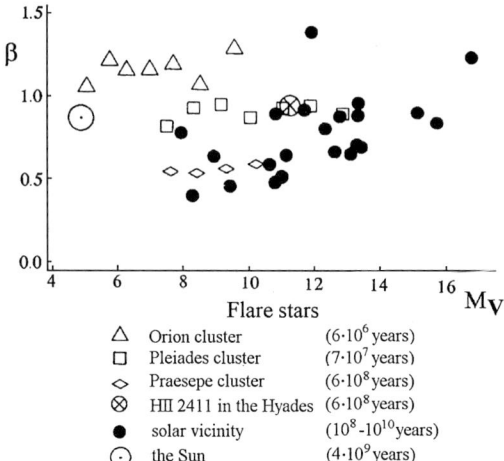

Fig. 40. Spectral indices of flare energy spectra (Gershberg, 1989)

solar vicinity there are objects of various age, including those older than flare stars in Praesepe. Confirming the dependence of β on age, spectral indices of flare stars in the solar vicinity demonstrate the greatest scatter in Fig. 40 and expand to the region of values lower than β values characteristic of Praesepe. However, it should be noted that the β invariance of flare stars in each of the considered clusters in the luminosity range to $4^m - 5^m$ is a rather unexpected result, because it means that the character of flare activity, in general, is insensitive to the changes in the internal structure of stars. At the same time, for example, such an important parameter of the internal structure as the thickness of the convective zone in this luminosity range should definitely vary noticeably, and one could try to associate it with the correlation of β and M_V observed on flare stars in the solar vicinity. In the framework of a purely observational approximation the existence of this correlation and the correlation of β with the cluster age suggest that, on average, less bright flare stars in the solar vicinity are younger objects than brighter stars.

The correlation of β and the age of the revealed flare stars in clusters is violated in Fig. 40 by spectral indices of two individual energy spectra of flares: the star HII 2411 in the Hyades and the Sun, whose spectral indices in this figure correspond to much younger stars. The cause of the infringement can be interpreted using the research results for the energy spectra of solar flares in the region of soft X-ray emission obtained by Kasinsky and Sotnikova (1989, 1997). They analyzed such energy spectra from 1972 to 1993 and revealed a clear change of some characteristics of flare activity of the Sun with the 11-year cycle phase. In particular, it follows from their data that as the phase shifts for one year there is a close correlation of β with the Wolf numbers: $r(\beta_{+1}, W) = 0.84$. There is quite good agreement with the results obtained by Kurochka (1987) for optical flares on the Sun in 1978–79 $\beta = 0.80$,

Kasinsky and Sotnikova found for X-ray flares $\beta = 0.66$ and 0.68 in 1978 and 1979, respectively. But throughout the two cycles β_X covers the range from 0.47 near the epoch of minimum solar activity to 0.93 near maximum activity. This means that during the epoch of minimum activity, when rare and strong flares dominate in the flare ensemble, the value of β of the Sun is older than the oldest cluster Praesepe in Fig. 40 and is much younger, as in the Pleiades, during the epoch of the maximum, when β approaches unity and the contribution of frequent low-energy events sharply increases. Based on photographic observations of HII 2411, Szecsnyi-Nagy (1986) suspected the cyclicity of its activity with a two-fold increase of the number of flares during the maximum phase.

For the active flare dwarf EV Lac there are no sufficiently consistent data. Mavridis and Avgoloupis (1987) from observations in 1974–79 in the B band found a significant decrease of the spectral index β during low flare-activity seasons when the total radiative energy decreased two-fold. Alekseev and Gershberg (1997a) analyzed the observations carried out in 1986–95 in the U band and found cyclicity in the changes of β with a characteristic amplitude of ± 0.1, but did not detect any change of the level of flare activity. On UV Cet variations of β on short time intervals were suspected by Lacy et al. (1978).

Thus, the data on individual energy spectra of flares in Fig. 40 can be influenced by appreciable effects of physical variability of these spectra.

The mentioned statistical study of X-ray flares on the Sun (Kasinsky and Sotnikova, 1989) is based on the consideration of more than 23 000 events. The total number of X-ray flares recorded on dwarf stars by now does not reach a hundred and fifty and there are inconsistent data again. According to Pallavicini et al. (1990a), there is a power-energy spectrum of flares with $\beta \sim 0.7$ in the sample of about thirty X-ray stellar flares on red dwarfs detected by EXOSAT. But, as stated above, based on EUVE observations, Audard et al. (2000) concluded that for flares with $E_{EUV} > 10^{32}$ erg $\beta > 1$ for F–G dwarfs and for K–M stars $\beta < 1$ was more probable.

C. Mean flare energies. The power differential spectrum follows from the power cumulative energy spectrum of flares

$$dN = \text{const} \times dE/E^{1+\beta} , \tag{39}$$

which allows the average energy of recorded flares to be estimated:

$$\bar{E} = \int_{E_{\min}}^{E_{\max}} E dN \bigg/ \int_{E_{\min}}^{E_{\max}} dN = E_{\min} \frac{\beta \left[(E_{\max}/E_{\min})^{1-\beta} - 1 \right]}{(1-\beta)\left[1 - E_{\max}/E_{\min})^{\beta}\right]} . \tag{40}$$

This relation is especially evident at $\beta = 1$

$$\bar{E} = E_{\min} \frac{\ln(E_{\max}/E_{\min})}{1 - E_{\min}/E_{\max}} . \tag{40'}$$

In other words, the average energy of flares recorded on a star is equal to the product of the minimum energy of flare detected on the star using a certain observation technique, and a slowly varying function of the ratio E_{min}/E_{max}. For the most intensely studied flare stars, on which many tens or even more than one hundred flares were recorded, this ratio is close to one hundred, and for absolutely brighter stars it is apparently a little higher. Assuming for weak flare stars $\beta = 1.4$ and $E_{min}/E_{max} = 50$, and for bright stars, $\beta = 0.4$ and $E_{min}/E_{max} = 200$, we obtain

$$\bar{E} = (2.8-17)E_{min}. \tag{41}$$

Consideration of extensive observational series of AD Leo carried out by Pettersen et al. (1984a) and flare stars in the Pleiades by Haro et al. (1982) demonstrated that the average amplitudes of recorded flares fell into the range of values expected from (41) (Gershberg, 1985).

Thus, the average energy of flares in a rather large sample is determined first of all by the energy of an extremely weak flare detected on the given star. The extremely low energy is determined in turn by the observational techniques, thus the average energy (or the average amplitude) – as opposed to the conventional concept – cannot be the measure of flare activity of stars.

D. Dependence of the frequency of observed flares on stellar luminosity. Analyzing photoelectric observations of eight flare stars by Moffett (1974), Mirzoyan (1981) discovered an increasing average frequency of flares with the increase of absolute stellar magnitudes. Let us consider this question in more detail on the basis of more complete data.

On the x-axis in Fig. 38 one can see the cumulative frequencies of flares, thus the abscissa of the rightmost point before the break of each spectrum – to an accuracy of an insignificant extent in the logarithmic scale of frequencies of the observed energy spectrum after the break – corresponds to the average frequency of all recorded flares. There is a reliable positive correlation between the average frequencies and absolute stellar magnitudes: $r(\log \nu, M_B) = 0.81 \pm 0.10$. Figure 41 presents the results of a comparison of these values and those obtained from the observations in violet rays (Gershberg, 1985). The equations of straight lines of the linear regressions are as follows:

$$\lg \nu_U = (0.14 \pm 0.04)M_U - 2.3 \pm 0.6,$$
$$\lg \nu_B = (0.14 \pm 0.03)M_B - 2.8 \pm 0.4. \tag{42}$$

Therefore, the statistical dependence of the average frequency of all flares observed on flare stars on their absolute luminosity can be presented by the relation

$$\nu \propto L^{-0.35 \pm 0.09}. \tag{43}$$

The observed dependence is doubtless, and the appreciable dispersion of points in Fig. 41 and appreciable probable errors of the numerical factors corresponding to it in (43) may be caused by the heterogeneity of the initial observational data providing a nonuniform set of frequencies of the breaks

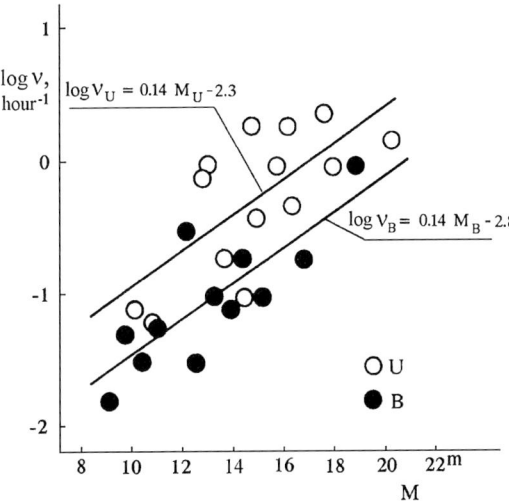

Fig. 41. The average frequency of recorded flares vs. absolute luminosity of flare stars (Gershberg, 1985)

of the flare-energy spectra, and by physical nonuniformity, for example, with respect to age. Finally, the dispersion of points in Fig. 41 may be due to the fact that flare stars are often components of binary systems, which can contain stars of different brightness and different levels of flare activity, therefore the dispersion in the observed dependence $\log \nu$ on M should manifest itself even if there is a functional relation between these parameters for individual flare stars.

The decrease of average frequency of recorded flares with the increase in stellar absolute luminosity is due to observational selection: on brighter stars the threshold of detection of flares is higher. However, using the observed energy spectra of flares, one can show that for the same sample of flare stars the effect of observational selection should lead to the relation

$$\nu \propto L^{-0.96\pm0.13} \tag{44}$$

(Gershberg, 1985). A significant difference between (43) and (44) means that at the transition from absolutely weak to absolutely brighter flare stars the average frequency of flares decreases not as fast as expected because of an increasing detection threshold. Apparently, this effect can be connected with the increase in the surface area of brighter stars and the availability of a greater number of active regions producing flares. Indeed, using the data of Pettersen (1976, 1980) on absolute luminosities and the sizes of single flare stars, one can obtain the statistical relation between the values

$$L \propto R^{5.0\pm0.6} . \tag{45}$$

Then, for the multiplier that distinguishes (43) from (44), we obtain

$$L^{0.6\pm0.2} \propto R^{3.0\pm1.4} \propto 4\pi R^2 \times R^{1.0\pm1.4} \ . \tag{46}$$

The last factor $R^{1.0\pm1.4}$ is too uncertain for any conclusions to be made on the level of specific flare activity per unit surface of a flare star; but the influence of the size of stellar surface on the frequency of flares is rather probable.

Thus, the dependence of average frequency of recorded flares on absolute stellar luminosity is determined both by the dependence of the detection threshold on stellar luminosity, and real distinctions of average frequencies of flares on the stars of various luminosity, and – contrary to the widespread belief – the true average frequency of flares is higher on brighter stars. The erroneous opinion arose because the higher frequency of recorded flares found observationally on low-luminosity stars was not considered from the point of view of inevitable effects of observational selection (Petit, 1970; Gurzadian, 1971a). This essential effect of observational selection was not taken into account by Ambartsumian (1978) in calculations of the distribution of average frequencies of flares from the chronology of opening of flare stars in the Pleiades cluster.

Finally, we note that since statistical studies suggest that true flare frequencies and their maximum luminosities on brighter flare stars are higher than on less bright stars, red dwarfs should be considered as the most explicit carriers of the considered type of activity not because such activity is the most advanced, but only because it is the most noticeable on such low-luminosity cool stars.

E. On the physical sense of the power dependence $\tilde{\nu}(E)$. As stated above, the formal analysis of the observed energy spectrum of flares leads to physically significant conclusions on some statistical properties of the flare activity of stars. It is obvious that interpretation of the discovered power character of the dependence $\nu(E)$ can provide additional information on the nature of flares. The first attempt to analyze energy spectra of flares was undertaken by Pustil'nik (1988). He concluded that if a stellar flare was a final result of dissipation of nonpotential excesses of magnetic field in the active region and optical emission madeup a certain share of the total energy released during the dissipation, the flare radiation of the flare should be described by the relation

$$E(l) \propto \int_V \left(\frac{\Delta H^2}{8\pi}\right) dV \propto \begin{cases} H^2 l^3 & \text{at } l < z \ , \\ H^2 l^2 & \text{at } l > z \ . \end{cases} \tag{47}$$

Here, l is the characteristic size of the region of magnetic-field dissipation and z is the characteristic scale of heterogeneity of the field with height; $l < z$ and $l > z$ suggest volumetric and layered dissipation of nonpotential excesses of magnetic field, respectively.

Simple dimensional considerations make it possible to obtain the following expression from (47)

$$\tilde{\nu}(E > E_0) \propto \begin{cases} E_0^{(s-2)/3} & \text{for volume dissipation}, \\ E_0^{(s-2)/2} & \text{for layer dissipation}. \end{cases} \quad (48)$$

Here, s determines the dependence of turbulent velocity on the size of the turbulent element l

$$v_{\text{turb}} \propto l^s. \quad (49)$$

In the existing theories of stellar convection, the value of s varies in a rather narrow range: $s = 1/2$ in the case of magnetic-field convection, $s = 1/3$ for Kolmogorov's turbulence, and $s = 1/4$ for acoustic and Alfven turbulence. If the strength of the dissipating magnetic field H is assumed independent of l, for the given values of s (48) yields the sought power dependence $\nu(E)$ with spectral indices within 0.5–0.9. A weak negative correlation $H \propto l^{-1/3}$ would allow this range of spectral indices to be shifted to (0.85, 1.4).

Kasinsky and Sotnikova (1988) considered the model with a dissipating kinetic part of the energy of turbulent plasma motions. The following expressions were valid instead of (47)

$$E(l) \propto v_{\text{turb}}^2 l^3 \propto l^{2s+3} \quad (50)$$

and

$$\tilde{\nu}(E) \propto E^{(s-2)/(2s+3)}. \quad (51)$$

Thus, this model also yields the power dependence $\nu(E)$, but with systematically lower spectral indices, from 0.4 up to 0.5.

However, it should be noted that both proposed schemes include nonobvious assumptions: the structure of the considered size lives until the first flare and that its lifetime is $t(l) = l/v_{\text{turb}}$. However, the existence of homologous flares on the Sun shows that in the structures that generate flares such processes can develop repeatedly. On the other hand, if the lifetime of these structures is not determined by turbulence, but rather by the diffusion of magnetic field, i.e., the lifetime of the structure $t(l) = l^2/\nu_{\text{m}}$, where ν_{m} is the magnetic viscosity, the following expression is valid

$$\tilde{\nu}_{\text{flares}}(l > l_0) \propto l_0^{-3}, \quad (52)$$

and we obtain the power spectrum with $\beta = 1$ or $3/2$ for volumetric or layered dissipation of the magnetic field, respectively.

Lu and Hamilton (1991) showed that in the case of a general approach to coronal magnetic fields on the Sun as to the self-organizing complex system close to the critical state, and within the context of the notion on observed solar flares as a superposition of numerous elementary processes of reconnection in these fields, one should expect the energy distribution of flares with a spectral index of 0.4. But within the framework of the notion the spectral index of the energy spectrum should not depend on the level of solar activity, which contradicts the results of Kasinsky and Sotnikova (1988, 1989). Recently, Podlazov and Osokin (2002) simulated numerically flares within the framework of the updated avalanche concept proposed by Lu and Hamilton and found $\beta = 0.37$, which is still much lower than the appropriate values in the energy spectra of flare stars.

Litvinenko (1994) showed that within the framework of the traditional scheme of origin of flares from reconnection of magnetic flux tubes, one could obtain the distribution of probabilities of the occurrence of flares of certain energy that for high-energy flares – more than 10^{26} erg – yields the differential energy spectrum of power type $dN \propto E^{-7/4}dE$, i.e., $\beta = 0.75$ – and a softer spectrum for low-energy flares. Probably, an important confirmation of the situation in the region of sufficiently strong stellar flares were the above results of Audard et al. (1999, 2000).

The notions of current sheets, developed recently, led to the concept of essentially nonuniform dynamic structures of turbulent layers containing numerous clusters of the regions with switching over normal and sharply reduced conductivity. The passage of current through such a nonstationary medium is described by percolation, that is filtering through a stochastic network. In the process, the dissipation of current differs quantitatively from the case of electron–ion collisions. Based on the concept of fractals, the percolation theory also leads to a statistical power dependence of global properties of the system on the properties of its separate parts with exponents of 0.6–0.9 for two-dimensional systems and 1.5 for three-dimensional systems (Pustil'nik, 1997, 1999).

Thus, the power character of the dependence $\tilde{\nu}(E)$ follows from the consideration of stellar flares as processes proceeding in separate structures of turbulent magnetized plasma. This statement remains valid despite the significant progress in the actual content of these ideas.

2.3.1.3 Total Energy of Flare Emission

Simultaneous observations of stellar flares in three bands of the UBV system (Lacy et al., 1976) showed that there was a distinct statistical relation between total emission of flares in these bands

$$E_U = (1.20 \pm 0.08)E_B = (1.79 \pm 0.15)E_V , \tag{53}$$

which is valid in the energy range 10^{27}–10^{34} erg. It follows from (53) that

$$E_{UBV} = 2.4E_U = 2.9E_B = 4.3E_V . \tag{54}$$

For a wider $UBVRI$ system covering the whole optical range, similar relations can be obtained from long-term observations of one of the brightest and most active flare stars EV Lac (Alekseev and Gershberg, 1997a). As stated above, in 1986–95 227 flares on this star were recorded in the Crimea in the

$UBVRI$ system. Among them only those flares were selected for which all color parameters of flare emission at maximum brightness were determined with an error not exceeding $0\overset{m}{.}15$. Eight flares were selected and for all of them the brightness amplitude ΔU at maximum exceeded $1\overset{m}{.}8$, and the average values of color indices of flare emission at maximum brightness were

$$\langle U - B \rangle = -0\overset{m}{.}87 \pm 0\overset{m}{.}0.15, \quad \langle B - V \rangle = 0\overset{m}{.}04 \pm 0\overset{m}{.}01 ,$$
$$\langle V - R \rangle = 0\overset{m}{.}48 \pm 0\overset{m}{.}06, \quad \langle V - I \rangle = 0\overset{m}{.}81 \pm 0\overset{m}{.}21 . \tag{55}$$

(Earlier, from extensive UBV observations of eight flare stars for flares of different amplitudes, Moffett (1974) found average color indices of intrinsic flare radiation $\langle U - B \rangle = -0\overset{m}{.}88 \pm 0\overset{m}{.}31$ from 153 flares and $\langle B - V \rangle = 0\overset{m}{.}34 \pm 0\overset{m}{.}44$ from 77 flares. Ishida et al. (1991), from 127 flares on YZ CMi, AD Leo, and EV Lac, estimated $\langle U - B \rangle = -0\overset{m}{.}98 \pm 0\overset{m}{.}36$ and $\langle B - V \rangle = 0\overset{m}{.}24 \pm 0\overset{m}{.}30$, from 59 flares with the amplitudes of $\Delta U > 1\overset{m}{.}5$ $\langle U - B \rangle = -1\overset{m}{.}03 \pm 0\overset{m}{.}23$ and $\langle B - V \rangle = 0\overset{m}{.}14 \pm 0\overset{m}{.}20$ and from 17 flares with $\Delta U > 2\overset{m}{.}5$ $\langle U - B \rangle = -0\overset{m}{.}98 \pm 0\overset{m}{.}17$ and $\langle B - V \rangle = 0\overset{m}{.}05 \pm 0\overset{m}{.}13$. Pooling together the data from different publications, Grandpierre and Melikian (1985) estimated from 276 flares $\langle U - B \rangle \sim -1\overset{m}{.}0$ and from 174 flares $\langle B - V \rangle \sim 0\overset{m}{.}4$). Using the absolute calibration of the scale of stellar magnitudes, from (55) one can obtain

$$L_{UBVRI} = 1.6 L_{UBV} = 4.2 L_U = 3.8 L_B = 7.6 L_V . \tag{56}$$

Differences in the appropriate coefficients in (54) and (56) by 10–20% are caused both by the real dispersion of colors of flare emission at maximum brightness and small systematic changes of these colors during the development of flares. However, such distinctions are negligible for the estimates of the total flare energy.

It is natural to use (54) or (56) after the estimate of total flare emission in one of the photometric bands with the help of the energy spectrum of flares found for this band. The total energy of flare emission calculated from (37) depends essentially on the accepted limiting values of flare energy: on $E_{\max}$ for energy spectra with $\beta < 1$ and on $E_{\min}$, for $\beta > 1$. The calculations of Shakhovskaya (1979) showed that if first the total energy of optical emission of flares was calculated on the assumption that on each star $E_{\max} = 3 \times 10^{35}$ erg, the value corresponded to the strongest flares recorded in clusters, and $E_{\min} = 2 \times 10^{27}$ erg corresponded to the weakest flares on the faintest flare star in the solar vicinity CN Leo, and secondly the values $E_{\max}$ and $E_{\min}$ really recorded on each flare star were then used, the first estimate of total flare energy would exceed the second estimate of the same value by 2 to 40 times. Thus, even using an average of these two estimates of total energy of flare emission, one cannot exclude the error of the average by an order of magnitude.

The total energy of flare emission is usually specified as the ratio of time-averaged observations of the intensity of flare emissions to a certain constant

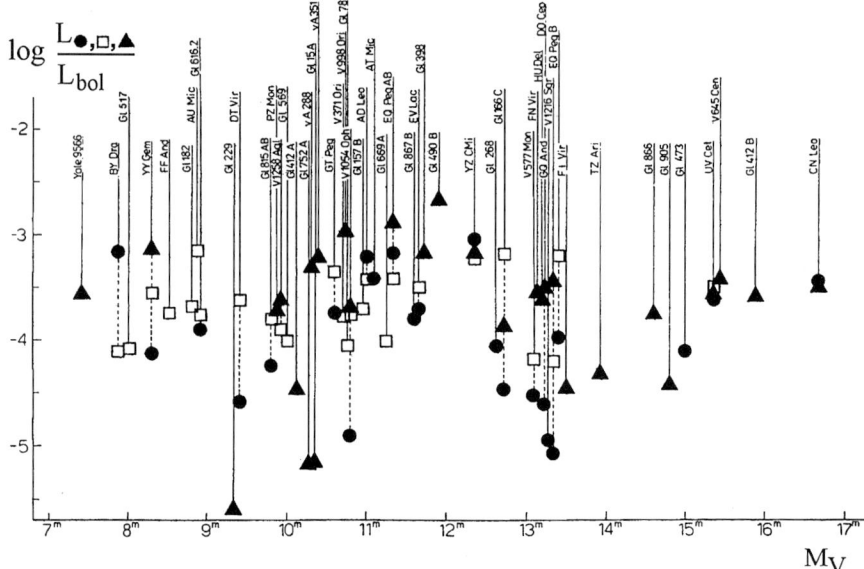

Fig. 42. Energy losses of flare stars for optical flare emission (*filled circles*), emission of quiet chromospheres (*squares*) and coronae (*filled triangles*) vs. bolometric luminosity of flare stars of different luminosity (Gershberg and Shakhovskaya, 1983)

intensity related to the stellar radiation rather than in absolute units. Thus, initially the sum of equivalent durations of flares to the time of monitoring $\Sigma P/T$ was often used, which apparently was equal to the ratio of the total flare energy in a certain photometric band to permanent stellar radiation in the same band. For the most active flare stars this dimensionless ratio varies from several thousandths to 3–5 hundredths. However, in the U and B bands regularly used in flare recording, flare stars emit only an insignificant portion of energy, therefore the ratio $\Sigma P/T$ determines only the visible pattern of flare activity. More suitable, from a physical viewpoint, is the ratio of total energy of all optical emission of flares to bolometric luminosity. The ratios for 23 flare stars presented in Fig. 42 are reproduced from the paper by Gershberg and Shakhovskaya (1983). In plotting the diagram, the values of limiting flare energies really recorded on each flare star were used as E_{max} and E_{min}. According to these data, the ratio of the time-averaged emission of flares to bolometric luminosity of flare stars is within 10^{-5}–10^{-3}. No dependence of this ratio on the absolute luminosity of stars was found.

As to the above concept of heating of stellar coronae by numerous microflares, it should be noted that the method of energy-spectrum evaluation does not provide confident estimates of the contribution of low-energy flares to the general flare emission: at $\beta < 1$ when formally their contribution is insignificant, there is no confidence that the spectral index found for strong

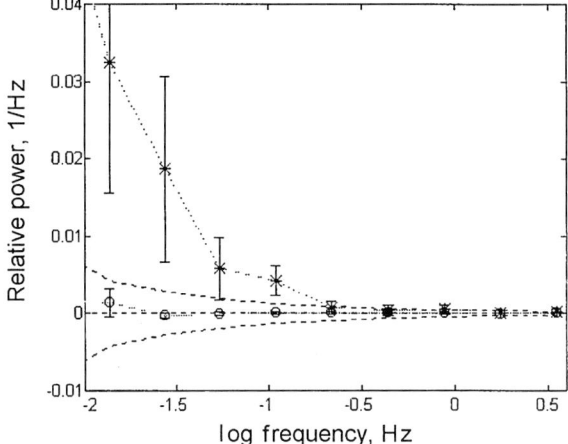

Fig. 43. Relative intensity of the variable component of brightness in the U band within 0.05–3.5 Hz: quiescent state of EV Lac (*asterisks*) and reference stars (*circles*) (Zhilyaev et al., 2001)

flares is valid for low-energy flares, and at $\beta > 1$ the lack of knowledge of the value of $E_{\min}$ brings essential uncertainty. The recent results obtained by Zhilyaev et al. (2001) with statistical photometry should be cited (Fig. 43): within the frequency range of 0.05–3.5 Hz the relative brightness of the variable component of EV Lac reaches 0.25% of the stellar brightness in the U band, which is almost an order of magnitude lower than the average intensity of individually recorded flares. This estimate is much stricter than the one obtained earlier by Beskin et al. (1995).

2.3.2 Estimates of Total Energy of Flares

The data on energy losses of stellar flares, which differ from the losses for optical radiation, are rather poor and sketchy.

It is known that the energy of radio emission of flares is much lower than the energy of their optical radiation $E_{\mathrm{radio}}/E_{\mathrm{optics}} = 10^{-5}$–$10^{-2}$ (Lovell, 1971).

The ratio E_X/E_{optics} differs from unity apparently by not more than one order of magnitude in both directions (Katsova and Livshits, 1986); probably the span of the estimate is partly explained by its essential variability during the flare (Kahler et al., 1982; Haisch, 1983). According to Haisch (1983), $L_X/L_{\mathrm{trans}} \sim 20$.

The contribution of radiation in the far ultraviolet to the total flare emission is rather uncertain. Based on ROSAT WFC, IUE and optical observations of the flare on BY Dra of 1 October 1990, Barstow et al. (1991) concluded that EUV radiation contained an essential part of the total flare emission.

Audard et al. (1999, 2000) carried out a statistical study of EUV flares. To carry out photometric observations of two young G dwarfs, 47 Cas and EK Dra, in the deep-photometry mode EUVE, they applied an algorithm of statistical identification of flares from the fidelity test of every readout and identified 28 flares, which occupied about 60% of the total duration of observations. They showed that the power function represented well the energy spectrum of these flares and found that in the energy range of 3×10^{33}–6×10^{34} erg the spectra had spectral indices of 1.2 ± 0.2, i.e., numerous weak flares made a decisive contribution (up to 90%) to the total EUV radiation of flares on these stars. Then, Audard et al. (2000) similarly investigated eight other stars, and in joint consideration of the whole sample of 10 F2–M6 objects found a certain change of spectral index of the energy spectrum from 1.3 for F–G through 0.9 for K to 0.8 for M dwarfs.

In March 2000, a wide cooperative program was arranged to monitor AD Leo: photometrical observations by three observatories, spectral studies by two observatories, ultraviolet observations by STIS HST and EUVE, and radio observations from Jodrell Bank (Hawley et al., 2003). The program made it possible to estimate the contribution of different components to the total flare emission. In analyzing four flares of different amplitude and duration in all cases the main contribution was attributed to continuous emission, at the impulsive phase the contribution of ultraviolet lines achieved several per cent of the total radiation and exceeded the contribution of optical lines, whereas at the stage of smooth decay the optical lines provided up to 10–20% of the total radiation and exceeded the contribution of ultraviolet lines.

If $E_{\text{optics}} \sim E_{UV} \sim E_X$ and $< E_{EUV}$, as a rough approximation one can accept $E_{\text{bol}} \sim (5-6) E_{\text{optics}}$.

Thus far, no direct data on the energy of stellar flares spent for particle acceleration have been obtained. As to the kinetic energy of the matter of flares, some preliminary estimates can be made on the basis of available spectral data.

It is known that in the spectra of flares on red dwarf stars some shifts of emission lines for several tens of km/s and appreciable broadening of their profiles of the lines – up to 10–15 Å at the level of half-intensity – were found (Greenstein and Arp, 1969; Gershberg, 1978). The fact that in the initial stages of flares, emission lines often have asymmetric profiles with a red wing extended noticeably more than the blue one (Gershberg and Shakhovskaya, 1971; Kulapova and Shakhovskaya, 1973; Bopp and Moffett, 1973; Bopp, 1974b; Byrne, 1989; Houdebine et al., 1993a) indicates that the motion of matter plays an essential, maybe even dominating, role in the observed variations of profiles. Thus far, only one detailed study of flare kinematics has been undertaken for the flare on AD Leo of 28 March 1984 (Houdebine et al., 1993a). However, the upper and lower estimates of kinetic energy of this event differ by two orders of magnitude. From the analysis of ultraviolet spectra, Byrne (1993a) concluded that the kinetic energy of matter in flares exceeded their radiation, whereas Gunn et al. (1994b) estimated the excess of emission energy over the kinetic energy from the spectra in the optical range of the flare on AT Mic as two orders of magnitude. In ten other flares on red dwarf stars the analysis of the observed profiles of emission lines near maximum brightness made it possible to present them as a superposition of different Gaussians (Byrne, 1989; Robinson, 1989; Eason et al., 1992; Abdul-Aziz et al., 1995): one narrow profile of instrumental width and one wide profile or the sum of several narrow Gaussians. This testifies to the kinematic heterogeneity of flares: apparently the narrow unshifted components are kinematically similar to the quiet chromosphere, whereas other components – widened and displaced – correspond to the motion of matter with velocities of several hundreds of kilometers per second. The removal of matter from a stellar surface at about 400 km/s was found from displaced components of the emission line HeII λ 4686 Å in the spectrum of active EV Lac (Abranin et al., 1998b). These emissions were noticeably delayed with respect to flare maxima; thus, they were interpreted as the analogs of solar coronal ejection of matter into the interplanetary space.

Now let us focus on solar flares.

It is known that many characteristics of recorded stellar flares are very close to those of relatively rare solar white-light flares. Neidig and Kane (1993) considered the data on eight flares of this type and found that in these events the ratio of the power of nonthermal electrons with an energy of more than 50 keV, which was responsible for hard X-ray emission, to that of optical emission varied within 0.5–22, and the ratio of intensities of optical emission and soft X-rays in the range of 1–8 Å varied within 4–100.

According to the results of Kurochka (1987), the maximum energy of the total optical and ultraviolet emission of hydrogen plasma of solar flares in 1978–79 calculated from H_α emission was equal to a few 10^{32} erg. According to Lin and Hudson (1976), in the strongest solar flares in August 1972 the recorded optical and ultraviolet radiation reached 5×10^{31} erg that made up from 8 to 33% of the total energy losses including the radiation in the whole electromagnetic spectrum, losses for acceleration of fast particles and motion of matter. Based on these estimates, we conclude that the total energy of the strongest solar flares apparently does not exceed 10^{33} erg. We note that, according to Hudson (1978), the energy spectra of radiation of solar flares in soft X-rays, near 20 keV and in the microwave range have a power character with spectral indices of 0.80–0.85, which practically coincides with the optical and ultraviolet data. For proton fluxes from solar flares, the spectral index noticeably differs: $\beta = 0.15$, but the contribution of these losses to the total energy of flares is rather insignificant. The same is valid for hard X-ray emission of solar flares for which, according to Dennis (1985), $\beta = 0.40$. On the other hand, as searches for solar "superflares" showed (Lingenfelter and Hudson, 1980), during 19 and 20 cycles of solar activity, during the last 7000 and even 10^7 years there were no flares on the Sun in which the energy of hard particles exceeded the flare of 4 August 1972 by more than an order of magnitude.

As to flare stars, Fig. 38 shows that on the brightest flare stars in the solar vicinity – BY Dra and AD Leo – in the strongest flares in the B band more than 10^{34} is emitted, maybe even 10^{35} erg. On flare stars in the Pleiades and in Orion where the technique of simultaneous observations of several hundred flares stars enables recording of very rare events on brighter stars, the maximum value of E_B approaches 3×10^{35} erg. Taking into account (56) and assuming that in stellar flares the same ratio between radiative and nonradiative losses of energy as for the strongest solar flares is valid, the total energy of the strongest stellar flares of considered type should reach 3×10^{36} erg.

Thus, to interpret the observations, the theory of stellar flares of the considered type should explain the energy release up to 10^{33} erg in one event on the Sun and at least up to 3×10^{36} erg on the brightest UV Cet-type stars. But there are reports of the existence of extremely rare flares, whose energy is higher than the above limits by two orders of magnitude (Schaefer et al., 2000; Batyrshinova and Ibragimov, 2001).

2.4 Dynamics and Radiation Mechanisms of Flares in Different Wavelengths

Though flares recorded on red dwarf stars of the UV Cet-type are extremely diverse, available observational data make it possible to present a preliminary physical picture of these short-term phenomena. To sort the data logically, let us use the data on solar flares.

It is known that the so-called precursors of flares are observed on the Sun. One of the indisputable precursors is the activation of prominences: shortly before a flare, formerly quiet filament structures, which are seen as hanging bright structures in the corona and are found as dark filaments in rays of the red line of hydrogen on the solar disk, come into motion with velocities of tens of km/s in the place where the flare then occurs. These motions are connected with incipient reorganization of the magnetic field of the appropriate active region as prominences "hang" on magnetic force lines. There are observations in which the reorganization is recorded directly.

A confidently recorded onset of a solar flare is concerned with the occurrence of nonthermal X-ray emission. The spectrum of this radiation and its polarization features unambiguously evidence that this radiation is nonthermal and is caused by fast particle fluxes in the lower corona. The question on the mechanism of formation of the fluxes and the source of their energy is still disputed. However, most existing phenomenological models of these phenomena start from the observed phase, since strong fluxes of fast particles allow one to understand the subsequent thermal X-ray, ultraviolet, optical, and radio emission of various natures that appears further as the flare develops. Thus, it is reasonable to start the consideration of observations of stellar flares from the discussion of X-ray and radio emission from the corona. Then, we consider the results of ultraviolet observations that provide data on the luminescence of the transition region and the upper chromosphere, and the most numerous data on optical emission of flares, which is formed in the stellar chromosphere and probably encompasses the upper layers of the photosphere.

Widely using the research results for solar activity as guiding the studies of stellar activity, one should not forget about the essential quantitative distinctions between the phenomena. For example, the average flux F_X from a unit surface of dMe stars in the quiescent state is higher than the appropriate flux from the Sun at least by 3 orders of magnitude and by hundreds of times at the maximum of the 11-year solar cycle; the time-averaged optical emission of flares differs by 3–4 orders of magnitude. The qualitative difference of solar and stellar flares is that radiative losses in the optical, ultraviolet and X-ray regions in the latter are of the same order of magnitude, whereas in the former they differ essentially. Therefore, starting research of stellar activity from simple scaling of solar phenomena, one should be ready to discover qualitatively new processes on stars.

2.4.1 X-ray Emission of Flares

If optical flares on red dwarf stars were discovered occasionally, the accompanying bursts in hard radiation were purposefully sought for a long time.

Already in 1967, the data on the level of cosmic noises recorded from the Crimea and Gorky were considered in studying the geophysical effects of solar activity. These records were used to find sudden absorption of the noises during recorded optical flares on red dwarfs, the absorptions occur under additional ionization of the earth's atmosphere by sporadic hard radiation (Gershberg et al., 1969). This effect was not found, therefore the upper limit of the ratio of luminosities of stellar flares in two wavelength ranges was estimated very roughly as $L_X/L_B < 10^5$.

Within the framework of certain physical notions, X-ray emission of stellar flares for the first time was precalculated by Grindlay (1970). He considered a number of models, which explained radiation of the bursts recorded by then on the UV Cet at meter wavelengths by synchrotron radiation of monoenergetic relativistic electrons or electrons with a power-energy spectrum. As a result, he estimated the efficiency of the relativistic particles in the generation of X-ray emission due to the inverse Compton effect and nonthermal bremsstrahlung emission. He found that the latter mechanism was more effective for radiation in the range of energies $E > 10\,\text{keV}$. Then, Gurzadian (1971b) estimated the expected X-ray emission of flares, reasoning from his own concept (Gurzadian, 1965) of stellar flares as a result of the occurrence of a large number of subrelativistic (~ 1 MeV) electrons above the surface of a cool star and their interactions with photospheric radiation. Gurzadian also concluded that bremsstrahlung emission of fast electrons was most effective in the generation of X-ray emission. But, according to Edwards (1971), the level of X-ray emission of stellar flares predicted by Grindlay should lead to such a high total radiation of flare stars in the Galaxy that it would essentially exceed the observed X-ray background. Tsikoudi and Hudson (1975) analyzed the data obtained by the OSO-3 X-ray telescope during 82 optical flares on four red dwarf stars, but did find the effects of flares. Thus, they established another rigid restriction on X-ray emission of stellar flares $L_X/L_U < 300$, which also contradicted Grindlay's model and to a lesser degree, the Gurzadian model. However, according to Kahler and Shulman (1972) and Crannell et al. (1974), the contradiction is completely eliminated if the nonthermal models of Grindlay and Gurzadian are rejected and the expected X-ray emission of stellar flares is estimated by scaling the known data on solar flares.

For the first time, X-ray emission of a flare on a red dwarf was recorded by the Astronomical Netherlands Satellite (ANS) during monitoring of YZ CMi on 19 October 1974 (Heise et al., 1975). The satellite was equipped with two instruments produced by the Utrecht Laboratory for Space Research for recording soft (0.2–0.28 keV) and medium (1–7 keV) X-rays. Another instrument produced in the Cambridge Center for Astrophysics (USA) was designed for detecting hard X-rays (1–30 keV). In soft and medium X-rays, the emission burst on YZ CMi occurred practically simultaneously, but in soft X-rays the flare was observed for six minutes, whereas in medium X-rays – only for one and a half minutes. The Cambridge instrument did not detect the flare. The luminosity at flare maximum $L_X^{\max}$ in the specified energy ranges reached 2.5×10^{29} and 3.6×10^{30} erg/s, and full radiation E_X was 4×10^{31} and 2×10^{32} erg, respectively. The maximum recorded flux in soft X-rays was 0.14 quanta/cm^2 s. (Since flare stars are the objects nearest to the Sun, distances to them are determined rather precisely from trigonometrical parallaxes and interstellar absorption that depends on the number of neutral atoms on the ray of sight can be neglected. Thus, in calculating the absolute luminosity L_X from the observed flux F_X an error can be caused by absolute calibration of the equipment and the deviation of radiation from isotropicity, the latter is practically excluded for thermal emission.) During the X-ray flare the star was not monitored in other wavelength ranges, and this flare was only recorded for 5 h, 5–12 min at each revolution.

The flare on UV Cet of 8 January 1975 was recorded in soft X-rays by ANS and in the optical range from the Earth (Heise et al., 1975). The optical flare was rather strong – $\Delta U \sim 6^m$ – and lasted more than six minutes, but X-ray emission of the flare was recorded from optical maximum only for 48 s. At the moment of maximum the recorded flux was 0.21 quanta/cm^2 s, which corresponded to the flare luminosity of 6×10^{28} erg/s. According to the observations, the ratio $L(0.2 - 0.28\,\text{keV})/L_U \leq 0.03$ was valid for this flare. These values appeared to be less than all these precalculated, the closest estimate to the real value was that of Kahler and Shulman. On the whole, ANS observations of UV Cet lasted for more than five hours, over this time three other weak optical flares were recorded, but they did not have appreciable X-ray emission.

2.4 Dynamics and Radiation Mechanisms of Flares in Different Wavelengths

The first successful observations of stellar flares in X-rays stimulated Mullan (1976a) to develop a purely thermal quantitative model. According to the model, hot coronal gas arising in the stellar atmosphere during a flare determines its further development: flare X-ray emission is caused by thermal radiation of gas, optical emission arises due to heating of underlying layers through heat conductivity. In the model, the ratio L_X/L_{opt} near maximum brightness should be determined only by the ratio of radiative losses and losses for heat conductivity of hot coronal gas. The expected value of this ratio in flares on red dwarfs should be noticeably lower than on the Sun and make a few thousandths or hundredths.

The concept of thermal X-ray emission of stellar flares put forward by Mullan has now become conventional.

During the joint USSR–USA manned space flight of the Soyuz–Apollo satellites the flare star Proxima Cen was observed in soft X-rays in the range of 0.065–0.28 keV for 1100 s on 21 July 1975 and for 78 s on 22 July 1975 using the telescope produced at the University of California (Haisch et al., 1977). During the second observational session, the signal, which noticeably exceeded both the background and the signal recorded the day before, was registered. Unfortunately, the X-ray observations were not accompanied by optical monitoring. American researchers attributed the signal to a flare and postulated its purely thermal nature. In the context of this hypothesis quantitative characteristics of high-temperature components of the stellar flare were estimated for the first time. These estimates used a simple relation connecting specific luminosity of optically thin plasma of unit density and determined chemical composition $l_\lambda(T)$ with the luminosity of the whole plasma body $L_\lambda = \int l_\lambda n_e^2 dV$; for homogeneous medium the relation is as follows

$$L_\lambda = l_\lambda(T) \int n_e^2 dV = l_\lambda(T) EM , \qquad (57)$$

where EM is the volume emission measure of the source. To use (57), one should determine the plasma temperature from independent considerations. But if it is somehow fixed, from the found values $L_\lambda(T)$ and EM one can easily estimate the total X-ray emission in the range of sensitivity of the detector

$$L_X = EM \int l_\lambda(T) d\lambda . \qquad (58)$$

According to Haisch et al. (1977), if on 22 July 1975 thermal emission of the optically thin plasma was recorded, at 2.5–100 MK the emission measure of the radiating volume was 3×10^{52}–1×10^{53} cm^{-3}, and its full luminosity throughout the X-ray range varied from 6×10^{29} to 3×10^{30} erg/s. Similarly, they considered X-ray flares on YZ CMi and UV Cet and estimated the full X-ray luminosity as 8×10^{30} and 2×10^{30} erg/s and emission measures as 5×10^{53} and 1×10^{53} cm^{-3}, respectively, for a temperature of radiating gas in the flares of 10^7 K. (In the described calculations the values of $l(T)$ were used for the standard chemical composition, found by Kato (1976) and Raymond et al. (1976); today, usually the Raymond-Smith and MEKA models and modifications of the latter are used.)

Quantitative parameters of the above stellar flares and those recorded later in different X-ray experiments are summed up in Table 13. The data are not quite uniform, since they were taken from original publications that appeared at different times and were not adjusted to the development of stricter analysis methods. Moreover, for burning T_b and decay T_a times of flares some publications present the full duration of these phases, and the others, the times of an e-fold increase of brightness to the maximum level and an e-fold decrease after the maximum. In some case, the values L_X^{max} and E_X were calculated for Table 13 from the published values of luminosity and total energy in the whole X-ray range. Therefore, Table 13 is not suitable for strict statistical consideration but is convenient for approximate estimates.

In late 1975, the first simultaneous monitoring in optical, radio, and X-ray wavelength ranges was carried out (Karpen et al., 1977). Observations in the range of 0.15–50 keV were executed from SAS-3 using the X-ray telescope produced at the Massachusetts Institute of Technology. The X-ray monitoring of YZ CMi continued during 15 optical and 7 radio flares but did not reveal hard emission. The three strongest optical flares yielded the ratio $L(0.15$–$0.8\,\text{keV})/L_B < 0.3$. This value is between the values predicted by Kahler and Shulman (1972) and Mullan (1976a).

In May 1977, Proxima Cen was observed within the framework of similar cooperative radio–optical–X-ray studies from SAS-3 (Haisch et al., 1978). During the X-ray observations 22 optical flares were recorded from the Earth, but no reliable X-ray flares were revealed. Like before, for the strongest optical flare the ratio $L_X/L_{opt} < 0.08$ was obtained.

Important results on hard flare emission from the UV Cet-type stars were obtained by the American High-Energy Astrophysics Observatory HEAO-1. Between August 1977 and January 1979 the satellite carried out three all-sky surveys in soft (0.15–3 keV), medium (2–20 keV), and hard (2–60 keV) X-rays. Each source was in the field of sight of the HEAO-1 instruments for 10 s at each revolution, every 30 min within a week. In October 1977, two X-ray flares on AT Mic were recorded. The flare on 25 October 1977 was so strong that the for first time an X-ray spectrum of the flare was recorded (see Fig. 44) (Kahn et al., 1979). In the spectrum within the energy range of 0.2–20 keV thermal radiation of an optically thin plasma was confidently found at 31 ± 7 MK and the iron emission line K_α with an excitation energy of 6.7 keV, which at the particular temperature really had to be excited effectively. The total X-ray luminosity of this flare was 1.6×10^{31} erg/s, the emission measure of 1.4×10^{54} cm^{-3}, and total X-ray emission of at least 5×10^{32} erg. If one assumes that the spectrum of the second, weaker flare on AT Mic of 27 October 1977 was the same, the following parameters are valid for it: total X-ray luminosity of 4.6×10^{30} erg/s and $EM = 4.0 \times 10^{53}$ cm^{-3}.

2.4 Dynamics and Radiation Mechanisms of Flares in Different Wavelengths 245

Table 13. Qualitative parameters of stellar flares

Star	Date	Experiment	Range (keV)	T_b (min)	T_a (min)	L_X^{max} (erg/s)	E_X (erg)	T (M K)	ME (cm^{-3})	h (cm)	n_e (cm^{-3})	Description and Analysis of Experiment
YZ CMi	19.10.74	ANS	0.2 – 0.28	6		2.5×10^{29}	4.2×10^{31}	10^*	5×10^{53}			Heise et al., 1975
			1–7		1.5	3.6×10^{30}	1.9×10^{32}					
UV Cet	8.1.75	ANS	0.2–0.28		0.8	6×10^{28}	3×10^{30}	10^*	1×10^{53}			Heise et al., 1975
Prox Cen	22.7.75	Soyuz–Apollo	0.065–0.29		>1.3	1.5×10^{29}		2.5–100*	5×10^{52}			Haisch et al., 1977
YZ CMi		SAS-3	0.15–50									Karpen et al., 1977
Prox Cen		SAS-3	0.15–50					10^*	$<5 \times 10^{51}$			Haisch et al., 1978
AT Mic	25.10.77 I	HEAO-1	0.15–18		>0.5	7×10^{30}	$>2 \times 10^{32}$	30	1.4×10^{54}			Kahn et al., 1979
	II							40				Connors et al., 1986
AT Mic	27.10.77	HEAO-1	0.15–18			2×10^{30}		30^*	4.0×10^{53}			Kahn et al., 1979
AD Leo	22.11.77 I	HEAO-1	0.15–2.5			5×10^{29}		30^*	1.1×10^{53}			Kahn et al., 1979
	II					7×10^{29}		30^*	1.4×10^{53}			Kahn et al., 1979
EV Lac	25.12.77	HEAO-1	0.5–20	<30	~22	2.9×10^{29}		10–30*	3×10^{52}	5×10^9	$(4-10) \times 10^{11}$	Ambruster et al., 1984
EV Lac	30.12.77	HEAO-1	0.5–20		<78	2.2×10^{30}	4×10^{33}	10–30*	2×10^{52}			Ambruster et al., 1984
EQ Vir	13.1.78	HEAO-1 Einstein	0.5–20			7.4×10^{30}						Ambruster and Wood, 1984, 1986
Wolf 630	28.2.79	IPC	0.5–4.8	11	>6	4×10^{29}		32	8×10^{51}			Johnson, 1981, 1983b
WX UMa	24.5.79	IPC		8		1×10^{28}						Johnson, 1981, 1983b
Prox Cen	6.3.79	IPC	0.2–4	5	25	7.4×10^{27}	$>6 \times 10^{30}$	12–17	1×10^{51}	10^{10}	10^{11}	Haisch et al., 1980, 1981
Prox Cen	7.3.79	IPC			50							Haisch et al., 1980, 1981
YZ CMi	25.10.79	IPC	0.2–4	1	10	3×10^{28}	3.6×10^{31}	20	2×10^{51}	10^{10}	3×10^{11}	Kahler et al., 1982
YZ CMi	27.10.79	IPC		7			4×10^{28}					Kahler et al., 1982
AD Leo	13.5.80	IPC	0.2–3.5		160	3×10^{29}	2×10^{33}	12	1×10^{52}	1.3×10^{10}		Favata et al., 2000a
TZ Ari	14.7.80	IPC		5	>7	3×10^{28}			2×10^{51}			Johnson, 1981, 1983b

(continued)

Table 13. cont.

Star	Date	Experiment	Range (keV)	T_b (min)	T_a (min)	L_X^{max} (erg/s)	E_X (erg)	T (M K)	ME (cm^{-3})	h (cm)	n_e (cm^{-3})	Description and Analysis of Experiment
Prox Cen	20.8.80	IPC	0.2–4	34	130	1.3×10^{28}	2×10^{31}	27				Haisch et al., 1983
HD 27130	19.9.80	IPC+MPC	0.2–10		>200	10^{31}	2×10^{34}	50	6×10^{53}	2×10^{10}	4×10^{11}	Stern et al., 1983
BD+14° 690		IPC	0.2–4	16	30	2×10^{30}						Stern and Zolcinski, 1983
vA 500		IPC	0.2–4	16	30							Stern and Zolcinski, 1983
vA 288		IPC	0.2–4		~400							Stern and Zolcinski, 1983
Gl 669	27.3.79	IPC	0.2–4	<3	10	$>10^{29}$	10^{31}	24				Harris and Johnson, 1985
Gl 669B	19.8.80	HRI	0.2–4	<8	<8							Harris and Johnson, 1985
Gl 34B	4.2.80	HRI	0.2–4									Harris and Johnson, 1985
Gl 338A	27.4.80	HRI	0.2–4	<12								Harris and Johnson, 1985
V1216 Sgr	24.3.81	IPC	0.2–4	<2	5	5×10^{27}	10^{30}					Agrawal et al., 1986
HZ 1733	8.2.81	IPC EXOSAT	0.2–4		17	6×10^{28}	6×10^{31}	20	3×10^{51}			Caillault and Helfand, 1985
1E0419.2+ +1908	21.8.83	LE	0.02–2.5	40	80	6×10^{29}	6×10^{32}	20*	3×10^{53}			Smale et al., 1986
π^1 UMa	31.1.84	LE+ME	0.1–10	8	24		2×10^{33}	30	7×10^{52}	2×10^{10}	7×10^{11}	Landini et al., 1986
Wolf 1561		LE	0.05–2	10	30	5×10^{29}	5×10^{32}					Pallavicini et al., 1990a
BY Dra	24.9.84	LE	0.05–2	15	30	1×10^{30}	6×10^{32}	$\lesssim 10^*$	1.2×10^{52}			de Jager et al., 1986
EXO 040830- -7134.7	11.10.84	LE	0.05–2			1.2×10^{30}	7×10^{33}					Van der Woerd et al., 1989
YY Gem	14.11.84 I II	LE+ME	0.05–6	35 10	65 25	4.7×10^{30} 1.3×10^{30}	6×10^{33} 4×10^{32}	from 64 to 24	6×10^{52}			Pallavicini et al., 1990ab Pallavicini et al., 1990ab
Gl 867 A	18.11.84 I	LE	0.05–2	8	8	4.4×10^{29}	6×10^{31}					Pollock et al., 1991
Gl 867AB	II	LE	0.05–2	10	15	5.1×10^{29}	1×10^{32}					Pollock et al., 1991

2.4 Dynamics and Radiation Mechanisms of Flares in Different Wavelengths

Table 13. cont.

Star	Date		Band	Range									Reference
UV Cet	6.12.84	I	LE	0.05–2	<2	4	2.0×10^{28}	2×10^{30}					Pallavicini et al., 1990a
		II	LE		<2	5	1.9×10^{28}	4×10^{30}					Pallavicini et al., 1990a
		III	LE		5	10	2.5×10^{28}	5×10^{30}					Pallavicini et al., 1990a
		IV	LE		<3	10	3.0×10^{28}	5×10^{30}					Pallavicini et al., 1990a
EQ Peg	7.12.84	I	LE	0.04–2	5	15	2×10^{29}	2×10^{31}					Haisch et al., 1987
		II	LE		45	20	4×10^{29}	2×10^{32}	26	1×10^{52}	1×10^{10}	2×10^{11}	Haisch et al., 1987
			ME	1.5–5.5	20	40	7×10^{28}			5×10^{51}	6×10^{9}		
		III	LE		5	10	1.5×10^{29}	2×10^{31}					Haisch et al., 1987
		IV	LE		10	15	3×10^{29}	4×10^{31}					Haisch et al., 1987
Prox Cen	2.3.85		LE		10	25	4×10^{27}	2×10^{30}					Pallavicini et al., 1990a
YZ CMi	4.3.85	I	LE	0.02–2	25	30	1×10^{29}	7.8×10^{31}					Doyle et al., 1986
		II	LE	0.02–2	10	40	1×10^{29}	1.4×10^{32}					Doyle et al., 1988b
Wolf 630	8.3.85		LE	0.05–3	1	30	1.4×10^{29}		22	8×10^{51}			Tagliaferri et al., 1987
			ME	2–7	1	30							
VB 8 AB	8.3.85		LE		4	60	7.3×10^{28}	8×10^{31}		6×10^{50}		1×10^{11}	Johnson 1987; Tagliaferri et al., 1987, 1990
AT Mic	24.5.85		LE		3	30	6.5×10^{29}	1.3×10^{33}	39	5×10^{52}			Pallavicini et al., 1990a
UV Cet	4.8.85	I	LE		<5	10	2×10^{28}	5×10^{30}					Pallavicini et al., 1990a
		II	LE		5	15	2×10^{28}	5×10^{30}					Pallavicini et al., 1990a
		III			5	10	2×10^{28}	9×10^{30}					Pallavicini et al., 1990a
EQ Peg	6.8.85	I	LE	0.04–2		120	2×10^{30}	4×10^{33}	from 42 to 18	from 2×10^{53} to 2×10^{52}			Pallavicini et al., 1986, 1990a
			ME	1–20									
		II	LE		10	60	3.4×10^{29}	3×10^{32}					Pallavicini et al., 1986, 1990a
Wolf 630	25.8.85		LE	0.05–2	5	15	4.6×10^{29}	9×10^{32}	from 48 to 29	3×10^{53}	9×10^{8}	1×10^{12}	Doyle et al., 1988a
			ME	2–7	5	25		4×10^{32}					Tagliaferri et al., 1987

(continued)

Table 13. cont.

Star	Date	Experiment	Range (keV)	T_b (min)	T_a (min)	L_X^{max} (erg/s)	E_X (erg)	T (M K)	ME (cm^{-3})	h (cm)	n_e (cm^{-3})	Description and Analysis of Experiment
EV Lac	13.10.85	LE			120	1.3×10^{29}	1.3×10^{32}					Ambruster et al., 1989a
	15.10.85	LE		10	75	1.3×10^{29}						Ambruster et al., 1989a
YZ CMi	19.11.85	LE		8	20	8.2×10^{29}	7×10^{31}					Pallavicini et al., 1990a
AD Leo	15.12.85	LE		10	60	4.8×10^{29}	1.5×10^{33}					Pallavicini et al., 1990a
BD+48°1958A	16.12.85	LE+ME				6×10^{29}	2×10^{32}		3×10^{51}		2×10^{11}	Rao et al. 1990
UV Cet	23.12.85 I	LE		<3	20	1.0×10^{29}	6×10^{31}	10^*			5×10^{11}	de Jager et al., 1989
		ME		<3	30			40			2×10^{11}	
	II	LE		<3	10	1.0×10^{29}	3×10^{30}					Pallavicini et al., 1990a
	ROSAT											
CC Eri	10.7.90	PSPC	0.2–2	60	220	7×10^{29}		28–12				Pan & Jordan, 1995
HD 147365	30.7.90	PSPC		360	360	4×10^{29}	1×10^{34}					Güdel et al., 1995b
47 Cas	19.8.90	PSPC	0.1–2.0		<90	4.3×10^{30}		25^*		4×10^{10}	5×10^{11}	Güdel et al., 1995b
47 Cas	21.8.90	PSPC	0.1–2.0		90	1.2×10^{31}	5×10^{34}	25^*	4×10^{53}			Güdel et al., 1995b
AZ Cnc	19.10.90	PSPC	0.1–2.4		500							Fleming et al., 1993
Prox Cen	8.10.90	PSPC	0.1–2.4		300							Fleming et al., 1993
CN Leo	25.11.90	PSPC	0.1–2.4		200							Fleming et al., 1993
EK Dra	11.90	PSPC	0.4–2.4		60	1×10^{30}	4×10^{33}			$(1–2) \times 10^{10}$	2×10^{11}	Güdel et al., 1995a
EQ 1839.6+8002	14.2.91	LAC GINGA	1–37	4	20	10^{31}	10^{34}					Pan et al., 1997
VB 8	25–27.2.91	PSPC			>100			2 and 10	5 and 5×10^{49}			Springfellow, 1996
												Reale & Micela, 1998

2.4 Dynamics and Radiation Mechanisms of Flares in Different Wavelengths

Table 13. cont.

Object	Date	I/II	Instrument									Reference
AD Leo	8.5.91	I	PSPC									Springfellow, 1996
		II	PSPC	60	160	2×10^{29}	2×10^{32}	12	1×10^{51}	$<1.7 \times 10^{10}$		Favata et al., 2000a
UV Cet	31.12.91		PSPC					10			10^{11}	Stepanov al., 1995
UV Cet	2.1.92	I	PSPC	0.1	0.1							Schmitt et al., 1993a
		II	PSPC	0.1	5							Schmitt et al., 1993a
EV Lac	13.7.92		PSPC					8 and 35	8 and 23×10^{51}			Sciortino et al., 1999
G 102-21	23.9.93		PSPC	0.2–2.4		16	5×10^{29}	$>5 \times 10^{32}$	5 and 16	$(8-50) \times 10^{51}$		10^{12}	Micela et al., 1995
YY Gem	26.10.93		ASCA	0.4–7	1.6	70			11–3				Gotthelf et al., 1994
Castor AB	26.10.93		ASCA	0.4–7									Gotthelf et al., 1994
H II 2034	11.8.90		PSPC	0.5–1.8	<160		3×10^{31}		10				Schmitt et al., 1993b
H II 2147			PSPC		<180	<1000	2.5×10^{31}		4 and 14	2 and 11×10^{53}			Gagné et al., 1994a,1995
H II 1516			PSPC		<80	80	1.6×10^{31}		13	2×10^{54}		$>1.3 \times 10^{11}$	Gagné et al., 1995
HCG 97			PSPC		<60	80	1.6×10^{30}						Gagné et al., 1995
H II 174			PSPC		90	220	3.2×10^{30}						Gagné et al., 1995
H II 191			PSPC		16	80	3.2×10^{30}						Gagné et al., 1995
H II 212			PSPC		<70	100	2×10^{30}						Gagné et al., 1995
HCG 143			PSPC		10	25	2.5×10^{30}						Gagné et al., 1995
H II 345			PSPC		<70	110	5×10^{30}						Gagné et al., 1995
HCG 181			PSPC		16	80	3×10^{30}						Gagné et al., 1995
H II 1100			PSPC		50	160	3×10^{30}						Gagné et al., 1995

(*continued*)

Table 13. cont.

Star	Date	Experiment	Range (keV)	T_b (min)	T_a (min)	L_X^{max} (erg/s)	E_X (erg)	T (M K)	ME (cm^{-3})	h (cm)	n_e (cm^{-3})	Description and Analysis of Experiment
XRNumber 191		PSPC		<75	180	8×10^{30}						Gagné et al., 1995
H II 2244		PSPC		<80	130	4×10^{30}						Gagné et al., 1995
HE 421		PSPC		<70	<750	6×10^{30}	$>6 \times 10^{33}$					Prosser et al., 1996
AP 20		PSPC		<150	<840	1×10^{31}	$>2 \times 10^{34}$					Prosser et al., 1996
AP 108		PSPC		<70	<840	4×10^{31}	$>3 \times 10^{34}$					Prosser et al., 1996
Prox Cen	18–20.3.94	ASCA	0.5–12					44 and 7				Haisch et al., 1995
EQ Peg	23.6.94	ROSAT HRI	0.1–2.4	2	10	1×10^{30}	2×10^{32}	27	3×10^{52}		4×10^{10}	Katsova et al.,2002
UV Cet	5/6.01.95	ROSAT+ASCA	0.1–2.4		80	3×10^{27}	2×10^{31}					Güdel et al., 1996
UV Cet	6/7.01.95	ROSAT+ASCA	0.5–10		70	6×10^{27}	3×10^{31}					Güdel et al., 1996
HD 197890	20.4.95	ASCA	0.4–10			2×10^{30}		9 and 37	6 and 12×10^{52}			Singh et al., 1999
Gl 890	19.11.95	ASCA	0.4–10		40	4×10^{29}		9 and >23	2 and 0.2×10^{52}			Singh et al., 1999
AD Leo I	4.5.96	ASCA	0.5–10		10	9×10^{28}	1×10^{32}	20	1×10^{52}	$<10^{10}$		Favata et al., 2000a
II		ASCA	0.5–10		20	3×10^{29}	8×10^{31}	48	5×10^{52}	4×10^{9}		Favata et al., 2000a
III		ASCA	0.5–10		60	7×10^{28}	2×10^{32}	38	2×10^{52}	7×10^{9}		Favata et al., 2000a
AU Mic	13.6.96	RXTE	2–15	<13	1000			20 and 93				Gagné et al., 1998
EQ Peg	2.10.96	RXTE	2–10									Gagné et al., 1998
AD Leo	23–24.4.97	SAX						4, 11 and >100	2, 3 and 0.3×10^{51}			Sciortino et al., 1998
VB 10	19.10.97	PSPC			<19	3×10^{27}						Fleming et al.,2000
EQ Peg	3.12.97	SAX	0.1–5	300	330							Landini et al.,2001

(*continued*)

2.4 Dynamics and Radiation Mechanisms of Flares in Different Wavelengths

Table 13. cont.

Star	Date	Instrument									Reference
EV Lac	7–8.12.97	SAX						3.2, 8.3 and 25	5, 4 and 8×10^{51}		Sciortino et al., 1999
YY Gem	4–5.11.98	SAX	0.1–10								Tagliaferri et al., 2001
AD Leo	2.4–16.5.99	SAX							10^{51}		Güdel et al., 2001
LP 944-20	15.12.99	Chandra	0.1–4.0	90		1×10^{26}	2×10^{29}	3			Rutledge et al., 2000
ER Vul	29/30.3.01	Chandra		30				2–30			Brown et al., 2002
YY Gem	29/30.9.00 I	XMM-Newton+Chandra		30		8×10^{29}		10–40	$3 \times 10^{51-51}$	10^{9}	Stelzer et al., 2002
YY Gem	29/30.9.00 II	XMM-Newton+Chandra		45		5×10^{29}		10–46	$2 \times 10^{51-52}$	10^{9}	Stelzer et al., 2002
Castor A	29/30.9.00 I	XMM-Newton+Chandra		19	30		1×10^{32}				Stelzer & Burwitz, 2003
Castor A	29/30.9.00 II	XMM-Newton+Chandra		12	36		1×10^{32}				Stelzer & Burwitz, 2003
Castor B	29/30.9.00 I	XMM-Newton+Chandra		9	16		6×10^{31}				Stelzer & Burwitz, 2003
Castor B	29/30.9.00 II	XMM-Newton+Chandra		9	12		6×10^{31}				Stelzer & Burwitz, 2003
UV Cet AB	26/27.11.01	Chandra	0.8–10			2×10^{26}					Audard et al., 2003
Prox Cen	12.8.01 I	XMM-Newton	0.15–2.5			2×10^{28}		2–4	10^{50}	2×10^{10}	Güdel et al., 2002a
Prox Cen	12.8.01 II	XMM-Newton	0.15–2.5			4×10^{28}	2×10^{32}	2–4		-4×10^{11}	

* Coronal temperatures are taken from independent considerations.

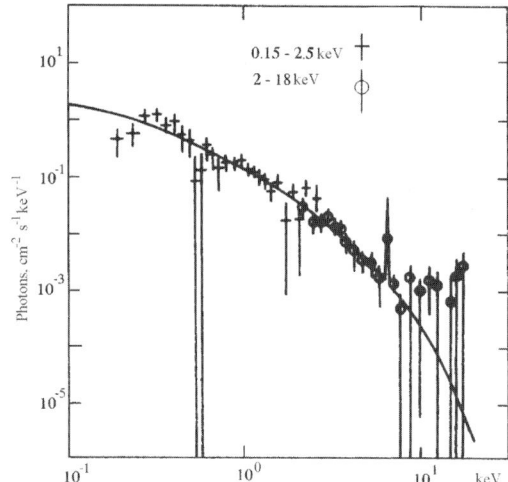

Fig. 44. X-ray spectrum of the flare on AU Mic of 25 October 1977 (Kahn et al., 1979)

During similar HEAO-1 observations of the flare star AD Leo on 22 November 1977 two rather weak X-ray flares were revealed. Their parameters cited in Table 13, on the assumption that they had the same spectrum as the flare on AT Mic of 25 October 1977, are rather close to the appropriate values of the flare on AT Mic.

HEAO-1 observations were not accompanied by optical monitoring. But using the statistical properties of optical flares obtained for the stars observed by Kunkel (1975a), Kahn et al. concluded that in these flares $L_X/L_{\rm opt} > 1$. This contradicted Mullan's thermal model, in which heat conductivity from the corona downward was considered as a determinative of excitation of an optical flare, but was in agreement with the observations of the strong solar flare of 4 August 1972, which yielded much more complete and authentic data than previous estimates from many weaker solar flares.

Ambruster et al. (1984) found from HEAO-1 data two flares on EV Lac (25 and 30 December 1977) with X-ray luminosities of 3×10^{29} and 2×10^{30} erg/s, respectively. There are suspicions that on 25 December 1977 it was the decay stage of a strong flare that was recorded, and three and four hours after the burst on 30 December 1977 there was still increased X-ray luminosity. Since the time interval between these bursts was equal to 1.25 rotational periods of the star it is rather probable that both flares took place in the same active region. The amplitude of the X-ray flare of 30 December 1977 was very high: $A_X \sim 54$, higher by an order of magnitude than that of earlier recorded X-ray flares on red dwarfs. At 10–30 MK the measured flux corresponded to $EM = 3 \times 10^{52}$ and 2×10^{53} cm^{-3} for the first and second flares, respectively. The authors estimated for X-rays $(\Sigma P/T)_X = 0.5$, which was higher by one to one and a half orders of magnitude than the appropriate ratio for the U band.

2.4 Dynamics and Radiation Mechanisms of Flares in Different Wavelengths

But one should take into account the qualitative distinctions of these ratios in the two considered intervals of the spectrum: in the optical range the stellar photospheric emission is used as a photometric standard, which physically is not connected to flares, whereas in X-rays physically homogeneous emissions of hot gas of the disturbed and quiet stellar corona are compared. Though two flares are not sufficient for statistical conclusions, we note that the total energy of X-ray emission recorded on EV Lac bursts by one order of magnitude surpassed the total energy of flare emission in the U band, which was estimated from the available statistics for optical flares on this star.

Similar analysis of HEAO-1 data allowed Ambruster and Wood (1984) to find a flare on EQ Vir on 13 January 1978.

Finally, Connors et al. (1986) thoroughly analyzed the entire databank, about 40 000 HEAO-1 scans obtained over one and a half years to find fast phenomena (1–5 s). They analyzed the records in medium X-rays, which are less subject to the effect of interstellar absorption than soft X-rays, and found five new bursts. On the basis of statistical consideration of the data they concluded that annually about 20 000 flares of duration from 1 to 30 min occurred in the whole sky, which provided fluxes within 2–20 keV of more than 10^{-10} erg/cm^{-2} s on the Earth. Most probably, the majority of bursts are concerned with flares on red dwarf stars.

The next important step in the studies of X-ray emission of stellar flares was the launch of the Einstein Observatory. By this time, an up-to-date concept on the upper solar atmosphere and the governing role of local magnetic fields in its formation were formed. According to these concepts, the magnetic flux tubes connecting solar spots of various polarity form steady discrete arches, plasma loops, in which plasma can move along but not across the axes (see Fig. 15). The arches reach extreme heights of the order of the solar radius and to a considerable extent determine both the properties of the quiet corona and the development of solar flares.

Already during the epoch of pure optical observations two characteristic types were distinguished from the visible picture in H_α rays in an immense variety of solar flares: rather fast and less strong compact flares and longer and stronger two-ribbon flares. Further studies showed that the basic distinction of these types was related to the topology of local magnetic fields: compact flares developed in individual closed magnetic loops, which remained invariant throughout the flare process, whereas two-ribbon flares involved the whole arcades of loops, a strong initial disturbance in them led to breakage of initial loops; the flux tubes of a new open magnetic configuration then closed again, the field energy reduced, and the portion of energy released provided long additional heating of flare matter.

In most solar flares, impulsive and gradual phases can be distinguished. The impulsive phase is characterized by a short-lived burst of hard X-ray emission, which comes from a few small sites at the base of coronal arches

Table 14. Parameters of solar flare arches after Pallavicini et al. (1990a)

	Weak Flares (Compact)	Strong Flares (Two-Ribbon)
Luminosity at maximum (erg/s)	10^{26}–10^{27}	10^{27}–10^{28}
Temperature (MK)	10–30	10–30
Emission measure (cm^{-3})	10^{48}–10^{49}	10^{49}–10^{50}
Electron density (cm^{-3})	10^{11}–10^{12}	10^{10}–10^{11}
Loop height (cm^{-3})	10^{9}	10^{10}
Field strength (G)	100–300	50–100

placed close on both sides of the neutral line of magnetic field. This suggests bombardment and heating of the chromosphere by fluxes of energetic particles. The gradual phase of flares begins simultaneously with the impulsive one, but lasts longer; it has a smooth light curve, radiation occurs in the range of soft X-rays and proceeds from the rather extended diffuse region. The radiation is attributed to hot plasma, which rises from the arch base, propagates over a large area, and slowly spends its energy for radiation and heat conductivity. Finally, one should bear in mind that the described types of flares are seldom realized in a pure form in active solar regions: as a rule, in real flares the characteristics of one or the other kind dominate.

Table 14 lists typical parameters of the arches of weak and strong solar flares obtained in observations.

Let us return to stellar flares.

Rewriting (18) in the following way

$$T_{\max} = 6.3 \times 10^{-3} (n_e h)^{1/2} , \qquad (59)$$

where n_e is the density at $T_{\max}$, Natanzon (1981) showed that flares on YZ CMi and UV Cet recorded by ANS, the flare on Proxima Cen detected by Soyuz–Apollo, and two flares on AD Leo recorded by HEAO-1 satisfied this relation as well. It should be noted, however, that the estimates of the parameters of stellar flares involved in (59) from the first rather limited X-ray observations required additional assumptions: the decay times of flares were determined for the heat conductivity of hot gas, cross sections of arches with flares covered the same part of stellar surface as was estimated for the total area of active regions on two flare stars – AD Leo and EQ Peg – from optical observations during different seasons, the vertical extension of flares was equal to the scale of heights of the appropriate stellar chromosphere. These assumptions are not obvious, but, nevertheless, it is clear that a detailed discussion of the local processes occurring in the atmospheres of flare stars is impossible without allowance for the structure of the atmospheres.

On 28 February 1979, the Einstein Observatory recorded the first stellar flare in the Wolf 630 AB system (Johnson, 1981). This flare and those of 24 May 1979 in the BD+44°2051 AB system and of 14 July 1980 on TZ Ari

2.4 Dynamics and Radiation Mechanisms of Flares in Different Wavelengths

stars were detected as a byproduct of the survey undertaken by Johnson in X-rays for a large sample of objects near the Sun (Johnson, 1981, 1983a). Since each object was observed only for about half an hour and there was no optical support, the data on these flares are rather limited. The flare on Wolf 630 AB was recorded from the beginning of burning till the beginning of decay for 11 min and, since the duration of 12 recorded optical flares on this star varied from 8 to 250 s, either X-ray emission was found during a flare that was very slow for this star or it lasted noticeably longer than the optical one. The rise and decay of the flare were synchronous in the soft and harder regions of the spectrum, but the flare amplitude was higher in harder X-rays. Spectral analysis of these observations of Wolf 630 AB yielded, at a flare temperature of 32 MK, X-ray luminosity at maximum reached 4×10^{29} erg/s, and $EM = 8 \times 10^{51}$ cm^{-3}. The X-ray flare on BD+44°2051 AB lasted about eight minutes and its luminosity at maximum was about 1×10^{28} erg/s. The flare on TZ Ari lasted about 10 min, but till the end of observations of this star that lasted for about 20 min its X-ray emission remained at the level that was noticeably higher than that of the quiescent state; the luminosity of the flare in a maximum was 3×10^{28} erg/s and $EM = 2 \times 10^{51}$ cm^{-3}.

On 6 March 1979, the Einstein Observatory recorded the first stellar flare within special monitoring of Proxima Cen (Haisch et al., 1980). On time averaging of 5 min at the stage of burning one readout was obtained, at luminosity maximum, two, and in the decay phase, three; an e-fold decrease of brightness took 20 min. As to the time characteristics, the flare was rather close to typical strong solar flares. During this event, for the first time the change of spectral structure of X-ray emission was confidently recorded, and the analysis of this radiation resulted in the estimates presented in Table 15.

Table 15. Parameters of the X-ray flare on Proxima Cen of 6 March 1979 (Haisch et al., 1980, 1981)

	L_X (0.2–4 keV) (erg/s)	T (MK)	EM (cm^{-3})
Preflare state	1.5×10^{27}	4	
Burning	6.0×10^{27}	17	0.8×10^{51}
Maximum brightness	7.4×10^{27}	12	1.3×10^{51}
70 min after the maximum	1.8×10^{27}	6	

Thus, the maximum temperature was achieved at the phase of burning of the flare, i.e., prior to maximum X-ray luminosity, and the value of maximum temperature and the time shift between temperature and luminosity maxima were close to the appropriate values typical of solar flares.

During the flare on Proxima Cen simultaneous observations of the star were arranged for the first time: in X-rays, in the ultraviolet from the IUE satellite, in the optical range by four telescopes of three Australian observatories, and in the radio range by the 64-m Parkes radio telescope. But the flare was recorded only in X-rays. Haisch et al. (1981) concluded that the thermal radio emission, which should be expected from hot plasma responsible for the revealed hard emission, should have an intensity below the detection of the equipment used, while the rather probable nonthermal radio emission of the flare could be directed and thus did not hit the Earth. The absence of accompanying phenomena in optical and ultraviolet ranges can be explained by spatial separation of the regions of formation of flare radiations, which is typical of the loop model and shielding of the bottom parts of arches, where optical and ultraviolet flare radiation is formed, with stellar limbs. However, another interpretation is possible. Estimating the upper limit of optical emission from observations, we get $L_X/L_{\mathrm{opt}} > 1$. On the other hand, the recorded X-ray emission can be provided by 3–4 loops similar to those in which strong solar flares of importance 3 are observed: loops with a height of 10^{10} cm, cross-sectional radius of 10^9 cm, and an average density of 10^{11} cm^{-3}. For such characteristics of hot gas the ratio of radiative losses to those for heat conductivity definitely exceeds unity. Thus, we should consider an alternative to the model proposed by Mullan (1976a): hot coronal structure relaxing at the expense of radiative cooling. In this case, it is natural to expect $L_X/L_{\mathrm{opt}} > 1$, and the time of radiative cooling of such gas for the obtained estimates of its temperature and density should coincide with the characteristic time of flare decay. The parameters h and n_{e} of the EV Lac flare of 25 December 1977 (Ambruster et al., 1984) summarized in Table 13 were also obtained on the assumption that the flare decay time was determined by the time of radiative cooling of relaxed coronal plasma.

The successful observations of the flare on YZ CMi of 25 October 1979 (see Fig. 27) from the Einstein Observatory were mentioned above. This event lasted for about 10 min and for the first time the development of a stellar flare was traced in optical, radio, and X-ray ranges. The burst in optical and X-ray emission started almost simultaneously, but the maximum of optical luminosity was achieved much earlier than that of X-ray emission. Figure 27 shows the flare light curve in X-rays and calculated the temperature and emission measure of hot gas: at maximum, the temperature reached 20 MK and $EM = 4 \times 10^{51}$ cm^{-3}. As before, calculations were based on the adjustment of the distribution along the energy of flare quanta to known distributions $l_\lambda(T)$ for thermal radiation of homogeneous hot plasma at various temperatures and standard chemical composition.

In the YZ CMi flare of 25 October 1979 no delays of the emission measure maximum with respect to the temperature maximum was noticed, as in the Proxima Cen flare of 6 March 1979, this is also typical of solar flares. However, one cannot exclude that the lack of delay was due to fast burning of the YZ CMi flare and averaging of the X-ray data over appreciable time intervals. If the cross section of a flare plasma loop is assumed equal to the area of this flare measured in optical observations, 10^{19} cm^2, and the loop height is taken equal to 10^{10} cm, an electron density of a few 10^{11} cm^{-3} follows from the found estimate for the emission measure. In this case, the time of radiative cooling of hot gas, as in the Proxima Cen flare of 6 March 1979, is close to the time of the observed flare decay, which agrees with the ratio found for total emission energies $E_X/E_U \sim 2$. On the other hand, Kahler et al. (1982) found that the ratio L_X/L_{opt}, which was widely used earlier for the diagnostics of flare radiation, underwent essential changes during the YZ CMi flare of 25 October 1979: from 0.1 near optical maximum to 1 and more at the decay phase of optical flare, and the average value of about 1.5.

During the 3-day cooperative observations of YZ CMi on 27 October 1979 one more weak X-ray flare was recorded. An essentially quantitative conclusion following from the data on this event confirms the estimate E_X/E_U found for the flare of 25 October 1979 (Kahler et al., 1982).

On 20 August 1980, during cooperative observations with IUE, a rather strong flare that continued for more than 2 h was recorded on Proxima Cen by the Einstein Observatory (Haisch et al., 1983). Due to slow development and rather high luminosity, the light curve of the flare was confidently constructed and the temperature changes of radiating gas were found (see Fig. 28). The maximum temperature of 27 MK was achieved 2–3 min prior to the maximum luminosity, as during the flare of 6 March 1979, and the maximum luminosity within 0.2–4 keV was 1.4×10^{28} erg/s. For the estimated temperature of hot gas this value corresponds to full luminosity at a maximum of 2×10^{28} erg/s and a total energy of X-ray emission of 4×10^{31} erg. An hour and a half before the strong flare the X-ray activity of the star, several overlapping weak bursts, was observed for half an hour. The bursts were accompanied by a slightly increased temperature of radiating gas. Immediately after the maximum of X-ray luminosity the e-fold decay lasted for about 20 min, then the rate of brightness decay and gas cooling slowed down, and secondary bursts of similar amplitude and duration to those that occurred prior to the strong flare were observed on the light curve.

Near the maximum brightness of the Proxima Cen flare of 20 August 1980 a distortion of the X-ray spectrum was observed for several minutes in the low-energy region, which corresponds to the occurrence of cold neutral matter with the number of particles of about 10^{20} cm^{-2} in the ray of sight. It is natural to attribute this effect to the passage of a strong prominence above the flare, and, as Haisch et al. (1983) noted, the long decay phase and luminosity of this stellar flare resembled two-ribbon solar flares, which are characterized by prominences in the active regions. Finally, the ratios of total emission energies in the transition-region emission lines and in the Ly_α to E_X in this stellar flare were estimated as 0.05–0.06, which was lower by an order of magnitude than similar ratios in the well-examined solar flare of 5 September 1973. However, later Byrne and McKay (1989) analyzed this flare again using the method of differential emission measure, which takes into account the contribution of many weak lines to the general radiative losses, and the ratios of transition-region emission lines sensitive to electron density. As a result, they removed the discrepancy and found that volumes and luminosities of the regions of corona radiating in the flare and the transition region were comparable.

On 19 September 1980, Stern et al. (1983) recorded a strong X-ray flare on the star HD 27130 (= BD+16°577) in the Hyades by the Einstein Observatory using IPC and MPC sensitive to higher-energy emissions. The earlier stellar survey of this cluster ranked this star as a strong source of X-ray emission. The star was observed during five revolutions, and only during the first of them was it in a quiescent state. Flare burning was missed: when the star came out of the Earth shadow its radiation was already at the level of 10^{31} erg/s, which is 35 times higher than its quiet-state luminosity. During the initial stage of observations the flare emission corresponded to a temperature of about 50 MK, $EM = 4 \times 10^{53}$–10^{54} cm^{-3}, and the decay rate, e-fold decrease, was about 40 min. Full X-ray emission of the flare within the range of 0.2–10 keV exceeded 3×10^{34} erg. By that time, extensive X-ray observations of solar flares by Skylab had already been analyzed (Moore et al., 1980) and led to the conclusion that near the maxima of such flares the time of an e-fold decay τ_d, the time of radiative cooling τ_rad, and the time of cooling due to heat conductivity τ_cond were approximately equal

$$\tau_\mathrm{d} \sim \tau_\mathrm{cond} = 3 \times 10^6 k n_\mathrm{e} h^2 / T^{5/2} \sim \tau_\mathrm{rad} = 3kT/n_\mathrm{e} \int l_\lambda(T) \mathrm{d}\lambda \ . \tag{60}$$

As stated above, at known (or set) temperature the emission measure of a flare is determined from the measured luminosity L_X; then the condition (60) combined with EM allows one to estimate the average density of flare plasma, the height of magnetic loops, the product of their number and the squared loop thickness to height ratio. Assuming that (60) is valid for the HD 27130 flare, Stern et al. (1983) from found values of EM and temperatures estimated the characteristic electron density and volume of the flare: $n_\mathrm{e} \sim 4 \times 10^{11}$ cm^{-3} and $V = 4 \times 10^{30}$ cm^3. Thus, in the decay rate, temperature, and density of this event have close analogs among X-ray flares on the Sun and on UV Cet-type stars, but in total energy and, especially, volume it stood out. Probably, since HD 27130 is a binary system of G and K dwarfs with a period of 5.6 days, the recorded flare can be related to X-ray bursts on RS CVn-type stars, which, as is well known, involve much greater volumes than flares on UV Cet-type stars, and their physical nature can differ from that of solar flares. It should be noted that the X-ray luminosity of HD 27130 in the quiescent state is close to the lower limit of those values corresponding to RS CVn-types stars, but unlike the latter, HD 27130 does not contain the components that left the main sequence. It is unknown whether another important characteristic of RS CVn-types stars, fast and synchronous rotation and revolution of one or both components, holds in this system. In addition to this huge flare on HD 27130, during the survey of the Hyades in soft X-rays, Stern and Zolcinski (1983) recorded X-ray bursts in the spectral binary G0 V system BD + 14°690 and on the K dwarf vA 500, as well as slow – about eight hours – flare decay on the flare dMe star vA 288. During analogous observations of the Pleiades Caillault and Helfand (1985) detected flares on HZ 1136 and HZ 1733, the

2.4 Dynamics and Radiation Mechanisms of Flares in Different Wavelengths

former was similar to the flare on HD 27130, the latter, to many X-ray flares on M dwarfs.

Using the high-resolution imager (HRI) installed at the Einstein Observatory, whose sensitivity was lower than that of IPC and that did not provide information about the spectrum but had a spatial resolution of 4 arcsec, Harris and Johnson (1985) observed four binary systems, including one or two dwarf M stars, from February till August 1980, one session each. During the observations they found flares on Gl 34B and Gl 338A at constant X-ray luminosity of Gl 338B and the flare on Gl 669B at constant luminosity of Gl 669A. The time characteristics of these bursts are summarized in Table 13. Harris and Johnson analyzed IPC data on Gl 669 obtained occasionally in 1979 during two observational sessions of another nearby source, and in one session they found a flare with a burning time of less than 3 min and longer decay: the maximum luminosity of the flare was over 10^{29} erg/s, the total energy, 10^{35} erg, and temperature, 24 MK.

Agrawal et al. (1986), on 24 March 1981, recorded by the Einstein Observatory an X-ray flare on the star V 1216 Sgr with a time of burning < 2 min and decay time ∼5 min. Its luminosity at maximum was 5×10^{27} erg/s and total energy $E_X(0.2-4\,\text{keV}) \sim 10^{30}$ erg. The development of the flare was similar to characteristic optical flares on this star.

Studies of X-ray flares on the UV Cet-type stars from the Einstein Observatory were reviewed by Haisch (1983). Supplementing (60) by the approximation

$$\int l_\lambda(T)\mathrm{d}\lambda \sim 10^{-26.2} T^{1/2} \qquad (61)$$

and the estimate for volume $V = h^3/100$, i.e., accepting the solar ratio of loop thickness to its height as 1/10, Haisch estimated anew the temperature, density, and linear size of loops in eight stellar flares. He obtained temperatures within 30–350 MK and systematically higher – by a factor of 1.5 to 5 – than those found directly in X-ray spectra of flares. The density varied within 8×10^{11}–10^{13} cm^{-3} and the linear sizes, from 2×10^9 to $>6 \times 10^{10}$ cm, the latter estimate is for the flare on HD 27130. For four of eight flares – on Proxima Cen of 6 March 1979 and 20 August 1980, on TZ Ari of 14 July 1980, and on V 1216 Sgr of 24 March 1981 – the direct determination of temperature from the spectrum and from (60) and (61) differ by no more than a factor of 2–2.5. They are similar to solar flares of moderate power, but their electron density is higher by approximately one order of magnitude, which results in large EM and L_X and a shorter decay time. Simple scaling of the remaining three flares – on YZ CMi of 19 October 1974 and 25 October 1979 and on HD 27130 of 19 September 1980 – did not yield a "scaled" solar flare. The status of the flare on Wolf 630 of 28 February 1979 in this sample remained uncertain.

Assuming that flare loops persist due to magnetic field pressure, which should exceed the gas pressure, Haisch estimated the lower limit of field strength from the found values of plasma density and temperatures: it was

260   2 Flares

over 400–1600 G in the flares, comparable with solar flares, and over 700–9000 G in the others. For such fields, the characteristic time of development of instability of magnetic structures for scales h appeared to be of close order of magnitude to the times of burning of the appropriate flares.

Further, upon comparing the full stock of thermal energy at the maximum of flares $E = 2n_\mathrm{e}(3kT/2)h^3/100$ and the total emission energy of radiation, which for plasma at temperatures of 10–100 MK observed within 0.2–4 keV differed from directly measured E_X only by some tens of per cent, Haisch (1983) found that for flares on Proxima Cen of 6 March 1979 and 20 August 1980 and probably the flare on Wolf 630 of 28 February 1979 the value of E was higher by one and a half or two times than E_X. This means that in these two or three events, in addition to the initial pulsed energy release, one can suspect on-going heating during the decay, as in strong two-ribbon flares on the Sun.

As many as 22 flare stars were observed by EXOSAT, about 30 flares were recorded on 11 of them. Furthermore, flares were found on three stars, which occasionally came into the field of sight of the satellite instruments. Figure 45 shows the light curves of four flares recorded by EXOSAT.

The first X-ray flare on a red dwarf star was recorded by EXOSAT by Smale et al. (1986). In studying the field in the vicinity of T Tau on 21

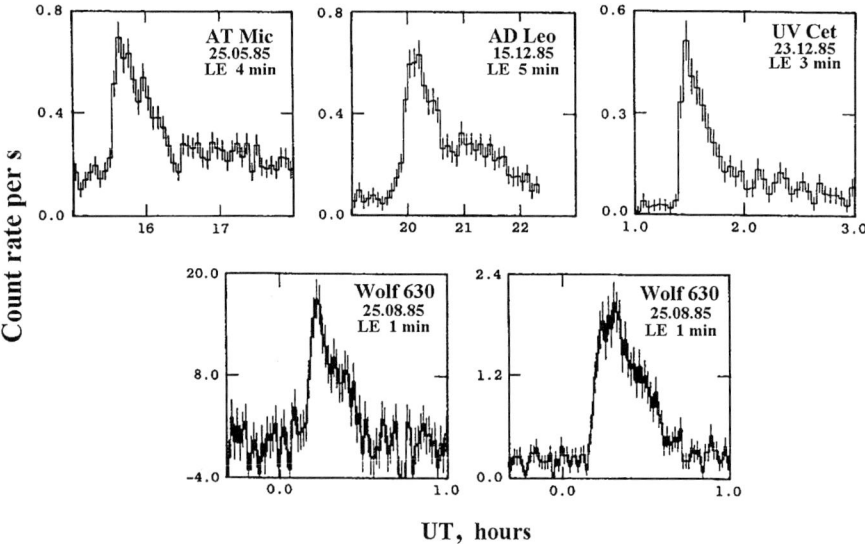

Fig. 45. EXOSAT light curves of X-ray flares on four dMe stars; the observation range and the time of averaging of readouts is specified below the date (Cheng and Pallavicini, 1991)

2.4 Dynamics and Radiation Mechanisms of Flares in Different Wavelengths

August 1983 they detected the event on the object that was later named 1E0419.2+1908 and was a typical UV Cet star. The flare was observed by both soft X-ray telescopes, its burning and decay time were about 40 and more than 80 min, respectively. Based on the estimate of the distance to the star, assuming that the temperature of the coronal plasma was 4 and 20 MK in quiescent and flare states, from the measured flux they estimated the amplitude of the X-ray flare as $A_X \sim 40$, maximum luminosity as $L_X \sim 6 \times 10^{29}$ erg/s, the emission measure as 3×10^{53} cm^{-3} and the total energy – to a factor of 3 – as $E_X \sim 10^{33}$ erg. This object, as an X-ray source, was found also in two images from the archive of the Einstein Observatory, which showed the star in quiescent and active states. The active state of the star was manifested in a slightly increased level of X-ray emission and slightly increased hardness. Quantitative parameters of this and successive flares recorded by EXOSAT are cited in Table 13.

On 31 January 1984, both EXOSAT instruments recorded a strong flare on the single G0 V star π^1 UMa, which is an order of magnitude older than the Sun, has several times higher rotational rate, and displays strong emission of the chromosphere, a transition region and corona in the quiescent state (Landini et al., 1986). At burning over about 8 min the maximum of flare luminosity in ME was achieved approximately 12 min earlier than in LE, and the maximum hardness of this range was recorded at the burning phase. The spectrum averaged over the whole flare in the band of 1–8 keV was presented within the framework of the model of optically thin plasma by Landini and Monsignori Fossi (1984) as thermal radiation with $T \sim 30$ MK and $EM \sim 7 \times 10^{52}$ cm^{-3}. These estimates did not contradict LE flare luminosity. If the flare decay was determined only by radiative losses, $n_e \sim 7 \times 10^{11}$ cm^{-3} and $h \sim 7 \times 10^9$ cm $\sim 0.1 R_*$, the area of this flare did not exceed 0.1% of the stellar surface. The total thermal energy of the flare plasma was close to E_X, which suggested a lack of supporting heating in the decay phase. From the energy properties this flare is close to the strongest events on dMe stars and exceeds the strongest solar flares by two orders of magnitude. Though its volume is close to that of solar two-ribbon flares, optical monitoring did not reveal this flare on π^1 UMa because of insignificant contrast with the photospheric radiation.

The flare on Wolf 1561 was recorded occasionally in studying another X-ray source (Pallavicini et al., 1990a).

The flare in BY Dra system on 24 September 1984 was recorded during joint X-ray, optical and radio monitoring observations (de Jager et al., 1986). The development of the X-ray flare was similar to the gradual phase of a solar flare. The lack of a signal in medium X-rays made it possible to accept $T < 10$ MK and estimate the emission measure as 1.2×10^{52} cm^{-3}. De Jager et al. believed that a rather short optical flare that occurred at the beginning of the long X-ray flare corresponded to an impulsive phase during which photospheric layers of the star were burnt out and evaporated. Within the model, they estimated the density of coronal plasma as $n_e \sim 2.3 \times 10^{11}$ cm^{-3}.

The X-ray source EXO 040830-7134.7 that was later identified as a dMe star was found in soft X-rays in the field of a dwarf nova during 30 sessions and seven times higher emission was recorded from it on 11 October 1984 (van der Woerd et al., 1989).

EXOSAT observations of the α Gem system made it possible to distinguish the red dwarf system YY Gem from the brighter pair α Gem AB and record two flares in the YY Gem system in the LE and ME regions on 14 November 1984 (Pallavicini et al., 1990b). The peak of the first, stronger and longer flare in ME occurred somewhat earlier than in LE, and the flare emission in ME was almost half as long. On the basis of the analysis of this radiation the temperature of the flare plasma was estimated as 64–24 MK and EM as 5×10^{53}–4×10^{52} cm^{-3}. This was the strongest X-ray flare recorded by EXOSAT on red dwarfs.

Another visual pair resolved by EXOSAT was Gl 867 AB. The distance between its components FK and FH Aqr is only 25 arcsec, thus only the methods of maximum likelihood used to analyze X-ray images of the system made it possible to distinguish quiet radiation from the coronae of the components and conclude that X-ray flares occurred on both of them: the first X-ray burst recorded during four-hour monitoring on 18 November 1984 was due to the flare on FK Aqr, and the second, the superposition of two flares on each of the components at a 13-min interval (Pollock et al., 1991).

On 6 December 1984, over less than four hours, four flares of similar power were recorded on UV Cet: the time of burning and the decay time varied from < 2 to 5 min and from 4 to 10 min, respectively. The luminosity at maxima was $(2-3) \times 10^{28}$ erg/s and the total emission in X-ray range was $(3-7) \times 10^{32}$ erg (Pallavicini et al., 1990a).

On 7 December 1984, Haisch et al. (1987) detected four flares in the EQ Peg system by EXOSAT. The second flare that lasted for more than an hour and was confidently recorded in the LE and ME regions is the most interesting: though the light curve in ME looked as usual with a fast rise and smoother decay, burning in LE was very slow and took approximately twice as long as in ME, and even longer than the decay in LE. Pallavicini et al. (1990a) believed that the unusual form of this light curve was caused by the superposition of two events in LE. But Haisch et al. explained this feature by temperature heterogeneity of the flare plasma from the data about the temperature gradient along arch axes known from the observations of solar flares and considerations on the radiation of soft X-rays at different stages of such flares. However, the observations of stellar flares did not allow a detailed model to be elaborated, and quantitative estimates of plasma parameters in the flare on EQ Peg were carried out using the traditional homogeneous model. The hardness of radiation in ME determined as the ratio of fluxes $F(4-7\,\text{keV})/F(1.5-4\,\text{keV})$ was maximum at the beginning of the flare and then smoothly decreased. A similar conclusion follows from spectral analysis of the emission in ME: the temperature of the flare plasma reached 26 MK at the very beginning of burning and went down to 14 MK during the decay phase. The total emission in the

2.4 Dynamics and Radiation Mechanisms of Flares in Different Wavelengths

LE region was 5×10^{32} erg. If the decay was caused by radiative cooling, $n_e \sim 2 \times 10^{11}$ cm^{-3}. LE and ME measurements yielded a practically identical emission measure of 5×10^{51}–10^{52} cm^{-3}. If this flare occurred in the arch structure, its height, density, and emission measure were closer to the appropriate parameters of solar two-ribbon flares rather than compact flares. Three other flares recorded during the 8-h observational session were shorter and their energy release in the X-ray range was lower by 5–10 times (Pallavicini et al., 1990a).

In observations of YZ CMi on 4/5 March 1985 two rather slow X-ray flares were recorded by Doyle et al. (1986). During these events, the optical monitoring was incomplete. But between the detected X-ray flares there was a rather strong optical burst with $\Delta U \sim 1\overset{m}{.}2$ and a total energy of optical emission of about 6×10^{30} erg, which did not manifest itself in X-rays. This result contradicts the earlier revealed close relation between the variations of UV Cet emission in H_γ line and in soft X-rays (Butler et al., 1986). Doyle et al. (1988b) considered three probabilities for optical flares without X-ray emission: reconnection in low loops where, due to a high density of matter, all allocated energy quickly relaxed and no hot coronal plasma occurred; the absorption of X-ray emission by overlying cold matter, as on 20 August 1980 (Haisch et al., 1983) near the maximum of flare on Proxima Cen; deeper heating of the chromosphere by penetrating proton rather than electron beams.

The Wolf 630 system including at least seven members was repeatedly observed by EXOSAT. In processing the 13-h session of 8 March 1985 using the initial standard program, Johnson (1987) found a flare in soft X-rays with $A_X \sim 30$, a duration of about 10 min and increased luminosity during the next 40 min in the binary star VB 8 involved in the Wolf 630 system and containing an M7 dwarf. It is essential that both components of VB 8 AB have too low masses for maintaining thermonuclear burn of hydrogen, and optical flares in it were not recorded before. Table 13 cites the parameters of this flare obtained later by Tagliaferri et al. (1987, 1990) with the improved program for analyzing EXOSAT data.

Using this program, Tagliaferri et al. (1987) reprocessed the data of the session of 8 March 1985 and, in addition to the flare on VB 8, found a flare on Wolf 630 in LE and ME. But the most complete data were obtained during the session of 25 August 1985 from a stronger flare on itself Wolf 630, which was also found in LE and ME (Doyle et al., 1988a; Tagliaferri et al., 1987). This event occurred after the star was in a quiescent state for about seven hours. The maximum of this flare in ME took place 30 s before the maximum in LE. The temperature of the flare plasma reached 48 MK during the burning phase and reduced to 29 MK during the decay phase. The maximum hardness of the spectrum took place two minutes before the maximum of X-ray brightness. If one assumes that the flare plasma did not obtain additional energy from outside after the maximum and its cooling was due to radiative losses only, the rate of flare decay would correspond to the density $n_e \sim 10^{12}$ cm^{-3} and its volume would be 3×10^{28} cm^3. If the flare was related to N_1 loops and the loop radii amounted to the α share of their height, the found volume corresponds to $h \sim 2 \times 10^9 (\alpha^2 N_1)^{-1/3}$ cm. If cooling losses dominate over heat conductivity in the flare, $h \sim 2 \times 10^{10} (\alpha^2 N_1)$ cm, and at $\alpha = 1/10$, as on the Sun, this value is lower by an order of magnitude than the previous one. Since the intermediate mode of cooling is the most probable, the heights of flare loops should make a few 10^9 cm, i.e., about 10% of the stellar radius. Parallel optical observations revealed a rather good correlation of soft X-ray emission with flare emission in the H_α line: though stellar spectra were obtained with a time resolution of 14 min, they proved the coincidence of the moments of maxima and the increase of equivalent width of the line from 2.1 Å in the quiescent state to 3.2 Å at flare maximum. Excessive radiation in the line lasted approximately twice as long as the X-ray emission, but its total energy amounted to about 4% of the flare emission within the range of 0.05–7 keV. The same E_{H_α}/E_X ratio is observed in compact solar flares.

Observations of UV Cet on 4 August 1985, 6 December 1984, and 23 December 1985 revealed several individual flares of similar power on the background of rather unstable radiation (Pallavicini et al., 1990a).

On 6 August 1985, EXOSAT LE and ME detected one of the strongest and longest X-ray stellar flares in the EQ Peg system (Pallavicini et al., 1986, 1990a). Its duration was similar to that of two-ribbon solar flares, about 2 h, but its energy was two orders of magnitude higher than that of solar flares. The flare spectrum at different phases is well represented by free–free radiation of the coronal plasma plus the emission in the iron line of 6.7 keV. Reduced hardness of the spectrum was revealed during the decay and cooling of the radiating plasma from 42 to 18 MK. The decay rate of the flare in the case of the prevailing radiative losses and the absence of additional heating, corresponded to the density $n_e \sim 6 \times 10^{11}$ cm^{-3}, which is typical of solar flares.

The nine-day cooperative program aimed at studying EV Lac in October 1985 resulted in two successful sessions of EXOSAT observations (Ambruster et al., 1989a). On 13 October 1985, noticeably increased X-ray emission was observed for more than two hours. Comparison of X-ray light curve and optical photometry suggests that there were two long – an hour or an hour and a half – X-ray flares that, however, were not accompanied by significant changes in stellar brightness in optical wide bands, and two fast X-ray bursts, one of which was identified as an optical flare with a amplitude $\Delta U = 1^m\!.8$, which was accompanied by appreciable but short amplification of H_β, CaII, and MgII emissions. Upon termination of this two-hour X-ray activity within the next three hours X-ray emission was similar to that of a constant source. During a 12-hour session on 15 October 1985 on the background of obvious but weak variability of X-ray emission with a characteristic time of more than one hour a rather strong X-ray flare was recorded, which had an abrupt start, but an optical burst with an amplitude of only $\Delta U \sim 0^m\!.4$ corresponded to it. As the flare decayed, there was one more optical burst of equally low amplitude. There was also a flare with $\Delta U = 1^m\!.5$ with a simultaneous fast X-ray burst. As on the night before, these rather fast events were accompanied by amplified emission in H_α and CaII but without appreciable changes of the intensity of CIV λ 1550 Å.

Considering the results of EV Lac observations, Ambruster et al. (1989a) noted the lack of close correlation of optical and X-ray emission of flares: rather weak bursts in X-rays corresponded to appreciable optical flares and weak optical flares to strong X-ray activity. At the same time, there is a close relation between the events found photometrically in the U band and chromospheric emission. Recalling a similar situation with YZ CMi observations on 4 March 1985 (Doyle et al., 1988b) and the X-ray flare on Proxima Cen of 3 August 1979, which were not accompanied by strengthened optical emission (Haisch et al., 1981), Ambruster et al. advanced the idea on isolated magnetic regions in stellar atmospheres and cited some solar data supporting the idea.

Kundu et al. (1988a) carried out simultaneous EXOSAT and VLA observations: flares recorded on UV Cet, EQ Peg, YZ CMi, and AD Leo had a rather weak correlation of activity in X-rays and in the microwave range.

Rao et al. (1990) analyzed EXOSAT data for BD+48°1958A, which was supposed to be a flare star because of the variability of the structure of H_α emission, and found evidence of a long – about 2 h – weak X-ray flare with rather hot flare plasma.

During one of the most successful cooperative observations of UV Cet on 23 December 1985 a flare of very high optical amplitude, $\Delta B \sim 5^m$, was recorded by LE and ME EXOSAT (de Jager et al., 1989). The preflare LE flux corresponded to a temperature of 4 MK and an emission measure of 3.3×10^{50} cm^{-3}. At maximum, which was achieved in 140 s, the LE radiation increased by a factor of 20 and 20 min after the maximum was still twice as high as the preflare level. Radiation in ME increased at maximum to a level of 3σ and was observed for about 30 min. Quantitative analysis of the data resulted in a conclusion about a significant heterogeneity of flare plasma being responsible for X-ray emission in this event. If LE and ME radiations were caused by the same plasma body, for the recorded radiation in soft X-rays its temperature should be 1 MK, i.e., lower than the coronal temperature in the quiet star, while for the radiation recorded in medium X-rays, it should be 40 MK. If one assumes that emission in both ranges was caused by plasma at 40 MK, the emission measure for LE should be an order of magnitude higher than for ME.

The results of EXOSAT X-ray observations of flare stars were reviewed by Pallavicini et al. (1990a). About 30 X-ray flares recorded on some fifteen stars allowed for statistical conclusions. First, a rather high frequency of the recorded flares in the considered sample of stars should be noted – on average, one event every 8–10 h. Secondly, there is a clear positive correlation of this frequency with the level of X-ray emission of quiet stellar coronae. Thirdly, the range of characteristic times varied from minutes to hours, and total energy of X-ray emission of flares, from 3×10^{30} to 10^{34} erg. Thus, the temporal characteristics of stellar and solar flares are identical, the energy of X-ray flares on red dwarfs is higher by orders of magnitude than the energy of solar flares. In spite of the significant difference, two types of flares occur on stars: pulsed, similar to solar compact flares, with burning in minutes and decay in ten minutes, and with a long decay – up to an hour and more, similar to solar two-ribbon flares. Stellar flares in soft X-rays, similar to solar compact flares, are usually smoother and longer than optical bursts accompanying them. Though flares of different energy of both specified types can occur on the same star, as on the Sun, there is a tendency: flares with higher $L_X^{\max}$ and E_X are recorded on stars with higher quiet X-ray emission. The energy spectrum of X-ray flares constructed by Pallavicini et al. (1990a) has power characteristics with the spectral index of 0.7, as in the optical range. Flare plasma is quickly heated to 40–50 MK and cools down to 20–10 MK. Emission measures of the recorded stellar flares are equal to 10^{51}–10^{53} cm^{-3} and essentially exceed solar values. The ME emission reaches a maximum somewhat earlier than LE and is noticeably shorter. Pallavicini et al. did not confirm the concept of microflares and gave another interpretation to the X-ray observations that underlies the concept. Earlier, Collura et al. (1988) obtained similar results in analyzing EXOSAT observations of 13 dMe stars: quiet X-ray emission of the stars could not be presented by the superposition of microflares with characteristic times of hundreds of seconds, and the spectral index of the energy spectrum of individually recorded flares was close to 0.52. They concluded that if microflares played an essential role in the heating coronae, their distribution should have been noticeably different from the spectral index.

In the late 1980s–early 1990s, when the results of EXOSAT that had completed its operation were intensely studied together with the first data from ROSAT, new approaches to the analysis of X-ray stellar flares were advanced.

The relations (60) underlying some of the above estimates of the parameters of stellar flares are approximately fulfilled in many solar flares, but do not have a distinct physical substantiation. This shortcoming was removed by van den Oord and Mewe (1989), who proposed for flare decay a model of quasistatic cooling of flare loops as a series of equilibrium states of a magnetic structure. They showed that if there was power dependence of $\int l_\lambda(T) d\lambda$ on temperature, a physically exact model was realized at $\tau_{\mathrm{rad}}/\tau_{\mathrm{cond}} = 0.18$ and $\int l_\lambda(T) d\lambda \sim T^{0.25}$. In this case, radiative losses prevailed, but one should not neglect heat conductivity, while simple power relations described the change of temperature and emission energy with time. Comparing the relations with observations, one can estimate the density of matter in flares, loop heights and all other parameters, as in analyzing (60). Basically, the model of quasistatic cooling can also include additional heating at the flare decay stage, but it is not clear whether the estimate of flare parameters remains unambiguous. It is obvious that ideologically this model is close to compact flares, but originally van den Oord and Mewe applied it to a very strong flare on Algol recorded by EXOSAT, which does not match the simple morphological classification of solar flares.

2.4 Dynamics and Radiation Mechanisms of Flares in Different Wavelengths

Considering the results of the ROSAT investigation of X-ray stellar flares, one should remember that the satellite had two operating modes. During sky surveys, each object was recorded for 20 s every 96 min during five days, which enabled recording of long and strong flares. At long individual pointings with considerably increased sensitivity, weak and fast flares could be recorded.

ROSAT was used to continue checking the concept of microflares: within two weeks Schmitt (1993) collected about twenty half-hour records of UV Cet, but their consideration did not reveal the expected effects of microflares. Some of these X-ray observations were accompanied by monitoring in the U band. Two contiguous optical flares on 2 January 1992 with $\Delta U \sim 1\overset{m}{.}2$ were accompanied by X-ray bursts. The first purely impulsive optical flare was about 12 s long and was accompanied by a short X-ray burst. The second optical flare had a very fast impulsive phase, but it was followed by a gradual five-minute phase, and in X-rays there was a fast single burst, which was followed by long X-ray luminosity three minutes after. A similar delay of soft X-rays after an impulsive beginning of optical flare was recorded in the flares on the Sun, UV Cet, and BY Dra (de Jager et al., 1986, 1989). Both X-ray bursts on UV Cet took place 30 s after optical pulses, and they could be low-energy tails of hard X-ray emission concerned with initial beams of energetic particles (Schmitt et al., 1993a), but the reason for the 30-second delay remains unclear.

In studying late Me dwarfs with ROSAT, Fleming et al. (1993) recorded several X-ray flares on three stars. On 19 October 1990, an already mentioned strong flare with $A_X \sim 160$ was recorded on AZ Cnc during six revolutions around of the Earth. Within two days, when Proxima Cen was accessible within the RASS program, they recorded a flare during four revolutions and two flares, with individual readouts and apparently the decay of a slow flare, whose maximum was missed. On 25 November 1990, a flare on CN Leo was recorded on three consecutive revolutions.

Pan and Jordan (1995) carried out two cycles of ROSAT PSPC observations of the flare star CC Eri within the range of 0.1–2.4 keV with a spectral resolution of 2.1. They found its variability at times from several minutes to several hours. On 10 July 1990, they recorded a rather slow flare with a burning time of about an hour and a decay of about 3.7 h, an amplitude $A_X \sim 2$ and maximum luminosity $L_X(0.2-2\,\text{keV}) \sim 7 \times 10^{29}$ erg/s. There was an additional burst on the descending branch of the flare an hour and a half after the maximum. The spectra of CC Eri in a quiescent state and during the flare were presented within two-temperature models, and in both states various algorithms inevitably suggested the existence of a coronal component with a temperature of about 10^7 K. The time of development of the flare of 10 July 1990 was similar to two-ribbon solar flares, but its energy was an order of magnitude higher compared to the latter. To analyze the "pure flare spectra" obtained by subtraction of the stellar spectrum out of the flare from the observed spectra of active star, Pan and Jordan applied the analytical theory of Kopp and Poletto (1984) and Poletto et al. (1988) on magnetic reconnection in two-ribbon flares. They presented the descending branch of the light curve of the flare on CC Eri within the framework of this model and found that during the decay phase the temperature of the plasma decreased from 28 to 12 MK, and EM – from 42×10^{51} to 1×10^{51} cm^{-3}. Depending on the specified extension of the flare on the stellar latitude from 33° to 5°, its area varied within 8–0.2% of the stellar hemisphere, the magnetic-field strength within 250–1500 G, an initial rise speed of the reconnection point within 14–2 km/s, the maximum height of rise of this point within 2×10^{7}–4×10^{9} cm, and ranges of electron density within $(13,1)\times10^{10}$–$(220,14)\times10^{10}$cm^{-3}. Thus, the model of the flare on CC Eri with continuous energy release should have either rather high density, or much greater volume than large two-ribbon solar flares. Within the framework of the flare model with only pulsed energy release, parameters obtained by Pan and Jordan are close to those in the previous models where the total flare area was close to 8% of the stellar hemisphere.

Within the RASS program the star F0V 47 Cas, which probably belongs to the Pleiades moving group, was observed every 96 min during 3.5 days. On the background of clear flux oscillations with a period of about a day that were visible only in the soft part of the recorded range and were apparently caused by the rotation of the star with an asymmetrical corona, Güdel et al. (1995b) detected, on 19 and 21 August 1990, flares at the maxima of periodic oscillations. The first was visible only in the hard part of the recorded range. The estimates showed that the observed pattern could be related to the large active region on the star with coronal temperature of about 2 MK. At maximum brightness the luminosity of flares was 4.3×10^{30} and 1.2×10^{31} erg/s. If the temperature of the flare plasma is assumed to be 25 MK, the total energy of X-ray emission of the second flare would be about 7×10^{34} erg. This event is one of the strongest among stellar flares. If the duration of the first flare is assumed to be due to radiative relaxation, at $EM\sim4\times10^{53}$ cm^{-3} and $n_e\sim(2-4)\times10^{10}$ cm, and the size of the flare loops is about the size of a loop in the cool component of the preflare corona.

Springfellow (1996) published light curves of AD Leo for 8/9 May 1991 and of VB 8 for 25–27 February 1991. Over four hours of AD Leo observations one strong flare was found with $A_X\sim3$ and the decay to an FWHM level of eight minutes and two weaker flares. On the light curve of VB 8 there was one long flare of about 100 min in total.

Under spectral analysis of a strong flare on EV Lac recorded by ROSAT PSPC on 13 July 1992, Sciortino et al. (1999) found that a hot component with $T\sim36$ MK should be introduced in addition to isothermal MEKAL models that enabled presentation of the spectrum of the quiet star; MEKAL models are modifications of MEKA (Mewe et al., 1995b).

A byproduct of ROSAT PSPC observations of Gl 213 stars was the detection of a high activity level of the M3 dwarf G 102-21 (Micela et al., 1995). Later, this X-ray source was found in the archives of the Einstein Observatory. The ROSAT data for G 102-21, eight sessions within 3.35 days, were analyzed within the range of 0.20–2.0 keV. In the records of 23 September 1993, a strong flare with $A_X \sim 8$ and $L_X^{\max} \sim 5 \times 10^{29}$ erg/s was found. Representation of the flare spectrum within the framework of the two-temperature model yielded $T_1 = 4.8$ and $T_2 = 16$ MK, $EM_1 = 8 \times 10^{51}$ and $EM_2 = 51 \times 10^{51}$ cm^{-3}. The high-temperature coronal component appeared to be necessary for the models of quiet corona as well. The value of T_1 was practically equal for the flare and quiet corona, whereas the EM_1 of flares was twice that in the quiescent state. The values of physical parameters of this flare found resembled those of compact solar flares and, if it was located in one loop, the field strength in it should have been 630 G.

The young G star EK Dra that has just reached the main sequence was in the sight of PSPC ROSAT for five days during the sky survey. A flare invisible in softer rays of 0.1–0.4 keV was recorded on it on 23 November 1990 in the range of 0.4–2.4 keV. The characteristic time scale of the flare was about an hour, $L_X^{\max} \sim 1 \times 10^{30}$ erg/s and $E_X \sim 4 \times 10^{33}$ erg. If the decay of the flare was caused by radiative cooling, $n_e \sim 2 \times 10^{11}$ cm^{-3} and $h \sim (1-2) \times 10^{10}$ cm (Güdel et al., 1995a). Similarly, on 30 July 1990 a slow flare whose burning and decay took about six hours was recorded on the F star HD 147365 (Güdel et al., 1995c).

Of particular interest for the general picture of the considered activity are completely convective objects with masses lower than 0.3 solar masses. Fleming et al. (2000) carried out additional long-term observations of the M8 star VB 10 (= Gl 752 B), which had not been found in ROSAT surveys or two-hour individual monitoring. Upon six-hour observations they found an extremely weak image of the star in the image of the field obtained throughout the monitoring. However, detailed analysis proved that all of the 10 photons composing this image were recorded on 19 October 1997 during a 19-min session. Assuming that the star was recorded during the flare, Fleming et al. estimated its luminosity at maximum as $L_X = 3 \times 10^{27}$ erg/s and the stellar luminosity out of the flare as less than 2×10^{25} erg/s. In the full duration and the part of total monitoring the X-ray flare took, this event was close to the UV flare recorded on this star on 12 October 1994 by HST (Linsky et al., 1995). Another important similarity of the two flares is that the level of quiet radiation of the star in both wavelength intervals was lower by 1–2 orders of magnitude than the radiation in flares. Based on this fact, Fleming et al. concluded that on this star the solar-type corona with $T \sim 10^6$ K could be absent because of an essential difference of thermodynamic conditions in the photosphere.

During the ROSAT sky survey, Schmitt et al. (1993b) found a very strong flare on HII 2034, quickly rotating K2 dwarf in the Pleiades with $v \sin i > 50$ km/s and $P_{\rm rot} \sim 8.^h3$: at least a ten-fold increase of the flux was recorded in one of the scans, whereas the whole duration of the event was no more than 3 h, and the greater part of the recorded photons was in the hard region of the recorded wavelength range. Sciortino et al. (1994) found flares with $A_X \sim 10$ on three stars in the Pleiades: a flare with fast burning and decay in tens of minutes on the very weak star Hz 892; an event on HCG 307 similar to a compact solar flare, and a flare with slow burning and slow decay on HCG 144 similar to a two-ribbon solar flare.

In individual pointing images, Gagné et al. (1994a) recorded a flare on the fast G8 rotator HII 2147 in the Pleiades and estimated the parameters of the two-temperature model of the flare: $T_1 = 3.6$ and $T_2 = 14\,\mathrm{MK}$; $EM_1 = 2 \times 10^{53}$ and $EM_2 = 11 \times 10^{53}\,\mathrm{cm}^{-3}$. In a more detailed analysis of these data, Gagné et al. (1995) found 11 other flares with $L_X > 10^{30}$ erg/s, which were among the strongest X-ray flares. These results confirmed the conclusion of Schmitt (1994): strong flares occurred on stars of all late types from G to M and in a wide range of rotational rates $v \sin i = 9\text{--}45\,\mathrm{km/s}$. The data on a flare on a K star, HII 1516, also enabled the parameters of this event to be estimated within the framework of the simple model of quasistationary cooling: $n_e > 1.3 \times 10^{11}\,\mathrm{cm}^{-3}$, $T \sim 13\,\mathrm{MK}$, and $EM \sim 2 \times 10^{54}\,\mathrm{cm}^{-3}$.

Pye et al. (1994) noted at least 10 members of the Hyades with variable X-ray brightness whose flare spectra were recorded by PSPC. X-ray flares occurred with the greatest probability during observations of VB 141 and VB 190.

In studying three regions of the α Per cluster in the mode of individual pointing, three slow flares with $L_X^{\max} \sim 10^{31}$ erg/s and $E_X > 10^{34}$ erg with burning from 1 to 2.5 h and decay within many hours were detected on three stars of the cluster (Prosser et al., 1996).

During the ROSAT sky survey, Güdel et al. (1994b) recorded a strong flare on the G star Gl 97, which at $v \sin i \sim 4\,\mathrm{km/s}$ had high luminosity $L_X \sim 10^{29}$ erg/s. If the flare light curve is approximated by a triangle whose vertex is at maximum brightness, the duration of this event in FWHM is about 100 min and $E_X \sim 2 \times 10^{33}$ erg, which is higher than the strongest solar flares by an order of magnitude.

During five-day ROSAT HRI observations of AT Mic, McGale et al. (1994) found fast oscillations of X-ray flux with a characteristic time of about 20 s and an amplitude of about 10% imposed on smooth variations with a period of about 8 h and $A_X \sim 2$.

Schmitt (1994) summed up the ROSAT studies of stellar flares completed by early 1993 both during the sky survey and in individual pointings on selected objects. The basic conclusion from the observations within the RASS program is that it enabled X-ray flares on late stars of all spectral types to be revealed. In addition to the above flares on three Me dwarfs (Fleming et al., 1993) and the fast rotator K0 V HD 197890 (Bromage et al., 1992), Schmitt cited the 40-day light curve of the F5 dwarf 36 Dra, on which among about 700 measurements two flares with $A_X \sim 4$ and 10 are clearly seen, and the three-day light curve of B star in π Lup binary system with a strong flare at the end of the observation period. A stronger flare on 36 Dra had $L_X \sim 2 \times 10^{30}$ erg/s and $E_X \sim 10^{33}$ erg, whereas the flare on π Lup, 3×10^{31} and 10^{35}, respectively. On the three-day light curve of EV Lac obtained during the first half of the monitoring – on 18/19 December 1990 – several fast bursts are seen, all the

2.4 Dynamics and Radiation Mechanisms of Flares in Different Wavelengths

second half (19–20 December 1990) took a long time – about a day – increased brightness and one very strong ($A_X \sim 25$) and fast flare. The hardness of the spectrum calculated from PSPC data during this state was definitely higher than before it. Spectral analysis of these observations resulted in the two-temperature model of a quiet corona with temperatures of 1.8 and 5 MK and the one-temperature flare model with $T = 25$ MK. Schmitt simulated this flare on EV Lac within two different models: quasistatic cooling by van den Oord and Mewe (1989) and a two-ribbon flare by Kopp and Poletto (1984). Within the framework of the first model, the light curve of the flare was presented at $T = 30$ MK, $EM = 1.5 \times 10^{52}$ cm^{-3}, $n_e \sim 3 \times 10^{10}$ cm^{-3} and the length of the coronal loop of 6×10^{11} cm or $10 R_*$, i.e., a structure of rather low density and large extension. Within the two-ribbon model, an appreciable dependence of the ascending branch of the flare on the chosen parameters was kept, whereas during the decay phase this dependence was very weak, and observations could be presented by theoretical curves with $n_e = 3 \times 10^{12}$–5×10^{10} cm^{-3} and the strength of the magnetic field was within 2–5 kG. Comparison of the calculations did not allow selection of one of the models, though the fact that the loop size was ten times greater than the stellar radius raised suspicion. Consideration of the summary seven-hour ROSAT monitoring of the UV Cet star in the mode of individual pointing confirmed the flickering in the spectrum of brightness intensity found by Pallavicini et al. (1990a) with EXOSAT and advanced its region to 10 MHz, white noise occurred at higher frequencies.

Apparently, the first stellar flares in the range of medium X-rays from 2 to 18 keV were recorded from Ariel V: Rao and Vahia (1987) identified eight flare stars as fast variable objects found by the satellite, at maximum brightness the luminosity of these stars varied from 2×10^{30} to 3×10^{31} erg/s.

During GINGA LAC observations of the radio galaxy 3C390.3, within 2–36 keV on 14 February 1991, a flare was recorded on the dM4e star EQ 1839.6+8002 (Inda et al., 1994). The flare fell onto the last 37 min of the 14-h monitoring of the radio galaxy field, over four minutes it reached the maximum with $A_X \sim 10$ and then smoothly decayed approximately over 20 min. The large-angle counter (LAC) onboard GINGA, performed spectrophotometry in 48 energy channels with a 16-s time resolution. Pan et al. (1997) analyzed in detail the unique observations. The flare flux at $E > 20$ keV was negligible, the rate of development of the flare within 8–20 keV was noticeably higher than within 2–6 keV. Pan et al. (1997) divided the whole recorded interval into 24 small intervals and adjusted the model of isothermal optically thin plasma, 1T RS model, the "pure spectra of the flare" of each of them was adjusted by varying the temperature and emission measure of the flare plasma. Results of the analysis are presented in Fig. 46.

The plots show a 2–3-minute delay of $EM^{\max}$ behind $T^{\max}$, similar dynamics of the temperature and a flare intensity within 8–20 keV and similar dynamics of EM and intensity in the softer range. The total energy of the

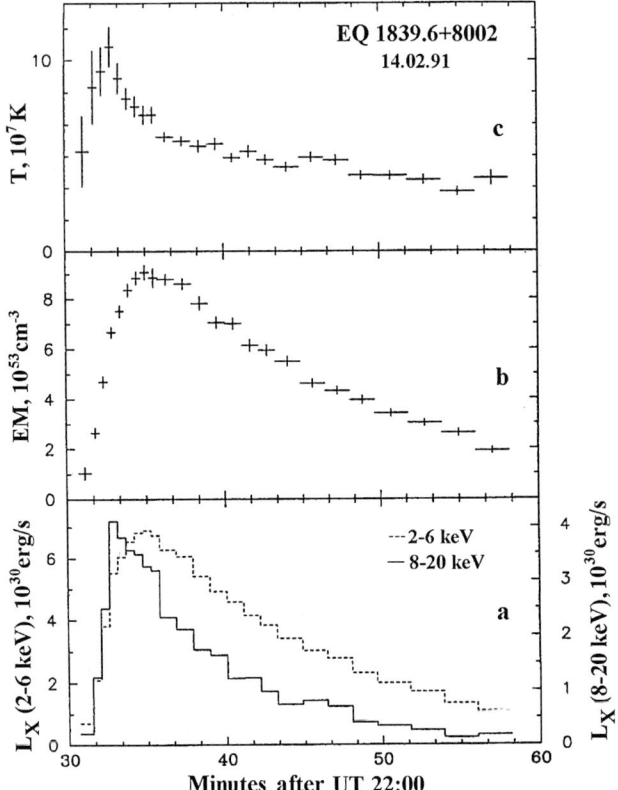

Fig. 46. Flare on EQ 1839.6+002 of 14 February 1991 recorded by GINGA in medium and soft X-rays: (**a**) light curves within 2–6 and 8-20 keV; (**b**) development of the emission measure; (**c**) dynamics of flare plasma temperature (Pan et al., 1997)

X-ray emission of this flare was about 10^{34} erg, while its luminosity at maximum was 10^{31} erg/s, which corresponds to $L_X/L_{\text{bol}} \sim 0.25$. Using the method proposed by Cheng and Pallavicini (1991) to analyze flares on the (T, EM) plane, Pan et al. (1997) distinguished three phases of the flare: a burning phase when the temperature and emission measure grew simultaneously, the phase of continued growth of EM at reducing temperature, and the decay phase when both parameters decreased. They carried out preliminary analysis of each phase on the assumption of constant mass or size of radiating plasma and found that in both cases the beginning of the second phase corresponded to an essential decrease of heating after which EM grew due to evaporation, but slight heating took place at the decay phase as well. The key parameters of the flare are as follows: the length and section of coronal loops $(1-4) \times 10^{10}$ cm and 2×10^{19} cm^2, respectively, and $n_e \sim (0.4$–$1.7) \times 10^{12}$ cm^{-3}. Comparison of the parameters of this flare with other flares on M dwarfs showed that here L_X, E_X, and EM were higher by 1–2 orders of magnitude, $T^{\max} \sim 10^8$ K was several times higher, while n_e, loop length, and magnetic flux tube strength were the same as in typical X-ray flares on M stars. In the archives of the Einstein Observatory, Pan et al. found flares on EQ 1839.6+8002 of 10 April 1980 with $A_X \sim 23$ and on 7 November 1985 with lower amplitude.

Advancement of observations of stellar flares in medium X-rays opened qualitatively new opportunities: since many heavy elements play an essential role in this range, the adjustment of theoretical models to the observed spectral distributions enables one to assess the chemical composition of radiating plasma. These qualitatively new results were obtained by ASCA and BeppoSAX in X-rays and by EUVE in extreme ultraviolet.

The system α Gem was monitored for one day within 0.4–7 keV by SIS ASCA. On 26 October 1993, a flare on YY Gem with $A_X \sim 4$, following a weaker burst that occurred approximately 10 h before, and a strong flare on Castor AB with $A_X > 10$ were recorded (Gotthelf et al., 1994). The summary quiescent-state spectrum and the spectra of the flare on YY Gem were analyzed using the RS and MEKA models of radiation of hot equilibrium plasma. (It should be noted that representation of an X-ray spectrum in the framework of the MEKA model yields higher temperatures, a much higher EM_1/EM_2 ratio and much lower values of EM_1 and EM_2 than the RS model for the same spectrum (Sciortino et al., 1999).) The spectrum of flare maximum on YY Gem could be presented by means of rising temperature of the hottest component of the quiet corona, and the flare spectrum of Castor AB, by means of the two-temperature RS model with $T_1 = 3$ and $T_2 = 11$ MK.

The star Proxima Cen was monitored for 50 000 s from SIS and GIS ASCA on 18–20 March 1994. Over this time, two flares took place on it in the range of 0.5–12 keV with amplitudes of about 5 and 10 (Haisch et al., 1995). Within the framework of the two-temperature model the active stellar corona was composed of components with temperatures of 7 and 44 MK, whereas no hot element was found in the quiet corona.

Using ASCA, Singh et al. (1999) carried out X-ray spectroscopy of flares on two rapidly rotating late dwarfs HD 197890 (Speedy Mic) and Gl 890 and revised the observational results obtained for YY Gem system by Gotthelf et al. (1994). The flare on HD 197980 of 20 April 1995 was recorded at the end of a 13-h observational session: its luminosity increased approximately twice over 33 min and reached 2×10^{30} erg/s. The flare on Gl 890 was found at the end of a 27-h observational session on 19 November 1995: at $A_X \sim 2$ the exponential decay lasted for about 40 min. The spectra were analyzed using MEKAL and VMEKAL models accounting for variations of chemical composition of the radiating plasma. Isothermal and two-temperature coronae and coronae with a continuous dependence of EM on temperature (CEM) set by Chebyshev's polynomials were considered. The spectrum of the flare on HD 197890 in the range of 0.4–6 keV was better represented by the two-component corona with $T_1 = 8.6$ and $T_2 = 37$ MK, $EM_1 = 6 \times 10^{52}$ and $EM_2 = 12 \times 10^{52}$ cm^{-3}. As compared to the quiescent state, the greatest – two-fold – amplification in the flare was noticed in EM_2. Attempts to present the quiet corona and flare with identical content of heavy elements within the 2T models failed, but the solar content was presented within the CEM model and in the 2T model with a five-fold depletion of heavy elements. The spectrum of Gl 890 during the flare was presented by 2T or CEM models of coronae for the solar content of elements, but this representation noticeably improved at a multiple of 3–16 decrease of the content of heavy elements. Gotthelf et al. (1994) analyzed the flare on YY Gem of 26 October 1993 within the MEKA model using SIS data only. Singh et al. (1999) repeated the analysis using the MEKAL model and the data of all ASCA instruments and found that 2T models for the solar content of heavy elements well presented the observations, but introduction of the third component noticeably improved the representation, and a decrease of the content of heavy elements by a factor of two yielded the best result. Singh et al. found that emission measures of hot components of HD 197890 and YY Gem coronae and the temperature of the hot component of Gl 890 corona increased considerably during flares.

The results of the analysis of the quiet corona of AD Leo obtained by Favata et al. (2000a) in observations from three X-ray observatories were reported above. Now let us consider their conclusions about flares on this star. They examined the flare of 13 May 1980 with $A_X \sim 7$ recorded by IPC of the Einstein Observatory, two flares on 8 May 1991 at an interval of 12 h with $A_X \sim 5$ and 3 found by ROSAT and three flares with $A_X \sim$ 3–10 recorded by SIS ASCA. ASCA observations were run on 2–4 May 1996: during the first day the star was in a quiet state, then with intervals of 5.6 and 1.9 h rather strong flares occurred, then till the end of observations the star was active with appreciable oscillations of X-ray luminosity. All the flares were uniformly analyzed using the hydrodynamics approach with obvious allowance for the supporting heating of radiating flare plasma. An empirical relation of the characteristic decay time with the inclination of the flare track on the $\log n_e - \log T$ plane was established for each instrument separately, since this relation depended on the used range of energies. The fullest results were obtained for the second flare of 1996: its descending branch was split into seven intervals, each of them was analyzed independently. Within the framework of the model this flare had essential heating during the decay phase, thus its light curve at this phase was determined by the temporal dependence of heating, rather than by self-decay of the loop. The estimate $h \sim 4 \times 10^9$ cm is seven times lower than R_* and tens of times lower than the pressure scales of heights. At a maximum luminosity of about 2×10^{29} erg/s and total energy $E_X = 3 \times 10^{31}$ erg a field of 0.6 kG was sufficient to maintain such a loop and 1.4 kG were sufficient for its complete emission. The decay phase of the weaker third flare involved supporting heating as well, and the size of its loop was similar. Decay heating was also found in the first flare recorded by ROSAT

2.4 Dynamics and Radiation Mechanisms of Flares in Different Wavelengths

PSPC. The ROSAT light curve of the longest second flare on AD Leo of 8 May 1991 was divided into five intervals 1.7 to 2.3 h long, which were analyzed independently. The decay of this flare was completely determined by the time dependence of additional heating of the radiating plasma. Thus, only the upper limit of the loop size could be estimated: $h < 17 \times 10^9$ cm. Analogously, the light curve of the flare recorded by the Einstein Observatory on 13 May 1980 was divided into intervals of 10–26 min. Their analysis suggested essential supporting heating and a largest loop of 13×10^9 cm. Thus, in all 6 flares the loop sizes fell in a narrow interval and were appreciably less than the stellar radius. The maximum temperature in loops was within 10–50 MK, but E_X covered a wide range of values due to differences in the times of flare decay. Comparison of these results with the conclusions of Cully et al. (1997) for the EUV flare on AD Leo obtained on the assumption of an absence of supporting heating revealed an essential difference in the estimates of the loop sizes, which were overestimated by an order of magnitude. An important conclusion was made by Favata et al. for the loop thickness: the thickness-to-height ratio is 0.3, not 0.1, as on the Sun. Favata et al. (2000a) concluded that all considered flares developed in similar compact coronal loops and their decay was determined by the process of supporting heating rather than self-decay of flare loops. Otherwise, loops would be five times larger. For the standard small size of loops the transition from the level of solar activity to that of dMe stars should not be determined by the filling factor (which is limited to values 1.5–2 orders of magnitude higher than the solar level) but rather by pressure in loops, i.e., densities of flare loops. This pattern is in agreement with the models of a quiet corona of AD Leo with $f \ll 1$ and with thick loops obtained earlier by Giampapa et al. (1996) and Sciortino et al. (1999).

During many-hour monitoring of EV Lac by ASCA GIS on 13–15 July 1998, Favata et al. (2000b) recorded three flares, and the second flare with a maximum close to 13 July 1998 at 20:20 UT was one of the strongest stellar X-ray flares: its amplitude reached 300 at full duration of about four hours and a luminosity at maximum of $L_X/L_{\text{bol}} \sim 0.25$. For several minutes its X-ray luminosity was comparable with photospheric luminosity. Figure 47 shows the light curve of the flare with a time resolution of one minute and the spectra of the flare within each of the time intervals: (1) preflare state, (2) and (3) ascending flare branch, (4) maximum, and (5–9) decay. The results of quantitative analysis of spectra at separate intervals are presented in the bottom diagrams. Within the two-temperature model, quiet X-ray emission with luminosity $L_X = 3 \times 10^{28}$ erg/s was presented by $T_{1,2} = 9$ and 22 MK and $EM_{1,2} = 2.4 \times 10^{51}$ and 0.8×10^{51} cm^{-3}. Flare radiation was presented within the one-temperature model: a maximum temperature of 73 MK was achieved during the premaximum time and the greatest $EM = 7.4 \times 10^{53}$ cm^{-3} occurred at the moment of flare maximum. On intervals 3–6 a 2–3-fold increase of the content of heavy elements and the subsequent return to the preflare level were found. Physical parameters of plasma radiating in the flare were

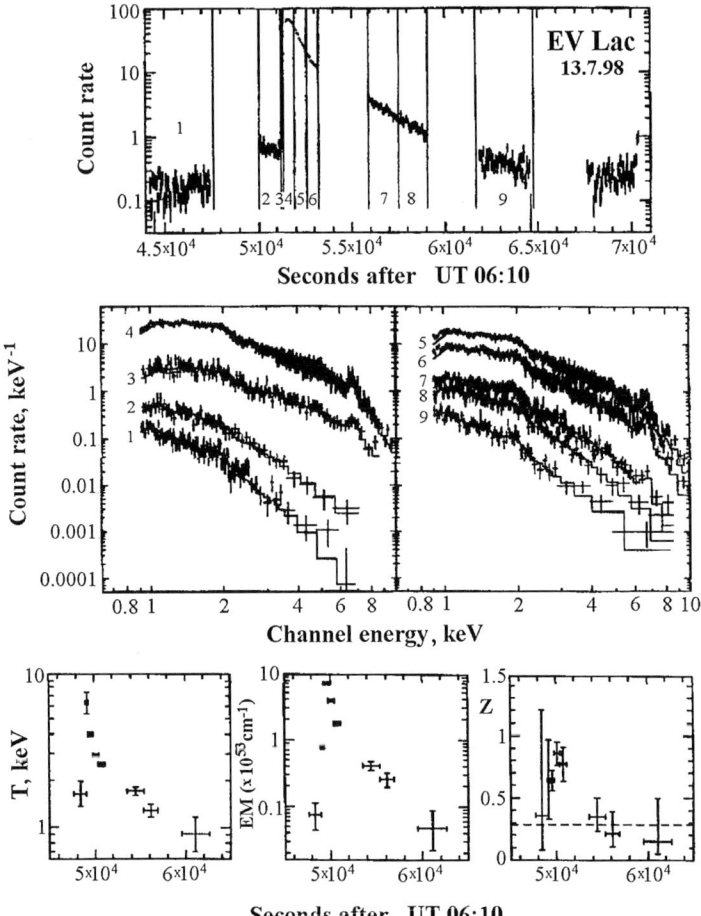

Fig. 47. The flare on EV Lac of 13 July 1998 recorded by ASCA. Top down: the light curve; the spectra on the separated sections of the light curve; results of spectral analysis: the dynamics of temperature, emission measures, and the content of heavy elements (Favata et al., 2000b)

estimated by Favata et al. from their decay character using both models: in the approximation of quasistatic cooling and in the hydrodynamic scheme of decay of an arcade of flare loops with long heating. In the first model, the character of flare decay corresponded to a flare loop of 5×10^{10} cm ($\sim 2R_*$) and a density of matter of 6×10^{11} cm^{-3}. In the second analysis scheme, the drift of the flare was considered during its decay on the $\log n_e - \log T$ plane, which, as stated above, made it possible to find additional heating. In this case, as for the above flare on AD Leo, the light curve of the flare was determined not by the self-cooling rate but by the time variations of additional heating. This approximation yielded the following parameters for the flare on EV Lac:

the true temperature of the flare at maximum was 150 MK, a loop height of 1.3×10^{10} cm ($\sim 0.5 R_*$), a plasma density in the loop of $(2$–$0.2) \times 10^{12}$ cm^{-3}, and a pressure of $(8$–$0.9) \times 10^4$ dyn/cm^2. The obtained loop size is large, but not too large, similar loops occur even on the Sun.

Numerous conclusions within the framework of the hydrodynamic model of flares about the rather small size of coronal loops $h < R_*$ led Favata et al. (2000a) to the conclusion on the nonreastic nature of the free relaxation model that regularly yielded $h > R_*$.

Covino et al. (2001) analyzed ROSAT and ASCA X-ray observations of LQ Hya. On 5 November 1992, ROSAT recorded a flare with $A_X > 10$ and a decay of about four hours, the maximum hardness of its X-ray spectrum was achieved approximately 2–3 min after the maximum brightness. The spectrum of this flare was presented by 2T model at metallicity of $Z \sim 0.1$. During the flare, the cool component of coronal radiation practically did not change, and the size of the flare loops was estimated as $1.5 R_*$. The data on LQ Hya flare of 7 May 1993 recorded by ASCA were presented by the 2T model with $Z = 0.2$–0.6 for different elements.

BeppoSAX LECS and MECS observations with a spectral resolution of 5 and 12 in the ranges of 0.1–7 keV and 1.5–7 keV, respectively, yielded additional data on flares in medium X-rays.

Landi et al. (2001) observed EQ Peg for 19 h and on 3 December 1997 recorded a flare in the range of 0.1–5 keV that lasted for many hours. Within the 2T MEKAL model the recorded radiation was presented as a sum of components with temperatures of 6.7 and 15 MK.

Within a day, three flares with amplitudes of 4–5 and a time of exponential decay of 13, 67, and 73 min were recorded on AD Leo. The star was active for 30–50% of the monitoring time. Two strong flares with a decay time of 75 and 23 min and several weaker bursts were revealed over 9 h on EV Lac. Spectral analysis of these flares showed that 2T models were insufficient irrespective of the accepted chemical composition of the radiating plasma. The flare spectra were successfully presented by 3T MEKAL models with metallicity of the flare plasma decreased by 3–5 times on AD Leo and by a factor of 2, on EV Lac. Evidences of the existence of at least two types of loops were obtained: hundreds of dominating low-temperature compact loops with a size less than $0.1 R_*$ and a total area of bases less than 1% of the stellar surface and tens of extended loops responsible for the hottest radiation with very small total area of bases (Sciortino et al., 1999).

During five-day observations of YY Gem in November, 1998 Tagliaferri et al. (2001) recorded three flares. The hardness of X-ray emission was the greatest during the weakest flare. Spectral analysis of the third flare using the 2T MEKAL model yielded the following parameters of flare plasma: temperatures of components of 9 and 40 MK and increase of metallicity to $Z = 0.8$ from 0.2–0.4 in the quiescent state.

2 Flares

In 1999, AD Leo was observed during 45 days within the framework of the cooperative program that involved BeppoSAX and EUVE, VLA radio systems and optical telescopes from Pennsylvania and the Crimea (Güdel et al., 2001a). For about seven days the star was highly active with almost continuous oscillations of brightness with an amplitude of 1.5–2. In the mode of deep photometry about a dozen EUV flares were recorded with a decay time of up to 2–3 days. Over a shorter period LECS and MECS recorded about ten X-ray flares, which were partially coincident with EUV events. On the other days EUV activity coincided with the bursts of microwave radiation and with the strong optical flare of 30 April 1999. From EUVE data, Sanz-Forcada and Micela (2002) found radiating matter with $\log T \sim 7.1$ in the flares on AD Leo in addition to radiation of quiet corona with DEM maxima of about $\log T \sim 6.3$ and 6.9.

The X-ray astronomy satellite RXTE with a sensitivity of 2–15 keV was used for multiwavelength observations of AU Mic and EQ Peg on 12–15 June and 2–6 October 1996, respectively (Gagné et al., 1998). At the beginning of the flare on AU Mic of 13 June 1996 that was thoroughly tracked in EUV over a day, with a time resolution of eight minutes, there was an individual X-ray burst with $A_X \sim 20$. The spectrum of this flare was presented within the 2T MEKAL model with $T_1 = 20$ and $T_2 = 93$ MK, $EM_1 = 5.0 \times 10^{52}$ and $EM_2 = 0.6 \times 10^{52}$ cm^{-3}, whereas for the quiet star the 1T MEKAL model with $T = 18$ MK appeared to be sufficient. But the flare spectrum could be equally well presented also by the thermal MEKAL model + power spectrum corresponding to nonthermal energetic electrons. Thus, the observations did not reveal the nonthermal component of X-ray emission of the AU Mic flare, though this could be expected from the light curve. Further, at least five strong bursts on EQ Peg were recorded within the range of 2–10 keV, and their spectra could be presented by purely thermal models and the combination of thermal and nonthermal components. Significant similarity of light curves in the optical U band and in medium X-rays was found in the EQ Peg flare of 2 October 1996 recorded in all four ranges of the cooperative observations.

In ACIS Chandra observations, Rutledge et al. (2000) recorded on 15 December 1999 a flare on the rapidly rotating brown dwarf LP 944-20, which was not detected in X-rays out of the flare. The maximum flare luminosity was 10^{26} erg/s, which is lower by an order of magnitude than in the flare on the M8 dwarf VB 10 (Fleming et al., 2000). Within the RS model of the flare plasma they estimated the flare temperature as about 3 MK.

During LETGS Chandra observations of the L 726-8 system, Audard et al. (2003) found much higher flare activity of the B component. On the B component, UV Cet itself, they distinguished flares with two types of light curves: symmetric with slow burning and slow decay, which occurred on the A component as well, and flares with fast burning of about 80 s and a decay of about

40 min. Apparently, this is due to the appreciable difference of flare activity of these stars in the radio range as well. At the same time, the spectra of quiet components were rather similar; their analysis led to the conclusion that appreciable emission measures occurred at 3–6 MK.

In XMM-Newton observations of Proxima Cen on 12 August 2001, Güdel et al. (2002, 2004), recorded X-ray emission by EPIC. A spectrum with a resolution of 300 within the range of 0.35–2.5 keV was recorded by RGS in the region of emission of OVII of about 22 Å and OM recorded stellar brightness in the U band. Thus, for the weakest recorded flare $L_X(0.15 - 10\,\text{keV}) = 2 \times 10^{26}$ erg/s, which corresponds to an average solar flare. Some flares were preceded by short bursts in the U band. The strongest flare that followed a strong burst in the U band (see Fig. 25) lasted for more than five hours, its maximum luminosity was $L_X = 4 \times 10^{28}$ erg/s and $E_X = 2 \times 10^{32}$ erg. In convolving of the optical light curve with the decay constant of 200 s its correlation with X-rays noticeably increased. From the OVII and NeIX triplets, Güdel et al. estimated the electron density of the flare plasma. In combination with the measured emission measures they obtained the following estimations of mass: $5 \times 10^{14}, 2 \times 10^{15}, 3 \times 10^{14}$, and 2×10^{15} g at the moments of primary flare maximum, its decay, secondary maximum and its decay, respectively (see Fig. 25). The volume of hot plasma amounted to $7 \times 10^{26}, 2 \times 10^{28}, 4 \times 10^{26}$, and 5×10^{28} cm^3. The density and mass increase during the decay phase of the relatively cold component of the flare plasma were interpreted as a result of cooling of the hotter component arising at the flare maximum. Based on the estimates, Güdel et al. concluded that the evaporation of the chromosphere played the main role in the development of the flare. Analyzing the same flare, Reale et al. (2004) estimated the characteristic size of flare loops in two independent ways and obtained close values: 10^{10} cm $\sim R_*$. According to their estimates, no effect of final optical thickness was found in FeXVII lines of the flare. Analyzing the light curve obtained over 18.5 h of monitoring, they considered the whole ensemble of flares that took place on it.

During XMM-Newton observations of the α Gem system, when for the first time its A and B components were resolved, high flare activity of both components was found by Güdel et al. (2001b). On the B component the frequency of flares was somewhat higher and their decay was faster than on A. Using the 2T VMEKAL algorithm, a flare coronal plasma was modeled for a strong flare on YY Gem of 25 April 2000. As initial parameters, the 3T model obtained for the quiescent state was used. The resulting parameters of flare plasma were as follows: $T_1 = 9$ and $T_2 = 37$ MK, $EM_1 = 0.4 \times 10^{52}$ and $EM_2 = 3.4 \times 10^{52}$ cm^{-3} at appreciable iron enrichment.

During the simultaneous Chandra and XMM-Newton observations of YY Gem on 29/30 September 2000, within 11.5 h, two X-ray flares were recorded with luminosities of 2 and 8×10^{29} erg/s at maxima (Stelzer et al., 2002). In comparing the high-dispersion spectrum of the stronger flare with the appropriate spectrum of the quiet star before the flare, an appreciable strengthening of emission lines was revealed in the region of 16 Å. In constructing the models of coronal plasma of these flares from EPIC XMM-Newton spectra, two more components with increased iron content were added to the above three-component model of the quiet corona. As in earlier studies, the temperature of the flare plasma reached maximum during burning of flares, while the emission measure did so at the beginning of the decay.

Simultaneous Chandra and XMM-Newton observations of the α Gem AB allowed Stelzer and Burwitz (2003) to divide the components, to find high flare activity of each of them with a characteristic time between flares of 3–5 h, to attribute each of the recorded flares to one or another component, and to conclude that flares occurred on invisible components of these bright stars.

In XMM-Newton/EPIC observations of the EQ Peg system, Robrade et al. (2004) for the first time divided the components of this close pair, established the X-ray flare activity of each of them, and determined that the A component was brighter, but flares were more frequent on the B component.

In the close pair ER Vul comprising two G dwarfs with an orbital period of 0.69 days, Brown et al. (2002) recorded a 30-min X-ray flare.

Let us briefly sum up the results.

X-ray flares on active dwarfs are nonstationary relaxation of optically thin gas at an initial temperature of tens of megakelvins. They occur on stars of F–M spectral classes with rather strong coronae. Gas results from primary energy release in a flare, and its relaxation is due to heat conductivity, radiative losses, long heating at the decay phase, or combinations of these processes. In the strongest flares this hot gas fills magnetic loops of sizes up to the stellar radius, its total radiative energy in X-rays reaches 10^{34} erg, while the weakest flares depend on the opportunity of detection of such events with a low S/N ratio, since, as a rule, stellar flares are recorded on the background of quiet luminosity of coronae. Some stellar flares are considered as analogs to solar compact flares, others to solar two-ribbon flares. However, this classification is not always certain and unequivocal. Feldman et al. (1995) found that the correlation between temperature and emission measure of solar flares, which covers the ranges of $T = 5$–35 MK and $EM = 10^{47}$–10^{51} cm^{-3}, toward higher temperatures and EM was continued by the active stars AU Mic, Algol, π Peg, and UX Ari.

The relatively simple mechanism of thermal radiation of optically thin gas made it possible to determine a number of its characteristics in stellar flares: electron density, dimensions of magnetic loops, strength of magnetic fields, and even the variation of chemical composition of matter during flares.

2.4.2 Radio Emission of Flares

Radio emission of flares in the solar atmosphere is the field of solar physics that has plenty of data at its disposal. This is due to the fact that many magneto-

2.4 Dynamics and Radiation Mechanisms of Flares in Different Wavelengths 281

hydrodynamic disturbances arise and propagate in low-density plasma sited in magnetic fields. On the other hand, various electromagnetic radiations induced by these disturbances can be observed in a wide wavelength range from millimeters to tens of meters. By the late 1950s, numerous radio data on solar flares were cumulated. Fast growth of the interest in flare red dwarfs naturally encouraged attempts to find similar phenomena on these stars.

After successful detection of stellar flares their research was stimulated by the progress of radioastronomical equipment and in its turn stimulated the development of plasma astrophysics.

On 28 September 1958, Lovell started observations of flare stars on the then largest 250-ft radio telescope in Jodrell Bank. During the first night two radio bursts were recorded. Over one and a half years data on 474 h of radio monitoring of UV Cet, YZ CMi, AD Leo, EV Lac, and BD+19°5116 AB at 100, 158 and 240 MHz were gathered. Radio observations were conducted with additional detectors displaced from the axis to control the atmospheric noise. Over this time, 13 bursts that were not in the comparison channel were recorded, but the absence of parallel optical control left doubts as to the stellar nature of these events. In autumn 1960, simultaneous radio and optical observations were started with a network of five Baker–Nunn cameras mounted by NASA to monitor artificial satellites. The time resolution of photographic observations was 2 min. In superimposing the epochs of the obtained radio-emission records at the moment of maxima of 23 recorded rather weak optical flares a distinctly increased radio emission was found in the interval from 2 min before to 8 min after the summed up optical maximum (Lovell et al., 1963).

On 25 October 1963, the Jodrell Bank reflector simultaneously recorded a flare on UV Cet at 240 and 408 MHz. The brightness maximum at greater frequency took place two minutes after the optical maximum, while the maximum brightness at lower frequency, a minute later. Comparison of these features with the characteristics of solar radio flares showed that, based on the emission delay with respect to the optical maximum, the total duration of the event, frequency drift rate, and spectral index, this radio flare on UV Cet could be ranked between solar radio bursts of II and III types (Lovell et al., 1964).

A more distinct time relation of optical and radio flares was established using the results of photoelectric monitoring. According to Lovell and Chugainov (1964), in three flares on UV Cet, one flare on YZ CMi, and one flare on EV Lac the events in both ranges began almost simultaneously, but the maximum of radio brightness was reached 6–8 min after the optical maximum. This conclusion was confirmed by the most successful cooperative observations in October 1963, when the data of 132-h radio monitoring, 55-h observations with Baker–Nunn cameras, and 61-h photoelectric and visual observations by Soviet astronomers were collected (Lovell and Solomon, 1966).

For one and a half years since autumn 1960, UV Cet, Proxima Cen, V 371 Ori, and V 1216 Sgr were observed from Australia with optical support from amateur astronomers (Slee et al., 1963a). For more than 1000 h observations were conducted on the north–south baseline of the interferometer near Sydney at 15 and 3.5 m and for 52 h on the 64-m reflector in Parkes at 75 and 20 cm. As a result, a flare on UV Cet and two flares on Proxima Cen were recorded in the radio and optical ranges and five flares on UV Cet, Proxima Cen and V 371 Ori were recorded without optical support. Optical monitoring was visual, the duration of the optical flares accompanied by radio bursts was 40–60 min at $\Delta m_\text{vis} = 0.3 - 1^m$, the flux density of radio signals was up to 16 Jy, though several of the brightest optical flares were not accompanied by radio bursts.

The strongest was the flare on V 371 Ori observed on 30 November 1963 at both frequencies at Parkes and at 15 m with an interferometer. In the latter case, a half baseline observed the star, while the other, the control region of $3°$ from the star (Slee et al., 1963b). A flare at 75 cm lasted for about 15 min, its rather symmetric light curve had a maximum three minutes after the optical peak, but on the descending branch one could see numerous 6–20 s weakenings of brightness. The optical monitoring was executed by the Baker–Nunn camera and visually by a team of amateurs: $\Delta m_\text{vis} \sim 0\overset{m}{.}6$. At 20 cm and 15 m radio emission was recorded several minutes earlier, just before the beginning of the optical burst. The main feature of this stellar flare was an enormous radio-emission energy of 10^{31} erg, which was higher than that of earlier recorded stellar flares by two orders of magnitude and that of strong solar flares in this range by 6 orders.

Slish (1964) and then Zheleznjakov (1967) proved that recorded essentially nonthermal radiation of the flare on V 371 Ori with a brightness temperature to 6×10^{21} K in decameters could not be due to noncoherent synchrotron radiation of relativistic electron, but probably was a result of coherent synchrotron emission under conditions close to the upper layers of the solar corona.

In May–August 1965, UV Cet, V 371 Ori, V 1054 Oph, and EQ Peg were monitored by the Parkes antenna at 153, 408, and 1410 MHz (Moisseev et al., 1975). Optical monitoring was executed photoelectrically in the Crimea, by the Baker–Nunn camera, and by a team of amateurs in Australia. Over 60 h of joint observations seven optical flares were recorded and for three of them radio emission was recorded at 153 MHz; no bursts were observed at other radio frequencies. The optical flare on UV Cet of 28 August 1965 had $\Delta m_{pg} = 1\overset{m}{.}5$, and burning and decay times of three and four minutes, respectively. Five minutes after optical maximum a seven-minute radio flare started with a flux density of 5 Jy at maximum. The flare on V 371 Ori of 22 August 1965 with $\Delta B \sim 0\overset{m}{.}15$ lasted for eight minutes and a radio burst of 3.5 Jy began simultaneously with the optical flare. The optical flare on V 1054 Oph on 9 May 1965 had $\Delta m_\text{vis} \sim 0\overset{m}{.}3$ and radio flux of 7 Jy at maximum.

In October 1967, during cooperative observations, UV Cet was monitored in Australia by the Parkes radio telescope at 150 and 2650 MHz and by the Culgoora radio heliograph at 80 MHz (Higgins et al., 1968). On 3 October

2.4 Dynamics and Radiation Mechanisms of Flares in Different Wavelengths   283

1967, with an interval of two hours two strong optical flares were recorded, both had high activity at 150 MHz. During a stronger flare that started at 14:36 UT, the maximum flux at 150 MHz was 25 Jy and at 80 MHz, 37 Jy. At 150 MHz, 4 more radio flares simultaneous with optical bursts were found, but they were not confirmed at 80 and 2650 MHz.

During three nights in December 1963 and in one night in November 1968 Slee et al. (1969) carried out simultaneous radio and optical observations of the Orion nebula containing a large number of flare stars. Radio observations were carried out at the Parkes antenna at 136/150 and 408 MHz, optical monitoring, using the chain method by a Schmitt camera with a time resolution of eight minutes. Over 34.1 h of joint observations nine flares with $\Delta B > 0\overset{m}{.}6$ and duration from 22 to 70 min were recorded. The maximum radio emission delayed for 1–19 min with respect to the optical peak, but in two flares it outstripped the optical maximum for 4 and 15 min. The duration of radio flares varied from 1.5 to 50 min. Slee et al. assumed the existence of two different types of radio flares with respect to their duration. In the strongest radio flare that occurred at 13:16 UT on 20 December 1963 and corresponded to two practically simultaneous optical flares in different parts of the nebula, the maximum flux at 136 MHz reached 230 Jy, while one of the longest flares at 16:24 UT on 21 December 1963 had smooth burning and decay and lasted for more than 40 min; close to maximum brightness the spectral index was equal to -1.65. For the strongest flare, Slee et al. estimated the full electromagnetic radiation as 5×10^{35} erg, radio emission amounted to 1% of this value.

One of the most interesting radio flares was recorded on 19 January 1969 on YZ CMi during 12-h radio monitoring from Jodrell Bank (Lovell, 1969). At 1:59 UT a fast optical burst with $\Delta m_{vis} = 1\overset{m}{.}7$ was recorded by Andrews in Armagh; five minutes after the sharp maximum a fast decrease was replaced by a smooth decay, which proceeded until the star disappeared from sight in Northern Ireland. But at 2:49 UT Kunkel in Chile began photoelectric monitoring and tracked the slow decay of the flare in the U band during the following 3.5 h. The 4.5-h optical flare was recorded successfully at Jodrell Bank at 240 and 408 MHz. Records at different frequencies differed noticeably due to the frequency drift of radiation, the time change of the spectral index, and some technical reasons. The flare intensity at maximum reached 30 Jy, and in the middle part, where the estimates of the spectral index were the most authentic, it varied from -1 to -2.5, which in general agreed with similar estimates for the values recorded in Australia (Higgins et al., 1968). Assuming that the frequencies of the recorded radio emission corresponded to the frequencies of plasma oscillations, Lovell estimated the density of matter in the corona as 7×10^8 and 2×10^9 cm^{-3} for 240 and 408 MHz, respectively. He concluded that the long luminosity of plasma required an input of additional energy during flare decay and estimated the total energy of radio emission of the flare as 3×10^{29} erg, which is lower than its optical radiation by five orders of magnitude. Lovell noted the closeness of the frequency drift to the corresponding value in solar events of type II. The flare brightness temperature was 10^{15} K at meter wavelengths, but this estimate could be increased by four orders of magnitude if the flare was a local rather than global event in the stellar atmosphere. In total, up to 1969, within the program of radio-optical observations of flare stars at the Jodrell Bank radio telescope 36 campaigns were run and radio monitoring data for 4000 h were accumulated (Lovell, 1964, 1971).

On 11 October 1972, the upgraded Jodrell Bank radio telescope, at a frequency of 408 MHz, and the 30" telescope of the Stephanion Observatory in Greece recorded a strong flare on UV Cet in the B band (Lovell et al., 1974). Figure 25 shows the light curves of the flare in both wavelength ranges. One can see the considerable delay of radio emission and its noticeably long duration as compared to the optical flare. The flare intensity at maximum reached 12 Jy. As the strong flare on YZ CMi of 19 January 1969, it was used by Kahn (1974) to construct the model of a disturbed atmosphere.

In autumn 1972, the 37-m radio telescope of the Vermilion River Observatory was used to observe EV Lac at 170 MHz over three nights (Webber et al., 1973). Optical support was provided by the 102-cm reflector of the Prairie Observatory (Illinois) and the 51-cm Palomar telescope. On 2 October 1972, both optical telescopes recorded a flare with $\Delta U \sim 0\overset{m}{.}14$ and duration of about 25 s, two and three minutes later there were radio bursts of about 25 Jy that decayed completely in nine minutes. At the same time using the same radio telescope, Tovmassian et al. (1974) observed the Pleiades with support from optical telescopes in Tonantzintla and Palomar. Over eight hours of joint observations one radio flare with an intensity of about 35 Jy was found, this result was confirmed by optical observations. But an attempt to find radio emission of flare stars in the Pleiades with the help of the Ooty radio telescope (India) failed (Sanamian et al., 1978).

Between October 1973 and January 1974, Spangler et al. (1974b) monitored YZ CMi, AD Leo, Wolf 424, and V 371 Ori by the 305-m Arecibo radio telescope at 196, 318, and 430 MHz for more than 70 h. The sensitivity of the telescope was an order of magnitude higher than that of its predecessors. On the first three stars, 26 flares were recorded with amplitudes of tenths of Jy and duration of ten seconds, the average frequency of the events was comparable with that of optical flares – from several hours to ten hours for one flare. For the flares on YZ CMi and Wolf 424 recorded at 196 and 318 MHz, the spectral index was estimated as -2.5 and -2.9, respectively. Then during 15 nights in January–February 1974 Spangler and Moffett (1976) carried out simultaneous observations of YZ CMi, AD Leo, and Wolf 424 by the Arecibo radio telescope and the 91-cm and 76-cm telescopes of the McDonald Observatory. Over 57.8 h of joint monitoring the radio fluxes of 13 radio events were detected directly from the records, one of the events definitely was not accompanied by an optical flare. In analyzing the moments of 62 optical flares, 15 more radio flares were found. Figure 26 presents the light curves of a flare on Wolf 424 recorded during this campaign. On the whole, the observations revealed a considerable diversity of light curves at two radio frequencies and

2.4 Dynamics and Radiation Mechanisms of Flares in Different Wavelengths

distinctions of their light curves from the light curves of relevant optical flares. Most probable was the delay of radio burst with respect to the optical one for 0–5 min. Of 10 radio flares recorded at both frequencies four started simultaneously, in three flares a frequency of 318 MHz and in three of 196 MHz were in the lead. The recorded radio flares did not show a preference to strong optical flares, and one of 3 strong optical flares on AD Leo definitely was not accompanied by a radio event. All this evidences the coherent mechanism of a directed radio emission of stellar flares. By the end of 1974, the list of thus-studied flare stars was supplemented by BD+16°2708, Ross 867, G 25–16, EQ Peg, and G 3–33, the total duration of radio monitoring exceeded 400 h and the number of recorded events exceeded 70, and the range of spectral indices extended from -2– -10 (Spangler et al., 1975).

In observations of AD Leo by the Arecibo radio telescope on 1 April 1974 Spangler et al. (1974b) obtained the first radiopolarimetric data: all four Stokes parameters were measured at 430 MHz. A burst was recorded with an intensity of 0.5 Jy and the duration at the levels of 1/2 and 1/10 of the maximum intensity was 12 and 40 s. The degree of circular polarization at maximum was 56% and increased to 92% by the end of the flare, and the linear polarization was 21% at the maximum. By analogy with decimeter bursts on the Sun, Spangler et al. used the gyrosynchrotron mechanism of emission of moderately relativistic electrons in magnetoactive plasma to interpret the observations.

During a three-day cooperative campaign of observation of YZ CMi in November–December 1975 radio observations were run on eight radio telescopes at 12 frequencies in the range from 38 to 6300 MHz. For the first time radio–optical–X-ray coverage was arranged over 1/3 of the campaign duration (Karpen et al., 1977). In total, 11 radio events were recorded. During three of the longest events lasting from half an hour to an hour and a half no optical monitoring was arranged, the duration of all other radio bursts did not exceed one minute. The greatest number of events was recorded by the Ooty radio telescope at 327 MHz. The burst at 16:12 UT on 1 December 1975 was recorded by the Culgoora telescope at 160 MHz and from Parkes at 5000 MHz. Some 100 s after the radio flare at 20:44 UT on 1 December 1975 there was an optical flare with $\Delta B \sim 0\overset{m}{.}4$. The flare at 7:05 UT on 3 December 1975 was recorded simultaneously at 196 and 318 MHz and in the optical range.

Nelson et al. (1979) summarized the data of radio–optical observations of 15 flare stars obtained in 1972–76 by the Culgoora radio heliograph at 80 and 160 MHz in the mode that detected only long events, but not fast bursts. Over 110 h of observations supported by amateur astronomers 21 optical flare and 19 radio events were recorded independently, variability was found on 11 objects. The greatest activity was displayed by AD Leo: 30 and 58 radio flares were recorded over 34.5 h of observations at 80 MHz and 67.7 h at 160 MHz, respectively. The least active were AT and AU Mic, on which over 10.8 and 24 h of observations at 160 MHz no radio events were recorded. At the same time, no optical flares were found on AD Leo and AU Mic, while on AT Mic three flares were recorded. Only four pairs of radio and optical flares were contiguous: a five-minute flare on UV Cet of 21 July 1975 with $\Delta m_{pg} = 0\overset{m}{.}3$ and $F_R = 1.4$ Jy, two flares on BD + 16°2708 on 18 May and 17 June 1975 with $\Delta m = 0\overset{m}{.}3$ and $0\overset{m}{.}8$ and duration of 70 s and >35 min with $F_R = 1.1$ and 1.9 Jy, and a 15-s flare on V 1054 Oph of 10 August 1975 with $F_R = 0.8$ Jy. In all cases, radio observations were executed at 160 MHz and F_R was determined by integration over half an hour. Three of the four optical flares were short pulsed events, after which long – up to 33 min – radio flares were observed, which resembled solar bursts of types II and IV and noise storms after strong solar flares. Thus, most optical flares did not cause detectable radio emission and radio flares were frequently recorded in the absence of optical flares. Nelson et al. assumed that this could be explained by the fact that both were recorded at the detection threshold or, as on the Sun, optical flares were not always accompanied by radio flares on meter wavelengths. For eight radio flares, significant circular polarization was found from 40 to 61% and more. If the size of the radiation source was $\sim 0.1 R_*$, the brightness temperature was $\sim 10^{15}$ K. Robinson, one of the authors of this publication, showed that coherent synchrotron or coherent cyclotron radiation matched such a temperature, but both yielded heavily collimated beams, which could not be observed for more than two hours. To maintain such a radiation, a continuous supply of relativistic electrons to the medium is required.

Davis et al. (1978) carried out interferometric observations of YZ CMi at 408 MHz on the Jodrell Bank radio telescope and the 25-m Defford antenna. Over 48 h two flares were found: on 15 December 1977 with an amplitude of > 0.04 Jy and on 18 December 1977 with an amplitude of 0.12 Jy. The energy of the latter was lower by two orders of magnitude than that of the flare of 19 January 1969.

In January 1975, Moffett et al. (1978) carried out a 33.9-h radio–optical monitoring of YZ CMi, AD Leo, and Wolf 424 to detect centimeter flare emission. Radio observations at 1420 MHz were conducted on the Arecibo radio telescope. The optical monitoring was carried out at the McDonald Observatory and on the 152-cm telescope of the Mount Hopkins Observatory. Radio flares were suspected near two of the 41 optical flares, but 12 of the 14 radio events were not accompanied by optical flares.

In May 1977, during 3-day cooperative studies of Proxima Cen radio observations were carried out at 6 cm at the 64-m Parkes antenna and at 13 cm at the 25-m antenna in Johannesburg. Optical monitoring was carried out at the 76-cm telescope of the South African Observatory and the 41-cm telescope in Cerro Tololo, Chile. Observations in Parkes were carried out during four optical flares, one of them was strong. Radio observations at 13 cm covered the time of 26 optical flares, but no events in the radio range were detected. Hence, it was concluded that $L_R/L_{\mathrm{opt}} < 2 \times 10^{-5}$ (Haisch et al., 1978).

During cooperative observations on 6–8 March 1979 Proxima Cen was radio monitored at 6 cm from Parkes (Haisch et al., 1981). Over 24 h six flares were recorded with ΔU varying from $0\overset{m}{.}6$ to $4\overset{m}{.}5$; independently 12 radio flares with F_R varying from 7 to 12 mJy were found; but only three pairs appeared to be close in time. Radio emission was not revealed during the strong X-ray flare. The most confident correlation was between flares on 8 March 1979 at 17:30 UT: they started and reached the maximum of 8 mJy and $\Delta U \sim 4\overset{m}{.}5$ simultaneously. Three radio flares were not accompanied by optical events five minutes before and after and seven radio bursts were not accompanied by optical flares. On the basis of the rather modest sample Haisch et al. suspected that there was a tendency of a five-minute delay of radio bursts and a lack of correlation between the amplitudes of the events.

A cooperative campaign of multiwavelength observations of YZ CMi arranged in October 1979 suggested radio monitoring from six radio telescopes at 12 frequencies from 275 to 7875 MHz (Kahler et al., 1982) over 40 h. One strong flare was recorded on 25 October 1979 (see Fig. 27). A confident signal from the flare was obtained at the Jodrell Bank radio interferometer at 408 MHz: radio emission began 17 min after the impulsive beginning of the optical and X-ray flare, reached a maximum of 60 mJy, and then lasted for more than half an hour, while at other frequencies it decayed. Green Bank observations at 515 MHz detected one more radio event, probably a narrow-banded one, since it was not recorded at 430 and 1428 MHz. If the radiation source was of the order of a stellar diameter, the brightness temperature of the flare of 25 October 1975 was 2×10^{12} K. The temperature and the delay with respect to the optical flare are similar to solar bursts of type II caused by plasma oscillations and bursts of type IV in a continuum caused by gyrosynchrotron emission of quasirelativistic electrons. But on the Sun, the delay of bursts of type II usually takes about two minutes, which makes the gyrosynchrotron nature of radio emission of the flare on YZ CMi more probable.

Slee et al. (1981) described radio–optical observations of AT Mic and a flare that occurred on 25 October 1980. Radio observations were carried out at 6 cm in Parkes and photographic monitoring, by the 26″ refractor of the Mount Stromlo Observatory with a five-minute resolution. During serial measurements of the star and control region with an integration time of 100 s an accuracy of about 3 mJy was achieved, and the flux at maximum was $F_R = 20$ mJy. Burning of the radio flare took less than 100 s, its decay lasted for about one hour. The optical flare started almost simultaneously with the radio burst and was of almost the same duration. This is a unique case of a very close correlation of radio and optical events. Three years later in Australia, new radio-optical observations of AT Mic (Nelson et al., 1986) were organized. During 3 nights the radio monitoring was carried out at 5 GHz in Parkes and at 843 MHz in Molonglo, optical observations were conducted in U, B, and J bands by the telescopes in Mount Stromlo and Perth. Radio observations covered the time of 10 optical flares, but the signal $> 3\sigma$ was recorded only during one of them at 15:30 UT on 3 August 1983 with $\Delta B = 0\overset{m}{.}4$. On the other hand, no optical flares were found during two radio bursts. But both flares on AT Mic – of 25 October 1980 and 3 August 1983 – were recorded in two ranges and had appreciable durations, whereas two flares without optical events were short radio bursts.

In the early 1980s, a new stage in studying stellar flares started with the launch of the Very Large Antenna (VLA) designed for recording waves of centimeter lengths.

In spring 1980, YZ CMi, Wolf 424, and BD + 16°2708 were observed at 20 cm with VLA. Over 22 h the only event on YZ CMi was detected with a maximum luminosity of 2×10^{14} erg/s. Hz and duration of about two hours, and strong circular polarization (Gibson and Fisher, 1981). The obtained light curve can be interpreted as three bursts separated by 45- and 41-min intervals, or as one event with 30-min burning and 12-min decay. The observations were continued in autumn 1980 with the optical support from the 200″ Palomar reflector and the 60″ Mount Wilson telescope. On 19 December 1980, at 6 cm Fisher and Gibson (1982) recorded a flare on UV Cet with a fast rise and an exponential decay, immediately after which another flare began at 21 cm. It is not clear whether it was an analog to solar drift bursts or an independent event. During three nights they monitored UV Cet, YZ CMi, and CN Leo for 4 h and recorded 10 or 11 radio events. All the events lasted for about 20 min and did not have on impulsive beginning. At the same time 36 optical flares were recorded but they had no correlation with radio events, though there were rather strong optical flares: $\Delta B = 0^{m}\!.88$ on UV Cet, $1^{m}\!.43$ on YZ CMi, and $1^{m}\!.04$ on CN Leo. Based on the lack of correlation of radio and optical emission of flares and high circular polarization of radio emission of some flares, Fisher and Gibson concluded that gyrosynchrotron or another coherent mechanism was responsible for the centimeter radiation of stellar flares.

Six-hour VLA observations of the system L 726-8 on 16 October 1981 at 6 cm were fulfilled with separate recording of circular polarization components (Gary et al., 1982). At 6:52 UT the right-polarized component of L 726-8 increased from 4 to 16 mJy, whereas the left-polarized component remained invariant. The light curve of the flare constructed with a resolution of 10 s demonstrated five or six quasiperiodic oscillations with a characteristic time of about 56 s (Fig. 48), over which the degree of circular polarization varied from 40 to 82%. Gary et al. considered various opportunities for the occurrence of radiation: modulation of a radio source by an external agent and variations of the mechanism of energy release. In the first case, these could be the oscillations of magnetic flux tubes, but they modulate incoherent radiation and consequently could not account for high brightness temperature and high circular polarization. The second variant can be realized in electron-cyclotron maser, assuming that the conditions required for this mechanism – sufficient strength of magnetic field, availability of free energy for electrons

2.4 Dynamics and Radiation Mechanisms of Flares in Different Wavelengths

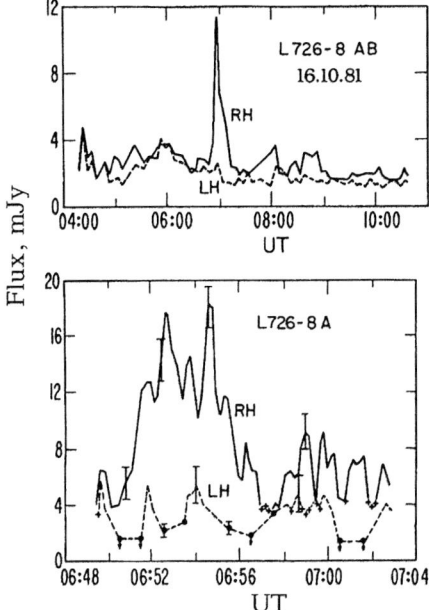

Fig. 48. The flare on L 726-8 A of 16 October 1981 recorded in right- and left-circular polarization at 6 cm, RH and LH, respectively (Gary et al., 1982)

and free emission at the second harmonic of the gyrofrequency – are met in the atmosphere of dMe stars. According to Gary et al., this mechanism operates at B $\sim$ 900 G, brightness temperature of 10^{14} K, and f $\sim 10^{-4}$; but the mechanism of 56-s oscillations remained obscure.

In the course of observations of AD Leo on 1 February 1983 at 1400 MHz from Arecibo with a higher time resolution, Lang et al. (1983) recorded a very interesting flare. Its burning phase lasted for 18 min, and the decay phase was six minutes, whereas the degree of circular polarization did not exceed 15%. But on the ascending branch numerous fast bursts with 100% left-circular polarization occurred within three minutes, the burning time of these bursts did not exceed the time resolution of the instrument – 200 ms, the flux density achieved 130 mJy, and the brightness temperature was 10^{13} K. The event was interpreted as the emission of an electron-cyclotron maser at the second gyrofrequency in a longitudinal field of B $\sim$ 250 G.

During the next days the cooperative program recorded significant variation of AD Leo activity (Gary et al., 1987). On 2 February 1983, simultaneously with an optical impulsive a flare with $\Delta U \sim 0^m\!.7$ and duration of about three minutes VLA was recorded at 6 cm a radio flare with an essential and variable polarization that lasted for about an hour and a half. On 5 February 1983, numerous independent flares were recorded at 20 and 6 cm over 7 h, and one of the flares, seen only at 6 cm, was 100% left polarized. According to Gary et al., a weak correlation of flare activity at 20 and 6 cm can mean various mechanisms of flare generation or various regions of their exit from the corona.

On 15 July 1985, Lang and Willson (1986b) carried out 1.5-h observations of AD Leo at 1415 MHz using the Arecibo telescope with a maximum time resolution for the instrument of 5 ms. They detected two events 50 and 25 s long with intensities of 30 and 10 mJy and a high degree of left-circular polarization. During the second half of the 50-s event quasiperiodic brightness oscillations occurred with a characteristic time of 3.2 s and the strongest component of these oscillations, in its turn, was composed of five separate bursts with a characteristic time of 32 ms, a flux density of 70 to 400 mJy, a rise time of not more than 5 ms and 100% circular polarization. Components next to the strongest one did not display such a structure. At such a fast burning the size of the radiation sources could not exceed 1.5×10^8 cm, the filling factor was estimated as 2.5×10^{-5} and the brightness temperature as $\sim 10^{16}$ K. The radiation source could be caused by electron-cyclotron maser or plasma oscillations. In the first case, the longitudinal field should be 250 G and $n_e \sim 6 \times 10^9$ cm^{-3}. In the second case B $\ll$ 500 and 250 G and $n_e \sim 2 \times 10^{10}$ and 6×10^9 cm^{-3} for the first and second harmonics of plasma oscillations, respectively. Quasiperiodic oscillations and fast bursts, according to Lang and Willson, could be caused by radial pulsations of coronal loops induced by trapped energetic particles or their pulsed source.

On 3 February 1983, VLA microwave observations of YZ CMi were carried out within a wide cooperative program (Rodonò, 1986a; van den Oord et al., 1996). At 6:08 UT an optical flare occurred with $\Delta U = 0\overset{m}{.}84$ and before it completely decayed at 6:12 UT there was an optical burst with $\Delta U = 3\overset{m}{.}83$. This optical maximum coincided with a burst at 6 cm, which reached a maximum of 5 mJy in seven minutes and lasted for about half an hour; significant circular polarization was absent. At 6 cm, at a time resolution of 10 s, the fine structure of the light curve was not seen, and at 2 and 20 cm no radio emission was recorded. If the radiation source was equal to the stellar size, its brightness temperature was about 10^9 K. The analysis of the data sets by van den Oord et al. resulted in the conclusion that there was optically thick synchrotron radiation of electrons with an energy of megaelectron-volts in the magnetic field of about 100 G.

On 5 October 1983, VLA recorded the activity of Gl 182 at 2 cm (Rodonò, 1986a). Quiet radio emission of this star was not revealed, thus the light curve of the flare, constructed with a minute time resolution had three symmetric peaks at 10 to 6 mJy with rise and fall times of about one minute. The flare was not visible at 6 and 20 cm.

On 25 October 1983, a 3.5-h VLA recording of the radiation of YZ CMi at 6 cm revealed an interesting phenomena (Pallavicini et al., 1985). At first there was a fast radio burst with an amplitude of about 6 mJy and duration of about 30 s on the background of quiet radiation of 2.5 mJy. Then 83 and 75 min after there were gradual 20- and 30-min flares. All these brightness variations

2.4 Dynamics and Radiation Mechanisms of Flares in Different Wavelengths 291

occurred on the background of growing circular polarization. Pallavicini et al. found analogs of all these events in solar centimeter-range radio emission. Observing this star on 10 December 1984, Lang and Willson (1986a) found two slow one-hour flares with an intensity of ∼20 mJy, the events at 1515 MHz preceded for about an hour similar events at 1415 MHz. Identification of the appropriate events at these frequencies means narrow-band radiation $\Delta\nu/\nu \ll 0.1$ with slow frequency drift.

The flare on AD Leo of 28 March 1984 was studied in the wide wavelength range from ultraviolet to microwaves (see Fig. 29). At 3:22 UT there was an optical burst with $\Delta U = 2\overset{m}{.}1$, and after several additional rises the flare faded 22 min after the beginning. Near the optical maximum there was a radio burst of up to 32 mJy at 2 cm, the radio emission at 6 cm started simultaneously, but had a smooth rise and reached a maximum of 10 mJy 10 min later (Rodonò, 1986a). The microwave radiation decayed simultaneously with the optical radiation. This flare was not found at 20 cm, which probably suggests its localization in the lower corona.

In early 1985, Kundu et al. (1987) observed UV Cet, AT and AU Mic with VLA at 20 and 6 cm and recorded microwave flares on all stars. In the binary systems AT Mic they found flare activity on both components at both frequencies. During one-hour observations on 5 February 1985 over 13 min an increase of radio emission from 3.6 to 5.6 mJy from the northern star of the pair was detected at 20 cm. Over the next 12 min the value decreased to the initial level. On 22 March 1985, the southern star displayed variations of weakly polarized radiation at 20 cm from 10 to 7 mJy. On 6 February 1985, during one-hour VLA observations of UV Cet, Kundu et al. recorded a smooth rise of the radio emission from 1.7 to 3.2 mJy at 20 cm that lasted for 10 min. Twenty minutes later there was a sharp jump to 10 mJy. Circular polarization increased to 40%, while radio emission at 6 cm was invariant at the level of 2.5 mJy. There was a weak 100% polarized flare on the main component of the system L 726-8 A. Two-hour VLA observations of AU Mic on 22 March 1985 recorded the moment of maximum at 26 mJy and subsequent decay of a strong radio flare at 20 cm: over 85 min its brightness halved and circular polarization was kept throughout the observations at the level of 70–90%. But this strong flare was not visible at 6 cm (Kundu et al., 1987).

On 22 March 1985, Jackson et al. (1989) carried out one-day VLA observations of AD Leo, EQ Peg, UV Cet, Wolf 630, YY Gem, and YZ CMi at 20 and 6 cm. In most cases they found some kind of flare activity. The most interesting was the following result: at a time resolution of 10 s at the burning phase of a flare on AD Leo the sign of circular polarization changed from $V = 4$ mJy at 1465 and 1515 MHz to -13 and -7 mJy, respectively; at the maximum the flare reached 18 mJy. There was a single increase of intensity from 0.6 to 1.4 mJy at 6 cm with a subsequent decrease to 1.0 mJy, but no peculiarities in polarization were found. Another important conclusion drawn from these observations was that optically identical components of binary systems could essentially differ in the radio range.

In August–December 1985, Kundu et al. (1988a) carried out simultaneous VLA and EXOSAT radio and X-ray observations of UV Cet, EQ Peg, YZ CMi, and AD Leo to detect a correlation between flare radiation in two wavelength ranges, 6 and 20 cm. Each star was observed continuously for 7–11 h with a time resolution of 6.7 s. AD Leo was monitored continuously in each microwave region by a half of VLA-antennas, other stars, by all antennas that were switched every 5 or 10 min from 20 cm to 6 cm and back. During nine-hour observations of the system L 726-8 on 4 August 1985 two right-polarized flares with intensities of 10 and 35 mJy were found at 20 cm on UV Cet and L 726-8 A, respectively. At the same time, bursts with an intensity of several mJy were recorded at 6 cm, but their duration was longer than that of flares at 20 cm. EXOSAT simultaneously revealed increased activity, but there was no correlation between the peaks of radiation in the two ranges. It is noteworthy that in the frequency range of 1390–1450 MHz the flux from the flare on L 726-8 A was approximately twice that at 1490–1540 MHz, and the general character of the light curves differed appreciably: at low frequencies a maximum was clearly recorded, whereas at higher frequencies it was absent. This could be due to the drift of radio-burst frequencies, but then one should consider the flare radiation as a narrow-band one with a width of about 100 MHz (White et al., 1986).

In observations of EQ Peg on 6 August 1985 at the beginning of the session a strong X-ray flare was recorded, then at 20 cm there was a many-hour activity at a level of 3–5 mJy and at 6 cm at a level of 10–12 mJy. But it is possible that this microwave activity started already on 4 August 1985, when the system was monitored in the radio range for half an hour and over this time a strong flare with an intensity of ∼50 mJy was recorded on EQ Peg B at 20 cm (Kundu et al., 1988a).

During the observations of YZ CMi on 19 November 1985 from 10:00 to 13:30 UT at 20 cm a 100% left-polarized flare was found with an intensity of 14 mJy; at 6 cm a flare started at 9:24 UT and decayed by 13:00. An X-ray flare started at 8:30 UT and faded in an hour and a half. There was no increased X-ray emission during the strong radio flare at 20 cm.

During 11-hour observations of AD Leo on 15 December 1985 a 100% left-polarized 2.5-h flare with an intensity of 80 mJy was recorded, but it was not noticed at 6 cm or in X-rays. This flare was visible at 1415 MHz, but did not manifest at 1515 MHz, which could not result from the frequency drift, since the star was monitored long before the flare and after it. Details less than a minute were not visible on the light curve of the flare. Such a long and strong polarized radiation has no analogs in solar phenomena (White et al., 1986; Kundu et al., 1988a).

Thus, no distinct correlation between the flare radiation in the microwave range and in soft X-rays was found, at least distinct maxima in one range have no analogs in another range.

2.4 Dynamics and Radiation Mechanisms of Flares in Different Wavelengths

On 23 June 1986, Willson et al. (1988) carried out three-hour sessions of VLA observations of AD Leo and YZ CMi. Two flares were recorded on AD Leo at 6 cm, while no activity was recorded at 20 cm. The first flare had a pulsed light curve with burning shorter than 10 s, a full duration of some minutes, and 100% right-circular polarization. Two hours later there was a flare with a complex light curve and a duration of about 20 min and 100% left-circular polarization. On YZ CMi, a radio flare at 20 cm with rather symmetric light curve was recorded, its total duration was about 20 min, the polarization was low and there were no attributes of narrow-band emission, but immediately after its decay a slow radio flare started at 6 cm with strong left-circular polarization.

Bastian and Bookbinder (1987) obtained on VLA the first dynamic spectra of stellar flares in the microwave range. The working band of 50 MHz centered at 1415 MHz was divided into 15 narrow bands, and recording was conducted in each of them. VLA antennas were divided into two groups for independent recording of right- and left-polarized components. The whole system worked with a resolution of 5 s. On 28 June 1986, five-hour monitoring of UV Cet yielded unique data. On the background of nonpolarized radiation smoothly changing within 2–18 mJy, which was higher than the regular level of 1 mJy, for an hour and a half Bastian and Bookbinder recorded two flares that had completely different parameters. The first event was 100% left polarized. Its light curve was rather simple, with 10-s burning and a twice longer decay. The maximum intensity was about 220 mJy. The event was rather broadband, since light curves in all 15 narrow frequency bands were identical. The second event consisted of two bursts separated by several minutes with a 3-min burning phase and a 1.5-min decay phase. It had a time structure up to 5 s. The radiation was 70% right polarized. The first and second bursts achieved 80 mJy and 100 mJy, respectively. Essential differences in the structure of the event at close frequencies were found within the 41-MHz band. If the source was about stellar size, the brightness temperatures were estimated as 3 and 2×10^{11} K. Bastian and Bookbinder analyzed the conditions at which both cyclotron maser instability and plasma radiation, which originally arise as monochromatic effects, can provide narrow-band emission.

On 3 July 1986, a dynamic spectrum of a flare on UV Cet was obtained by Jackson et al. (1987a). Observations at about 20 cm were carried out at 1385 and 1502 MHz by 13 VLA antennas and at 1435 and 1652 MHz by the other 14 VLA antennas that were switched every 10 min from 20 to 6 cm and back. Near 6 cm, emission at four close frequencies was studied. The session continued for six hours with a time resolution of 6.7 s. Within this period there was a weak (up to 2 mJy) 100% right-polarized flare with a duration of about 80 min on the star L 726-8 A. On UV Cet, at 20 cm a 100% polarized flare was detected at about 11:00 UT with an intensity of $\sim$10 mJy and a nonpolarized flare at about 14:00 UT with an intensity of $\sim$4 mJy. At 6 cm there were weak flares with an intensity of about 2 mJy at 11:45 and 14:20 UT. If the events were considered as uniform bursts at these two frequencies, one would suggest that there was a frequency drift toward high frequencies, i.e., the motion of a disturbing agent toward the stellar surface. The dynamic spectrum displayed a rather complicated pattern of frequency and time changes of microwave radiation.

In November 1987, the studies were continued in Arecibo by Bastian et al. (1990). Since the Arecibo radio telescope is six times more sensitive than VLA, it enabled separation of bursts of terrestrial origin with lower reliability, but in the dynamic spectra the noise was efficiently rejected based on the spectrum type. During three-hour monitoring of AD Leo near 1415 MHz with a resolution of 20 ms a 4.5-min burst with a distinct peak of about one second was recorded. The whole burst had 100% right-circular polarization, and a one-second peak and its nearest environment displayed a rather complex structure with a lifetime of separate details to ten of milliseconds and an appreciable change of light curves in 2.5-MHz bands from 1395 to 1435 MHz, corresponding to a spectral index of about 12. The one-second peak at maximum achieved 940 mJy, the maximum level for stellar flares at this frequency, and with regard to the time structure the temperature brightness exceeded 6×10^{15} K. Six minutes after the decay of this burst another 100% right polarized burst with duration of about 6 min took place on the star, the burst peak was several seconds long. But unlike the previous burst, this peak consisted of ten oscillations with a quasiperiod of 0.7 s and coherent frequencies within the whole frequency band of 40 MHz. The oscillations started 20 s before the maximum of the peak, and its decline was much more abrupt than burning. Bastian et al. observed YZ CMi for two hours with a resolution of 20 ms at 430 MHz and found a single 100% left-polarized burst with a duration of 6.5 s and intensity up to 1.6 Jy. As in the case of AD Leo, the light curve of the burst had short – up to 50 ms – decays with a frequency drift of -250 MHz/s. Other details evidence both negative and positive frequency drift. If the size of the source was determined by the time scale of 50 ms, its brightness temperature was $\sim 3 \times 10^{16}$ K. But this value can be only the lower limit: first, the propagation velocity of disturbances can be only 0.1 of the speed of light and, hence, the linear size of the source is 10 times smaller. Secondly, 50 ms is the resolution of the instrument rather than the real burning time, which leads to an overestimation of the size of the source. Bastian et al. concluded that the decisive role in the selection of the nature of the observed coherent radiation was the ratio of the frequency of plasma oscillations to the electron-cyclotron frequency: if this ratio was less than unity or close to it, instability of the electron-cyclotron maser could occur in the medium, but its appreciable broadband character required a heterogeneous medium. If the ratio was over or close to 3, the conditions for plasma oscillations were created.

In November 1987, Güdel et al. (1989) carried out simultaneous observations of AD Leo, UV Cet, YZ CMi, and Wolf 630 by Effelsberg, Jodrell Bank, and Arecibo radio telescopes at 1665, 1666, and 1415 MHz, respectively. On

2.4 Dynamics and Radiation Mechanisms of Flares in Different Wavelengths

7 November 1987, all three telescopes recorded a sharp burst of right-polarized microwave radiation, all telescopes recorded identical time, brightness, and polarization characteristics. The total duration of the radio event was 40 s, but its light curve consisted of numerous pulsations of different amplitude with duration up to 125 ms and intensity up to 670 mJy. Thus, the event is one of the brightest observed at 18 cm. The obtained data allowed one to suspect narrow-banded radiation $\Delta\nu/\nu < 0.1$ and a frequency drift with a velocity of 240 MHz/s. The value is close to similar values in solar decimeter pulsations, the time division of pulses is analogous as well, but solar events are weaker than the burst on AD Leo by four orders of magnitude.

White et al. (1989a) carried out two observational series of all known flare stars accessible to VLA and located at a distance of up to 10 pc from the Sun, which earlier were not identified as microwave-radiation sources. In the first series, each star was monitored for 20 and 15 min at 20 and 6 cm, respectively. In the second series, monitoring was slightly shorter. As a result, they found microwave radiation from Wolf 47, Gl 234 AB, Wolf 424, Wolf 461, DT Vir, Ross 867, DO Cep, Gl 867 AB, and EV Lac. Together with earlier known sources they made about 40% of the known nearby flare stars and, naturally, this percentage would increase if the exposures were longer, because in all cases apparently a flare rather than quiet microwave radiation was recorded. An extremely strong 95% polarized flare was recorded on DO Cep at 6 cm: it was almost invisible at 20 cm and was essentially narrow-banded, since the light curves differed markedly at frequencies with $\Delta\nu = 450$ MHz. Comparing the fact of detection of microwave radiation with other global stellar parameters, White et al. noted the maximum for M4e–M5.5e stars and the reduced occurrence of this radiation before and after the maximum. The data of this survey suggest that the sources of microwave radiation coincide with the strongest X-ray sources among flare stars and that flares are more often found at 20 cm, while quiet radiation is found at 6 cm.

In January 1987, Lang and Willson (1988) carried out four-hour VLA observations of YZ CMi. Like Bastian and Bookbinder (1987), they divided the 50-MHz band into 15 narrow bands and with a resolution of 10 s conducted serial monitoring of the right- and left-polarized components of radiation. On the background of practically constant radiation of the right-polarized component, the left-polarized one displayed significant variations of brightness with an intensity up to 18 mJy and with a characteristic time of minutes. Detailed consideration of eight local maxima on the light curve showed that four of them contained significant differences between radiations in narrow bands, i.e., microwave emission was narrow-banded with $\Delta\nu/\nu \sim 0.02$, whereas in four other local minima the effect of narrow-bandedness was not noticed: $\Delta\nu/\nu > 0.03$. Interpreting the recorded radiation in the context of an electron-cyclotron maser, Lang and Willson estimated the magnetic field as $B \sim 260$ G and noted that since radiation was limited to a narrow cone with the opening angle from 70° to 85° with respect to the field flux tubes, one should expect its narrow-bandedness to be from 0.01 to 0.1, and the sharp field gradient results in the broadening of the band, while the small size of the source results to its narrowing.

During the strongly polarized flare of 29 March 2001 in the close binary system ER Vul that lasted for about an hour and was recorded by VLA at 20 cm, Brown et al. (2002) found only a very weak response with HETGS Chandra.

On 15 July 1989, using the upgraded Arecibo radio telescope Lecacheux et al. (1993) recorded a four-second burst on AD Leo at 6 cm. At a resolution of 20 ms and a maximum intensity of the burst of 400 mJy the brightness temperature of the source was $10^{10}(R_*/r_s)^2$, where r_s is the size of the radio source. The burst radiation was 20% polarized and concentrated mainly within the band of 135 MHz.

On 24 November 1990, in the course of a 14-h session of observations of the young star AB Dor by the Australian Telescope Compact Array (ATCA) Lim (1993) recorded two very strong circularly polarized flares on the nearby star Rositter 137 B, a very young fast rotator, and their polarization was opposite to that of quiet stellar emission.

On 6 October 1988, Spencer et al. (1993) observed YZ CMi at 6 cm using the broadband interferometer (BBI) at Jodrell Bank. They recorded flare activity as a set of short-lived bursts with a frequency of 5–6 per hour and an intensity of ∼15 mJy on the background of slowly varying emission of about 1.4 mJy. There was a burst with an intensity of up to 42 mJy and total duration of several minutes that consisted of a number of fast pulsations with a duration of less than 20 s. The total energy of the burst was higher than that of similar events on the Sun by four orders of magnitude.

In 1990–93, Abada-Simon et al. (1994, 1997) for more than 81 h observed nine known flare stars at 1.4 and 5 GHz with the Arecibo telescope. Using certain criteria to reject bursts of nonstellar origin, they selected only 12 bursts on AD Leo detected after 40 h of radio monitoring from the initial list of more than fifty radio events at 1.4 GHz. The duration of bursts varied from several seconds to about one and a half minutes, four bursts lasted for 3.5 min, the maximum intensity achieved 70 mJy, half of the bursts were 100% circularly polarized, the others were not polarized. One of the bursts consisted of one right-polarized peak and two left-polarized peaks. Over five hours of monitoring at 5 GHz no bursts were found. Upon more detailed analysis of the strongest and 100% right-circularly polarized burst on AD Leo of 13 February 1993 Abada-Simon et al. found that during this flare the emission frequency repeatedly increased and decreased with the drift module of 1–5 MHz/s and formed arches on the dynamic spectrum with a duration of 10–20 s. The limitation of the detection band did not allow determination of extreme values of emission frequencies, but sometimes it filled the whole 50-MHz band. In considering the central 5 s of the burst with 20-ms resolution, clear differences in the adjacent bands with a width of 10 MHz with separate peaks up to

350 mJy were ascertained, they increased "from zero" during the next 20-ms intervals. Such peaks corresponded to brightness temperature of more than 10^{15} K. These characteristics, according to Abada-Simon et al., conform to the maser mechanism with a magnetic field of about 500 G.

On 31 December 1991, the 100-m Effelsberg telescope at 4.75 GHz recorded a radio flare on UV Cet. The flare started five minutes after the maximum of optical flare with $\Delta U \sim 1^m$ that lasted in the B band for about six minutes. Optical observations were performed at the 80-cm telescope of the Wendelstein Observatory using the five-channel $UBVRI$ photometer with a resolution of 20 ms (Stepanov et al., 1995). The radio burst was more than 75% left-circularly polarized. When 12 narrow 3.1-MHz bands were separated from the full 50-MHz band, significant differences were found in the light curves at the radio flare maximum and in its nearest vicinities: at a frequency lower than 4751 MHz the flare practically was not visible, but it definitely went below the detection limits of 4725 MHz. In the range of 4725–4744 MHz the light curve of the flare constructed with a resolution of 0.125 s had a total duration of about one minute, a maximum intensity of 250 mJy and numerous narrow peaks with an intensity of up to 150 mJy. The characteristic brightness temperature reached 3×10^{12} K. Comparing the mechanisms of the electron-cyclotron maser and plasma oscillations, Stepanov et al. concluded that in the case of the maser, the magnetic field should be ~850 G and the electron density less than 7×10^{10} cm^{-3}, but above 10 MK there would be problems with an exit of radiation from the medium. In the case of plasma oscillations, $n_e = 3 \times 10^{11}$ cm^{-3} and B > 200 G. This mechanism explains the fine structure of the dynamic spectra. In both cases, the radiation source size should be about 10^9 cm.

Using the same instruments, Stepanov et al. (2001) recorded a microwave burst on AD Leo of 19 May 1997 with a flux of about 300 mJy, a duration of about one minute, and 100% right polarization. In the decay phase they found quasiperiodic oscillations with a characteristic period of about 2 s, which most probably were due to magnetohydrodynamic oscillations of the flux tube containing the radiation source. Stepanov et al. identified the burst with $T_{br} > 5 \times 10^{10}$ K with plasma oscillations at the fundamental frequency of the source with $n_e \sim 2 \times 10^{11}$ cm^{-3} and a field of about 800 G. Only rather nonuniform corona with two orders of magnitude lower average density than that in the source could pass this radiation through without appreciable absorption. Acceleration of electrons by an electric current at the loop base could initiate plasma oscillations. Later, using the Wigner–Ville transformation technique to obtain the dynamic spectrum of low-frequency pulsations, Zaitsev et al. (2004) reanalyzed the observations within the framework of the model of free oscillations of coronal magnetic tubes. They estimated the loop height as $\sim 1.4 \times 10^{10}$ cm $> R_*/3$, the electric current in the loop as 4.5×10^{12} A and its total energy as 5×10^{33} erg, which is 2–3 orders of magnitude higher than the appropriate solar values.

Güdel et al. (1995a) investigated microwave radiation of the very young solar-type star EK Dra at 6 and 3.6 cm. Analysis of 36 one-hour sessions of VLA observations (September–October 1993) revealed the variability of radio emission at 3.6 cm. From its average intensities the intervals of "strong flux" were established when, on intervals of 10–180 min the intensity achieved 0.4–0.5 mJy that was 5–6 times higher than the quiet level, but polarization in this case was low. In the case of a "weak flux", such flare-like events did not occur. If the size of the radiation source is equal to the solar radius, the brightness temperature should be 2×10^7 K. Measurements at 6 cm conducted during high-intensity radiation at 3.6 cm did not reveal such events.

As stated above, Lim and White (1995) observed four quickly rotating G–K dwarfs in the Pleiades at 3.6 cm with VLA. On HII 1136, the fastest G rotator in the cluster, they found a low-polarized radio flare, whose emergence was accompanied by a one-hour increase of intensity up to 1 mJy. The flare maintained this state for 2 h and then exponentially decayed with a characteristic time of 1.4 h. On the descending curve there was a short 100% circularly polarized burst.

Summarizing the properties of the not numerous microwave flares on F–G–K stars, Benz (1995) noted their regular distinction from flares on M dwarfs: the former were low-polarized and could be caused by gyrosynchrotron emission; they were longer and less frequent; their radiation was apparently incoherent and required more energy for the same intensity, whereas highly polarized mostly narrow-band radio flares on M dwarfs were caused by the coherent mechanism.

To determine the sizes of radiation sources, Benz et al. (1995) observed EQ Peg B and AD Leo at 18 cm using the intercontinental interferometer system VLBI. The system included antennas of VLA, Arecibo, Effelsberg, Jodrell Bank, Green Bank, and Owens Valley observatories. On 15 March 1990, a radio flare was recorded on EQ Peg B with an intensity of 8 mJy at maximum, duration of about one hour, and right-circular polarization to 75%. The size of the radiation source was estimated as less than 1.8 stellar photospheres. During two observational sessions of AD Leo in September 1991 slow oscillations of the intensity of radio emissions at a level of 0.6 and 1.8 mJy were recorded, as well as 50% right-circular polarization. Based on the analysis of the obtained data, the upper limits of the sizes of the radiation sources were estimated as 1.9 and 3.7 stellar photospheres. The estimates of the sizes correspond to the lower limits of brightness temperatures of 4×10^{10} K on EQ Peg B and $2 \times 10^{9-10}$ K on AD Leo.

One of the critical moments of the phenomenological model of solar flares consists in the consideration of the fluxes of accelerated particles that appear during the initial energy release – the causal process is not clear as yet – quickly heat up the chromosphere, which leads to its evaporation and an increase in thermal radiation from the corona. One of the most forcible experimental confirmations of the idea is the so-called Neupert effect: the intensity of flare thermal radiation from the corona at each moment of time is proportional

2.4 Dynamics and Radiation Mechanisms of Flares in Different Wavelengths 299

to the integral from the nonthermal radiation initiated directly by accelerated particles since the flare beginning till this moment. For the Sun, the measure of thermal coronal radiation is thermal X-ray emission. Hard nonthermal X-ray emission of electrons with energies of hundreds of kiloelectron-volts in the form of gyrosynchrotron emission is usually considered as a measure of nonthermal emission, though originally nonthermal radio emission was studied as such a measure. To detect the stellar analog of the Neupert effect, Güdel et al. (1996) observed the system L 726-8 AB in January 1995 at 6 and 3.6 cm by VLA and in X-rays by ROSAT and ASCA. During two nine-hour sessions they recorded a half dozen weakly polarized radio flares: one on L 726-8 A and the rest on UV Cet. Upon detailed consideration of their radio and X-ray emissions the authors found that three or four pairs of such radio and X-ray events to some extent satisfied the criterion of the Neupert effect: radiation in both ranges began practically simultaneously, fast growth of X-ray emission occurred at the moment of maximum radio emission, while maximum X-ray emission took place when radio emission decreased to the preflare level. Güdel et al. showed that one could expect regular distinctions between the Neupert effect on stars and on the Sun. They called the events "candidates for the Neupert effect".

During four sessions in 1990–94 Lim et al. (1996) studied microwave radiation of Proxima Cen at 20, 13, 6 and 3.5 cm using ATCA. They recorded flare radiation only during one session on 31 August 1991 at 20 cm. The flare lasted for about six minutes, reached a maximum of 20 mJy, and its radiation was 100% left-circular polarized. The absence of signal at other wavelengths testified to the narrow-bandedness of the flare.

In October 1996, EQ Peg was observed within the framework of the wide cooperative program that involved VLA, EUVE, and RXTE, and the McDonald Observatory for studies in the optical range (Gagné et al., 1998). On 2 October 1996, for nine hours EQ Peg activity was recorded in four wavelength ranges, six or eight radio flares at 3.5 cm with an intensity of 3–4 mJy and duration up to 30–40 min were recorded. After one of the strongest optical flares with $\Delta U \sim 2^m\!.5$, simultaneously recorded by EUVE and RXTE, a radio burst started with an appreciable delay, and the strongest radio flare simultaneous with the flare in medium X-rays was recorded by RXTE. Based on the results of these and previous observations, Gagné et al. concluded that the strongly polarized radio flares occurred irrespective of activity in other ranges, whereas low-polarized radio events were often correlated with flares at other wavelengths.

On 29 April 1999, during 10-hour VLA monitoring of AD Leo a radio flare was recorded at maximum EUV brightness of the flare (Güdel et al., 2001a).

In 1984–85, Jackson et al. (1990) studied flare activity of known dMe stars at decameters from the Clark Lake Radio Observatory: 12 objects were monitored at seven frequencies from 30.9 to 110.6 MHz over 143 h. Over this time meaningful signals were suspected only from AD Leo at the beginning and the end of the flare of 15 December 1985, which was recorded at 1415 MHz by VLA. According to Jackson et al., this means that flares, as a rule, do not reach the decameter wavelength range.

On 1–6 September 1992, UTR-2, a radio telescope near Kharkov, with optical support from Crimea, Sicily, and Greece monitored EV Lac in the decameter wavelength range (Abranin et al., 1994; Abdul-Aziz et al., 1995). UTR-2 is the world largest radio telescope for this range designed as a flat, fixed system of dipoles with an electrically controlled orientation diagram. Simultaneous observations at 25 and 20 MHz with a time resolution of 0.2 s were carried out for 5 h about midnight. Some special methods were used to suppress noise. Over 30 h more than 30 single and group radio bursts from 1–3 to 10 seconds were detected, one burst lasted about four minutes. During radio monitoring numerous optical flares were detected. Eight radio bursts could with great probability be related to optical events. It should be noted that optical flare activity varied considerably from one night to another, stellar activity in the radio range varied in parallel. For example, groups of radio bursts were observed only during the night between the 1st and 2nd of September, when optical activity was also maximum. Radio bursts were more often recorded at 25 than at 20 MHz, and the effect of frequency drift was not found. The bursts with the greatest probability identified with optical events had an intensity of 150–900 Jy at 25 MHz and 200–535 Jy at 20 MHz. These values correspond to brightness temperatures within 5×10^{17}–3×10^{18} K. In a number of characteristics the recorded radio bursts were similar to type III solar bursts but their intensity was higher by 3–4 orders of magnitude. Melnik (1994) thoroughly considered the mechanism of an electron-cyclotron maser as applied to dMe stars and concluded that it could explain all features observed on EV Lac – narrow-banded radiation, an absence of frequency drift, and the duration of bursts – and admitted brightness temperatures up to 10^{21} K.

Upon some technical and methodical improvements the campaigns of 1993 and 1994 dealt again with the decameter radiation of EV Lac. But in 1993, 25-h radio observations revealed no events that could be confidently compared with an optical stellar flare. In 1994, over 33 h 18 radio bursts were recorded, but after strict selection only one was with the greatest probability identified as an event of stellar origin: the radio flare of 26 August 1994 at 22:08 UT at 20 MHz with a duration of about 10 s and intensity of ∼150 Jy. Its beginning coincided, to an accuracy of 3 s, with the optical flare with $\Delta U \sim 0\overset{m}{.}5$ (Abranin et al., 1998ab).

In December 1994, UTR-2 conducted radio-optical observations of YZ CMi with optical support from Crimea and Greece (Abranin et al., 1997). Over 33 h 12 radio bursts with durations from 5 to 20 s were recorded. Eight of them occurred during photometric monitoring in the U band from the Crimea, but none of them coincided with an optical flare. The only flare recorded in the B band from Greece took place 160 s after the radio burst, its brightness temperature was $10^{16}(R_*/r_\mathrm{s})^2$ K.

2.4 Dynamics and Radiation Mechanisms of Flares in Different Wavelengths

Thus, the radio observations of stellar flares yielded extremely important data on nonthermal radiation and, consequently, on nonequilibrium processes occurring in stellar coronae that could be responsible for this radiation. Though thermal X-ray emission and nonthermal radiation of flares arise in the same coronal regions, their role in flares is incomparable: X-rays radiate the main portion of the energy obtained by coronal gas from the initial energy release and probable additional heating during the flare, whereas the radio range is mainly an information channel that carries a negligible amount of flare energy, which makes it possible to investigate the independent coronal component, accelerated particles. It is natural that in the wide wavelength range from millimeters to tens of meters different technical means were used, thus the basic conclusions of these studies should be summarized separately for several subranges.

In the meter range, the radiation of stellar flares was studied for several decades. These studies revealed high brightness temperatures of sources, high spectral indices, and a high degree of circular polarization of radiation. These properties suggested coherent and directed radio emission of flares. As to the concrete mechanisms, first the concept of plasma oscillations excited by intense motions of flare plasma prevailed. In combination with regular delays with respect to optical events, the properties of stellar flares found were compared with the much weaker solar bursts of types II and IV. Today, only the data on the strongest flares are used from the whole body of early experimental results, because, as a rule, they have good optical confirmation, whereas the reliability of the allowance for the ground noise is doubtful and there could be confusion with nearby radiation sources.

The most effective were the studies of stellar radio flares in the centimeter wavelength range, where radio events display considerable variation of time, brightness, and polarimetric characteristics, wide-bandedness and a high degree of correlation with optical flares. It suffices to remember that the time change of microwave flares expands from that typical of optical light curves with a fast rise, slower decay, and a total duration of tens of minutes to separate bursts and even sharp peaks lasting for tens of milliseconds. If, in the 1960s–1970s, many hundreds of hours of monitoring were needed to detect a radio flare in the meter-wavelength range, modern telescopes record microwave flares in hours. As to physical models of such events, two coherent mechanisms are actively discussed: plasma oscillations and instabilities of an electron-cyclotron maser. The main point of the discussion is the fact that the parameters of the radiating medium obtained within both models – plasma density and strength of magnetic fields – are rather similar. Thus, it is not clear whether these models are alternative or additive.

The richest data on radio flares in centimeter waves yielded dynamic spectra with considerable diversity of spectra and time characteristics. Frequency drift of both signs, fast decays, and quasiperiodic pulsations of these radio bursts resemble solar decimeter bursts.

Except for the unique flare on V 371 Ori of 30 November 1963 recorded simultaneously in decameter and meter waves and in the optical range, no decameter range flares were unequivocally associated with optical events: about twenty radio events with different reliability are compared to contiguous optical flares, both individually and at the level of general activity on certain nights. As to the physical model of such radio events, there are competing maser and plasma mechanisms, which basically can provide brightness temperatures up to 10^{20}–10^{21} K.

Finally, the available pattern of stellar radio flares differs from the solar one not only by the orders of magnitude higher intensity of processes, but also by such qualitatively new components as a high-temperature plasma with $T > 10^7$ K and constantly replenished resources of relativistic particles that are absent on the Sun.

2.4.3 Ultraviolet Emission of Flares

Information about ultraviolet emission of flares on active dwarfs is less abundant than on X-ray flares. Ultraviolet emission, as X-ray emission, is accessible only from space vehicles, but the flare activity of stellar coronae was investigated by wide-angle instruments during sky surveys and in studying the vicinities of some objects within other programs, whereas flares in the chromospheres and transition regions require long monitoring with individual pointing.

As in a number of the above cases, ultraviolet spectra of solar flares serve as initial data in considering the appropriate stellar spectra, however, essential differences between the spectral resolution of stellar and solar spectra require certain caution (Doyle and Cook, 1992).

In monitoring of Proxima Cen on 6 March 1979 with IUE and the Einstein Observatory, an X-ray flare with appreciable amplitude was recorded. Simultaneously, a spectrogram was obtained in the range of 1150–1950 Å. However, no changes as compared to the spectrum of the quiet star were revealed (Haisch et al., 1981).

IUE observations of Gl 867 A without photometric support detected an additional continuum in the range of 1200–1950 Å, approximately twice as large an intensity of the emission lines of CII λ 1335 Å, SiIV λ 1396 Å, and CIV λ 1550 Å, and less amplified other lines in one of three spectrograms obtained with hour exposures (Butler et al., 1981).

In 1980, simultaneous observations of flare stars and RS CVn-type variables in ultraviolet and other wavelength ranges were carried out by researchers from Armagh, Boulder, Palo Alto, Socorro, Caltech, and Catania. Participants of this cooperation performed most of the further observations of active dwarfs in the ultraviolet (Rodonò, 1986a).

In monitoring of Proxima Cen on 20 August 1980, IUE and the Einstein Observatory recorded a strong X-ray flare with a duration of about one hour. Spectrograms of the flare in the short-wavelength range recorded by IUE are shown in Fig. 28. The CIV, NV, HeII, CII, and CI lines were many times stronger near the maximum brightness and SiIV, CIII, AlIII, and CI emissions appeared, which were not seen in the spectra of the quiet star (Haisch et al., 1983). The total emission energy of the transition region in the flare was estimated as 10^{30} erg, 20 times lower than the X-ray emission of this flare. However, later, when Byrne et al. (1987) developed the technique of differential emission measures and estimating electron density from the ratio of line intensities to analyze IUE spectra, Byrne and McKay (1989) revised the observations of this flare and found that the characteristic density in the transition-region was 4×10^{10} cm^{-3}, the volumes of flare X-ray and ultraviolet emissions were comparable, radiative losses for Ly_α and the transition-region lines were 1.5×10^{31} erg and were close to $E_X = 3.5 \times 10^{31}$ erg.

Bromage et al. (1983) during 17.5-h IUE monitoring detected strong flares on UV Cet, AT Mic, EV Lac, and EQ Peg, while usually, on average, one ultraviolet flare is detected over a day of monitoring. Flares on AT Mic and EQ Peg with maximum amplification of the CIV emission were monitored without photometric support, and rather strong optical flares on UV Cet and EV Lac demonstrated rather weak amplification of the emission. This fact confirmed that there was no simple dependence between the optical continuum and ultraviolet line flare emission, which could be connected with the occurrence of these radiations in spatially different regions. The strong continuum in the range of 1700–1900 Å in the flare on AT Mic could be presented as blackbody radiation with a temperature of 13 000 K. In the same wavelength range a weaker continuum was found in a flare on EQ Peg; the availability of CIII λ 1176 Å emission in the absence of CIII λ 1909 Å and SiIII λ 1892 Å testified to a density of matter above 10^{10} cm^{-3}.

Baliunas and Raymond (1984) recorded a flare in the system EQ Peg AB on 2 September 1981: IUE provided the spectrum of the system as a whole, while the Whipple Observatory recorded the spectrum in the visible range of EQ Peg B, in which strengthening of H_α and H_β emissions was found during the flare. The intensity of CIV emission increased approximately by a factor of three, the SiIV, NV, HeII, and CII lines appreciably amplified, in the range of 1700–1900 Å the continuum grew above the detection threshold. In the long-wavelength range obtained for the decay phase, MgII emission was amplified approximately by 10%, and FeII blend within 2600–2650 Å, by 20%. A maximum of helium emission in this flare took place earlier than in other transition-region lines close to the beginning of strengthening of hydrogen lines, whereas the flare on AU Mic of 4 August 1980 observed without photometric support but with shifting of the stellar image along the slit during three hours had another time change of emission: HeII and SiII lines remained bright enough after appreciable fading of CIV, and maximum intensity of CI,

CII, SiII, and HeII took place approximately 20 min after CIV maximum (Butler et al., 1983).

Using an improved technique to extract data from IUE telemetry and their absolute calibration, Bromage et al. (1986) thoroughly analyzed the observations of a flare on AT Mic of 19 September 1980 and reprocessed the archives of observations of flares on Gl 867 A of 11 September 1979, Proxima Cen of 20 August 1980, and EQ Peg B of 2 September 1981. They found that on both spectrograms obtained during one-hour observations of AT Mic on 19 September 1981, emission lines were appreciably stronger, and on the first spectrogram the strengthening was more intense: fluxes in CIV, SiIV, CIII, OI, and CII lines increased from 7.5 to 2 times as compared to the quiescent state. Absolute fluxes in these lines, averaged over 30-min exposures, were 4–5 orders of magnitude higher than those in the strong solar flare of 15 June 1973 recorded by Skylab, and the increase was even higher for high-temperature lines. Considering the data for the four stellar flares, Bromage et al. noted that in a solar flare the strengthening of the Ly_α line was much greater than in stellar flares, which could be a consequence of the large optical thickness of the line in stellar atmospheres. On the first spectrogram for the flare on AT Mic of 19 September 1980, a flare continuum was seen in the whole range of 1150–1970 Å, in the second, only in the range $\lambda > 1500$ Å and about 1900 Å its intensity was four times lower than on the previous spectrogram. The continuum in the first spectrogram of this flare was rather flat and similar to that recorded in the flare on Gl 867 A. The continuum in the second spectrogram was close to that of the flare on EQ Peg, and both were much higher than the continuum of the flare on Proxima Cen of 20 August 1980. The continuum of the solar flare was 4–5 orders of magnitude lower in the region of 1900 Å and lower by one more order of magnitude in the region of 1300 Å.

Using the same technique, Phillips et al. (1988) analyzed the spectra of UV Cet obtained on 17 September 1980. In a 55-min spectrogram obtained in the range of 1150–1950 Å, during which a five-minute optical flare occurred, they found strengthening of the CIV line by 30%, while in the spectrogram of the range of 1950–3200 Å obtained almost half an hour later, the MgII emission was strengthened by the same extent.

Upon more detailed analysis of AU Mic observations on 3–6 August 1980, during which in the range of 1150–1950 Å a dozen flares with $\Delta U < 1^m$ were recorded, Butler et al. (1987) noticed that as the energy of flares E_U varied within two orders of magnitude, the fluxes of the strongest ultraviolet emissions of CIV, CII, and HeII changed by not more than a factor of two. In different flares these emissions behaved in a different way: CIV and HeII emissions changed nonsynchronously, some flares with appreciable amplitude ΔU were not accompanied by significant changes of ultraviolet emissions and vice versa.

Of a dozen BY Dra spectra obtained in October 1981 in the range of 1150–1950 Å, two were presumably obtained during stellar flares: CIV, SiIV,

2.4 Dynamics and Radiation Mechanisms of Flares in Different Wavelengths

CII, and NV lines were stronger by some tens of per cent as compared to the other spectra; there was a tendency of growth of this increase as the formation temperature of the lines increased (Butler et al., 1987).

Based on the observations of BY Dra and AU Mic, Butler et al. (1987) constructed the curve of column emission measures for these stars in quiescent and active states. They found that the EM of these stars in the active state was higher by 1.5 than the appropriate EM for a quiet star, but all the EM of the stars were an order of magnitude higher than the appropriate value of the quiet Sun at about 3×10^4 K and by two orders of magnitude than that of about 2×10^5 K, at which they are much higher than even the EM of a solar flare.

In the spectrogram of YZ CMi obtained with a 40-min exposure that covered almost the whole strong flare of 3 March 1983 with $\Delta U = 3^m.8$, the strengthened lines are well seen: CIV by a factor of 5.6, SiIV by a factor of 10, HeII by a factor of 5, and NV, CII, and SiII by a factor of 3–4. The total luminosity of the lines in the flare achieved 2×10^{29} erg/s and 3×10^{28} erg/s in the continuum in the range of 1150–1950 Å. Comparison of the curves of radiative losses in a quiet star and during the flare showed a shift of the transition region to the depth of the atmosphere during the flare (Rodonò, 1986a; van den Oord et al., 1996).

On 4–5 October 1983, two ultraviolet flares were recorded on Gl 182. In the first and second one-hour exposures of 4 October 1983, CIV lines were four and two times stronger. The spectrum obtained later in the range of 1950–3200 Å had a 30% stronger MgII emission. The unique spectrogram in the range of 1150–1950 Å of the flare of 5 October 1983 revealed a 20-fold amplification of CIV and SiIV lines, the highest recorded on dMe stars, other emission lines increased by 7–14 times, the line Ly_α by one and a half times (Mathioudakis et al., 1991). From the ratios of intensities of the lines CIII λ 1176 Å and SiIV λ 1396 Å to the intensity of the line CIII λ 1908 Å electron densities were estimated during two exposures of the flare of 4 October 1983 and one of the flare of 5 October 1983 as 1.5×10^{10}, 1.0×10^{10} and 1.7×10^{10} cm^{-3}. Using the DEM technique, radiative losses of flares on 4 and 5 October were estimated as 8×10^{32} and 3×10^{33} erg within $4.3 < \log T < 5.4$ and 1×10^{34} and 6×10^{34} erg within $4.0 < \log T < 8.0$. In the flare of 5 October 1983, within 1150–1950 Å, a strong continuum was found with a total energy of 2×10^{33} erg, the ratio of this energy to that of CIV emission was close to the values found by Phillips et al. (1992) in flares on other stars. From absolute fluxes and estimates of n_e Mathioudakis et al. found the volumes of flare plasma in both flares on Gl 182 and then the radii of curvature of magnetic flare loops: 9×10^9 and 2×10^{10} cm in the events on 4 and 5 October, respectively.

In cooperative observations of AD Leo in early February 1983 the star was monitored by IUE and VLA, and in the optical range. In total, nine flares were recorded. The following conclusions were made from this small sample: a) CIV fluxes correlate with brightness in the U band; b) flares in the U band are much shorter than the duration of strengthening of HeII and radio emissions;

c) radio emission correlates with HeII, but does not correlate with CIV. Thus, it was concluded that HeII in flares was excited by coronal radiation (Gary et al., 1987). On the spectrogram obtained eight minutes after the fast flare on 2 February 1983 at 11:37 UT, which lasted in the U band for about three minutes, a two-fold strengthened line of HeII λ 1640 Å was found without any traces of strengthening of other transition region lines; the strengthened helium line was observed 20 min after and the maximum Ly_α took place 37 min after the optical flare (Byrne and Gary, 1989).

In the decay phase of the strong flare on AD Leo of 28 March 1984, the spectrum within 1950–3200 Å was obtained. On the spectrogram one can see a many-fold strengthened MgII doublet and FeII λ 2600 Å blend. Since during this exposure the amplitude of the optical flare was already low, probably the emissions were due to the slow decay of the strong flare with $\Delta U = 2^m.1$, which occurred 15 min before the beginning of the IUE exposure (Rodonò, 1986a).

During the very strong flare on AD Leo on 12 April 1985, whose amplitude reached $4^m.5$, and the total duration, according to ground-based photometric observations, was more than two hours, Pettersen et al. (1986b) obtained a spectrogram within 1150–1950 Å. The total exposure was 56 min, the first half of the impulsive phase, including its maximum, lasted 15 min, so the contribution of the quiescent state during the first 41 min was negligible. In this spectrum, the lines of Ly_α, CII, CIV, and SiIV, and the continuum in the region $\lambda > 1780$ Å were overexposed, but rough estimates by means of completion of profiles from the wings showed that in this spectrum the Ly_α line was 20 times stronger, lines CIV, stronger by 45 times, HeII, 14, and NV, stronger by nine times. In the spectrogram obtained within 1950–3200 Å, numerous unidentified details were seen, while the intensities at the blend limits of FeII λ 2630 Å and λ 2750 Å were an order of magnitude stronger than in the quiet star. It was concluded that the observed flare continuum resulted from imposing many metal lines and recombination continua. In any case, $UBVR$ measurements could be equally well presented by free–free radiation at 10 MK and blackbody radiation at 10 000 K, but none of these models matched the ultraviolet flare continuum (Hawley and Pettersen, 1991).

During the flare on EQ Peg of 7 December 1984, EXOSAT provided light curves in LE and ME, and IUE recorded three 10-min spectrograms within 1950–3200 Å that encompassed a slow LE ascending branch and flare maximum phase (Haisch et al., 1987). In all three spectrograms the emission of MgII was practically identical, 70% higher than the level of the quiet star. The emission of FeII blend increased from the first to the third spectrograms from two to four times as compared to the level of the quiet star. The continuum near 2950 Å behaved as FeII blend, whereas near 3100 Å the continuum regularly decreased from an 8- to 4-fold level of the quiet star. This could be due to the fact that many faint lines of iron and other ions fell into the band near 2950 Å, whereas the band of 3100 Å is a purer continuum that behaves as one developing in a flare in the U band.

Mathioudakis and Doyle (1989b) analyzed two IUE spectrograms of the system Wolf 630. In one of them obtained on 12 June 1981 within 1150–1950 Å, a flare was diagnosed from strengthened lines and the continuum. From the ratio of intensities of the lines of CIII λ 1176 Å and SiIV λ 1396 Å to CIII λ 1908 Å they estimated the electron density at the level of $\log T = 4.8$ as 1.5×10^{10} and 4.0×10^{10} cm^{-3} in the quiet star and during the flare, respectively. As in flares on BY Dra and AU Mic analyzed by Butler et al. (1987), the DEM curve in the flare on Wolf 630 was shifted toward higher values without any change in the inclination. From absolute fluxes in the lines and the estimate of n_e Mathioudakis and Doyle determined the radius of the magnetic flare loop as about 5×10^{10} cm, if a single loop was responsible for the flare. If the flare involved n loops, their characteristic radius should be $n^{1/3}$ times less. The spectrophotometric gradient of the flare continuum corresponded to free–free radiation at 25 000 K, and its total emission within 1250–1950 Å was 8×10^{31} erg, which is only half that of the total radiation of emission lines in the temperature range $4.3 < \log T < 5.4$.

We mentioned already that on 13–15 October 1985 EV Lac was monitored for 17 h by IUE and EXOSAT with photometric and spectral ground support. However, only a weak correlation of activity was found at different levels of the atmosphere (Ambruster et al., 1989a). The flare on 15 October 1985 at 6:25 UT with $\Delta U = 1\overset{m}{.}5$ was not accompanied by appreciable strengthening of the CIV lines, though the flux in H_β increased by 6σ, and in the CaII K line, by 2.3σ. The stellar chromosphere better correlated with variations in the U band than with coronal flares, and long X-ray activity on 13 October 1985 was not accompanied by essential changes in other atmospheric layers. This suggests magnetic isolation of the loops of the chromosphere and transition region. Another example of isolation is the X-ray flare on Prox Cen of 6 March 1979, which was not accompanied by variations in the ultraviolet, optical, and radio ranges. However, such independence is quite rare.

Phillips et al. (1992) compared the ultraviolet spectra of about a dozen stellar flares, in which a flare continuum was found within 1150–1950 Å, with solar flares. They found a linear correlation between flare emission in the CIV lines and in the continuum and concluded that the observed flare continuum was the recombinant luminosity of neutral silicon at the level of the temperature minimum of the stellar atmosphere ionized by the radiation of the transition-region lines.

In the ultraviolet flare of 26 December 1986 on the very fast rotating dM2e star Gl 890, Byrne and McKay (1990) found a several-times strengthened emission of CIV and a strong continuum, whose total energy within 1250–1950 Å was estimated at an order of magnitude higher than in CIV lines: 4×10^{32} and 3×10^{31} erg, respectively. Comparable energy of flare emission in the continuum was found in the flares on Gl 687 A and AT Mic (Bromage et al., 1986). Half an hour before this exposure the spectrum of Gl 890 within 1950–3200 Å was recorded, in which the MgII doublet and the blend of FeII λ 2600 Å were strengthened almost by a factor of 1.5, their total energies were 2×10^{31} and 1×10^{31} ergs, but a flare continuum was not found in this spectrum. In the spectrum of the same region obtained after the above shortwave spectrogram, the iron blend was still strengthened.

On the ascending branch of the strong X-ray flare of 14 November 1984 in the system YY Gem that lasted for more than three hours, the spectrogram within 1950–3200 Å displayed overexposed MgII emission. In the spectra of the 1150–1950 Å region obtained before and after the X-ray flare, one can see strengthened CIV, HeII, CII, and CI lines, but eclipses in the system complicated unequivocal interpretation of these data (Haisch et al., 1990a).

Enhanced emission of MgII found in the ascending branch and at the maximum of a moderate X-ray flare on Proxima Cen of 2 March 1985 lasted for about an hour and a half after the termination of the X-ray flare (Haisch et al., 1990b).

During a five-day monitoring of the young K2 dwarf HD 82558 (= LQ Hya) an ultraviolet flare was recorded on 30 October 1988 with triple amplification of the emission of CIV and a considerable continuum within 1250–1850 Å. Increased luminosity of CIV lasted for at least two hours (Ambruster and Fekel, 1990).

CC Eri was monitored by IUE on 2–4 November 1989. Over 48 h in two of twenty spectra obtained within 1150–1950 Å 2–3-times strengthened CIV emission and smaller strengthening of emissions of CII and SiII were revealed (Byrne et al., 1992a; Amado et al., 2000). E_{CIV} in these flares was a few 10^{31} erg, which is several orders of magnitude higher than the values typical of solar two-ribbon flares.

On 11 June 1991, the dM5.5e star Gl 866 was observed by IUE with good spectral and photometric ground support (Jevremovic et al., 1998a). The energy of two of five optical flares was comparable with the strongest solar flares and MgII emission was strengthened by a factor of 1.5–2.

A strong flare with a duration of more than three hours was recorded on LQ Hya on 22 December 1993 by IUE and ground-based telescopes (Montes et al., 1999). On the spectrogram within 1150–1950 Å obtained immediately before the optical flare, the lines of the transition region and continuum were stronger by a factor of 1.5, but chromospheric lines remained at the level typical of the quiet star. CIV lines were shifted by -250 km/s, SiIV and NV lines also displayed a blue shift but to a lesser extent, while chromospheric lines remained stationary. The spectrograms obtained for the range of 1950–3200 Å during the flare maximum showed that MgII emission was 2.5 times stronger, shifted by -40 km/s, and its FWHM was widened to 250 km/s. In another spectrogram obtained during smooth decay for the range of 1150–1950 Å, transition-region lines were 20 times stronger, while chromospheric lines were only 4 times stronger. The continuum recorded near 1500 Å displayed up to a 10-fold increase. Analysis of this radiation does not contradict the hypothesis that it is conditioned by SiI recombinations.

2.4 Dynamics and Radiation Mechanisms of Flares in Different Wavelengths

Table 16 presents stellar flares recorded on active dwarfs by IUE and in the subsequent ultraviolet experiments.

Summing up the IUE studies of stellar flares, one should note that the instrument recorded less than 30 flares on red dwarfs and the conditions of their observations, as a rule, were unique. Further, the stars could be observed by IUE with a time resolution of not less than 10–20 min, which appreciably exceeded the characteristic duration of most stellar flares. Therefore, these observations provide only the most general pattern of UV flares.

CIV emission is the most appreciable on quiet stars and in flares on active dwarfs, where its strengthening reaches one and a half orders of magnitude. As a whole, the level of strengthening of emissions in flares increases with their formation temperature. This law is satisfied for solar flares, where the transition region goes down to large densities and with growing EM its dependence on formation temperature becomes steeper.

There is no unequivocal relation between the amplitudes of optical flares and the effects occurring in the ultraviolet region, which may be due to spatial separation of these radiation sources.

Ultraviolet emission of flares usually lasts longer than that in an optical continuum.

During UV flares practically all emission lines are amplified, but to different extents, and their light curves do not coincide.

The HeII λ 1640 Å line falls out of the general law: it correlates poorly with CIV, and there are arguments in favor of coronal excitation of this line in flares.

In strong flares, the ultraviolet continuum occurs down to 1200 Å, but its physical nature has not been determined unequivocally. The most developed is the hypothesis about its recombination on silicon. When the continuum was recorded, its energy was comparable with ultraviolet lines.

In one of the most thoroughly studied flares that occurred on AU Mic on 19 September 1981, absolute fluxes in CIV, SiIV, CIII, OI, and CII lines averaged over a 30-min exposure were higher by 4–5 orders of magnitudes than those in the strong solar flare of 15 June 1973 recorded by Skylab. The excess was greater for high-temperature lines.

In stellar flares, strengthening of the Ly_α line was lower than in solar flares, which can be a consequence of the large optical thickness of the line in stellar atmospheres.

The continuum of a solar flare is 4–5 orders of magnitude lower near 1900 Å and still an order of magnitude lower near 1300 Å as compared to several stellar flares for which it was measured.

The total energy of flares in transition-region lines achieves $10^{31} - 10^{33}$ erg and is comparable with the appropriate values of E_X.

Smooth variations of the intensities of ultraviolet lines out of flares suggest that they are caused by numerous low-amplitude flares.

Table 16. Flares on active dwarfs in UV and EUV

Star	Date	Experiment	Strengthening CIV	Strengthening MgII	Other Characteristics	Description and Analysis of Observations
Gl 867 A	11 Sept 1979	IUE	× 2			Butler et al., 1981; Bromage et al., 1986
AU Mic	4 Aug 1980	IUE	× 2			Butler et al., 1983; Butler et al., 1987
Prox Cen	20 Aug 1980	IUE	× 7		×35 in X-rays	Haisch et al., 1983; Bromage et al., 1986; Byrne and McKay, 1989
UV Cet	17 Sept 1980	IUE	× < 1.5		$\Delta U = 2\overset{m}{.}4$	Bromage et al., 1983; Phillips et al., 1988
AT Mic	19 Sept 1980	IUE	×3.7			Bromage et al., 1983 and 1986
Wolf 630	12 June 1981	IUE	×2.8			Mathioudakis and Doyle, 1989b
EV Lac	3 Sept 1981	IUE	× < 1.5		$\Delta B = 0\overset{m}{.}7$	Bromage et al., 1983
EQ Peg B	3 Sept 1981	IUE	×9			Bromage et al., 1983
EQ Peg B	2 Sept 1981	IUE	×3		×2H_β	Baliunas and Raymond, 1984; Bromage et al., 1986
BY Dra	2 Oct 1981	IUE	×1.5			Butler et al., 1987
BY Dra	4 Oct 1981	IUE	×1.3			Butler et al., 1987
YZ CMi	3 Feb 1983	IUE	×5.6		$\Delta U = 3\overset{m}{.}8$	Rodonò, 1986a; van den Oord et al., 1996
AD Leo	2–5 Feb 1983	IUE				Gary et al., 1987; Byrne and Gary, 1989
Gl 182	4 Oct 1983	IUE	×4			Mathioudakis et al., 1991

Table 16. (cont.)

Star	Date	Experiment	Strengthening CIV	Strengthening MgII	Other Characteristics	Description and Analysis of Observations
Gl 182	5 Oct 1983	IUE	×20			Mathioudakis et al., 1991
AD Leo	28 March 1984	IUE		×7.5	$\Delta U = 2^m\!.1$	Rodonò et al., 1989
YY Gem	14 Nov 1984	IUE		over-exposure	×4 in X-rays	Haisch et al., 1990a
EQ Peg	7 Dec 1984	IUE				Haisch et al., 1987
Prox Cen	2 March 1985	IUE		×2.5		Haisch et al., 1990b
AD Leo	14 Apr 1985	IUE			$\Delta U = 4^m\!.5$	Pettersen et al., 1986b; Hawley and Pettersen, 1991
EV Lac	13 Oct 1985	IUE		over-exposure	$\Delta U = 1^m\!.8$	Ambruster et al., 1989a
EV Lac	6 Feb 1986	Astron				Burnasheva et al., 1989
Gl 890	26 Dec 1986	IUE	×4.5			Byrne and McKay, 1990
LQ Hya	30 Oct 1988	IUE	×3			Ambruster and Fekel, 1990
CC Eri	3 Nov 1989	IUE	×3			Byrne et al., 1992a; Amado et al., 2000
CC Eri	4 Nov 1989	IUE	×2.5			Byrne et al., 1992a; Amado et al., 2000
BY Dra	1 Oct 1990	IUE+WFC	×2.4		$\Delta U \sim 0^m\!.7$ flare in EUV and soft X-rays	Barstow et al., 1991; Phillips et al., 1992
BY Dra	2 Oct 1990	IUE	×1.9			Phillips et al., 1992
YY Gem	3 Oct 1990	WFC				Bromage, 1992

(*continued*)

Table 16. (cont.)

Star	Date	Experiment	Strengthening CIV	MgII	Other Characteristics	Description and Analysis of Observations
HD 197890	17 Oct 1990	WFC			EUV burst	Matthews et al., 1994
AU Mic	18 Oct 1990	WFC				Bromage, 1992
EV Lac	17 Dec 1990	WFC+IUE				Bromage, 1992
EV Lac	20 Dec 1990	WFC+IUE				Bromage, 1992
AT Mic	3 May 1991				X-ray burst	McGale et al., 1994
AD Leo	9 May 1991	HST	×90			Bookbinder et al., 1992
Gl 866	11 June 1991	IUE		×2		Jevremovic et al., 1998a
35 EUV flares recorded by WFC within RASS						Tsikoudi and Kellett, 1997
AU Mic	3 Sept 1991	HST			flare in SiIII λ 1206 and in the Ly_α wing	Woodgate et al., 1992
AT Mic	1 July 1992	EUVE				Vedder et al., 1994; Monsignori Fossi et al., 1994a; Brown, 1994
AU Mic	15 July 1992	EUVE			the strongest flare in EUV: $A_{EUV} \sim 20$	Cully et al., 1993, 1994; Monsignori Fossi et al., 1994b, 1996; Drake et al., 1994a; Brown, 1994, 1996; Schrijver et al., 1996; Katsova et al., 1999a
Proc Cen	20 July 1992	EUVE				Vedder et al., 1994
EUVE 2056	3 Aug 1992	EUVE			$A_{EUV} \sim 10$	Mathioudakis et al., 1995b
HII 314	3 Sept 1992	HST	×3			Ayres et al., 1994

2.4 Dynamics and Radiation Mechanisms of Flares in Different Wavelengths 313

Table 16. (cont.)

Star	Date	Experiment	Strengthening CIV	MgII	Other Characteristics	Description and Analysis of Observations
AU Mic	9 Sept 1992	IUE + HST			flare in SiIV doublet lines	Linsky and Wood, 1994
YZ CMi	25 Feb 1993	EUVE				Vedder et al., 1994
AD Leo	2 March 1993	EUVE			$\Delta U \sim 0\overset{m}{.}6$ $A_{EUV} \sim 4$	Hawley et al., 1995; Cully et al., 1997
AD Leo	3 March 1993	EUVE			$\Delta U \sim 0\overset{m}{.}6$ $A_{EUV} \sim 2$	Hawley et al., 1995
EQ Peg	30 Aug 1993 I	EUVE			×3 EUV	Monsignori Fossi et al., 1995b
EQ Peg	30 Aug 1993 II	EUVE			×5 EUV	Monsignori Fossi et al., 1995b
EV Lac	Oct 199.93 I	EUVE + IUE	×6	× > 3		Ambruster, 1995; Brown, 1996
EV Lac	Oct 1993 II	EUVE + IUE	×4.4	×1.6		Ambruster, 1995
LQ Hya	22 Dec 1996	IUE				Montes et al., 1999
Gl 752 B	12 Oct 1994	HST				Linsky et al., 1995
YZ CMi	21 Dec 1994	HST			SiIV ×25	Robinson et al., 1996; Mathioudakis et al., 1999
YZ CMi	22 Dec 1994	EUVE			×3 EUV + radio	Robinson et al., 2001
YZ CMi	April 1996	HST				Mathioudakis et al., 1999
AU Mic	13 June 1996	EUVE			×5 EUV+X-ray	Gagné et al., 1998

(*continued*)

Table 16. (cont.)

Star	Date	Experiment	Strengthening CIV	Strengthening MgII	Other Characteristics	Description and Analysis of Observations
EXO 2041.8-3129	14–15 June 1996	EUVE				Gagné et al., 1998
EQ Peg	2 Oct 1996	EUVE			$\Delta U \sim 2\overset{m}{.}5$	Gagné et al., 1998
EUVE J1438-432	1 May 1997	EUVE			×16 EUV	Christian and Vennes, 1999
12 flares recorded by EUVE within RAP						Christian et al., 1999
AD Leo	2 Apr–16 May 1999	EUVE				Güdel et al., 2001
AU Mic	26 Aug 2000	FUSE				Redfield et al., 2002
AU Mic	26 Aug 2000	FUSE				Redfield et al., 2002
AU Mic	10 Oct 2000	FUSE				Redfield et al., 2002
EUVE J0613-23.9B	22 Oct 2000	EUVE		×200	DS/S	Christian et al., 2003
AU Mic	10 Oct 2001	FUSE				Redfield et al., 2002

All spectra of stellar flares were recorded by IUE with a spectral resolution of 350, whereas HST GHRS enabled a resolution of 2000, 25 000 and 80 000, covering the wavelength ranges of 285, 30, and 8 Å, respectively. Therefore HST provided more refined but less systematic data about the activity of considered stars in the ultraviolet.

According to high time resolution HST observations aimed at estimating the contribution of numerous weak flares to the observed emission spectrum from individually recorded events – Saar et al. (1994c) on AD Leo and Saar and Bookbinder (1998a) on HD 129333 and LQ Hya – the contribution reaches 10 or even 20%. Similar results were obtained by Ayres (1999) in observations of three G dwarfs in the α Per cluster and the Pleiades.

On 9 May 1991, observations of AD Leo were carried out with a spectral resolution of 2000 and a time resolution of 1 s alternately in the ranges of 1170–1450 Å and 1390–1670 Å with five-minute exposures. On each revolution four exposures were made and the whole experiment was run on four revolutions of the instrument. At the maximum of a strong flare the emission of CIV for 25 s

2.4 Dynamics and Radiation Mechanisms of Flares in Different Wavelengths

was strengthened by a factor of 90 and that of SiIV for 15 s, by a factor of 60; amplification of the HeII line was halved and in CI λ 1561 Å and λ 1656 Å lines it was not appreciable. Many previously unknown emissions were revealed in this spectrum, CIV, SiIV, and HeII lines displayed components shifted by +1800 km/s, but 25 s later the shift decreased to 600 km/s (Bookbinder et al., 1992; Byrne, 1993a). According to Byrne, the kinetic energy of this motion is 25 times higher than the CIV radiative energy.

In two-hour observations of AU Mic on 3 September 1991 with a spectral resolution of 10 000 and a time resolution of 0.4 s during a flare in the SiII λ 1206 Å line the only three-second emission increase was recorded in the red wing of the Ly_α line. In duration and shift from the line center the increase complied with the pattern expected from the descending flux of fast protons accelerated during the impulsive phase. The energy of protons was sufficient to excite flare luminosity of the transition region. But, in repeated 3.5-h observations carried out one year later the effect was not found (Woodgate et al., 1992; Robinson et al., 1993).

Three rapidly rotating G dwarfs in the Pleiades were observed on 3 September 1992 during three consecutive revolutions of HST around the Earth. Activity of HII 314 was discovered with confidence in the CIV lines at characteristic times of several minutes (Ayres et al., 1994). Over 89 min of general spectral monitoring on each revolution seven spectra were obtained in the range of 1150–1606 Å: in the spectra of the first revolution the line intensity was scattered, which obviously surpassed regular scattering, one of the spectra of the second revolution suggested a fast burst, and the spectra of the third revolution evidenced flare decay in the lines. The activity was not found on two other G dwarfs. In this connection, only smooth brightness variations with characteristic times of several hours and an amplitude of about 2 were found in the X-ray range on HII 314. Ayres et al. assumed that these variations in X-rays resulted from superimposing of numerous bursts with characteristic times corresponding to the found flares in CIV.

In 17 spectra in the region of the CIV doublet and 22 spectra in the region of the SiIV doublet obtained by Linsky and Wood (1994) on 9 September 1992 in high-spectral-resolution HST observations of AU Mic (162 s exposures), a fast flare was found on two successive SiIV spectra: on the first spectrogram one of the doublet lines was shifted by 40 km/s at FWHM = 260 km/s, in the second, by 20 km/s at FWHM = 430 km/s, and the flux increased by 3–4 times. Linsky and Wood attributed the initial red shift to the disturbance of the atmosphere by primary energy release in the corona and the subsequent blue shift, to evaporation of the chromosphere.

To find very fast brightness variations related to possible microflares, Robinson et al. (1995) carried out a high-speed photometric study of one of the weakest flare stars CN Leo (dM8e) by HST on 30 May 1993. The observations were run in the 645 Å wide band centered at 2400 Å, where the greatest contrast of flare and quiet star was expected. During four half-hour monitoring sessions with a time resolution of 0.01 s 32 flares with time structures

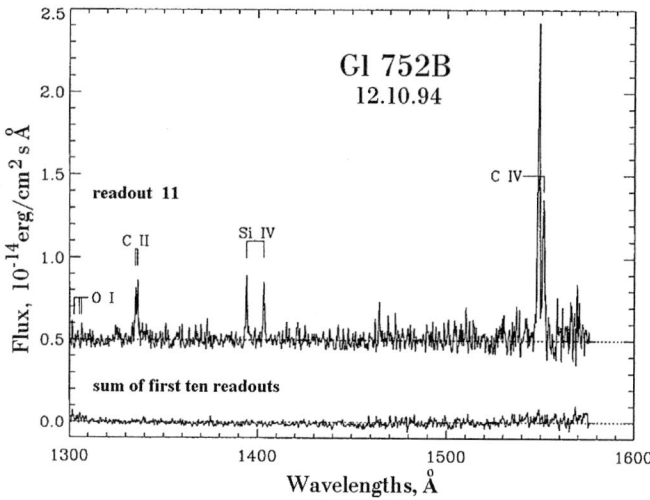

Fig. 49. HST ultraviolet spectra of Gl 752 B recorded on 12 October 1994: 10 spectra of the quiet star and a spectrum of the flare (Linsky et al., 1995)

of up to 0.1 s were recorded. The strongest of them had an amplitude of 18 and was about two minutes long. The weakest flares had $E_{UV} \sim 10^{27}$ erg and seldom lasted longer than several seconds. Strong flares resembled close groups of weaker events. The energy spectrum in the region of weak flares constructed from the 32 flares was appreciably below the spectra constructed earlier for the star from recorded optical flares.

On 12 October 1994, within one hour 11 spectrograms with a resolution of 22 000 were obtained for the star Gl 752 B (= VB 10, spectral type M8 Ve), which is intermediate between red and brown dwarfs. No emission lines were seen in the first 10 spectrograms. The eleventh spectrogram showed a strong spectrum of the transition region (Fig. 49). This fact proves the existence of high-level chromospheric activity even on low-mass stars (Linsky et al., 1995).

The 7.5-h monitoring of YZ CMi on 21 December 1994 with a spectral resolution of 1000 and a time resolution of 0.4 s over the range of 1150–1440 Å revealed 29 bursts with durations of 15–200 s. During one moderate-intensity burst the dependence of the line-emission amplification on the formation temperature was clear. But NV did not correspond to this dependence, while the coronal line of FeXXI disappeared during the flare. The flare continuum, as a rule, followed the line emission, but in some cases it grew sharply without appreciable amplification of lines (Robinson et al., 1996). Later in the spectrum of the strongest flare, Mathioudakis et al. (1999) found significant deviations of the intensity ratios of the components of the SiIV doublet (λ 1394/λ 1403) from 2:1, which suggests appreciable optical thickness in the λ 1394 Å line. A similar effect was found in the strong flare on this star that occurred in April 1996, simultaneously, a similar effect was found in the CIV doublet

2.4 Dynamics and Radiation Mechanisms of Flares in Different Wavelengths 317

and consistent estimates of optical thicknesses were obtained; based on these, the geometrical thickness of the transition region was estimated as several kilometers.

Observing the eclipse system CM Dra composed of dM4e stars with a time resolution of 5 s, Saar and Bookbinder (1998b) found numerous weak flares in the MgII line and in the adjacent continuum. Flares in the lines were weaker, less impulsive and longer than in the continuum. The total energy of flares in lines was at the level of 1% of the total radiation in these lines, at the same time they took up to 20–30% of the time.

From optical and ultraviolet observations of flares on AD Leo in March 2000 Hawley et al. (2003) constructed the light curves of the nine strongest ultraviolet lines of carbon, silicon, oxygen, helium, and nitrogen, and the six optical lines of hydrogen, helium, and the component of IR CaII triplet. In the strongest flares they found close to linear correlations of the intensities of CIV, SiIV, and NV lines with U-band radiation.

In studying 15 M dwarfs of different activity levels in close ultraviolet Fuhrmeister et al. (2004) found during a flare on LHS 2076 a strong forbidden coronal line of FeXIII λ 3388 Å and confirmed the existence and variability of this emission for CN Leo.

On spectral monitoring of EV Lac in the CIV λ 1550 Å line with a time resolution of 0.61 s from the space station Astron, a strong optical flare was recorded on 6 February 1986. In the smooth light curve of the flare in these lines 50 s after the beginning a very short burst was found (Burnasheva et al., 1989). The analysis of these observations led to a conclusion about a probable analogy of this burst with explosive phenomena observed in active regions of the Sun (Katsova and Livshits, 1989).

The wavelength range accessible for FUSE observations adjoins the short-wavelength range border accessible for HST.

Using FUSE, Redfield et al. (2002) recorded on 26 August 2000 and 10 October 2001 two six-minute flares on AU Mic. The first flare displayed a continuum increasing to short wavelengths, which was interpreted as free–free radiation of hot plasma, and the emission lines of CIII and OVI shifted to the long-wavelength range. The flares burned in the lines and in the continuum almost simultaneously. The amplitudes of flares in different lines were comparable, except for the emission of CIII λ 1176 Å: its amplitude in the first flare was almost six times higher than in the second.

HST enabled investigation of stellar flares with a much higher spectral resolution than IUE, and a time resolution of HST observations was similar to the time variations of flares, but spectral ranges of these observations were quite similar. ROSAT and HST were put into orbit almost simultaneously. ROSAT was equipped with a wide-field camera (WFC) that could record flares in the extreme ultraviolet from 60 to 1000 Å or 0.012–0.2 keV, i.e., the region adjacent to soft X-rays. EUVE was put into orbit later and operated in the same range of the extreme ultraviolet. These space observatories provided qualitatively new data on stellar flares. Observations in the extreme ultraviolet are important because this wavelength range contains numerous emission lines formed in the temperature range 10^5–10^7 K that enable the analysis of the state of radiating matter but are not observed in other regions of the spectrum (Monsignori Fossi et al., 1995a,b, 1996).

The first stellar flare recorded by WFC was the flare of 1 October 1990 on BY Dra, which was observed simultaneously by ROSAT and IUE, and in the optical range. The flare was recorded in all four wavelength regions. Using WFC data, Barstow et al. (1991) found that during the flare EM of hot plasma increased almost two-fold with respect to the level of the quiet star and reached 10^{52} cm^{-3}. If the observations of the flare on three revolutions suggest that its decay was about 4.5 h long and was caused by radiative relaxation, the density of the flare plasma was 10^{10} cm^{-3}, but in X-rays the decay lasted for about 10 h. The total energy of the flare in the range of 0.08–0.18 keV reached 7×10^{32} erg, which made up an appreciable share of the total emission of the flare.

On 17 October 1990, WFC revealed a flare on the single K0 dwarf HD 197890 within 90–210 Å. The duration of the flare was more than three hours with subsequent activity for 20 h. Further optical observations showed that this star had $v \sin i \sim 240$ km/s, thus, it is one of the fastest and youngest rotators that probably has not yet reached the main sequence and is similar to AB Dor (Matthews et al., 1994).

From the WFC sky survey data, Tsikoudi and Kellett (1997) studied the regions of 58 flare stars from the Pettersen list (1976) and found 28 EUV sources. In studying the fields of 67 stars with CaII emission they found 21 EUV sources. On 23 stars they detected 35 flares. Strong flares were found only on the most active flare stars: BY Dra, AU Mic, and EV Lac. Moderate-intensity flares were recorded on the well-known flare stars: YZ Cet, V 577 Mon, L 1113-55, CN Leo, Prox Cen, V 1258 Aql, and Gl 867. Weak flares were found on the known flare stars and stars for which the optical flare activity had not been established. As many as 19 of the 35 flares took place on 13 dMe stars. Except for these distinct several-hour-long flares, on a half of the examined objects Tsikoudi and Kellett found low-amplitude brightness variations with characteristic durations from 1–2 h to several days. Apparently, these slow variations did not correlate with individually recorded flares and were not caused by rotary or orbital modulation. Tsikoudi and Kellett called them milliflares. Comparing flare activity of late dwarfs from WFC data with other parameters of these stars, they established a positive correlation between the EUV number of flares and bolometric luminosity and negative correlation of

L_{EUV} with the rotational period. The dependence of flare activity on spectral class was similar to that found by Kunkel for flare luminosity in the U band (see Fig. 35): the maximum of this distribution falls on early K dwarfs. Finally, Kellett and Tsikoudi (1997) concluded that the flare activity of F–K dwarfs was more appreciable in EUV than in the optical range and that low-amplitude EUV "milliflares" were essential for heating stellar coronae.

The star AT Mic was used for calibration of EUVE. On 1 July 1992, all three spectrometers and a deep-survey photometer recorded burning of a strong flare. Based on these data, the DEM curve was plotted (Monsignori Fossi et al., 1994a).

During four-day EUVE monitoring of the star AU Mic on 15 July 1992 within 65–190 Å a strong flare was recorded by Cully et al. (1993) (Fig. 30). Its total energy in the extreme ultraviolet was 3×10^{34} erg and $EM \sim 6 \times 10^{53}$ cm^{-3}. The following weaker flare had $E_{EUV} \sim 2 \times 10^{33}$ erg and $EM \sim 3 \times 10^{53}$ cm^{-3}. These flares were 1–2 orders of magnitude stronger than the flare on BY Dra of 1 October 1990 recorded by Barstow et al. (1991) with WFC. Assuming that the decay of the strong flare of 15 July 1992 was due to radiative losses in the rapidly expanding plasma structure, from the light curve of the flare Cully et al. (1994) estimated the characteristic density of matter in it as $(4-6) \times 10^{11}$ cm^{-3} and the size as 5×10^{10} cm. With these parameters, the flare luminosity should be high due to the large volume of luminous matter, rather than its high density. Then, the flare would resemble more the processes on RS CVn-type stars rather than those on dMe. Monsignori Fossi et al. (1994b, 1996) analyzed independently simultaneous spectral data of three EUVE spectrometers and obtained different results: from the ratios of intensities of spectral lines of FeXXI they estimated the density of the quiet corona and near-flare maxima from 3×10^{12} to 2×10^{13} cm^{-3}. Using theoretical curves by Monsignori Fossi and Landini, from ten emission lines of FeIX–XXIV and the HeII λ 304 Å line Monsignori Fossi et al. (1996) constructed DEM curves for each of 7 intervals into which the whole monitoring interval of AU Mic was split. In the first and the last intervals, corresponding to the quietest state, the DEM curve reached maximum at about 8 MK. Then the curves were broken, whereas on the intervals with flares a significant amount of radiating plasma was found with a temperature above 50 MK. Synthetic spectra SW and MW EUV constructed using these parameters well presented the observations. According to these calculations, $L_{SW}^{max} = 5 \times 10^{29}$ erg/s, and this radiation was equal to 1/30 of the total radiative losses in the range of 1–2000 Å. The characteristic size of coronal loops should be about 3×10^9 cm and the filling factor was about 10^{-4}. Brown (1996) concluded that the maximum density occurred during the decay phase rather than at the maximum strength of flares. According to Schrijver et al. (1996), the high density of matter can even lead to appreciable optical thickness of the corona in resonance lines and scattering in these lines will lower the ratio of intensities of the lines and the continuum. Drake et al. (1994a) constructed the light curves of AU Mic in the lines of FeXXIV, XXIII, XVIII, and HeII λ 304 Å and found that the first two high-temperature lines of iron and the helium line sharply increased during the maximum of the flare of 15 July 1992, whereas the amplification of the FeXVIII line was appreciably weaker. Katsova et al. (1999a) assumed that the many-hour luminosity of high-temperature lines could be caused by additional heating of the flare plasma in the extended vertical current sheet.

The object EUVE J2056-17.1 was the brightest among the sources discovered during the sky survey. With the help of ground-based observations Mathioudakis et al. (1995b) identified it with an active dK7e-dM0e star. On 3 August 1992, within the range of 60–200 Å a strong flare was recorded with burning longer than an hour, an amplitude of 10, a luminosity at maximum of 1.3×10^{30} erg/s, a total duration of about one day, and a total energy of more than 10^{35} erg. Thus, the parameters of this flare were similar to the flare on AU Mic of 15 July 1992. An extremely strong line of lithium absorption corresponding to the content of this element in the stellar atmosphere $\log N(\text{Li}) = 2.4 \pm 0.4$ suggests that this star can be post T Tau or lithium is formed in it as a result of spallation during strong flares.

AD Leo was monitored on 1–3 March 1993 by EUVE with a good ground support. On 2–3 March, a high activity level was recorded (Hawley et al., 1995). First, on 2 March 1993, a flare with $\Delta U \sim 0\overset{m}{.}7$ was found, which practically was not seen in EUV, but three hours later a flare with $\Delta U \sim 0\overset{m}{.}8$ marked the beginning of a EUV flare with a total duration of eight hours. The total energy of this flare was $E_U \sim 5 \times 10^{32}$ erg and $E_{EUV} \sim 4 \times 10^{32}$ erg. Approximately a day later, a weaker EUV flare occurred on the star ($E_U \sim 3 \times 10^{32}$ and $E_{EUV} \sim 8 \times 10^{31}$ erg) that lasted for about four hours during which a rather fast burst with $\Delta U \sim 0\overset{m}{.}6$ and a subsequent long photometric activity were recorded. Comparison of light curves in different bands revealed a delay in the beginning of the flare in EUV with respect to U, while comparison of L_{EUV} and L_U during the active state supported the validity of the relation

$$L_{EUV}(t) \propto \int_0^t I_U \, dt \,, \tag{62}$$

which corresponds to the Neupert effect, if the emission in the optical U band is considered proportional to the flux of accelerated particles that emerged in the flare, and the EUV emission is proportional to the thermal radiation flare plasma heated by the particles. (Later, the relation (62) was confirmed in analyzing the flares on AD Leo of 10 March 2000 (Hawley et al., 2003).) Analysis of the EUV data within the theory proposed by Hawley et al. (1995) for burning and fast decay of flares in static coronal loops yielded the estimates for the flare of 2 March 1993: a loop length of 4×10^{10} cm, which exceeds the stellar radius, a loop section of 9×10^{19} cm^2, an average electron density of 3×10^{10} cm^{-3}, a maximum pressure of 180 dyn/cm^2, and an emission measure of 8×10^{51} cm^{-3}. Optical observations yielded the size of the flare 1×10^{18} cm^2 or 0.01% of the stellar surface. In the flare of 3 March 1993, the loop length was 1.5×10^{10}, its section was 2×10^{19} cm^2, the pressure was 280 dyn/cm^2, and $EM \sim 2 \times 10^{51}$ cm^{-3}. Later, Cully et al. (1997) thoroughly analyzed the spectra of AD Leo obtained by three EUVE spectrometers during the same observations averaged over four time intervals: during the first and second flares, during the decay of the first flare and in the quiet star before the first and after the second flare. Domination of emission lines of multiply ionized iron and the HeII λ 304 Å line was found in all averaged spectra, but low S/N ratio obtained in these observations did not enable consideration of "pure flare spectra". Cully et al. calculated DEM in two different ways and obtained consistent results. Further, DEM were calculated for the standard chemical composition of coronal plasma of AD Leo and for a ten-fold depletion of heavy elements $Z = 0.1$. The estimated sizes of coronal loops are $\sim Z^{1/2}$ and the average electron density in them is $n_e \sim Z^{-1}$. If in a quiet star DEM has a wide maximum about $10^{7\pm0.2}$ K, which is higher by an order of magnitude than DEM($10^{6.2}$ K) and can be caused by a set of coronal loops with different temperatures at the tops, then the DEM of flares has a maximum at $T > 10^7$ K, and in the decay phase this maximum goes down and shifts to lower temperatures, which is naturally associated with cooling of the coronal plasma and condensation of matter of the evaporated chromosphere. Spectral analysis suggests that during the flares mainly the high-temperature component of the corona changes, it yielded a somewhat lower temperature of the flare plasma as compared to a previous photometric consideration that underestimated the pressure and overestimated the filling factor of coronal loops. Thus, the conclusion on the decisive contribution of long loops to total coronal radiation and the range of density 10^9–10^{11} cm^{-3} was confirmed.

The system EQ Peg was observed by EUVE during 20 revolutions of the satellite in the DS/S mode: on 30 August 1993 two strong flares with amplitudes of about 3 and 5 and total durations of more than six and nine hours, respectively, were recorded in photometrical and spectral observations (Monsignori Fossi et al., 1995b).

On 9–13 September 1993, EUVE monitoring of EV Lac was conducted in the spectral range and in the photometric range of 65–190 Å. On 10 September 1993 at 6:10 UT, a rather strong flare was recorded almost from the very beginning. Burning of the flare lasted for several minutes, its EUV amplitude achieved 10, an initial decline was almost symmetrical to the rise to the maximum, then a smoother decay followed with an appreciable secondary burst approximately eight minutes after the main maximum. Using an exponential approximation of the descending branch of the flare the time of an e-fold decay was 19 min. The amplitude and duration of this flare are similar to those of two flares on this star recorded earlier by WFC. During the decay phase the star passed out of sight of the satellite, and on the next revolution there were no traces of the flare, though weak variations occurred on the EUV light curve. During the flare, approximately ten-fold-amplified lines of FeXVI–XXIV and HeII were found in the spectra in SW and MW ranges, the ratio of intensities of the FeXXII line sensitive to electron density yielded $n_e \sim 10^{11}$ cm^{-3} for the flare and 10^{13} cm^{-3} for the quiet star. A slightly weaker flare that took place on the same day at 22:30 UT was studied in detail by ground telescopes

(Abranin et al., 1998a). EUVE observations of EV Lac partly overlapped with IUE monitoring. Comparison of these data led to the conclusion that in the flare of 10 September 1993 at 6:10 UT CIV and MgII emissions were amplified by a factor of 6 and at least 3, respectively, whereas in the flare at 22:30 UT, by a factor of 4.4 and 1.6, respectively. The total losses for radiation in the earlier flare were $8 \times 10^{31}, 5 \times 10^{30}$, and $> 4 \times 10^{30}$ erg in the EUV range, CIV and MgII lines, whereas in the later flare they were $7 \times 10^{30}, 3 \times 10^{30}$, and $> 1 \times 10^{30}$ erg, respectively. After the flare at 6:10 UT the activity in MgII lines continued for two days, which can be associated with the visibility of a strong active region on the star, whose axial rotation period was about four days (Ambruster, 1995; Brown, 1996).

During EUVE observations in the DS/S mode, with individual pointings to a selected object using deep-survey photometers and spectrometers, other objects placed at a right angle to the object could be observed photometrically with a 20-times higher sensitivity than during the all-sky survey. In considering the observations of two dozen G–M dwarfs in October 1993–November 1994, EUV flares were found in the binary system of K dwarfs BD + 22° 669, on the dK3e star V 834 Tau, and the M0.5 star Melotte 25 VA 334 (Christian et al., 1998). During the observations within the "Right-Angle Program" on 1 May 1997 a strong flare was found on EUVE J1438-432 identified with one of dMe stars with 16-fold increase of brightness within 2.7 h, a subsequent e-fold decrease over 2.2 h, and full decay over 11 h. The total energy of flare was about 5×10^{33} erg, which is two orders of magnitude lower than in the strongest flare on AU Mic but close to the other events in EUV (Christian and Vennes, 1999). The results of the "Right-Angle Program" were summarized by Christian et al. (1999): 45% of EUV sources were identified with the stars of late spectral types and during these observations flares were recorded on approximately a dozen stars, including EUVE J0202+105, EUVE J0008+208, G 32-6, EUVE J0213+368, V 837 Tau, EUVE J0725-004, EUVE J1147050, EUVE J1148-374, EQ Vir, WT 486/487, EUVE J1808+297, and G 208-45. Later, the number of objects with flares recorded using this technique increased to 16, most of them were M dwarfs. The amplitudes of these flares varied within 1.5–16, burning times within 1–11 h, decay times within 3–18 h, luminosity at maximum within 8×10^{28}–4×10^{32} erg/s, and full energies within 3–500×10^{32} erg (Christian, 2001).

During cooperative observations in X-ray, UV, optical and radio ranges, AU Mic was monitored by EUVE on 12–15 June 1996: on 13 June 1996 a long (of about a day) flare was recorded with 13-min burning and 5-fold amplification of brightness in the range of 70–160 Å. In SW and a MW spectra EUV radiation was presented by the two-temperature model with $T_1 = 20$ and $T_2 = 80$ MK (Gagné et al., 1998).

In the course of similar multiwavelength observations of EQ Peg with EUVE on 2–6 October 1996, a flare was detected on 2 October 1996 in EUV with $\Delta U = 2^m\!.5$ (Gagné et al., 1998).

2.4 Dynamics and Radiation Mechanisms of Flares in Different Wavelengths

From EUVE data, Christian et al. (2003) discovered a strong flare on the dM3.5e star EUVE J0613-23.9B: a 200-fold amplification of brightness within the range 60–200 Å was primarily due to emission lines with a formation temperature above 10 MK. A strong Ly continuum in the range of 320–650 Å lasted less than 500 s, whereas high-temperature radiation continued for about eight hours and the total flare energy was $E \sim 3 \times 10^{34}$ erg. Within the semiempirical model the coronal density was estimated as $10^{14} - 10^{15}$ cm^{-3}.

On the basis of statistical analysis of EUVE observations of 10 F2–M6 dwarfs, Audard et al. (2000) concluded that the distribution of flares with respect to EUVE energies, as in the optical range, could be presented by a power function with spectral indices from 1.3 to 0.8. Values of these indices do not correlate with the period and velocity of axial rotation and Rossby numbers; but if we constrain consideration to only strong flares with $E > 10^{32}$ erg, a correlation between flare frequency and stellar luminosity L_X is found.

So, both stellar X-ray and ultraviolet flares have first of all higher intensity and total energy and in many cases longer duration than the appropriate solar flares. Unlike stellar radio flares, stellar ultraviolet flares did not show qualitatively new phenomena and processes. Among the features that differentiate stellar ultraviolet flares from solar ones, one should mention a less close correlation with optical and apparently X-ray flares. However, this can be due to different observational conditions. Certainly, quantitative analysis of these strong stellar phenomena requires cautious assumptions about small optical thickness of sources in the emission lines, about equal duration of luminosity of different lines, etc.

2.4.4 Optical Emission of Flares

In the optical wavelength range, solar flares are recorded first of all in spectral lines and only some rare and the strongest, the so-called white-light flares, display flare continuum. The sizes of such flares usually do not exceed 3 arcsec, which is 2000 km, their luminosity reaches 2×10^{29} erg/s and total radiation energies are 10^{30}–3×10^{31} erg, which corresponds to a flare on a dMe star with $\Delta U \sim 1^m$. In strong flares, the Balmer series is traced up to higher members, in the strongest, to H_{16}, the lines of neutral and ionized metals are excited, which testifies to the heating of deeper atmospheric layers. Absorption lines of neutral helium appear in strong flares at greater height, the lines in even stronger flares are replaced by emission. In the strongest events, the HeII λ 4686 Å line is recorded. Narrow emission cores of 1–2 Å and wide wings (to 10 Å) of hydrogen lines confirm the decisive role of the Stark broadening of these lines in solar flares. Often, the short-lived "blue asymmetry", amplification of the short-wavelength wing of the line, occurs. Usually after 1–2 min it is replaced by "red asymmetry" that lasts somewhat longer. CaII lines of solar flares lose the central absorption cores, but on the whole these lines change to a lesser degree than hydrogen lines. As a rule, the lines of metals in solar flares are rather narrow and only the strongest of them display self-absorption. In the rich variety of solar flares one can distinguish the events with limiting spectroscopic characteristics: flares with merging Balmer lines in the range of $\lambda < 4000$ Å, broad Stark wings of hydrogen lines, and strong emission HeI and HeII and metals, and flares without a Balmer quasicontinuum with narrow hydrogen lines, weak emission of HeI and metals and without HeII. The former group is associated with the impulsive phase and is related only to the small bright knots representing the bases of magnetic loops. The second group is associated with the gradual phase and large flare regions surrounding the bases of magnetic loops.

Considering solar optical flares, as an element of a general flare process in a wide range of the electromagnetic spectrum, one will find that a continuum and the lines of extreme ultraviolet that reflect the events occurring in the transition region appear simultaneously with hard X-rays, microwave bursts and white-light flare during the impulsive phase and several minutes prior to the maximum in H_α and in soft X-rays. The EUV burst lasts not more than several minutes and is much shorter than flares in H_α and soft X-rays, but longer than bursts in hard X-rays. Radiation from the chromosphere and transition region dominates in the early flare stage, while coronal radiation, dominates in the late stage.

It is often stated that flares on red dwarfs, on average, are an order of magnitude shorter than solar flares. However, one should bear in mind that the duration of solar flares is usually estimated from the time of H_α luminosity, while that of stellar flares is from the luminosity of continuum. Since the latter, with the greatest probability, corresponds to solar white-light flares, whose duration is an order of magnitude shorter than the luminosity of H_α, one should carefully use this statement in physical estimates.

Discovery of optical flares on red dwarfs and the first steps in their investigation are described in the Introduction and in detail in my early book (Gershberg, 1970a). Only the most general properties – the time and energy characteristics – of stellar flares in the optical range were considered in Sects. 2.2 and 2.3. Below, the features of the light curves of optical flares, their colorimetric and polarimetric properties, and the results of spectral studies are considered in more detail.

2.4.4.1 Light Curves

Light curves are among the major characteristics of optical emission from stellar flares, but they practically have no analogs in solar flares: if the light curve of a stellar flare is obtained directly during photometric observations, for each point of a similar curve of a solar flare one should carry out a labor-consuming procedure of two-dimensional integration over the solar disk. Certainly, now this can be easily done by computers, but 30–40 years ago, when solar-activity

2.4 Dynamics and Radiation Mechanisms of Flares in Different Wavelengths

studies were mainly optical, such facilities were not available, solar researchers did not believe that the curves were necessary and preferred filming of the development of flares.

Several hundred flare light curves on red dwarfs have been published since 1968 in almost three hundred issues of the Information Bulletin on Variable Stars (IBVS) of Commission 27 of the International Astronomical Union, and also in numerous publications by observers from Catania, Crimea, Byurakan, Armagh, Okayama, McDonald, South African observatories, etc. Examples of the curves are presented in Figs. 21, 22, 24–27, 29, and 31–33.

Light curves of flares on UV Cet-type stars, as a rule, are sharply asymmetric: a fast brightness burst turns into a smooth decay. This feature was found already in visual and photographic observations. The overwhelming majority of flares have a narrow and sharp maximum, but approximately 15% of flares have fast irregular oscillations near maximum. The descending branch of the light curve usually consists of two parts: an initial fast decrease, whose rate in absolute values is lower than that of the rise to maximum by a factor of 1.5–3, and the subsequent slow decay during which stellar brightness smoothly approaches the preflare level. The transition from the fast decrease to the slow decay is usually rather sharp and occurs at the level of 0.2–0.3 maximum flare brightness. Often, secondary brightness maxima are found on the descending branch of the light curve.

Despite a number of common properties, one can hardly find two flares with identical light curves, especially in the case of high time resolution observations. In observations with a low noise level a small preflare is sometimes found before the sharp rise to maximum: about a one-minute long smooth increase of brightness to the level of 20–35% of the amplitude of the future flare. There are authentic data about a slight brightness decrease below the level of the quiet star directly before the rise to the maximum.

As flare light curves were accumulated, repeated attempts were undertaken to find a correlation between various parameters of the curves. MacConnell (1968) considered 21 flares on AD Leo and 13 flares on BD-8°4352, recorded in the U band and suspected statistical dependences between amplitudes and flare burning and decay times. But Pettersen et al. (1984a, 1986a) did not establish such a correlation in 241 flares on AD Leo. Cristaldi et al. (1969) found a weak correlation of amplitude and the ratio of burning and decay times in flares on six red dwarfs. Gershberg and Chugainov (1969) from 90 light curves of flares on seven flare stars concluded that slower flares occurred on brighter stars. Using a greater number of initial data, Kunkel (1975a) expressed this conclusion in (28). Kiljachkov et al. (1979) found a weak correlation between the burning time and full durations of flares, and Sanwal (1995) suspected a faster burning of flares on lower-luminosity stars.

Correlations between various parameters of flare light curves were studied in detail by Shakhovskaya (1974a). In particular, she found a distinct correlation between the rate of initial fast decay of flares $-\Delta L_B/\Delta t$ and their luminosity at maximum L_{max}, where ΔL_B is the decrease of flare luminosity

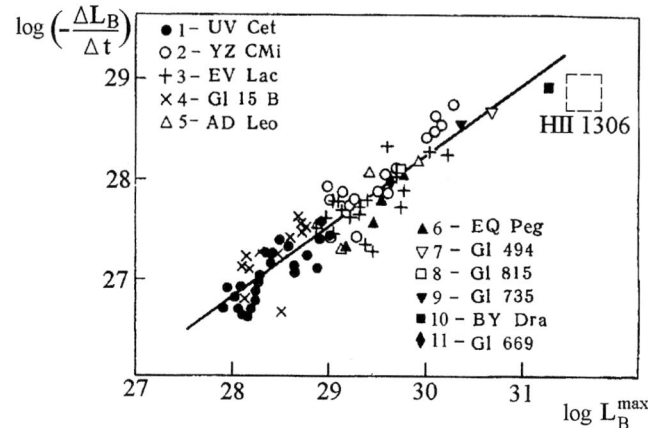

Fig. 50. The rate of initial decay of absolute flare luminosity after maximum brightness vs. absolute flare luminosity at maximum (Shakhovskaya, 1974a)

from the maximum to the start of slow decay and Δt is the duration of fast decay. More than 80 flares on 13 stars in the range of absolute luminosities of 8^m (Fig. 50) correspond to the found statistical relation with a correlation coefficient of 0.90. A similar but a less confident correlation was found for $-\Delta L_U/\Delta t$. The correlations are close to the relation

$$-dL/dt \sim L_{\max}^{3/4} . \qquad (63)$$

But the physical sense of this relation remains unclear.

A slow and smooth decay of most flares encouraged attempts to present the descending branch of the light curve analytically. Abell (1959), Roques (1961) and Chugainov (1961, 1962) used 1–2 exponents for this purpose. Later, Coluzzi et al. (1978) developed an interactive method for analyzing the light curves of flares, in which the curves were split into elementary components, each of which had a linear growth to the maximum and an exponential drop after maximum. The first physical model of flare decay suggested relaxation of a hot photospheric spot but did not provide satisfactory representation of the observations on appreciable extent of flares (Roques, 1961). An alternative model of relaxation of optically thin gas was proposed by Gershberg (1964): based on recombinations of hot gas and expansion into some dimensions, six of the 10 considered light curves of flares on EV Lac were presented. Then, this model was supplemented by the effect of gas cooling (Gershberg, 1967). However, the resulting characteristic rates of expansion were unrealistically high. A similar model was considered independently by Andrews (1965): assuming that the initial decrease after the maximum was caused by free–free emission of gas cooling down in a constant volume and recombination relaxation was responsible for the slow decay, he presented well the light curves of 15 flares. But Kunkel (1967) could use this scheme only for four of the

13 flares. Korovyakovskaya (1972) presented several flare light curves within the model of gas relaxation behind the shock-wave front, and Gershberg and Shnol (1972) included the effect of a nonstationary Ly_α radiation field in the calculations of theoretical light curves of relaxing plasma.

Within his model of "fast electrons" (see Sect. 2.5) Gurzadian (1969) presented the light curves of about 20 flares.

However, further spectral observations of flares on UV Cet by Bopp and Bopp and Moffett (1973) (see Fig. 24) and on some other flare stars by Moffett and Bopp (1976) showed that the content of optical emission from flares essentially varied during their development. When this fact was established, the attempts to construct the models of the whole light curves mainly terminated. (The only exception is the work by Kolesov and Sobolev (1990), in which the problem of construction of a light curve in a one-dimensional homogeneous stellar atmosphere was stated and solved analytically, but the quantitative comparison of the curves obtained with the observed flare light curves was not fulfilled.) However, clarification of the physical sense of individual characteristic details of light curves is still an urgent task. The most interesting results are considered below.

In the context of the idea of the decay of stellar flares through the relaxation of hot gas, Pettersen et al. (1984a) considered probable mechanisms of such relaxation and found that at the stage of initial fast decay the main role should be played by heat conductivity. Since the heat conductivity rate is proportional to electron density, flares on lower-mass stars with denser atmospheres, where one can expect a higher density of matter, should decay faster. These reasons were physical substantiation of the statistical relation (28), which is valid for the absolute values $M_V = 8$–17.

Analyzing about a hundred of the flares he recorded, Kunkel (1967) suggested considering their light curves as combinations of fast and slow components in different ratios. Independently, reasoning from the rate of flare burning and the character of descending branches of their light curves, Osawa et al. (1968) proposed to classify flares as fast bursts and flares with smooth decay. Oskanian (1969) independently proposed a more detailed classification of four morphological types of light curves. In the late 1960s, this partition seemed quite a formal procedure, its physical sense became clear later: according to Moffett and Bopp (1976), continuous radiation prevails in fast bursts, whereas in slow flares an essential contribution is made by emission lines. On the other hand, these could be analogs of solar impulsive and two-ribbon flares. Oskanian and Terebizh (1971a) noted an appreciable increase of the number of fast bursts in the total number of flares on UV Cet, the weakest among the stars considered by them. Avgoloupis (1986) analyzed statistically the light curves of 183 flares from the full sample and from each of Oskanian's morphological types separately and found that in the subsamples of separate types there were correlations between burning and decay times to the level of half-maximum and amplitudes, which in the whole sample were not significant.

A. Preflare brightness decrease and the "emission peak in the absorption saucer". In the late 1960s, during photoelectric monitoring of the brightness of flare red dwarfs in blue and green rays in the B and V photometric systems, respectively, Cristaldi et al. (1969) noticed that often stellar brightness slightly decreased prior to a flare and all flares on EV Lac took place in minima of slow oscillations of quiet brightness up to about $0\overset{m}{.}1$. Earlier, Roques (1961) revealed a similar effect in observations of YZ CMi. On 18 October 1968, a rather strong flare was recorded on UV Cet at the minimum of slow oscillations of stellar brightness (Chugainov et al., 1969). The flare on EV Lac on 9 October 1973 was observed in violet, blue, and red rays: a 15-s preflare brightness decrease by $0\overset{m}{.}1$ was recorded in red rays but was not noticed in violet and blue rays (Flesh and Oliver, 1974). Andersen (1976) recorded a similar situation on EV Lac on 7 October 1975: a preflare brightness decrease in the red region of the spectrum and less distinct decrease in blue rays. Moffett et al. (1977) recorded a preflare decrease of the H_β line in the flare on UV Cet of 6 January 1975. Rodonò et al. (1979) recorded very distinct preflare brightness decreases in the flare on YZ CMi of 5 January 1978 in blue rays using a two-channel photometer. This effect is seen extremely distinctly in the light curve of the flare on EQ Peg of 19 July 1980 recorded by Giampapa et al. (1982b) in violet rays, the preflare decrease lasted almost three minutes. The greatest preflare brightness decreases were recorded on BD + 22°3406 by Mahmoud and Soliman (1980) on 25 and 28 May 1980. On 15 June 1994, Ventura et al. (1995) recorded the longest preflare brightness decrease on V 1054 Oph that lasted for about 36 min and was visible in all UBV bands.

Shevchenko (1973) carried out the first statistical research of preflare brightness decreases of flare stars. He selected among published data 144 confidently recorded light curves of flares on four UV Cet-type stars. He concluded that half of the light curves did not demonstrate any preflare variations of brightness. In the other half, preflare brightening, i.e., a smooth increase of stellar brightness immediately before the sharp rise in the flare and a preflare brightness decrease were almost equally frequent. One could suspect that preflare brightness decreases occur mainly in the flares that are weaker than preflare brightness amplification. Cristaldi et al. (1980) analyzed the light curves of 277 flares on seven red dwarf stars recorded in blue rays in Catania in 1968–76 and found that 61% of the flares did not have preflare brightness variations, 30% had preflare brightening, and 9%, preflare brightness decreases. Simultaneously, it was shown that during preflare brightening the stars turned bluer, while during the preflare brightness decrease they reddenned.

The observations of Flesh and Oliver (1974) initiated the first theoretical model of preflare brightness decrease: Mullan (1975b) assumed that the observed brightness decrease in red rays was connected with a fast transition of strong emission of H_α into absorption. However, further spectral observations of flares did not support this model.

2.4 Dynamics and Radiation Mechanisms of Flares in Different Wavelengths

An alternative model proposed by Grinin (1973, 1976) was based on the fact that, according to calculations, on fast heating the atmosphere of cool stars should become less transparent, therefore the flux of radiative energy should decrease at the same time. The source of fast heating could be the impulsive phase, which by analogy with solar flares should begin with the appearance of short-lived hard radiation from the chromosphere. Developing this idea, Henoux et al. (1990) assumed that "black flares" should precede white-light flares on the Sun by 20 s.

In addition to this thoroughly developed model of abnormal relaxation of the atmospheres of cool stars after a temperature disturbance, Grinin (1980a) put forward an idea of suppression of radiation of active regions in the initial phase of stellar flare. This concept matches the interpretation of observations proposed by Rojzman and Kabichev (1985): almost always there is a smooth and slow brightening of stars before flares, which can not always be recorded photometrically, while the fast brightness decrease recorded directly before the fast rise is concerned with the short-time disappearance of an additional source responsible for the slow preflare rise. This scheme does not foresee the absolute decrease of stellar brightness below its normal level. This scheme has not been developed further.

As Grinin's model of abnormal relaxation of stellar atmospheres predicted the most appreciable preflare brightness decrease in the range of about one micrometer, since the mid-1970s this phenomenon has been studied mainly in the near-IR regions.

Bruevich et al. (1980) analyzed about 150 light curves of flares on three red dwarfs recorded in optical and near IR regions and found that practically all strong optical flares were accompanied by synchronous IR flares. A preflare decrease of stellar brightness was observed in almost 70% of cases. The study confirmed that the decrease was predominantly detected in rather weak optical flares. These conclusions were supplemented and refined by the observations of Tashkent researchers (Kiljachkov and Shevchenko, 1980).

During the strong flare on AD Leo of 28 March 1984 Rodonò et al. (1989) found a long brightness decrease in the IR K band at a rather insignificant preflare brightness decrease in this band. To explain the decrease of IR radiation of the star during this optical flare and the flare on UV Cet of 24 November 1985, Rodonò and Cutispoto (1988) used Gurzadian's concept (1977), according to which optical emission of flares is a side effect of eruption of a strong flux of relativistic electrons from the stellar interior: at the expense of Compton scattering they transformed a part of the infrared quanta radiated by the stellar photosphere into ultraviolet and optical ranges. This model predicted only a synchronous changes in the brightness antiphase in IR and in the optical range, but it did not explain the preflare decrease of brightness and there are serious doubts as to its physical validity on the whole.

During cooperative studies of EV Lac in 1991–93 its brightness was monitored simultaneously in the near IR K band from the Canary Islands and in all standard $UBVRI$ bands from the Crimea (Alekseev et al., 1994;

Abdul-Aziz et al., 1995; Abranin et al., 1998a). From six rather weak flares recorded simultaneously by both observatories in 1991–92, only during one burst in violet rays was the brightness decrease visible in the K band at the level of 2σ. In 1993, both observatories recorded a strong optical flare on September 10, but at the moment of optical maximum only a decrease at the level of 3σ was visible in the K band. However, several similar decreases were visible before the flare as well. In 1994–95, in parallel with observations in $UBVRI$ bands from the Crimea, the IR K band was monitored from Catania: in one of six flares recorded in optical rays a slight smooth decrease of IR brightness could be suspected 20–30 min after the optical maximum (Abranin et al., 1998b).

Hence, it follows that rather numerous photometric observations in the long-wavelength range did not yield a distinct and unequivocal picture of the preflare brightness decrease of flare stars. Of particular interest is a rather weak flare recorded on EV Lac during high time resolution observations by the $UBVRI$ system at the Crimean Observatory on 5 October 1996 by Shakhovskoy.

Figure 51 shows the light curves of this flare obtained by Zhilyaev et al. (1998) after processing the initial data by smoothing a Gaussian window with a width of 5 to 12.5 s. One can see two components of the flare: the traditionally recorded emission component, best seen in the U band, and the wide absorption component that is better detected in the long-wavelength range

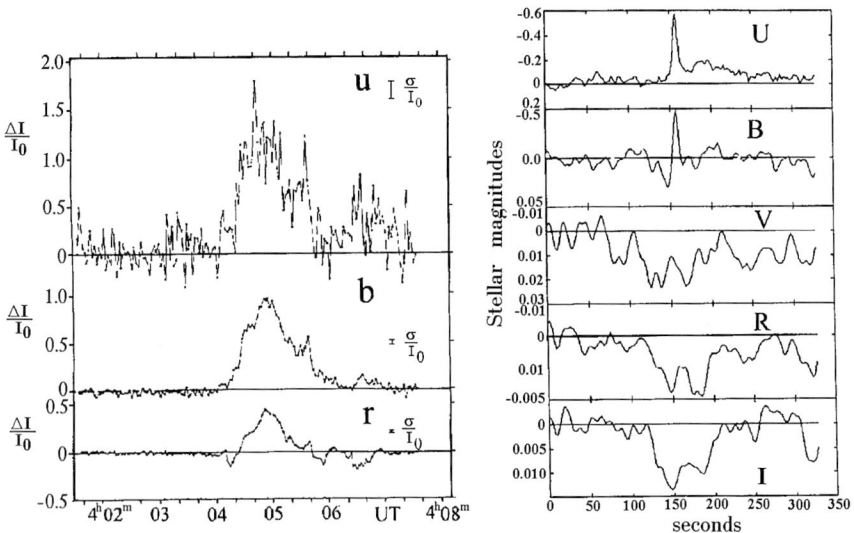

Fig. 51. Preflare decrease of EV Lac brightness or "emission peak in the absorption saucer": the flare of 9 October 1973 observed by Flesh and Oliver (1974) in violet, blue and red rays (*left*) and the flare of 5 October 1996 (Zhilyaev et al., 1998) in $UBVRI$ bands (*right*)

2.4 Dynamics and Radiation Mechanisms of Flares in Different Wavelengths 331

of the spectrum. This distinct two-component model, "emission peak in the absorption saucer", makes it possible to consider from a common point of view a number of photometric characteristics of stellar flares that seemed independent.

1. Since the late 1960s observers have noted that fast flares took place during slow minima of stellar brightness, because originally observations were carried out mainly in blue and green rays, B and V bands. Later, observers started monitoring in violet rays, U band, which increased the sensitivity to the emission component, but simultaneously reduced the probability of detection of the "absorption saucer".

2. The effect of stellar bluing during preflare brightening and reddening during preflare brightness decrease discovered by Cristaldi et al. (1980) suggested that these photometric variations were caused by various mechanisms and were substantially independent. Since the amplitudes of the absorption component are small, in strong flares the emission component with greater probability floods the absorption one, which results in predominant detection of the preflare brightness decrease in less bright flares noted by Shevchenko (1973), Cristaldi et al. (1980) and Bruevich et al. (1980).

3. The preflare brightness decrease on EV Lac of 9 October 1973 found by Flesh and Oliver (1974) was recorded with a high signal-to-noise ratio, had high amplitude, and is usually cited as one of the most illustrative examples of such phenomena (see Fig. 51). But on close consideration of the whole light curve of the flare one can suspect that after the decay of the emission component there was a final fragment of the absorption component, whose depth was comparable with the preflare brightness decrease. Thus, this flare with a "demonstration" preflare brightness decrease also matches the scheme of in "emission peak in the absorption saucer".

4. Coming back to the observed behavior of EV Lac in the K band during optical flares, one can suspect that in the two low-amplitude flares recorded in 1993 simultaneously in the Crimea and on the Canary Islands there were shallow "saucers" in the IR K band.

The above facts suggest that the "emission peak in the absorption saucer" scheme reflects essential properties of stellar flares and deserves special theoretical consideration. In other words, this scheme substantially changes the statement of the problem of the theory of the light curves of stellar flares: one should determine the decrease mechanism of photospheric radiation whose duration is comparable with or longer than the flare observed in violet rays rather than explain only short brightness decrease directly before the flare.

Abundant and diverse phenomena of solar activity provide prompting for interpreting many manifestations of the activity of UV Cet-type stars. But one can hardly find the analogs of rather long-living "absorption saucers" on the Sun. The disappearance of filaments recorded on the Sun at the very beginning of the development of flares proposed by Giampapa et al. (1982a) as a cause for the preflare decrease of ultraviolet stellar emission can hardly have an impact in the range of longer wavelengths. Though abundant solar observation data are available, there are no adequate observations on broadband stellar observations, during which the fluxes from the whole hemisphere containing the active region with spots and flares are recorded with high accuracy. But observers still have to check the hypothesis of Henoux et al. (1990) about "black" flares as precursors of white-light flares.

B. Fast burning. A characteristic feature of many stellar flares is extremely fast burning. Within the hydrodynamic model of stellar flares, Katsova and Livshits (1986) suggested that the duration of this phase is ultimately determined by the time interval required to shift the shock-wave front to the stellar photosphere at about one scale height of the stellar atmosphere. This suggestion was confirmed by the analysis of fast flares detected by the 6-m telescope (Beskin et al., 1988).

C. Secondary brightness maxima. Figures 22 and 29 show the light curves of flares on AD Leo of 18 May 1965 and 28 March 1984, where secondary maxima of brightness are distinctly seen. This feature is often observed in the light curves of strong flares. Andrews (1966a) noted that all secondary maxima found in four of the nine flares on YZ CMi took place approximately six minutes after the main maxima. The time interval between the maxima of light curves was rather typical of different flares on different stars with a total duration of tens of minutes (Gershberg and Chugainov, 1969). Apparently, small humps on descending branches and even practically horizontal sections of the light curves, called decay halts, are physically related to secondary maxima.

Mullan (1977) assumed that these details of light curves could be caused by cooling of expanding flare plasma owing to simultaneous action of heat conductivity and radiative losses, whose dependence on temperature was different.

Katsova and Livshits (1991) believed that secondary maxima were a response of the base of the second leg of the magnetic loop to the initial disturbance, which reached it through the whole loop from the first leg base, where the events that caused the primary flare maximum occurred.

However, in the flare on AD Leo of 2 March 1993 with $\Delta U \sim 0\overset{m}{.}6$ brightness remained at a noticeably high level for about three hours after the beginning of the decrease, thus the main flare energy was released in this state rather than, as usual, at about maximum (Hawley et al., 1995).

D. Short-period oscillations. On 28 November 1972 during monitoring of the star HII 2411 on the 207-cm Struve telescope of the McDonald Observatory Rodonò (1974) recorded a flare with an unusual light curve: its smooth brightness changes were superposed by high-frequency pulsations (Fig. 52). Recording was stopped twice to check whether the oscillations were due to technical problems, in so doing, the entrance diaphragm and the time of integration were changed. But the cause was not found. Shortly before this, in high-speed photometry studies at the same telescope, Moffett (1972) suspected a thin structure of the light curves of stellar flares, but upon thorough

2.4 Dynamics and Radiation Mechanisms of Flares in Different Wavelengths   333

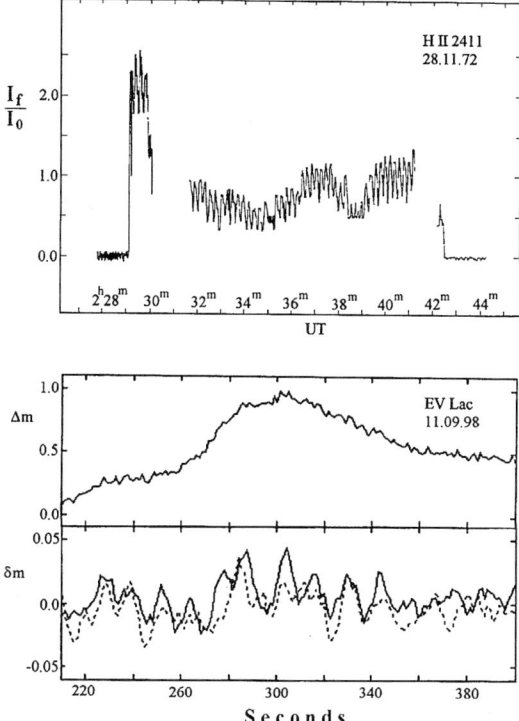

Fig. 52. Quasiperiodic brightness oscillations of stellar flares: *upper diagram*, the light curve of the flare on HII 2411 (Rodonò, 1974); *bottom diagram*, the initial light curve of the flare on EV Lac of 11 September 1998 at 21:55 UT and its high-frequency component; observations from the Crimea (*solid line*) and Greece (*dashed line*) (Zhilyaev et al., 2000)

examination of his results this structure was attributed to regular statistics of quanta (Gershberg and Shakhovskaya, 1973). As opposed to Moffett's data, where the light curve was composed of a set of symmetric parts of different duration, on the light curve of the flare on HII 2411 one could see regular details with systematically faster rise than decay, and with an average duration of 13.08 ± 0.06 s, which slowly increased by the end of the flare. Mullan (1976c) interpreted these oscillations as cyclotron waves propagating between spots on different sides of the equator. Zaitsev et al. (1994) considered them as magnetohydrodynamic oscillations of magnetic loops with a moderate quality factor.

Though there were telling arguments in favor of the reality of the oscillations discovered by Rodonò, his results remained unique for more than a quarter of a century, if we set aside similar data obtained by Rojzman and Kabichev (1985). Similar phenomena were found only recently in high-speed photometry of flares on EV Lac (Zhilyaev et al., 2000) carried out simultaneously at Peak Terskol, Caucasus, in the Crimea, by the Stephanion Observatory (Greece), and in Belogradchik (Bulgaria). Upon uniform processing of the observations it became clear that if the high-frequency component is filtered from the recorded light curves, low-amplitude quasiperiodic oscillations were well established. Figure 52 (bottom diagram) shows the initial light curve of the flare on EV Lac of 11 September 1998 and the residual curves obtained in the Crimea and in Greece. A good agreement of individual pulses obtained from observations at different observatories supports their validity. Zhilyaev et al. found the effect in three flares on EV Lac, quasiperiods of the found pulsations were up to 13 and 26 s, their amplitudes were about 2.5% in the B band and 10–15% in the U band.

Andrews (1989a,b, 1990a–c, 1991) tried to find short-period variations of brightness on a number of flare stars using various statistical methods. In analyzing the data for the quiescent state of V 1285 Aql in the R and I bands, he suspected quasiperiodic brightness variations with a characteristic time of about four minutes. Then he found a brightness variation in the U band of quiet stars with the following characteristic times: AU Mic – 25.4 s, V 1285 Aql – 39.3 s, V 645 Cen – 31.3 s, and V 1054 Oph – 24.7 s. In the records of brightness of AT Mic in the U band soon after the flare Andrews suspected quasiperiods of 13.2 and 7.9 s, whereas the latter apparently took place during the flare decay phase as well. The data of U photometry of YZ CMi provided several quasiperiods in the range from 7.1 to 92.9 s, which were supported at different time intervals. The analysis of brightness records of the star V 1054 Oph in the U band at eight time intervals before a strong X-ray flare revealed quasiperiods from 7.2 to 86.4 s, and many of them repeated at different intervals. During a strong flare on Gl 182, Andrews found quasiperiods at all the phases, the maximum characteristic time of variations was 59 s.

The physical sense of the statistically found quasiperiods in the light curves of active dwarfs remains unclear. They were interpreted using the effects of the terrestrial atmosphere, pouring out of the remains of the dust envelope on stars, astroseismology, and the Ionson–Mullan hypothesis (Mullan, 1984) on the interaction of convective motions with coronal magnetic loops at their base. The similarity of the found statistical quasiperiods with the above periods of brightness variations on HII 2411 and EV Lac is intriguing.

Later, Mullan et al. (1992b) found quasiperiodic variations of brightness with a characteristic time of about six minutes on AD Leo and with characteristic times from three to six minutes on Gl 549. To explain them, they developed the model of oscillating coronal loops.

Peres et al. (1993) suspected similar short brightness variations with a quasiperiod of two and four minutes shortly before the flare on FF And and variations with longer quasiperiods on V 1054 Oph and BY Dra. But, according to their quantitative estimates, the model of oscillating coronal loops by Mullan et al. (1992b) does not fit the energy considerations.

In photometric observations of YY Gem on 6 March 1988 Doyle et al. (1990c) within one night recorded four flares with time intervals of 48 ± 3 min.

2.4 Dynamics and Radiation Mechanisms of Flares in Different Wavelengths

They tried to explain this series of flares within the context of the events initiated by oscillating filament located above the active region. According to their calculations, if the filament is placed at small height, less than half the size of the active region, oscillations with a frequency of up to 1 Hz are possible.

From the power spectra of BY Dra Chugainov and Lovkaya (1988) detected the oscillations with periods of about 188 and 100 min with an amplitude of $0\overset{''}{.}002-0\overset{''}{.}005$ and with periods of 10–59 min with half the amplitude, and from direct measurements of V 1285 Aql brightness in the U band, oscillations with periods from 10 min to 1–2 h with amplitudes to $0\overset{''}{.}1$ (Chugainov and Lovkaya, 1992). Finally, high-accuracy measurements of radial velocity allowed Bouchy and Carrier (2002) to find p-modes of oscillations of

α Cen A with periods within 1.8–2.9 mHz and Carrier and Bourban (2003), oscillations on α Cen B with periods within 3–4.6 mHz, i.e., the analogs of five-minute solar oscillations.

E. Out-of-flare brightness variations. Many observers noted low-amplitude brightness variations in flare stars out of flares (Roques, 1958; Oskanian, 1964). Some of them were mentioned above. The most detailed research of these phenomena was done by Roizman (1983), Rojzman (1984) and Rojzman and Kabichev (1985). During photoelectric monitoring of EV Lac in the U band with regular measurements for comparison and reference stars, they found small variations of brightness on EV Lac at time intervals of hours and with amplitudes of up to $0\overset{m}{.}3$. With a lower amplitude this effect was observed in the B band. In many cases, such an increase in brightness preceded a flare or occurred on the nights when no flares were recorded (see Fig. 53), which enabled a conclusion on the similarity of these stellar phenomena to preflare slow brightening on the Sun. However, this analogy does not shed light on the physical aspects of the brightness variations.

Later, similar results were obtained by Mahmoud (1993b) in observations of EV Lac in the photometric B band.

2.4.4.2 Colorimetry

Sharp bluing of red dwarfs during strong flares was revealed in visual observations (Oskanjan, 1953). The rather "blue" color of the flare on the background of "red" photospheres results in fast growth of the flare amplitude toward shorter wavelengths and, hence, to a noticeable increase of frequencies of recorded flares. The first photoelectric colorimetric results were obtained by Johnson and Mitchell (1958) in observations of the flare on the star HII 1306 in the Pleiades – $\Delta U > 3\overset{m}{.}5, \Delta B \sim 1\overset{m}{.}6$, and $\Delta V \sim 0\overset{m}{.}7$ – and by Abell (1959) during the flare on AD Leo of 9 March 1959 – $\Delta U \sim 1\overset{m}{.}5, \Delta B \sim 0\overset{m}{.}3$, and $\Delta V \sim 0\overset{m}{.}1$. Noticeable differences in the amplitudes stimulated the use of colorimetric analysis in flare diagnostics. Basically, the analysis can be performed within two alternative models: one assumes that during the flare process the radiating body changes on the whole, the other assumes that a new source is

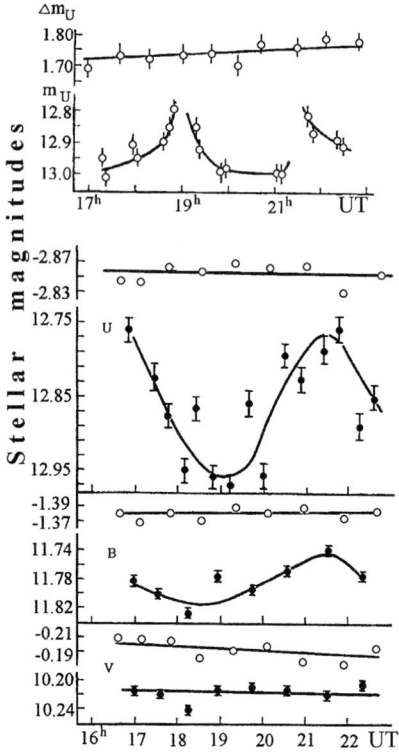

Fig. 53. Out-of-flare brightness variations on EV Lac: *top diagram* shows preflare increase and two descending branches of flares of 22 August 1980 in the U band (Roizman, 1983); *bottom diagram* illustrates the behavior of the star on 14 August 1981 in the UBV bands (Rojzman, 1984); *empty circles* above the light curves show the differences of stellar brightness of comparison and reference stars in the appropriate band

added to the initial one that remains invariable. The first model corresponds, for example, to pulsations of cepheids. On the other hand, spectral observations of flares showed that even during the strongest flares molecular bands were kept in the red region of the spectrum, i.e., flares are local phenomena, not covering the whole stellar surface, and the additive model can be used to analyze them. Apparently, the color characteristics of intrinsic flare radiation were determined first by Abell (1959): $U - B \sim -1^{m}\!.4$ and $B - V \sim 0^{m}\!.1$. A month later from the flare on AD Leo of 13 April 1959 Engelkemeir (1959) determined intrinsic flare radiation in the B and V bands. Assuming that the flare was caused by a hot spot on the stellar surface, from the color indices of main-sequence stars he estimated the effective temperature of the spot at the flare maximum as 11 000 K and its area as 0.11% of the stellar disk. Chugainov (1965) found systematic growth of the color index of $B - V$ flare radiation from $-0\overset{m}{.}2$ to $0\overset{m}{.}3$ as the flare decayed. Kunkel (1967) tracked the drift of flares in the two-color diagram $(U - B, B - V)$ and associated it with the increasing contribution of "burnt" photosphere to the total radiation of the flare. (The theory of "burn" in the approximation of small disturbance was proposed by Grinin (1973)). The first numerous colorimetry data on flares on red dwarfs were published by Kunkel (1970), Cristaldi and Rodonò (1973), Moffett (1974), Lacy et al. (1976). Later results are summed up in the relations (55).

One should bear in mind that a number of radiations of various nature are characterized by rather "blue" $U - B$ color indices, sharply distinct from this parameter of red dwarfs. Therefore, flare tracks on two-color $(U - B, B - V)$ diagrams, describing the drift of stars during flares, weakly depend on the specific mechanism of flare radiation, and colorimetric analysis based on the tracks is extremely uncertain. In other words, the analysis should be based of the intrinsic flare radiation rather than on the position of the star in the two-color diagram during the flare, otherwise erroneous results can be obtained. Thus, Gurzadian (1970), Moffett (1973), and Cristaldi and Rodonò (1975) compared observations with theoretical tracks "star + inverse Compton effect" and obtained satisfactory agreement. They thought it was an argument for the validity of the model. However, comparison of intrinsic colors of flares with the inverse Compton effect clearly proved the inconsistency of this model (Gershberg, 1978).

Colorimetric analysis disproved some other hypotheses on the nature of luminosity of stellar flares. Thus, Arakelian (1959) showed that proper colors of $B - V$ and $U - B$ flares on HII 1306 were much more "blue" than expected under synchrotron radiation in the wide range of the index of energy spectrum of relativistic electrons. Thus, synchrotron radiation cannot be considered as a basic mechanism of optical emission of flares on UV Cet-type stars. Later, this conclusion was made independently by Kunkel (1967). Klimishin (1969) estimated the color indices of a relaxed cloud of molecular hydrogen: $U - B \sim -1\overset{m}{.}0$ and $B - V \sim -0\overset{m}{.}5$, and this point on the two-color diagram is also rather far from the most probable position of stellar flares.

The first colorimetric models within the notions on fast occurrence and further relaxation of hot hydrogen structure above the photosphere of a cool star (Gershberg, 1964; Kunkel, 1967) did not yield a satisfactory presentation of observations (Cristaldi and Rodonò, 1975) and later were repeatedly revised. Strictly speaking, Kunkel (1967) made the first refinement by adding the radiation of heated photospheric section "burnt" by the flare to the luminosity of the hydrogen plasma.

Chugainov (1972a) observed three flares on EV Lac using a three-channel spectrocolorimeter of the Crimean Astrophysical Observatory with simultaneous recording of radiation in the ranges 3350–3650 Å, 4155–4280 Å, and 5120–5320 Å. Analysis of these data, which are analogous to UBV photometry with noticeably narrowed pass bands, suggests that flares at the brightness maxima do not lay on the lines of an absolutely blackbody or in the region of optically thin hydrogen plasma in the appropriate two-color diagram

On 3 August 1975, EV Lac was monitored using the 188-cm Okayama reflector; observations were carried out with a five-channel spectrocolorimeter in the wavelength range of 3300–6005 Å in 400 and 800 Å wide bands. Analysis of two strong flares recorded by Kodaira et al. (1976) made it possible to conclude that over about half an hour during the stronger second flare with $\Delta m_{UV} \sim 5\overset{m}{.}9$ and during several tens of seconds at the maximum of the first weaker flare with $\Delta m_{UV} \sim 1\overset{m}{.}9$ the energy distribution in the flare radiation was very flat, as, approaching decay, it became "redder". Kodaira et al. noted that their results were consistent with the data of Chugainov (1972a) and identified a flat spectrum with $m_\lambda \sim$ const with the radiation of hot hydrogen plasma at temperatures above 10^5 K.

Shmeleva and Syrovatsky (1973) showed that, depending on the rate of initial energy release, one of two temperature structures occurred in solar flares: with constant density (CDR) at very fast energy release or with constant pressure (CPR) at smoother burning. In this connection, developing the above concept of plasma heated to coronal temperatures at initial energy release as the basic component of stellar flares, Mullan (1976b) calculated $U - B$ and $B - V$ color indices for the structures. He concluded that in both cases the designed parameters were close enough to the observed parameters at brightness maxima of flares, though in CDR the basic contribution to optical emission was made by plasma with $T \sim 10^7$ K, whereas in CPR, it was by a plasma with $T \sim 20\,000$ K. Mullan attributed the sharp transition from fast decay to smooth decay with the reorganization of radiating plasma from CDR into CPR. He believed that the proposed model, as a whole, developed Andrews' scheme (1965) with prevailing continuous radiation at the beginning of the flare decay and line radiation, during slow decay, and explained two components of flares following Kunkel (1967) and Moffett and Bopp (1976).

On the basis of the above observations of flares on EV Lac of 3 August 1975, Kodaira (1977) developed the Andrews–Mullan idea about the decisive role of nonstationary high-temperature plasma in flares and proposed the following model: as a result of primary energy release, a structure of stellar size emerges with $n_e \sim 5 \times 10^{10}$ cm^{-3} and $T_e \sim 10^8$ K, at the base of the structure there is a smaller volume of denser and cooler plasma with $n_e \sim 5 \times 10^{13}$ cm^{-3} and $T_e \sim 10^5$ K; the latter is the source of optical and ultraviolet emission and the former is the energy pool, which relaxes in X-rays and heats up due to the heat conductivity of the latter.

Further, Mullan and Tarter (1977) considered the influence of flare X-ray emission on its position on the two-color $(U - B, B - V)$ diagram and found that the X-ray quanta falling on a star degraded as a result of multiple scattering and thus up to 10% of such quanta leaving the star fell into the $UBVR$ bands. This resulted in the shift of the flare on the diagram to the side corresponding to the observed drift. Thus, instead of Kunkel's model

2.4 Dynamics and Radiation Mechanisms of Flares in Different Wavelengths

of "hydrogen recombination emission + burn of photosphere" Mullan and Tarter proposed "luminosity of Shmeleva–Syrovatsky hydrogen structure + reflection of flare X-ray flux". Within this model, Schneeberger et al. (1979) presented the light curves of two weak flares on AD Leo recorded by them, and Worden et al. (1984) adjusted the initial decay of the flare on YZ CMi of 9 February 1979 with $\Delta U = 1^m\!.5$ for parameters close to those of a solar flare: $n_e \sim 10^{13-14}\,\mathrm{cm}^{-3}$ and $T_e = 20\,000$ K.

Using UBV observations, Lukatskaya (1972, 1977) suggested to study stellar flares based on their positions on the plane ($\Delta U/\Delta B, \Delta V/\Delta B$); however, this approach has not been widespread.

Using a five-channel $UBVRI$ photometer-polarimeter designed by Piirola (1975), Panov et al. (1988) in observations of flares on EV Lac established systematic variations of $U - B$ and $B - V$ color indices of flare radiation during the development of flares. But the most complete colorimetric analysis of stellar flares was executed using a similar photometer in the Crimea during long-term research of EV Lac. Independent color indices of the $UBVRI$ system were calculated for a set of known radiation sources: absolutely blackbody, optically thin and optically thick (in the Balmer continuum) hydrogen plasma of various density and temperature and with different probability of the escape of Ly_α of quanta from the medium, and radiation of the upper layers of the atmosphere of a red dwarf heated by a flux of fast particles. Then points were selected on the light curves of some strong flares, in which color indices were determined with sufficiently high accuracy. These observations were compared on four two-color diagrams with corresponding theoretical calculations. An example of such comparison is shown in Fig. 54 (Abranin et al., 1998a). The left diagram shows the light and color indices curves of the flare on EV Lac of 10 September 1993, numbered vertical straight lines at the top mark the points of this curve selected for further analysis and the appropriate color indices. Four two-color diagrams on the right mark the positions of the flare on these diagrams by numbered rectangles. The sizes of rectangles correspond to the accuracy of determination of color indices. One can see that on the $(U - B, B - V)$ diagram the emission at the moment of flare maximum (rectangle 3) was rather close to the curve of blackbody radiation, but on the other diagrams rectangles 3 are shifted from the curve toward hydrogen plasma. A premaximum radiation (rectangles 1 and 2) on the $(U - B, B - V)$ diagram is also located on the blackbody curve but at higher temperature and is also displaced toward the localization of hydrogen plasma on other diagrams. Selected points 4–7 on the descending branch of the flare are shifted from the blackbody curve on the $(U - B, B - V)$ diagram and, as seen from the $(B - V, V - R)$ and $(U - B, V - I)$ diagrams, the emission of hydrogen plasma dominates at this time. Thus, at any stage of flare development any of the considered radiation mechanisms alone is insufficient to explain the observed colorimetric characteristics of the flare in the whole range of the $UBVRI$ system, at least some combinations of these

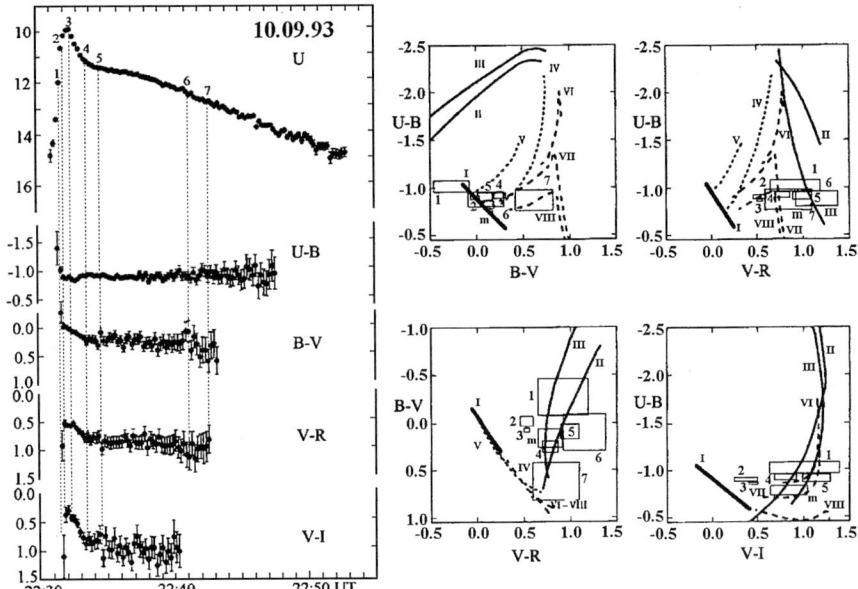

Fig. 54. The light curve of the flare on EV Lac of 10 September 1993, curves of color indices in the $UBVRI$ system and two-color diagrams for studying the nature of flare radiations (Abranin et al., 1998a)

mechanisms should be used. The most probable combination is short-lived blackbody radiation and longer luminosity of hydrogen plasma.

Figures similar to Fig. 54 were based on the Crimean $UBVRI$ observations of flares on EV Lac, which yielded similar results. Thus, Alekseev and Gershberg (1997a) from $U - B$ and $B - V$ color indices estimated blackbody temperatures of seven strong flares on EV Lac with $\Delta U > 1^m\!.8$ at maxima as 10 000 to 25 000 K, and from them, the sizes of flares as 1.1–25×10^{18} cm^2. Hawley and Fisher (1992) and Hawley et al. (2003) used blackbody approximation with $T \sim 9000$–$10 000$ K to describe the continuous radiation of flares on AD Leo.

It should be noted that the conclusion about insufficiency of one of the mechanisms for modeling optical flares was made by Bopp and Moffett (1973) based on spectral and photometric observations of flares on UV Cet. On the other hand, the conclusion by Lukatskaya (1977) on identical composition of the flare emission at maximum brightness and at the decay phase that resulted from her method of gradients contradicts the results of Bopp and Moffett and discredits the expediency of application of this method to stellar flares. Finally, the fact that one of the strongest stellar flares on G 102-21 was located, according to Pagano et al. (1995), in the region of radiation of hydrogen plasma on the $(U - B, B - V)$ diagram supports the validity of the method of colorimetric analysis of stellar flares. Nevertheless, application of

2.4 Dynamics and Radiation Mechanisms of Flares in Different Wavelengths

the technique of digital filtering to the flare on EV Lac of 11 September 1998 led to the conclusion on considerable color variations of flare radiation in the burning phase due to high-frequency brightness variations (Zhilyaev et al., 2003).

2.4.4.3 Polarimetry

Polarimetric observations of flares can provide meaningful astrophysical information only if the full cycle of measurements of polarization parameters is much shorter than the characteristic time of brightness variation and the telescope used is so large that statistical errors of the measured flux are much less than the measured values (Efimov, 1970). Unfortunately, many initial polarimetric observations conducted in the 1960s–1970s do not meet these requirements or do not contain exhaustive analysis of accuracy.

The first confident polarimetric results on stellar flares were obtained by Efimov and Shakhovskoy (1972) in observations with the Shajn telescope in the Crimea: they measured polarization parameters during the flare on EV Lac on 17 August 1969 in three points on the ascending branch of the light curve, at maximum and in 21 points on the descending branch. They found that flare radiation in the B band was not polarized to the accuracy of the measurement errors of $\pm 0.5\%$ caused mainly by the statistics of quanta of recorded flux. In other words, these measurements suggested that the polarization of stellar radiation during the flare was the same as during its absence. Similar conclusions were made by Karpen et al. (1977) based on the Crimean observations of the flare on YZ CMi of 2 December 1975 in the V band, by Pettersen and Hsu (1981) from polarimetric observations of the flare on AD Leo of 29 October 1979 in the U band using the 76-cm telescope of the McDonald Observatory, Eritsian (1978) from observations of several flares on EV Lac and AD Leo from Byurakan, and Tuominen et al. (1989) from the Crimean observations of the YY Gem system on 6 March 1988. But de Jager et al. (1986), based on the U band observations from the Crimea, suspected that the level of polarization of radiation of the flare on BY Dra of 24 September 1984 observed in the U band from the Crimea was 3σ.

The most reliable polarimetric results for stellar flares were obtained by Shakhovskoy using the Shajn reflector in the Crimea in the course of the cooperative program of EV Lac examination in 1989 and 1991 (Alekseev et al., 1994; Berdugin et al., 1995). Figure 55 shows the light curve of the strong flare of 14 September 1991 in the U band, the Stokes parameters of stellar radiation during the flares, and the ratios of the values of these parameters to their root-mean-square errors. The analysis of these data suggests that the degree of polarization does not exceed 2% for the time resolution of 10 s and 1% for a time resolution of 50 s. In weaker flares, these limits are accordingly higher.

In observations of the flare star BD + 26°730 with the 1.25-m telescope of the Crimean Astrophysical Observatory with a $UBVRI$ polarimeter Saar

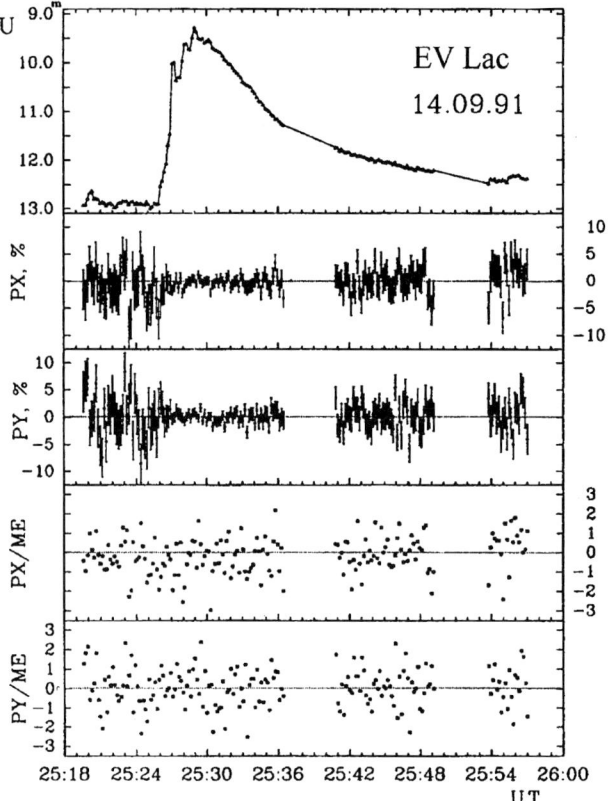

Fig. 55. Polarimetric observations of the strong flare on EV Lac of 14 September 1991, top down: the flare light curve in the U band; the Stokes parameters (the second and third charts); the ratios of the Stokes parameters to their errors (two bottom charts) (Alekseev et al., 1994)

et al. (1994b) detected some changes in the degree and angle of polarization plane over about three hours in the U and B bands. Upon considering several models of this phenomena, they decided that the most probable was the model concerned with the flare during which a flux of directed particles of $10^{9-10}\,\mathrm{erg/cm^2\,s}$ was generated.

Slower variations of polarization of active dwarfs are usually associated with the nonuniform magnetic structure of the stellar surface (see, for example, the results of Alekseev (2000) for MS Ser).

2.4.4.4 Spectral Studies

The optical spectrum of a flare on a UV Cet-type star differs heavily from the spectrum of its quiescent state and changes rapidly in the process of development of the flare. The main distinctive features of flares are strong continuous short-lived emission in the short-wavelength region of the spectrum and intensive emission lines with slower development. Figures 2, 22, 23, 24, and 27 present various spectral characteristics of optical stellar flares. Let us consider these characteristics in more detail with quantitative estimates and physical conclusions, whenever possible.

A. Continuous radiation, hydrogen emission, and Balmer decrement of flares. Strong continuous emission and strong emission lines of hydrogen were discovered as characteristic features on the very first spectra of stellar flares. During the first high time resolution observations with an image intensifier tube spectrograph at the 2.6-m Shajn reflector in the Crimea, the degrees of flooding by continuous emission of some absorption details – jumps of intensity at the heads of TiO bands and depths of the CaI λ 4227 Å line – were measured in the spectra of flares on AD Leo and UV Cet (Gershberg and Chugainov, 1966, 1967). However, the low accuracy of these measurements and small range of simultaneously recorded wavelengths prohibited reconstruction of the energy distribution in the continuous flare spectrum. But quantitative analysis of the equivalent widths of Balmer lines in the spectra of these flares suggested that the lines could be optically thin if they emerged at temperatures above 80 000 K or optically thick if they were formed at chromospheric temperatures (Gershberg, 1968).

Near the maximum of the flare on EV Lac of 11 December 1965 (Fig. 23) Kunkel (1967) recorded practically constant emission near the Balmer continuum from 3750 Å to 3500 Å. Together with the measured Balmer emission jump it was the first direct evidence of the involvement of hydrogen recombination emission of not too high a temperature in this emission.

Kunkel (1967) traced the Balmer series to H_{12} in the spectrum of the flare on EV Lac of 11 December 1965, to H_{14} in the flare on YZ CMi of 7 December 1965 and in the spectrum of EV Lac obtained during long-duration increased brightness of the star on 5 December 1965, and to H_{10} at the maximum of the flare on AD Leo of 11 December 1965. The latter corresponds to the density of radiating plasma of about 10^{15} cm^{-3}. In the spectra of flares on EV Lac and AD Leo of 11 December 1965 Kunkel (1967) estimated Balmer jumps as 4.6 and 5.7, which corresponds to temperatures of 23 000 and 19 000 K in a purely recombination spectrum of hydrogen plasma.

From the spectrum of the flare on YZ CMi of 5 December 1965 Kunkel (1967) obtained distinct evidence of faster decay of the continuum, brightness in the U band, as compared to the brightness decay of emission lines. The same conclusion followed from systematic growth of the equivalent width of the Balmer lines in the process of development of flares on AD Leo on 18 May 1965 (Gershberg and Chugainov, 1966) and from photoelectric measurements of equivalent widths of the H_β line in the flares on UV Cet and EV Lac executed by Chugainov (1968, 1969a). Special observations of the strong flare on UV Cet of 6 January 1975 carried out by Moffett et al. (1977) using a photometer and a scanner showed that the delay of the maximum emission in the H_β line was a real effect, which was connected not only with the growth of equivalent widths of the line because of a decrease of the continuum after the flare maximum. The delay was later confirmed many times (Kahler et al., 1982; Worden et al., 1984; Hawley and Pettersen, 1991; McMillan and Herbst, 1991), but Pettersen and Sundland (1991) suspected that it took place only during sufficiently strong flares.

The first Balmer decrement measured by Kunkel (1967) in five phases of the flare on EV Lac of 11 December 1965 at maximum brightness looked as follows: $H_\beta:H_\gamma:H_\delta:H_8:H_9:H_{10}:H_{11}$ = 1:1.24:1.48:1.22:1.17:0.94:0.80, and the upper members of the series weakened faster when the flare decayed. Kunkel presented the decrement within the framework of the chromospheric model with coherent scattering in lines at $n_e = 3 \times 10^{13}$ cm^{-3}, T = 20 000 or 25 000 K and optical thickness at the center of the H_α line of about 10^3; as the flare decayed the latter had to be replaced by 10^2.

The characteristics of hydrogen emission of the strong flare on AD Leo of 2 March 1970 recorded in the Crimea by Gershberg and Shakhovskaya (1971) were slightly different: the Balmer jump was close to 1.4 and the H_α and H_β lines were apparently always more intensive than the following members of the series, i.e., the inverse decrement was not observed. Small-emission Balmer jumps of close values were estimated by Chugainov (1972a) near maxima of three flares on EV Lac, and the inverse Balmer decrement was suspected in flares on YZ CMi (Gershberg, 1972b). Spectrocolorimetric observations by Kodaira et al. (1976) admit a small Balmer jump in the flare on EV Lac of 3 August 1975. According to the observations by Hambarian (1982), the Balmer jump in the spectrum of the flare in the binary system YY Gem on 5 February 1981 can be estimated as 1.5–2.

In observations of two flares on UV Cet of 14 October 1972 separated by 15 min, Bopp and Moffett (1973) obtained direct evidence of the decisive contribution of continuous emission to the optical flare radiation in maximum brightness and during the initial fast decay (Fig. 24). According to Moffett and Bopp (1976), the contribution of emission lines to the radiation of the UV Cet flare recorded in the B band on 11 November 1971 was 11% at the maximum and 16%, during the decay stage, to the flare on UV Cet of 12 November 1971 – 11 and 17%, and to the flare on EV Lac of 5 November 1971 – 5 and 28%, respectively. According to Giampapa (1983), within the range of 3600–4600 Å the contribution of emission lines to the total radiation of the quiescent state of UV Cet was 6%, at the beginning of the strong flare of 8 September 1979 – 18%, at maximum – 3%, during the decay phase it was about 30%. According to Doyle et al. (1988b), in the flare on YZ CMi of 4 March 1985 the contribution of emission lines to the total radiation in the U band was 10%. For the very strong flare on AD Leo of 12 April 1985 with $\Delta U = 4\overset{m}{.}5$ and a duration of about four hours Hawley and Pettersen (1991) found that within the range of 1200–8000 Å the continuum prevailed at all stages: the contribution of lines to the total radiation of the flare was 9%, about 4% during the impulsive phase and 17% during the gradual phase.

At maximum brightness of the stronger of two flares on UV Cet of 14 October 1972 Bopp and Moffett (1973) recorded the inverse Balmer decrement similar to that of Kunkel's model at $\tau \sim 10^2$, but its changes in the process of flare decay and the decrement of the previous weaker flare on UV Cet on the same day did not comply with Kunkel's calculations.

The data on the emission Balmer decrements of flares on UV Cet-type stars accumulated by the early 1970s were analyzed within the theory of radiation of optically thick hydrogen plasma with the velocity gradient of internal motions. Assuming the temperature of plasma to be equal to 15 000 or 20 000 K, all observational data were presented by models with an electron density from 10^{12} to 10^{14} cm^{-3} and the escape probability of Ly_α quantum of 10^{-5}–10^{-4} (Gershberg, 1974b). Later, Bruevich et al. (1990) elaborated the theory of the Balmer decrement without using the hypothesis about the velocity gradient of internal motions, but with allowance for multiple scatterings of quanta that allowed them to leave the medium in the line wings. Within this concept, Katsova (1990) presented the Balmer decrement of the flare on EV Lac of 11 December 1965 both at maximum and 15 min later, when the flat decrement was replaced by a steep one, as well as the flares on AD Leo of 4 March 1970, UV Cet of 17 September 1980, and YZ CMi of 4 March 1985. The observations were presented with an electron density of about 10^{14} cm^{-3}, somewhat lower temperatures and greater optical thickness $\tau_{Ly\alpha} \sim 10^6$ than in calculations within the concept of moving media. The transition from flat or inverse decrement at the flare maximum to a steep decrement at the decay phase was associated with the decrease of electron density of the radiating plasma. Within the framework of this model, Katsova estimated the area of the flare on YZ CMi of 4 March 1985 in the Balmer lines as 5×10^{18} cm^2, which exceeds by an order of magnitude the estimate of the area of the source of continuous radiation. In this flare, an appreciable decay of H_8 and H_9 at practical invariance of the ratio H_δ/H_γ was recorded (Butler, 1991). Grinin (1980b) showed that under the formation of the Balmer emission lines in the medium with considerable gradients of physical conditions combined with radiating interaction between high- and low-density zones an abnormally high inverse ratio I_{H_α}/I_{H_β} could occur.

Drake and Ulrich (1980) calculated the luminosity of flat hydrogen layer using the probability method for $n_e = 10^8$–10^{15} cm^{-3}, $T_e = 5000$–$40 000$ K, and $\tau_{Ly\alpha} = 10^4$–10^6. Using their calculations, from the Balmer decrement at the 15th minute of the flare on AD Leo of 28 March 1984 Butler (1991) estimated its electron density as $10^{14} - 10^{15}$ cm^{-3} assuming $T_e = 15 000$ K and $\tau_{Ly\alpha} = 10^4$. Within the Drake and Ulrich model, Jevremovic et al. (1998b) analyzed the Balmer decrement of three flares on the dM5.5e star Gl 866 recorded on 11 June 1991 during four-hour spectral and photometric observations with the South African Astronomical Observatory. To enable the best representation of relative intensities of the lines from H_β to H_{10}, the analysis involved the selection of four free parameters of the model: electron temperature, electron density, optical thickness in Ly_α, and temperatures of underlying surface.

As a result, for the flares with rather different light curves the appreciably varying parameters were obtained: $\log n_e$ = 12.5–14.9, T_e = 8000–20600 K, $\log \tau_{Ly\alpha}$ = 6.7–4.2, and temperatures of the underlying surface within 3800–7200 K. The models yielded the areas of flares from 0.1 to 5% of the stellar surface, or 8×10^{17}–6×10^{19} cm^{-2} and thickness from 0.5 to 1100 km.

In the strong but short flare on YZ CMi of 5 March 1985 Doyle et al. (1988b) found a nonmonotonic evolution of the Balmer decrement: in the range from H_δ to H_9 first it became flatter than at maximum brightness, then it steepened.

The Balmer decrement in the strong flare on AD Leo of 12 April 1985 was traced by Hawley and Pettersen (1991): it was rather flat at the flare maximum

$$H_\beta : H_\gamma : H_\delta : H_8 : H_9 = 1.20 : 1 : 0.89 : 0.66 : 0.53$$

and slightly steeper 1.30 : 1 : 0.72 : 0.50 : 0.36 8000 s later, but it did not reach the normal state.

On 26 March 1986, Doyle et al. (1990b) recorded a rather flat Balmer decrement in the spectrum of flares on Gl 375. A flat or even inverse decrement was found in the flare on YZ CMi of 18 May 1992 (Gunn et al., 1994a) and in flares on EV Lac of 10 and 11 September 1993 (Abranin et al., 1998a). But the Balmer decrement found by Phillips et al. (1988) in the flare on UV Cet of 17 September 1980 $H_\beta:H_\gamma:H_\delta:H_8:H_9$ = 2.00:1:0.69:0.47:0.42 approached the nebular decrement. In the flare on LQ Hya of 22 December 1993 the ratio H_α/H_β was close to 1.5 (Montes et al., 1999).

Petrov et al. (1984) examined the emission of H_β in the spectra of several flare stars with a resolution 0.7–1.0 Å using the 2.6-m Shajn telescope. During joint photometric monitoring they recorded two flares on UV Cet, three flares on AD Leo, and three flares on YZ CMi. Analysis of the data confirmed the conclusions about independent changes in the profiles and intensities of the Balmer lines in flares and the decisive contribution of the line emission to preflare brightening. It also revealed significant variations in the central intensity of H_β an hour before the flares.

Mochnacki and Zirin (1980) carried out spectral observations of five flare stars using a 32-channel spectrometer on the 5-m telescope. In the blue and visual regions of the spectrum, resolution was 160 Å, in red, 20 or 80 Å, whereas in the $\lambda > 5700$ Å region only separate sections of the spectrum were measured, thus the lower members of the Balmer series were measured separately, and starting from H_δ higher members were not separated. On 9 October 1979, they recorded a strong flare on YZ CMi with the time resolution of 10 and 30 s and on 10 October 1979, two fast flares on UV Cet. Analyzing the obtained data within the two-component model of Kunkel (1967), Mochnacki and Zirin found that at a similar Balmer decrement the Balmer jump in these flares was much smaller than in Kunkel's observations. They concluded that the emission from the heated photosphere dominated in these events even at maximum brightness. Assuming the radiation to be blackbody, from the measurements in the range of 4200–6900 Å they estimated the temperature of such radiation at maximum of the flare on YZ CMi as a value close to 9000 K, on UV Cet – 9500 and 7400 K; at measured absolute luminosities such temperatures correspond to the flares of 38×10^{17}, 2.9×10^{17}, and 3.3×10^{17} cm^2. According to Mochnacki and Zirin, fast burning of flares is connected with the increase of their areas, at maximum brightness the maximum of the blackbody temperature is reached and fast decay is caused by the decrease of this temperature.

At maximum brightness of the strong flare on YZ CMi of 25 October 1979 the energy distribution in the range of 4200–5900 Å was also presented by a blackbody temperature at 8500 K. In this case, the area of the flare was estimated as 1×10^{19} cm^2 or 0.5% of the stellar disk (Kahler et al., 1982). Katsova et al. (1991) approximated the optical continuum in the range of 3600–4600 Å at maximum brightness of the same star on 4 March 1985 by the blackbody radiation at 10 000 K and estimated the size of this six-minute flare with $\Delta U = 1^m.2$ as 5×10^{17} cm^2.

Giampapa (1983) compared the energy distribution in the continuous spectrum of the strong flare on UV Cet of 8 September 1979 with $\Delta U = 5^m$ in the range of 3700–4600 Å (Fig. 56) with several theoretical distributions: in the radiation of an absolutely blackbody and H^- at 7500 K, in the recombination spectrum of hydrogen at $T_e = 20\,000$ K and $n_e = 10^{12}$ cm^{-3}, and in free–free radiation of a hydrogen plasma at 50 and 100 kK. Comparison showed that both models of free–free radiation were suitable for representation of observations and there was a substantial divergence between them only in the range of $\lambda < 1800$ Å. Purely recombination luminosity occurs in the long-wavelength region of the range, but is insufficient for its "blue" part, similar situations with the luminosity of H^- and blackbody radiation exist, all of them require additional short-wavelength emission. The ion of H^- was used because it is considered as one of the probable radiators in solar white-light flares with a luminosity of 2×10^{29} erg/s and a total energy of 3×10^{31} erg. In these flares within 4000–6000 Å the spectrum is flat, the Balmer jump is 2–3, 90% of total radiation makes up the continuum and only 10% the lines. On the other hand, Grinin (1976) drew a conclusion on the strengthening of the emission of H^- in the field of the temperature minimum of stellar atmospheres under the effect of the flare impulsive phase. The last member of the Balmer series in the spectrum of the flare on UV Cet of 8 September 1979 was H_{15}, which corresponds to a plasma density of 10^{13} cm^{-3}, while the ratios of intensities H_{11}–H_{15} corresponded to Boltzmann distribution for the population of the levels, which means thermal excitation of this emission.

Using the 3.9-m Anglo–Australian telescope, Robinson (1989) recorded the flare on Wolf 424 of 4 April 1987 with the resolution of 1.5 Å with the exposures of 30 s. He found a flat spectrum near the maximum in the range of $\lambda > 3900$ Å and a smooth quasicontinuum in the short-wavelength region due to the superimposing of the Balmer lines. On the background of an almost

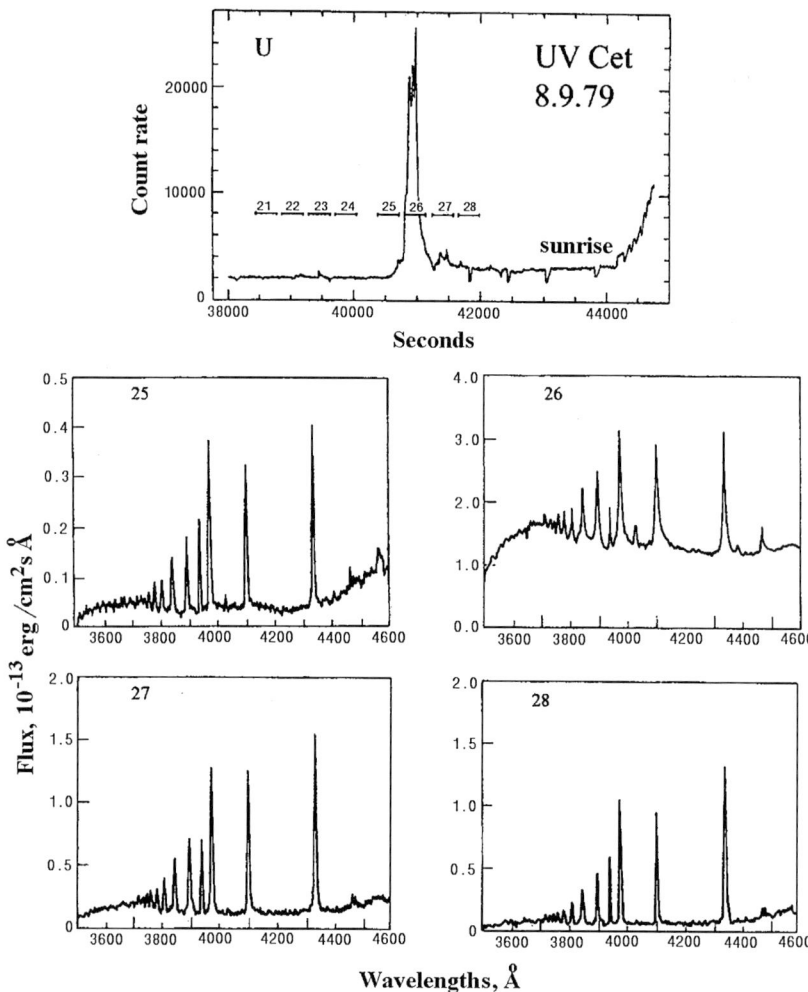

Fig. 56. The light curve of the flare on UV Cet of 8 September 1979 in the U band: seconds after 00:00 UT and energy distribution in its spectrum over the numbered time interval shown by horizontal sections are plotted along the x-axis (Eason et al., 1992)

monotonic decay of the flare continuum, Robinson found three essential bursts in hydrogen and calcium lines.

Zarro and Zirin (1985) obtained the spectra of YZ CMi in the range of 3600–4000 Å with a resolution of 3 Å using the 5-m telescope for the quiet star and during the flare of 19 February 1984. They showed that the flare continuum in the region $\lambda < 3800$ Å was formed as a result of superimposing of the upper members of the Balmer series.

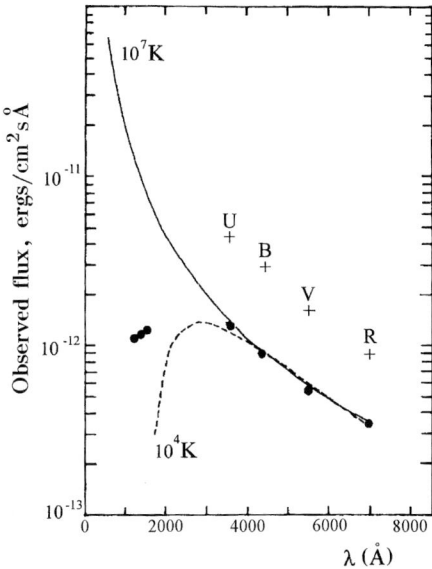

Fig. 57. Absolute fluxes in the $UBVR$ bands at the flare maximum and IUE spectrograms averaged over 15 min, *solid curve* shows the energy distribution in free–free radiation of plasma at 10^7 K, *dashed line* shows blackbody radiation at 10^4 K (Pettersen et al., 1986b)

The strong flare on AD Leo of 12 April 1985 with $\Delta U = 4^m\!.5$ recorded by Pettersen et al. (1986b) with IUE in the ultraviolet and $UBVR$ bands with a high-speed photometer of the 90-cm telescope of the McDonald Observatory allowed an important conclusion to be made on the nature of flare continuum in optical range. Figure 57 presents absolute fluxes in the $UBVR$ bands at the flare maximum and IUE spectrograms averaged over 15 min, the solid curve shows the energy distribution in free–free radiation of plasma at 10^7 K, the dashed line shows blackbody radiation at 10^4 K. One can see that in the optical range both distributions well represent the observations, but the ultraviolet points exclude the luminosity of plasma at coronal temperatures as the basic mechanism of optical emission of flare. At least a two-component model is needed to describe the flare continuum. Apparently, this is the only strong flare of the dMe star recorded from IUE and in $UBVR$ bands. Flare spectra obtained simultaneously in the optical range allowed the last member of the Balmer series H_{15} to be recorded.

De Jager et al. (1989) showed that at the maximum of the fast and strong flare on UV Cet of 23 December 1985 with $\Delta U > 6^m$, whose spectra were obtained at the 1.5-m telescope of the European Southern Observatory, the energy distribution in the range of 3500–7000 Å was well described by blackbody radiation at 16 000 K, which yields a flare size of $3 \times 10^{18}\,\text{cm}^2$, but as the flare decayed the blackbody distribution represented its continuum even less well.

Doyle et al. (1989) found that near the maximum of the flare on V 577 Mon of 28 February 1985 with $\Delta U = 3\overset{m}{.}8$ during several tens of seconds continuous emission in $BVRI$ bands was well presented by synchrotron, free–free, and recombinant radiation. But to represent the observed luminosity, the synchrotron needs too many luminous electrons, free–free radiation should exit from the volume comparable to the stellar one, whereas the recombination luminosity is 4000 times more effective.

McMillan and Herbst (1991) found the correlation between the equivalent widths of the emission line H_α and the degree of its variability and some evidences of the connection of strong flares on EV Lac with localization of spots.

During observations of the young fast rotator K2 LQ Hya by the 4.2-m Herschel telescope with the Utrecht echelle spectrograph in the range of 4840–7720 Å Montes et al. (1999) recorded on 22 December 1993 a strong flare that lasted for more than three hours. The first spectrogram revealed a strong continuum, which filled all the photospheric lines and was identified as blackbody emission at 7500 K from the degree of flooding. Simultaneously, the emission lines reached maximum at the moment when the continuum essentially weakened, and their decrease occurred at a decelerating rate.

Liebert et al. (1999) recorded a unique flare on the dM9.5e star 2MASSW J0149090+2951613 on 7 December 1993 in observations with the Keck telescope in the range of 6150–10 100 Å with a spectral resolution of 9 Å. They found an abnormally strong emission of H_α in the five-minute spectrum, which weakened at the three following 10-min spectra (Fig. 58). W_{H_α} achieved 300 Å, and its FWHM did not exceed the instrument width, whereas at the quiet star $W_{H_\alpha} = 8$–14 Å. In the short-wavelength part of the spectrum, the continuum was amplified by several times. The estimate of L_{H_α} showed that at the impulsive phase it was identical to or even higher than the bolometric luminosity of the star.

B. Calcium emission in the H and K lines is amplified in stellar flares later and to a lesser extent than hydrogen lines, and fades more slowly. Monitoring of YZ CMi revealed a many-hour increased luminosity of this emission after strong flares (Gershberg, 1972b). According to the estimate of Moffett and Bopp (1976), the delay of the CaII emission maximum with respect to the flare maximum varied from 4 to 52 min. Thus, they suspected that a greater delay is typical of stars of higher luminosity.

At the onset of the strong flare on UV Cet of 8 September 1979 the flux in the CaII K line decreased almost 3-fold, then increased to a 2–3-fold level of the quiescent state of the star, which could be due to the evaporation of the low chromosphere or further ionization of calcium by X-ray emission of the flare (Eason et al., 1992).

Houdebine et al. (1993b) found that flare CaII emission, to a considerable extent, was stipulated by radiative pumping by the emission of H_ε.

2.4 Dynamics and Radiation Mechanisms of Flares in Different Wavelengths 351

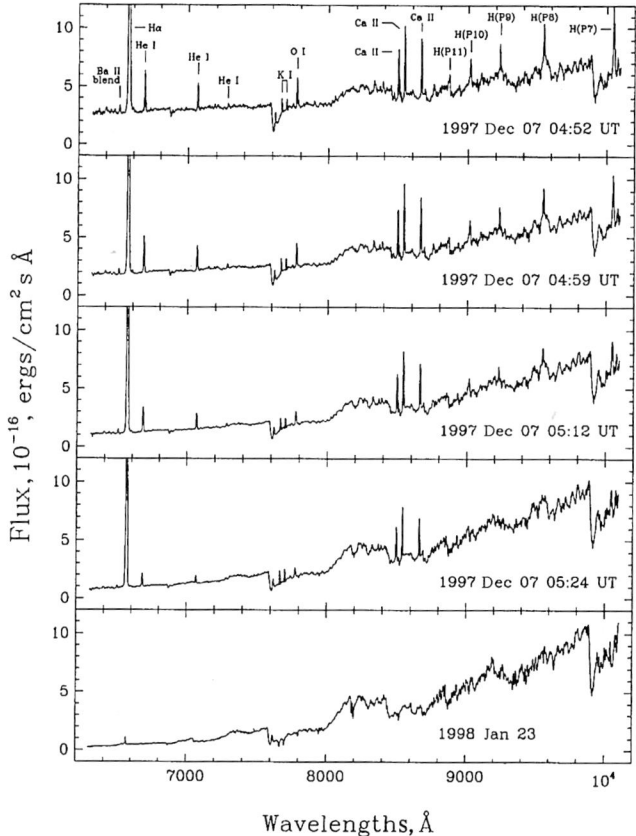

Fig. 58. Spectrograms of the flare on the dM9.5e star 2MASSW J0149090+2951613 of 7 December 1993 in the range of 6150–10 100 Å; the bottom diagram shows the spectrum of the quiet star (Liebert et al., 1999)

In the flare on the dM9.5e star 2MASSW J0149090+2951613 of 7 December 1993 IR triplet of CaII was recorded in emission (Fig. 58) (Liebert et al., 1999). In the flare on AD Leo on 13 March 2000 Hawley et al. (2003) discovered substantial delay of the lines of this triplet after the U-band emission.

We mentioned already that in observations of the active K2 star ε Eri with a four-slit spectrometer in the focus of the 2.5-m telescope of the Mount Wilson Observatory, Baliunas et al. (1981) recorded correlated variations of brightness of the H and K lines with amplitudes up to 7% and with characteristic duration of up to 15 min. These variations correspond to the flares with an energy of 2×10^{30} erg.

C. Lines of helium and metals. Emission of neutral and ionized helium was found by Joy and Humason (1949) in the first slit spectrogram of the flare on UV Cet of 25 September 1948 obtained with the exposure of 144 min. The emission lines of neutral helium λ 4026 Å and λ 4471 Å were recorded near the main maximum of the flare on AD Leo of 18 May 1965. By the secondary maximum phase (Fig. 22) they weakened, but during further flare decay, which was spectrographed in the green and red regions of the spectrum, other HeI lines were recorded (Gershberg and Chugainov, 1966). Alongside the lines of neutral He, at maximum brightness and during the fast decay phase of flares on UV Cet of 24 September 1965 the HeII λ 4686 Å line was found in its spectrum (Gershberg and Chugainov, 1967). But this is not a rule: in the flare on Wolf 359 of 20 March 1968 HeI lines were appreciably strengthened and the HeII line was absent (Greenstein and Arp, 1969). A similar situation with respect to helium occurred during the flares on UV Cet of 14 October 1972, from the stronger flare the emission of the MgI b line was detected for the first time (Bopp and Moffett, 1973). Moffett and Bopp (1976) found a vast diversity of helium emission in flares that were recorded with high time resolution, but never detected HeII emission.

In the weak flare on AD Leo of 30 March 1977 with $\Delta U = 0\overset{m}{.}4$ Schneeberger et al. (1979) recorded the behavior of the HeI line λ 5876 Å identical with H_α: an increase of the central intensity by 40% at constant width. During the decay stage in the spectrum of the strong flare on UV Cet of 8 September 1979 Eason et al. (1992) found strong lines of neutral helium λ 5876 Å with equivalent widths from 16 to 4 Å and λ 6678 Å with equivalent widths from 4 to 1 Å and numerous lines of FeI and FeII in the wavelength range of 5200–5400 Å; equivalent widths of D components of the sodium doublet varied from 3.8 to 1.5 Å.

On 2 September 1981, Baliunas and Raymond (1984) recorded about an hour-long flare on EQ Peg B simultaneously from IUE in the ultraviolet and in the range of 4700–7000 Å with a resolution of about 6 Å using the Reticon spectrograph. The intensity of H_β emission increased in the flare by a factor of 2, H_α by 30%, CIV lines almost by a factor of 4, HeII λ 1640 Å lines approximately 3-fold, with the amplification of ultraviolet emission being measured in the summary spectrum of the system. It is essential that the intensity maximum of helium lines took place approximately 15 min prior to the maximum of H_α and H_β. Assuming that recombinations in coronal plasma dominated in the emission of the HeII λ 1640 Å line, Baliunas and Raymond suggested the X-ray energy of the flare as 6×10^{32} erg and interpreted the flare as an analog of the gradual phase of a solar two-ribbon flare or as quickly cooling plasma of coronal temperature.

During spectral and photometric observations of several flare stars using the 2.6-m Shajn telescope, Petrov et al. (1984) looked for the emission of HeII λ 4686 Å in the spectra with a resolution of 0.7–1.0 Å, but did not find it in any of the six flares on AD Leo and YZ CMi recorded during this campaign.

Investigating radial velocities of K–M dwarfs in spring 1978, Mochnacki and Schommer (1979) carried out high-dispersion observations of YZ CMi with an image intensifier tube multichannel coude spectrometer of the 2.5-m telescope of the Mount Wilson Observatory. On 29 March 1978, over two

2.4 Dynamics and Radiation Mechanisms of Flares in Different Wavelengths 353

hours they obtained 11 spectra with emission lines of MgI b, FeI (15, 37) and FeII (42, 48, 49) in the range of 5090–5350 Å. On other dates these emissions were not visible. Radial velocities of the emissions did not show significant differences from the absorption lines, and in the course of observations the intensity of emissions weakened by approximately 30%. The components of MgI b of similar intensity suggested an appreciable optical thickness in these lines. Apparently, during this night a flare took place on the star, and the found emissions could have even longer lifetime than CaII lines.

In the maximum of the flare on AD Leo of 12 April 1985 Hawley and Pettersen (1991) identified seven HeI emission lines and ten FeI lines within the range of 3700–4433 Å, the emission peak in the center of the absorption line of CaI λ 4227 Å was observed for about an hour and a half after the flare started.

In the spectrum of the flare on Wolf 424 of 4 April 1987 Robinson (1989) noticed numerous weak emission lines of FeI and FeII, the line of MgII λ 4481 Å, and inverse emission of normally absorption lines of CaI λ 4227 Å and Mn I λ 4030 Å.

In the spectra of active EV Lac obtained during the cooperative observations in 1992 (Abdul-Aziz et al., 1995), one can easily see the lines of neutral and ionized helium λ 4471 Å and λ 4686 Å, respectively, three strong FeII emissions λ 4924, 5018 and 5169 Å and several weaker lines of iron and magnesium, some of which were mentioned earlier by Mochnacki and Schommer (1979).

The occurrence of the HeI lines was noticed in the flares on Gl 866 of 11 June 1991 (Jevremovic et al., 1998b), on AT Mic of 15 May 1992 (Gunn et al., 1994b), on Proxima Cen of 9 May 1994 (Patten, 1994), on LQ Hya of 22 December 1993 (Montes et al., 1999), on 2MASSW J0149090+2951613 of 7 December 1993 (Liebert et al., 1999), and on LHS 2065 of 12 December 1998 (Martin and Ardila, 2001).

Of particular interest are the continuous spectral observations of EV Lac with the Shajn telescope in the Crimea on the night of 31 August–1 September 1994 (Abranin et al., 1998b). The upper diagram in Fig. 33 shows the light curve of the star in the U band. One can see half a dozen characteristic fast flares of different amplitudes with durations of several minutes and long increased luminosity of the star, which started after the flare at 20:46 UT and lasted until the strongest flare on this night began at 23:24 UT. The next diagram presents the equivalent width of the H_β line. Close examination of the diagrams reveals the response of H_β emission to individual fast bursts of stellar brightness. But the main feature of the time course W_{H_β} is the dominating maximum concerned with an almost three-hour increased brightness of the star. The time course of W_{H_β} is rather similar to that of an equivalent-width blend of emission lines of FeII λ 5169 Å and MgI λ 5167/73 Å presented on the fifth diagram: the correlation coefficient r $(W_{H_\beta}, W_{\text{blend}}) = 0.94$. The third diagram shows the equivalent width of the HeI line λ 4471 Å. Comparison of this diagram with the first two shows that this line reacts to individual fast flares more explicitly than H_β, though as a whole the two lines correlate well with each other: $r(W_{H_\beta}, W\lambda_{4471}) = 0.76$. (Earlier, Hawley and Pettersen (1991) noted that the emission of HeI λ 4026 Å in the flare on AD Leo of 12 April 1985 preceded the Balmer lines.) The most interesting and unexpected results were obtained in the monitoring of the emission line of ionized helium λ 4686 Å (see fourth diagram). Formally, the line does not correlate with the above emissions: the appropriate correlation coefficients vary within 0.2–0.3. But thorough comparison of $W\lambda_{4686}$ with the light curve of the star detects that almost all spectra with increased luminosity of ionized helium were obtained 15–30 min after fast flares. The second feature of the emission line HeII λ 4686 Å is shown in the bottom diagram, where the summary spectrum of all active states of the star recorded during the campaign of 1994 is presented. One can see the splitting of the HeII emission: the long-wavelength component has a normal wavelength, whereas the short-wavelength one is displaced for -400 km/s. Among the spectra obtained on the night of 31 August –1 September, eight spectra had only a short-wavelength component, 2 spectra had both components, and one spectrum had only a long-wavelength component; these features are marked by $<, <>$, and $>$, respectively. The meaning of the observations is not clear. Probably, similar amplification of HeII lines took place in the flare on AD Leo of 2 February 1984, when in the spectrum obtained by Byrne and Gary (1989) from IUE with a 20-min exposure started eight minutes after the maximum of the fast flare, doubled intensity was recorded in the line of HeII λ 1640 Å arising in the recombination spectrum of HeII at the subsequent cascade transition after the radiation of the quantum of the line λ 4686 Å. Apparently, the short-wavelength component of this line can arise in the transition region from the chromosphere to the corona or in low corona under the formation of moving structures, whose analogs on the Sun yield transients into the interplanetary space.

Using the Utrecht echelle spectrograph with a resolution of 110 000 mounted on the 4.2-m Herschel telescope, Martin (1999) obtained on 6 June 1995 four spectra of the dM7e star VB 8 within 5400–10 600 Å. In one of them he found the evidence of a flare: an almost 2-fold increased W_{H_α} and the increased line width, an He D_3 emission line, and the emission cores of Na D lines and an amplified resonance K line λ 7699 Å. In the flare on LQ Hya of 22 December 1993 recorded by the same instruments, in addition to helium lines, emission Na D lines, the triplet MgI b and numerous lines of 42, 48, and 49 multiplets of ionized iron were found (Montes et al., 1999).

Considering the general properties of relaxing plasma with regard to variations of its temperature, expansion, ionization by soft X-ray, ionization and excitation of ultraviolet and optical continua, Houdebine et al. (1991) concluded that in spite of complicated geometry, expansion and radiative transfer, collision processes dominated in optically thin lines of flares with nonthermal heating only at the initial phase with not too high an amplitude and at $n_e = 10^{11} - 10^{14}$ cm^{-3}. For such models they calculated the functions of luminosity of optically thin lines of helium, hydrogen, and ionized calcium depending only on temperature and correlated the recorded luminosity maxima of the Balmer lines and K CaII with temperatures at the maxima of the appropriate luminosity functions. They constructed the cooling curves of flares on Proxima Cen of 24 March 1984, on UV Cet of 8 December 1984, and on AD Leo of 28 March 1984. The thus-obtained experimental cooling curves of flare plasma appeared to be rather close to theoretical Gurzadian curves (1984) for $n_e = 10^{11}$ cm^{-3}, but differed significantly from his curves for $n_e = 10^{10}$ cm^{-3}.

D. Broadening and shifts of spectral lines. Doubling of the width of hydrogen emission lines in the spectrum of the flare on UV Cet of 7 October 1957 was recorded by Joy (1958) in one of the first spectrograms of flares of this type. The line widths in flare spectra in separate phases of their development were first measured in observations of AD Leo in 1965 from the Crimea (Gershberg and Chugainov, 1966). During the flare of 13 May 1965 with $\Delta B = 1\overset{m}{.}5$ the FWHM of the H_α line grew from 5 to 7–8 Å at an instrumental width of 4.6 Å. During a stronger flare on 18 May 1965 (Fig. 22) H_α was observed only at the late decay stage, but the near-maximum brightness FWHM of H_γ and H_δ lines increased to 10–11 Å from 5–6 Å in the quiescent state. Several minutes after the maximum and termination of fast decay (see spectrograms 4 and 5 in Fig. 22) these lines started narrowing. During weak flares on AD Leo of 24 March and 16 May 1965 with $\Delta B \sim 0\overset{m}{.}2$ no changes in the H_α profile were noticed. Later that year the same equipment was used to observe UV Cet: six flares of different amplitude were recorded (Gershberg and Chugainov, 1967). The FWHM of H_β lines was 10–11 Å at the decay phase of the flare of 20 September 1965 with $\Delta V \sim 1\overset{m}{.}0$ and 12 Å in the weaker flare started 20 min later at the width of the instrumental contour of 7.8 Å. The strongest flares recorded during this campaign demonstrated the broadening of lines: the flare of 24 September 1965 with $\Delta V > 1\overset{m}{.}8$ near the FWHM maximum H_β line was 8.4 Å and H_γ line was 7.6 Å at the quiescent-state width less than 4.5 Å; in the flare of 26 September 1965 with $\Delta V = 1\overset{m}{.}4$ FWHM = 5.4 Å of H_α line at the same width in the quiescent state.

On the spectrogram of the flare on Wolf 359 of 20 March 1968, Greenstein and Arp (1969) traced with the strong Balmer continuum the Balmer series to H_{13} and found an increase in FWHM of H_β and higher members of the series from 8 to 15 Å. Gershberg and Shakhovskaya (1971) in the flare on AD Leo of 2 March 1969 found that the FWHM of H_γ, H_δ and $H + H_\varepsilon$ were 10–13 Å at the width of the instrumental contour of 6.2 Å and near the flare maximum there was "red asymmetry", the long-wavelength wing was longer than the short-wavelength one. During the decay phase this asymmetry disappeared, profile tops flattened, and a decrease of intensity in the center was outlined in some of them. Bopp and Moffett (1973) found practically the same width of $H_\beta, H_\gamma, H_\delta$, and $H + H_\varepsilon$ lines in the flares on UV Cet of 14 October 1972, but strong red asymmetry of structures was found only in the first of the two flares.

Kulapova and Shakhovskaya (1973) recorded FWHM of H_α lines corrected for the instrumental profile at maximum of the flare on AD Leo of 18 February 1971 with $\Delta B \sim 1\overset{m}{.}2$ equal to about 15 Å. The line profile had red asymmetry and during the flare decay phase the profiles of H_β, H_γ, and H_δ had flat tops with a small decrease in intensity in the center.

Joint consideration of observations of more than twenty flares from the Crimea and the McDonald Observatory in the late 1960s–early 1970s showed that there was no correlation between photometric and spectral characteristics of flares: between the flare amplitude and the width of H_α emission line, between the intensity and width of the line, the flare light curve and occurrence of helium lines, the relative contribution of the lines and the continuum.

Using an image intensifier tube echelle spectrograph of the 4-m reflector and a three-channel photometer of the 90-cm telescope of the Kitt Peak Observatory, Schneeberger et al. (1979) observed AD Leo: on 30 March 1977 two flares with $\Delta U = 0\overset{m}{.}40$ and $0\overset{m}{.}35$ and total durations of one and three minutes were recorded. In observations with a spectral resolution of 0.22 Å and a time resolution of 2 min the central intensity of H_α line in the second flare increased by 40%, but the line, whose width in the quiet star was 1.4 Å, remained invariant. Using the same instruments, Worden et al. (1984) recorded six short flares on YZ CMi with amplitudes up to $\Delta U = 1\overset{m}{.}5$, but did not find significant broadening of H_α or H_β in any of them. In the strongest flare of 9 February 1979 the time of the two-fold decay in the U band was 60 s, the total duration of the event was about 4 min, but the smoothly decaying afterglow of H_α lasted for about 2.5 h.

Eason et al. (1992) failed to present the profile of H_α emission in the spectrum of the flare on UV Cet of 8 September 1979 by the Stark curve and presented its as a superposition of two Gaussians, in which the wide Gaussian had a "blue" shift by 70 km/s at a Doppler half-width of 150 km/s. Eason et al. considered this result as a direct evidence of evaporation of plasma at the chromospheric density.

In observations of the flare on AD Leo of 31 May 1981 at maximum ($\Delta B = 0\overset{m}{.}62$) with a spectral resolution of 0.7–1.0 Å Petrov et al. (1984) found the wings of the H_β line up to ±15 Å with red asymmetry. Near the maxima of flares on YZ CMi of 30 January 1981 and 2 March 1981 with $\Delta B = 0\overset{m}{.}5$ and $0\overset{m}{.}7$, respectively, they found the wings of this line up to ±10 Å. Comparing the measured structures with the Stark theoretical profiles, Petrov et al. estimated the electron density as 10^{14}–10^{15} cm^{-3} in the mentioned flare on AD Leo of 31 May 1981 and 10^{14} cm^{-3} at the maximum of the second flare on YZ CMi.

In the spectrogram of the flare on AD Leo of 28 March 1984 with $\Delta U \sim 2^m$ recorded by the 3.5-m telescope of the European Southern Observatory, one can see that all emission lines are considerably broadened (Rodonò et al., 1986). In the spectrum of the flare on AD Leo of 12 April 1985 obtained at the 2.1-m telescope of the McDonald Observatory with a resolution of 3.5 Å, the width of the hydrogen wings at the base exceeded 20 Å at maximum and its broadening lasted for more than two hours, as long as the flare continuum lasted. H_{16} was the last distinct member of the Balmer series, and the broadening of calcium lines was very slight (Hawley and Pettersen, 1991).

In observations of the H_β line in the flare on YY Gem of 18 February 1986 Baliunas et al. (1986) found a doubling of the line width and a four-fold increase of its equivalent width in the spectrum of the B component of the system.

In the flare on YZ CMi of 4 March 1985 with $\Delta U = 1\overset{m}{.}2$ Doyle et al. (1988b) found broad wings at the maximum and especially during the fast decay of the flare in all the recorded Balmer lines. The lines H_γ and H_δ were symmetric, whereas H_6 and H_7 manifested red asymmetry. Before the flare, H_γ and H_δ were presented by a Gaussian with FWHM = 1.5 Å, but flare profiles could not be presented either by one Gaussian or by one or two Stark profiles. The profiles were presented by the sums of two Gaussians, which corresponds to the directed motions at ±250–300 km/s or to turbulence with a velocity of 500–600 km/s in the broad component and of 55 km/s in the narrow component. At these velocities, over the time of one exposure, the motion should throw an appreciable mass toward the corona and produce an X-ray flare. The event was not observed in the flare, thus Doyle et al. proposed to replace one long ejection by a chain of successive ejections in the loops of several adjacent flare nuclei, as is observed on the Sun.

During the six-minute flare on UV Cet of 17 September 1980 with $\Delta U = 2^m$, whose spectra were registered in the range of 3800–5100 Å with a resolution of 1 Å by the 1.9-m telescope in Southern Africa, Phillips et al. (1988) found broadening of the H_β and H_δ lines for about 1 Å, but with noticeable red asymmetry corresponding to directed motions with a velocity of about 100 km/s, which can be attributed to downward motion of chromospheric condensation. Similar red asymmetry of H_α profile was observed in the spectra of flares on EV Lac of 11 September 1986, 14 September 1986, and 10 September 1987 recorded from the Crimea (Gershberg et al., 1991a).

Robinson (1989) failed to present the wings of the Balmer lines near the maximum of the flare on Wolf 424 of 4 April 1987 by the Stark profiles.

Falchi et al. (1990) found broad wings of Balmer line profiles in the flare on V 1054 Oph of 15 June 1987 recorded spectroscopically by the 1.5-m telescope of the European Southern Observatory with a resolution of 6 Å.

Kinematics and dynamics of radiating gas in the flare on AD Leo of 28 March 1984 with $\Delta U = 2\overset{m}{.}1$ and duration of about 50 min (Fig. 29) were studied in detail from the spectra obtained at the 3.6-m telescope of the European Southern Observatory in the range of 3600–4400 Å with a resolution of 1.7 Å and an exposure of 60 s (Houdebine et al., 1990, 1993a,b). Before the flare, faint P Cyg components were found in the structure of CaII K lines, which were associated with dark filaments, typical precursors of solar flares. In the impulsive phase a blue wing extended to 80 Å was revealed in the H_γ line, which corresponds to the emission of matter at about 5800 km/s. This lasted for several minutes as long as the flare continuum weakened and practically disappeared. At the same time, a red shift was observed in the cores of the Balmer lines, in the lines of CaII K and HeI λ 4026 Å, which corresponds to downward motion of chromospheric condensations that could be initiated by the flux of energetic particles with a power of 9×10^{10} erg/cm^2 s. Later, oscillations of the line centroids with a period of 2.7 min and an amplitude of about 95 km/s were found, which Houdebine et al. attributed to a 2×10^9 cm long prominence with a magnetic field of 20 G. The estimates of Houdebine et al. showed that the kinetic energy was 5×10^{34} erg, which was 500 times higher than in solar coronal ejections (CME), and the amount of ejected matter was 40 times greater. It is possible that such events can lead to the loss of matter that has evolutionary significance for the star.

For several well-studied flares, Houdebine (1992) summed up the behavior of emission line widths. The FWHM of high members of the Balmer series during the impulsive phase achieve 20 Å and then monotonically decrease. In the gradual phase even at significant fluxes the broadening of the Balmer lines is insignificant. H_α and H_β display broadening less often than high members of the series. The intensity of lines correlates with the width directly neither during impulsive, nor gradual phases, but broadening was most appreciable near the maximum brightness in the U band, and in both phases the correlation of its broadening with the absolute flux in the U line is outlined. Helium lines are usually much narrower than hydrogen lines – by 2–3 Å. A slightly broadened CaII K line – about 1 Å – does not react to brightness variations during the flare. On the basis of a quantitative analysis of these data, Houdebine concluded that broadening of upper members of the Balmer series was caused by the Stark effect in the medium, whose density essentially exceeded the density in solar flares, while the broadening of the lower members was due to self-absorption. Under the decomposition of the Balmer line profiles from H_γ to H_9 in the spectrum of the flare on AD Leo of 28 March 1984 into two Gaussians, Houdebine obtained FWHM of the broad component from 22 to 17 Å and that of the narrow one, from 6.6 to 5.0 Å. He concluded that the narrow component was caused by a radiatively enhanced chromosphere around the flare, and the broad component, by the luminosity of the flare nucleus arising on deep penetration of particles into the layers with a density of up to $10^{15} - 10^{17}$ cm^{-3} during the first 30 s of the flare. The size of such nuclei was about 10^{18} cm^2, and its luminosity was detected only in the first flare spectrum. Broadening of the helium line λ 4026 Å could be caused by the Stark effect at a density of 4×10^{15} cm^{-3}, and the broadening of the CaII K line to 1.5 Å, by macroscopical motions. Later, comparative analysis of twenty flares on M dwarfs in the range of 3600–4500 Å by Houdebine (2003) confirmed the conclusions and revealed various correlations between the parameters of line and continuous emission during the impulsive and gradual phases of flares of different intensity.

The data on broadening and shift of emission lines in the spectrum of EV Lac were obtained at the 2.6-m Shajn telescope in the Crimea and the 6-m telescope in the Caucasus during a cooperative observation series. In the Crimea, observations of the H_α region with a resolution of 24 km/s near

2.4 Dynamics and Radiation Mechanisms of Flares in Different Wavelengths

the flare maxima on 28 and 29 August 1990 a red asymmetry to 130 km/s was found, and in the flare of 30 August 1990 a blue asymmetry to -100 km/s. The 6-m telescope observations showed that the flare decay on 29 August 1990 in the H_δ line occurred much faster than in H_γ and H_β (Gershberg et al., 1993). On 1 and 3 September 1992 spectral monitoring in the Crimea in the range of 4450–5500 Å with a resolution of 2.2 Å/pixel and with exposures of about 10 min detected small flares with appreciable amplitudes on EV Lac. During these events the FWHM of H_β practically did not change, but full widths of the line at the level of about 20% and 10% of the maximum noticeably increased: up to 12–15 Å wide wings appeared (Abdul-Aziz et al., 1995). Seven profiles of this line in an active star were presented by pairs of Gaussians of various widths: with the FWHM of narrow components from 4.0 to 5.7 Å and the FWHM of broad components up to 21 Å, and the centers of these components in pairs were spaced on the wavelength axis from $+1.3$ to -2.2 Å. Near the brightness maxima the broad component contributed more than half of the total flux of the line and during the decay this share vanished. Then, the same recorded profiles were presented by superposition of three to seven Gaussians with instrumental width and with various shifts along the wavelength axis. In this representation, the root-mean-square scattering of the velocities of components varied from 400 to 700 km/s. Similar results were obtained in the observations of the H_β line in the spectrum of EV Lac in 1994 (Abranin et al., 1998b). Thus, both representations of the observed profiles evidenced kinematic heterogeneity of the matter radiated in flares. Observations of the H_α line in the spectrum of EV Lac during the campaign of 1995 in the Crimea did not reveal an unequivocal correlation between stellar brightness in the U band and W_{H_α}. In many cases, the profile was appreciably asymmetric: the profiles were presented by pairs of Gaussians with the widths of narrow components within 1.2–3.1 Å and broad components within 4.2–5.8 Å and with a relative shift of their centers up to 60 km/s; broad components contained up to 79% of the total flux in the line (Abranin et al., 1998b).

During the strong flare on AT Mic of 15 May 1992, which was observed by the 1.9-m telescope of the South African Astronomical Observatory with a resolution of 0.9 Å and exposures of 2–3 min, strong blue asymmetry in the H_δ and H_8 profiles was found and for the first time, in CaII H and K lines (Gunn et al. 1994b). Radiation in the blue wings of the lines was comparable with that of the central components, and the shift along the wavelength axis corresponded to velocities from 200 to 700 km/s. The preliminary estimate of the density of evaporating matter as 10^{14} cm^{-3}, made with a certain assumption, led to the conclusion that the kinetic energy of the rising flux was much less than the flare radiative losses in the range of 3600–4200 Å. Several days later, the same equipment was used to observe YZ CMi with exposures from two to six minutes (Gunn et al., 1994a). As opposed to the described flare on AT Mic, at the maximum of the flare on YZ CMi of 18 May 1992 broad wings appeared only in the Balmer lines, in the CaII lines this effect was not observed and broadenings of hydrogen lines were symmetric. On decomposition of these profiles into the sums of two Gaussians the widths of the broad component corresponded to velocities of up to 250 km/s. Thus, Gunn et al. assumed that such components were formed by imposing many differently oriented ejections during the exposure. At maximum brightness of the flare on AT Mic of 21 August 1985 with $\Delta U \sim 4^m$ the widths of the $H_\gamma - H_9$ lines near the base increased to 25 Å. This broadening persisted as long as an enhanced continuum was seen, whereas CaII lines did not broaden in this case either (Garcia-Alvarez, 2002).

During spectral and photometric observations of the dM4e star RE J0241-53N from South Africa, Ball and Bromage (1996) recorded a strong flare with $\Delta U = 4^m$ that lasted for more than 2.5 h. Appreciable broadening of the Balmer lines increased in high members of the series and was equal to 8 Å in the H_δ profile. Using the profiles of H_8 and H_9 they estimated n_e as $\sim 2 \times 10^{15}$ cm^{-3}. The lines H_δ, H_8, H_9, and CaII K showed simultaneous Doppler shifts on slight preflare brightening of the star in the U band, and similar simultaneous shifts took place at slight brightenings during the gradual phase of the flare.

When purely flare line profiles of the flare on LQ Hya of 22 December 1993 were represented by two Gaussians, the FWHM of the narrow and broad components of the H_α emission was 59–69 km/s and 190–293 km/s, respectively. The broad component at the impulsive phase provided about 80% of the total flux in the line. The broad component of the H_β profile at the maximum contained 86% of the total flux in the line, and this emission decayed slightly faster than H_α; at maximum brightness both Balmer lines manifested a slight shift of the broad component toward short wavelengths, which was then changed to the shift of another sign (Montes et al. 1999). Simultaneous emission was also recorded for the helium D_3 and λ 6678 Å with broad wings.

In two flares on AD Leo recorded by HST STIS in March 2000, a shift of SiIV λ 1403 Å and C IV λ 1548 Å up to 30–50 km/s toward long wavelengths was detected, which Hawley et al. (2003) interpreted as a result of falling chromospheric condensations.

The set of listed data suggests that there is a significant variation of the behavior of emission line widths in various flares, and one should compare it with the known properties of solar flares, at which half-widths of H_α in bright compact nuclei reached 5–10 Å, whereas beyond these structures they did not exceed 1 Å. When the asymmetry is observed in profiles, a blue anomaly is recorded usually in the premaximum phase, and the red one reaches maximum values at the flare maximum.

2.5 Models of Stellar Flares

Though the mechanisms of radiation of stellar flares practically in all ranges of the electromagnetic spectrum are sufficiently clear, the integral theory of such processes has not been elaborated. As for solar flares, there is only general

2.5 Models of Stellar Flares

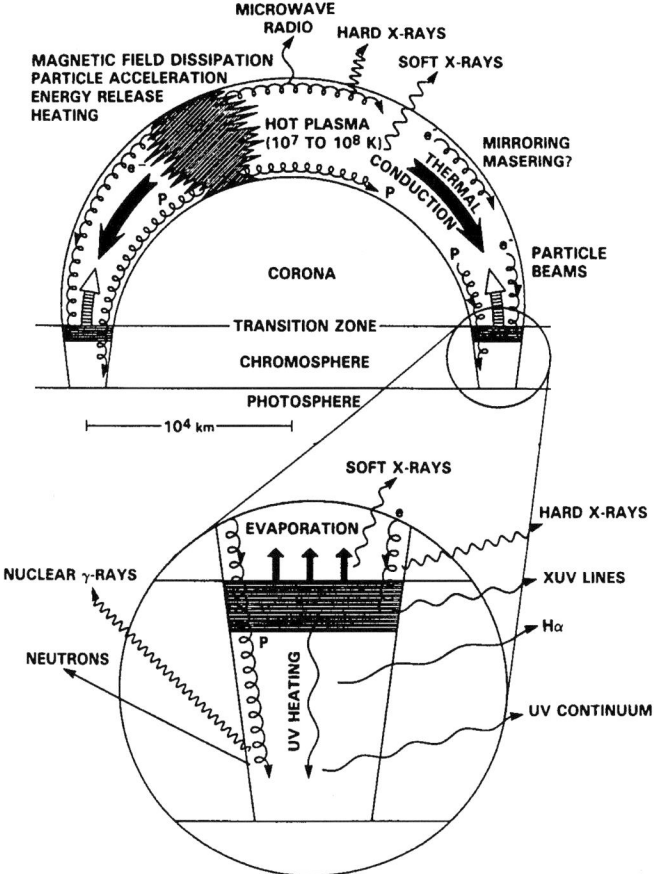

Fig. 59. Scheme of an impulsive solar flare after Dennis and Schwartz (1989)

understanding of the fact that the final source of their energy is magnetic-field energy and the initial phase of a flare consists of the formation of beams of high-energy particles. These beams are directed downward from the corona: going down along magnetic lines, they radiate microwave radio emission and bremsstrahlung hard X-rays, heat the chromospheric matter up to 10–20 MK, evaporating matter fills the magnetic loops and radiates soft X-ray. Since, with increasing temperature, radiative losses of plasma decrease, heating to tens of megakelvins occurs when heat conductivity becomes essential and final relaxation occurs in underlying dense layers. The scheme shown in Fig. 59 matches the vast majority of solar flares, though it still lacks physical fundamentals – the mechanism of flare triggering and the mechanism of formation of fast particles are unknown.

As opposed to the Sun, stellar hard X-rays cannot be recorded directly. Thus, only the features of the profiles in high-resolution spectra of one or two flares evidence fast-particle fluxes. Elaboration of the theory of stellar flares is also complicated by the lack of data on nonradiative components of their total energy and the absence of unequivocal links between the manifestations of flares in different wavelength ranges. The latter should be considered in more detail, though some comparisons of the kind were made above.

Above, we described the long discussion about the role of microflares in heating of stellar atmospheres concerned with the detection of correlation of quiet X-ray emission with the average flare emission and with Balmer emission out of strong flares. One should add that X-rays are emitted from an optically thin plasma of very high temperature, and hydrogen lines are radiated from the medium, which is one thousand times cooler and optically thicker in the discussed lines. Thus, X-ray emission comes from the total volume, while the Balmer emission arises from the surface. The Balmer lines were found right at the beginning or even in preflare brightenings, and soft X-rays – in the gradual thermal stage of flares. Their practical proportionality should be explained.

However, many individual flares do not satisfy the statistical relation between optical and X-ray emissions.

For example, the optical flare on YZ CMi of 4 March 1985 with $\Delta U = 1\overset{m}{.}2$ recorded by the 75-cm and 1.9-m telescopes of the South African Astronomical Observatory and EXOSAT was not accompanied by a simultaneous X-ray flare, there was no correlation of low-amplitude variations in the U band and in X-ray emission either (Doyle et al., 1988b). Above, we presented the description of the observations of EV Lac in October 1985, when rather complex relations of its activity were found in X-rays and in the optical range (Ambruster et al., 1989a). The X-ray flare on Proxima Cen of 6 March 1979 was not accompanied by ultraviolet activity. During simultaneous ultraviolet observations from IUE and ground photometric and spectral observations of UV Cet on 17 September 1980 Phillips et al. (1988) recorded a flare with $\Delta U = 2^m$, but with rather weak (only by 30%) amplification of CIV emission.

There are other examples of simultaneous observations in various wavelength ranges in the previous chapter. The observations were interpreted using the idea of anisotropic radio emission of flares, as well as localization of the bottom parts of flares behind the stellar limb, with the concept of development of these flares in the bottom layers of the atmosphere that are not connected with corona by magnetic fields – such magnetic isolated regions are known on the Sun – the hypothesis about absorption of hard radiation in the top layers of the atmosphere.

Though a general theory of stellar flares is still unavailable, we can critically consider the proposed general schemes and concepts of the processes.

In the 1960s, when the data on stellar flares were accessible only in optical range photometry and from the first results of radioastronomical studies, about ten models of flares were proposed. They were surveyed by Gershberg (1970a). Even for the appearance of ionized plasma above the photosphere of a cool star various schemes were discussed (Gershberg, 1968; Korovyakovskaya, 1972). Here, we shall briefly consider only those initial schemes that are somehow involved in further modeling of the flare activity of red dwarfs and discuss modern theoretical constructions in more detail.

In the first models of flares on red dwarf stars – the model of an asteroid falling onto a star proposed by Hertzsprung in 1924 (Oskanjan, 1964) and accretion of magnetized plasma proposed a quarter of a century later by Greenstein (1950) – flare stars were considered as a gravitating center only. However, the data available now on the dependence of the level of flare activity on the age of dwarf stars removes all doubts as to the decisive role of stars themselves, their rotation first of all, in the nature of this activity.

The concept of a flare as a hot spot was proposed by Gordon and Kron (1949) to interpret the first photoelectric observations. Over half a century it evolved essentially from a photospheric region cooling down due to blackbody radiation to a short-lived dynamic structure in the atmosphere and is now a natural element of the general picture of optical flares.

An alternative to the hot-spot model, the nebular model, was originally only phenomenological, but it was developed to determine the place and physical conditions of the appearance of chromospheric emission in flares. Today, its confrontation with the hot-spot model has been removed, since both components are included in the complex dynamic picture of optical flares.

Ambartsumian's concept (1954) about the eruption of prestellar matter with a large energy potential from the stellar depth initiated the hypothesis of Gurzadian (1965) about fast electrons and Papas' idea (1977, 1990) about electromagnetic solitons. But, if Papas derived only general equations of such solitons, Gurzadian elaborated the hypothesis: assuming that during a flare above the surface of a cool star a cloud of nonthermal electrons appears, he considered the expected consequences, and compared them to the observed properties of stellar flares. This hypothesis was repeatedly published by Gurzadian and drew the attention of many observers. However, it was subject to serious and all-round criticism – for example, by Haruthyunian et al. (1979), Gershberg (1980), Fomin and Chugainov (1980), and Mullan (1990). Leaving aside their forcible critical arguments, we will make only two remarks. First, the efficiency of fast electrons in the birth of quanta seen in the flare is 10^{-5}–10^{-6}; so this concept can hardly be considered as a solution, if it postulates the existence of an unknown energy source exceeding the energy of flares by many orders of magnitude to explain stellar flares. Secondly, if in the mid-1960s one could still argue about fast electrons that were ejected so far from stars that ionization losses were insignificant for them, which is the approximation underlying Gurzadian's theory, after the discovery and research of stellar coronae such reasonings make no sense.

Having considered a number of models, in which radiation of meter wavelength bursts recorded by the late 1960s on UV Cet could be explained by synchrotron radiation of relativistic electrons, Grindlay (1970) estimated the efficiency of these relativistic particles in the generation of X-ray emission due to the inverse Compton effect and nonthermal bremsstrahlung and found that the latter mechanism was more effective for radiation in the energy range $E > 10\,\mathrm{keV}$. However, as stated above, the level of X-ray emission of stellar flares predicted by Grindlay should lead to such a high total radiation of flare stars of the Galaxy that it would considerably surpass the observed X-ray background.

On the basis of simultaneous optical and radio observations of the flare on UV Cet of 11 October 1972 (Fig. 25), during which the radiation in these ranges was appreciably divided in time, Kahn (1974) proposed the model, which for the first time combined different aspects of the event. From the luminosity, mass, and size of the star he estimated thermodynamic characteristics of its photosphere: at convection up to a depth less than $\tau = 2/3$ an effectively radiating layer should be one kilometer thick, have a mass of $16\,\mathrm{g/cm^2}$, and an energy of convective motions capable of generating surface magnetic fields up to 900 G. A recorded optical flare on UV Cet had luminosity of about 2×10^{30} erg/s and a duration of about 10 s. Kahn estimated the total number and energy of relativistic electrons that could provide such luminosity in the lower corona through the synchrotron mechanism. Kahn simulated further development of the event with the radio emission at 408 MHz within the framework of the hypothesis about the emersion of a plasma bubble in the corona that was formed during reconnection of magnetic flux tubes. Assuming that the stellar corona was in the state of a stationary isothermal wind, Kahn estimated the size of the flare as about 2×10^9 cm, the magnetic-field strength in it as about 600 G, the total number of "optical" relativistic electrons as 10^{34} and the rise of the plasma bubble in the corona up to $3R_*$. Within the model, up to 40% of the energy produced by convective motions in the photosphere was spent for the maintenance of the corona. However, polarimetric and colorimetric observations showed that the contribution of synchrotron radiation to the optical flare was insignificant.

Grandpierre (1986, 1988, 1991) formulated and developed the convective model of stellar flares. According to this model, ascending convective cells acquire the speed of sound in subphotoheric layers, shock waves formed ahead of them generate all the observed effects of flares: shock waves cause heating and turbulence in the chromosphere, which are observed as its evaporation, and the fast particles generated by such waves cooperate with the overlying magnetic field tubes, generate electric fields, and produce bremsstrahlung, recorded as flare radiation. The distribution of accelerated particles along the flux tubes to the bases of loops yields the effects considered within the magnetic models of flares. But since traditional convective motions are subsonic, to obtain fast convective cells, Grandpierre postulated the existence of additional fast cells caused by the heterogeneity of thermonuclear burning in the stellar nucleus. However, solar-type activity was observed also in M6–M9 dwarfs, in which no thermonuclear burn occurs.

Apparently, all models of flares on red dwarfs that do not correlate with modern models of solar flares have lost their value now.

When the initial version of the nebular model with the relaxation of homogeneous plasma was found to be unable to present the observed colorimetric characteristics of optical flares, several attempts were undertaken to present these characteristics within the framework of the two-component model.

Mullan (1976a) constructed the model in which the basic component of a stellar flare was a significant volume of plasma of coronal temperature arising at the onset of a flare. Heat conductivity was considered as the basic heating mechanism of underlying layers of the stellar atmosphere. However, such a structure could not provide fast development of flares during the impulsive phase and better corresponded to the gradual phase. But, with the help of the model Mullan estimated the expected ratios of luminosities L_X/L_{optics}.

Kodaira (1977) made calculations within the two-component model to present the characteristics of the strong flare on EV Lac of 3 August 1975 (Kodaira et al., 1976). It was assumed that the hot component contained the main energy store of the flare, which due to heat conductivity was transported down and radiated in X-rays, whereas the cold component cools down due to the effective radiation in the saturated ultraviolet lines and the continuum and in the flat optical continuum recorded on EV Lac. However, further observations did not reveal the strong X-ray emission predicted by Kodaira corresponding to $T \sim 10^8$ K and the expected ultraviolet emission in the range of $\lambda < 1500$ Å.

Grinin and Sobolev (1977, 1988, 1989) stated and developed the idea about the higher energy of fast particles in stellar flares as compared to solar flares. They found that electrons with energies higher than 100 keV or protons with energies of 5 MeV in the total flux of 10^{12} erg/cm^2 can reach the layers with density to $10^{15} - 10^{17}$ cm^{-3} and directly heat the stellar photosphere and initiate continuous flare emission. In their model, deep atmospheric layers were heated simultaneously with fast particles reaching these layers and with optical and ultraviolet quanta generated in higher layers under the passage of fast particles. Later, Sobolev and Grinin (1995) calculated the Stark profiles of H_α, H_β, and H_γ within this model and found that profile wings can expand for 20 Å from the centers of lines. They also well presented the wings of the H_β line beyond 6 Å from its center in the spectrum of the flare on EV Lac of 1 September 1992 recorded from the Crimea. But, according to their calculations, H_α should be more broadened than H_β and H_γ, which apparently does not fit the observations.

Van den Oord (1988) compared the efficiency of excitation of flares by fluxes of electrons and protons and concluded that the efficiency of the latter was higher, since in this case fewer fast particles were required. Furthermore, one should not expect a correlation between optical and microwave bursts under excitation of optical flares by protons, which is supported by the observations.

Later, Grinin (1991) made a detailed review of the role of accelerated particles in solar and stellar flares. Having analyzed the observations, he concluded that in both cases the role of particles was important, but the sections of particle beams and their energy per unit surface in strong stellar flares were an order of magnitude higher than in solar white-light flares. According to his estimates, in stellar flares protons play a considerable or dominating role in the fluxes of accelerated particles, whereas in solar flares, electrons dominate. The lower limit of the energy spectrum of particles in stellar flares is several times higher than the appropriate value in solar flares. Extrapolating variations of the energy spectrum of accelerated particles in solar flares of different intensity to stars, one can expect that in strong stellar flares this spectrum is harder than in solar white-light flares, which shifts the maximum of energy release to deeper layers of the stellar atmosphere and on stars one can expect a higher energy ratio $(E_{\mathrm{optics}} + E_{UV})/E_X$ than on the Sun.

The most detailed analysis of heating and ionization of the chromospheres of red dwarfs by fluxes of protons with regard to the observed shift of Ly_α in the flare spectrum was made by Brosius et al. (1995): according to their estimates, the shift should proceed over 0.1–14 s.

Finally, Grinin et al. (1993) calculated the heating of fast electrons in stellar flares and estimated the expected color indices in the $UBVRI$ system depending on the characteristics of electron beams and the position of the flare on the stellar disk.

In the mid-1960s, the theory of solar flares was supplemented by the magnetohydrodynamic concept of flares: emerging of magnetic flux tubes from subphotospheric layers and reconnection of flux tubes. This process is considered inevitable in counter motions of a plasma with magnetic fields of opposite signs. But, this required enormous sizes of current sheets, where the reconnection should occur, and their insignificant thickness (Syrovatsky, 1976) seem somewhat unrealistic. Gershberg et al. (1987) noted that it was difficult to keep the store of free magnetic-field energy necessary for strong stellar flares in these atmospheric layers of red dwarfs. Later analysis of this mechanism showed that it could not effectively work during primary energy release, but it gives correct estimates of the expected density and temperature of plasma during the gradual phases of flares (Katsova and Livshits, 1988; Forbes, 1988). If a "basic" individual flare is within microflares with an energy of 10^{30} erg, instead of the region of normal flares with energies of 10^{31-32} erg, the problems connected to the release of a large amount of energy under reconnection of magnetic fields become less complicated. For $E > 10^{31}$, it is difficult to construct a model with release of this energy over seconds, but it is easier if the superposition of ten flares weaker by an order of magnitude is suggested.

The question of the primary energy source of solar flares was considered in detail by Mogilevsky (1980, 1986), who advanced the concepts of magnetohydrodynamic solitons. He showed that wave motions in nonlinear and nonstationary magnetized plasma in the convective zone resulted in the formation of magnetohydrodynamic solitons, which, unlike magnetohydrodynamic waves, propagated at super-Alfven velocities practically without dissipation and thus could carry considerable discrete portions of energy and matter away from subphotospheric layers. On flare stars, this mechanism can provide energy for the strongest stellar flares of the considered type (Gershberg et al., 1987). Arising in subphotospheric layers with a density of kinetic energy essentially exceeding that of magnetic energy, these structures quickly break up above the photosphere where the inverse ratio of energies occurs. Disintegration of solitons can generate the fluxes of fast particles necessary for most phenomenological models of flares.

The electrodynamic model of stellar flares dates back to the publication of Alfven and Carlquist (1967), who suggested consideration of a solar flare as a break in the circuit of a strong electric current, which can be caused by a local increase of resistance by many orders of magnitude. Zaitsev and Stepanov (1991) showed that the necessary conditions of the model for flare-energy release were the existence of a strong current, unsteadiness of the process, and the presence of neutral plasma component with different velocity. Thus, the process starts at the loop top and initiates a strong energy release in the chromosphere. The state-of-the-art of the model of an "electric circuit", or the circuit model, and competing electromagnetic models was reviewed by Zaitsev et al. (1994). The advantage of the circuit model over many models with current sheets consists in the fact that the interaction of current-bearing magnetic loops with a filament can provide sufficiently high energy release. As to the mechanism of particle acceleration in the flare, the uniform mechanism that can explain the variety of accelerating processes both in the model of current sheets and in the circuit model apparently does not exist. There is no agreement on the place of acceleration of particles: regions of the lower corona and upper chromosphere are disputed. The role of turbulence induced by fast particles in the distribution of fluxes of such particles, including the possible mode of strong diffusion, was considered in the above review. The studies of Melnik (2000) who discovered quasisoliton beam-plasma structures can bring essential changes to this field. Finally, an electrodynamic approximation connects short-period oscillations of red dwarfs with the fast magnetosonic oscillations of flare loops.

Hayrapetian and Nikoghossian (1989) and Hayrapetian et al. (1990) suggested that flares can result from the pinch effect, arising at emergency of magnetic tubes from the convective zone. The analysis of evolution of a tube, whose part is reconnected at a distance comparable with its transverse size, led to the conclusion about possible development of an isolated magnetic torus, unstable to constrictions, to formation of linear pinch and to the occurrence of a plasma of thermonuclear temperature. This model is an alternative to the model of reconnecting of magnetic flux tubes in current sheets.

Following the pioneering work by Kostjuk and Pikel'ner (1974), who were the first to show that the flux of fast electrons arising at the onset of a solar flare resulted in fast formation of a high-temperature plasma layer at some depth of the atmosphere, from which two hydrodynamic disturbances propagate upward and downward, Katsova et al. (1980) and Livshits et al. (1981) found that in denser stellar atmosphere at the total power of the flux of particles to 10^{12} erg/cm^2 s a downward disturbance should consist of a jump of temperature and a shock wave (Fig. 60).

The shock wave goes ahead of the temperature jump slowly moving downward, and due to strong radiative relaxation behind the wave front a jump of density by two orders of magnitude is formed. At thicknesses from 1 to 10 km, which is less by an order of magnitude than that on the Sun, this compression can reach an appreciable optical thickness in the continuum, which will provide appreciable continuous radiation of a stellar flare. In this downward disturbance one can expect a density of matter up to 10^{15} cm^{-3} and a temperature up to 10^4 K, in this case the profiles of Balmer lines should be broadened by the Stark effect and have red asymmetry, as is observed in many stellar flares. Physical parameters of this source of continuous radiation appeared to be close to those of the stationary model calculated by Grinin and Sobolev (1977), but now it is formed in an essentially nonstationary situation caused by short injection of the flux of fast particles and is displaced from the region of the temperature minimum in the stellar chromosphere.

The gasdynamic model was repeatedly and successfully used to interpret various observed facts. Thus, Katsova and Livshits (1986) showed that this model provided a natural explanation for some features of the light curves of impulsive stellar flares, in particular the fact that all structures on them were longer than hundreds of milliseconds. Analyzing the burst of CIV λ 1548/51 Å emission during the flare on EV Lac of 6 February 1986 recorded by Astron (Burnasheva et al., 1989), Katsova and Livshits (1989) interpreted it as the direct observation of the process of shock-wave formation with radiation accompanied by explosive evaporation of the chromosphere. The analysis of spectral and photometric observations of the impulsive flare on YZ CMi of 4 March 1985 within the framework of this model allowed Katsova et al. (1991) to estimate the area of the source of continuous optical radiation as $> 5 \times 10^{17}$ cm^2 and to explain the rather fast evolution of the emission of the Balmer decrement.

Later, Katsova et al. (1987) redesigned their initial calculation algorithm for the gasdynamic model by including separate consideration of electron and ion temperatures of plasma and recalculated the consequences of injection of a separate 10-s beam of fast electrons with an energy of 3×10^{11} erg/cm^2 s on

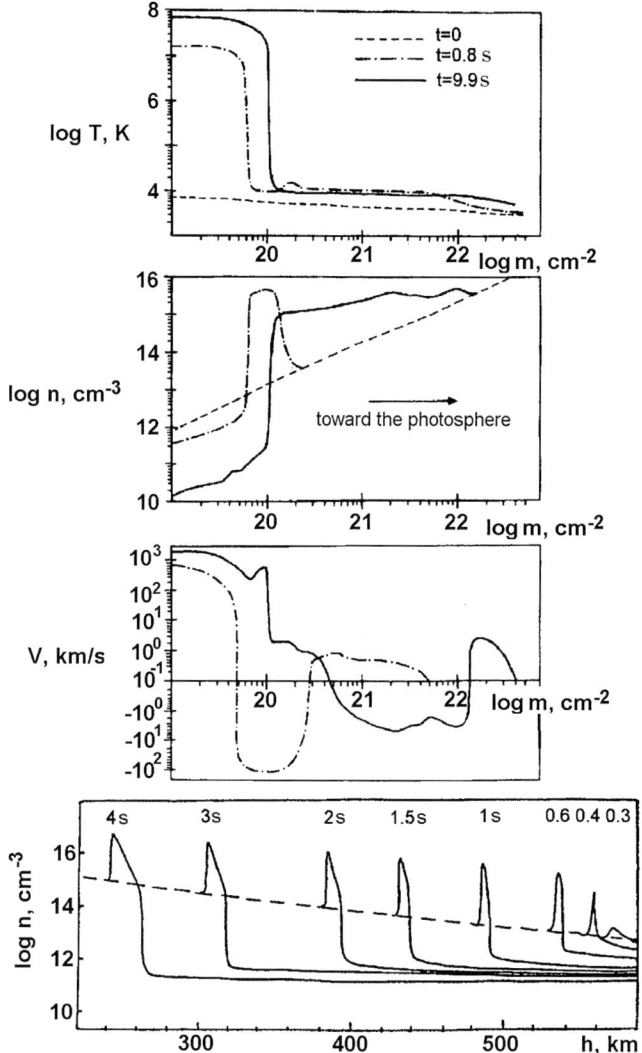

Fig. 60. Distributions of temperature, velocity, and density of matter in the gasdynamic model of a flare (Katsova, 1981b) and shock-wave propagation in the atmosphere of AD Leo (Katsova et al., 1997)

the magnetic tube throughout its height from the photosphere to the corona. Calculation showed that during the first 0.1–0.2 s the plasma was considerably heated at the level of the upper chromosphere and, as in previous calculations, two gasdynamic disturbances were formed and propagated up and down from the region of initial heating. In the disturbance going down, the resulting compression produces optical radiation. The compression has a nonuniform temperature structure that varies appreciably with time. The obtained model is applicable to flares that are not too strong. It agrees with the results of Cheng and Pallavicini (1991) for the top part of the tube and refines their results on the chromospheric level. The bottom diagram in Fig. 60 presents the calculation of the density profile of the downward disturbance in the atmosphere of AD Leo (Katsova et al., 1997), the velocity of this disturbance smoothly varies from 160 km/s 0.4 s after the beginning of the flare to 18 km/s 8 s later. The density of matter in this compression achieves 10^{16} cm^{-3} and conditions are close enough to LTE, the optical thickness becomes close to unity 2 s after the beginning of the flare and is kept at this level for several seconds. In the upward disturbance, the ion temperature at first considerably lags behind the electron temperature and only after 6 s do they equalize at the level of 30 MK, and the velocity of this disturbance changes from 200 km/s 0.28 s after the beginning of the flare to 1000 km/s after 8.5 s. The evaporating chromosphere fills the loop at a speed of 5×10^{18} particle/cm^2 s. If a flare is modeled as a sequence of fast injections, the proposed algorithm enables simultaneous calculation of optical, EUV, and soft X-ray emission of the heated tube. Katsova et al. found quite good agreement of the calculations with characteristic properties of flares in these wavelength ranges. With the help of this algorithm, Livshits and Katsova (1996) considered the conditions of occurrence of extreme ultraviolet emission during pulsed heating of the magnetic tube and found three different regions of formation of such radiation with different time characteristics. Then, Katsova et al. (1999b) considered the influence of stationary X-ray emission of the atmosphere on the condensation responsible for optical emission of the flare on AD Leo. They found that the allowance for this radiation noticeably improves the compliance with observations regarding the intensity of H_α and CaII K lines and continuum, since because of X-ray absorption a background source of continuous and line radiation is formed. After evaporation of the chromosphere the efficiency of fast particles falls in this place, so the pulsed heating is needed in a new place to restore such structures.

Katsova et al. (2002) analyzed the flare on EQ Peg of 23 June 1994 observed in X-ray and optical ranges: at its maximum several fast optical bursts took place that were followed by gradual decay. An X-ray maximum started four minutes after the optical one. The ratio of luminosities L_X/L_{optic} grew during the flare from 0.24 in the first peak to 15 in the transition to slow decay. Gasdynamic modeling of bursts enabled estimation of the size of the source of optical continuum as 7×10^{18} cm^2 and the amount of evaporated plasma in the corona. The calculated mass of coronal plasma and the observed rather slow decay of flares, suggest that this event combined the properties of compact and two-ribbon flares.

Using the formalism of Cram and Mullan (1979) proposed for calculations of the chromospheres of quiet stars, Cram and Woods (1982) calculated the structure of the chromosphere of a stellar flare, assuming that a quiet stellar atmosphere with $T = 3500$ K, $\log g = 4.75$, and with a solar abundance of elements was influenced by some heating mechanisms that changed the temperature variation with height. Calculations were executed for a plane-parallel geometry, hydrostatic balance, non-LTE ionization of hydrogen, and LTE ionization of other elements. From six constructed models, i.e., from the dependences of temperature and density of matter on column density, Cram and Woods calculated the equivalent width of H_α, H_β and H_γ and flare continuum in the range from 500 to 10 000 Å. They obtained different anticipated variations of the initial spectrum of the star. As the transition region shifts downward, which can be due to the influence of heat conductivity or in the gasdynamic model, a strong and narrow H_α emission with a weak central inversion should appear. The temperature increase in the upper chromosphere can occur because of the action of X-ray or ultraviolet emission or low-energy electron fluxes and lead to strong H_α emission with deep central inversion and a large Balmer jump. Temperature growth in the lower chromosphere and upper photosphere can arise under the effect of fluxes of high-energy electrons or protons, or hard X-rays, or dissipation of magnetic energy in the depth of the atmosphere. At very large energy release in the chromosphere the obtained model is localized on the two-color diagram near to observed flares.

Natanzon (1981) apparently was the first to apply the concept of stationary coronal loops proposed by Rosner et al. (1978) for the Sun to the analysis of stellar flares developing in such structures: from the then known X-ray observations of five flares on four stars he estimated a 5–6-fold temperature rise in stellar flares as compared to flares in solar loops, an excess of density of matter by many tens of times, slightly higher luminosity L_X, and an emission measure higher by 4–5 orders of magnitude.

Gasdynamic models of X-ray stellar flares also evolved from numerous models of solar X-ray flares, which from the beginning of the 1980s considered the response of coronal loops to various temperature variations. The coronal models considered the chromosphere as a boundary layer and the mass holder, and all attention was focused on the events occurring in the optically thin top part of the loop.

If we ignore the differences in the suggested structures of local magnetic fields, the basic qualitative difference in the energy of solar compact and two-ribbon flares will be that in the first case only initial pulse heating is suggested, while in long events of the second sort there is additional heating in the decay stage. A source of such heating, according to the solar model by Kopp and Pneumann (1976) and Kopp and Poletto (1984), is the process of reconnecting of magnetic flux tubes in the vertical current sheet formed at the break of initially closed loops during the primary strong energy release. The reconnection results in restoration of closed structures. Even for the one-dimensional case this model is described by a system of nonlinear equations with partial derivatives and supposes only numerical solutions. They yield the rate of energy release at each moment of time as a result of reconnection, density and temperature of the flare plasma, position and the speed of rise of the point of reconnection of the magnetic flux tubes. The resulting model depends on the product of the flare size and magnetic-field strength, thus absolute values of flare parameters can be obtained only by an independent estimate of one of the factors.

Poletto et al. (1988) applied the model of magnetic reconnection to analyze the decay phases of two long stellar flares: the light curves of the flare on EQ Peg of 6 August 1985 recorded by EXOSAT in the ME region and those of the flare on Proxima Cen recorded by the Einstein Observatory on 20 August 1980. Within the framework of this model the total energy of each flare and descending branches of their light curves were presented. Adding the Saar and Linsky (1985) estimates of the strength of photospheric magnetic fields on dMe stars, Poletto et al. estimated the area of the flare on EQ Peg as about 1% of the stellar surface at a loop height of $0.18 R_*$ and the density of flare matter from 6×10^{12} to 4×10^{11} cm^{-3} and on Proxima Cen as about 0.2% of the stellar surface at the loop height not higher than $0.09 R_*$ and a density of about 1×10^{12} cm^{-3}.

For the preliminary analysis of stellar flares attributed to the compact type, Kopp and Poletto (1990, 1992) proposed point-wise (zero-dimensional) analytical models that provided the time change of thermodynamic parameters of plasma averaged over the whole magnetic loop of the flare. Within the framework of these models they concluded that fitting of input conditions and loop geometry enabled flares with $EM = 10^{51} - 10^{53}$ cm^{-3} to be obtained, and that strong flares could arise only in large loops of up to 2×10^{10} cm.

Simultaneously with Poletto et al. (1988), Reale et al. (1988) carried out independent analysis of the flare on Proxima Cen of 20 August 1980 within the framework of the hydrodynamic model of solar compact flares modified for stellar events using the one-dimensional and one-liquid Palermo–Harvard program. As opposed to the previous model with reconnection of magnetic flux tubes, here pulsed heating of an originally static coronal magnetic loop from inside was suggested. By fitting model parameters, Reale et al. reproduced the observed flare light curve, the distribution of energy in its spectrum and the maximum temperature under energy dissipation over 700 s of 6×10^{31} erg in a semicircular rigid loop with $h = 5 \times 10^9$ cm, which is slightly less than the stellar radius and $n_e \sim 10^{11}$ cm^{-3}. Introduction of supporting heating enabled further reduction of the loop size. It is necessary to remember that Haisch et al. (1983) considered a rather slow decay of this flare as evidence of its similarity to solar two-ribbon flares.

To reveal the common properties of hydrodynamic models as applied to stellar analogs of solar compact flares irrespective of any actual events, Cheng and Pallavicini (1991) executed extensive numerical calculations of the models for dMe stars with the help of one-dimensional two-liquid programs aimed at determining the real area of allowable variations of the parameters involved in the calculations. At first, they constructed the equilibrium models of magnetic loops 2×10^9 and 8×10^9 cm of size expanding from the chromospheres to coronae under doubled solar acceleration of gravity, energy absorption providing loop balance, and a chromospheric temperature of 9000 K. Then, assuming five-minute energy release of different power at the loop top, they traced its further evolution. It appeared that the energy release resulted in fast heating of gas to 40–100 MK, fast – over several seconds – a shift of the temperature front downward, and heating of the chromosphere that caused its evaporation with a velocity of 400 to 2700 km/s, to lowering of the transition region and the compression of matter in the base of the loop. In the decay phase gradual or catastrophic cooling of gas occurs. If the energy release stops at the flare maximum, conditions for the formation of condensations are formed in cooling down plasma, the condensations with a velocity of several hundreds of kilometers per second reach the loop base. If the energy release proceeds after the flare maximum, the plasma cools down slowly and condensations are not formed. These calculations reproduce the observed X-ray flares, in general their intensity, light curves, and average coronal temperature. In particular, a flare with EM up to 10^{52} cm^{-3} can be obtained in a rather small loop with $h \sim 2 \times 10^9$, and the loops with $h \sim 8 \times 10^9$ cm are enough for the flare with $EM \sim 10^{53}$ cm^{-3}. Cheng and Pallavicini managed to reproduce the observed correlation of the calculated total energy during the development of flare E_X and EM in the range of 10^{50}–10^{53} cm^{-3}. As stated above, Katsova et al. (1997) refined these calculations for chromospheres.

Within the hydrodynamic model of compact flares, Reale et al. (1993, 1997) developed the technique of analyzing the decay phase for various gravity accelerations, loop sizes, maximum temperatures, and characteristics of additional heating. On the basis of extensive calculations of such models with the obvious account of supporting heating they found the empirical relation between the time of an e-fold decay of the flare brightness and the inclination track of the flare on the plane ($\log n_e$, T), where n_e was estimated from $EM^{1/2}$; the value of inclination also provided information on the presence or absence of supporting heating during the flare: at a slope of 1.5 there was no heating and at a slope of 0.5 this heating determined the rate of flare decay. In particular, their analysis showed that there was no such heating in the flare on Proxima Cen of 20 August 1980. In the Yohkoh observations of the Sun, Reale et al. (1997) tested the technique of quantitative analysis of stellar flares from their decay phase: comparison of direct measurements of the sizes of solar flares and the estimates with the help of the (EM, T) diagram yielded quite good agreement.

Continuing the studies of Reale et al. (1997), Reale and Micela (1998) elaborated the algorithm for direct estimate of the sizes of coronal loops from ROSAT PSPC data. The algorithm, without modeling of an actual flare, used earlier calculations of many hydrodynamic models of compact flares with the obvious account of supporting heating. They found that errors of such estimates were ~20% for 40 000 readouts and ~70% for 1000 readouts.

Application of the algorithm was illustrated by the example of flares on AD Leo and CN Leo. Using the parameters of the above flare on YY Gem of 29/30 September 2000, on the basis of the (EM, T) diagram, Stelzer et al. (2002) constructed the model of a flare loop with a length of 2×10^9 cm, the density increased to 5×10^{11} cm^{-3}, and noticeable heating during the decay phase. The short length of the loop – of about $0.6 R_*$ – points to the affiliation of the flare to one of the components of this binary system rather than to intercomponent medium.

Observations of a strong flare on Algol during an eclipse allowed consideration of two competing models of the decay of stellar flares – quasistatic cooling in long loops and heating in the decay phase – and established that the latter yielded more exact estimates of loop sizes. Therefore, an algorithm for analyzing flares with supporting heating was applied to a strong flare on EV Lac recorded by ASCA and to five flares on AD Leo recorded by different instruments. In all cases, rather compact loops with $h \sim (0.1 - 0.5) R_*$ and the essential effect of supporting heating were obtained. In this case, the light curve should be determined by the evolution of additional heating rather than by free cooling of flare plasma (Favata et al., 2001).

Reale (2002) compared all the above modeling schemes of flare-decay curves and provided the list of their applications to real events.

Within the hydrodynamic approach, Reale et al. (2004) thoroughly modeled the strong flare on Proxima Cen of 12 August 2001 (Fig. 25). They found that its burn, maximum brightness, initial decay and secondary maximum could be presented by the two-component model: a single pulse-heated loop responsible for the main maximum and an arcade of fewer hot loops of similar large size that flared half an hour later and were responsible for the secondary maximum; pulsed heating occurred at the level of loop bases, with smooth decay in their coronal parts.

Among flares that were the most successfully observed using different methods at different wavelengths, several events enabled rather detailed schemes of these processes to be obtained. Let us consider some of them.

The flare on BY Dra of 24 September 1984 with $\Delta U = 0\overset{m}{.}22$ was observed in the visible range and in soft X-rays by de Jager et al. (1986). Data analysis showed that they could be interpreted as "burning out of a well" 2×10^{17} cm^2 in the section with a bottom temperature above 25 000 K in the stellar atmosphere during five-minute impulsive phases by a beam of fast particles. Gas heated up to 10^7 K with an average density of 2×10^{11} cm^{-3} goes up covering the area of up to 10^{20} cm^2 and slowly relaxes radiatively during the gradual phase.

De Jager et al. (1989) analyzed a strong impulsive flare on UV Cet of 23 December 1985 with an amplitude of more than 5^m in ultraviolet, which was observed photometrically, in spectral, within the whole optical range, and in microwave ranges, and in X-rays. They arrived at the conclusion that the model of this event could be constructed analogously to solar impulsive flares with a vertically extended "cool" component – an optically thick source of visual radiation – about 700 km, which is higher by 1.5–2 orders of magnitude than on the Sun, a temperature of about 16 000 K and electron density of 10^{15} cm^{-3}, whereas the structures with temperatures of 40 and 10 MK and density of $(2-5) \times 10^{11}$ cm^{-3} were found in the hot component. On the whole, this supports the above model of "burnt wells".

The flare on YZ CMi of 3 February 1983 with $\Delta U = 3^{m}\!.8$ described earlier by Doyle et al. (1988b) was observed in optical, ultraviolet, and radio ranges. The analysis of the data by van den Oord et al. (1996) led to the conclusion on downward shift of the transition region, an effective blackbody temperature of optical radiation of about 9000 K, and an increase of emission measure in the whole range of temperatures from 10^4 to 10^7 K. Radio emission at 6 cm that appeared seven minutes after the optical maximum was interpreted as an optically thick synchrotron in the arcade of loops growing under the current sheet. The absence of emission at 20 cm was attributed to the absorption in the upper corona layers. In general, the flare was compared with a two-ribbon solar white-light flare with reconnection of magnetic flux tubes.

Fisher and Hawley (1990) considered the evolution of the parameters of coronal plasma averaged over a magnetic loop subjected to heating. Confining themselves to heating times longer than the time of passage of sound through the loop, they reduced the system of nonstationary nonlinear equations with partial derivatives to an ordinary differential equation, in which the average rate of cooling of a homogeneous-thickness coronal loop was described by its global parameters – size and column density. Without allowance for the role of fast particles, this equation enabled calculations for the flare on AD Leo of 12 April 1985: the change of average temperatures and emission measures of a loop at evaporation, self-similarity, and condensation stages distinguished by Fisher and Hawley at the gradual phase based on the value of the ratio of plasma heating and cooling rates. Then, Hawley and Fisher (1992) calculated five models of magnetic loops from the level of the photosphere to the transition region under strong flare energy release in the corona. The basic variable parameter of the designed models is coronal temperature at the loop top. At higher temperature, the transition region determined by the balance of heat conductivity from the corona and radiative losses of optically thin ultraviolet lines is narrowed and displaced downward to higher column densities. But the structure in the region of the chromosphere and temperature minimum is determined by the X-ray and ultraviolet emission field from the top of the loop rather than by heat conductivity. In these calculations of plasma temperatures, non-LTE functions of cooling on atoms of hydrogen and ions of calcium and magnesium were taken into account. The intensities of chromospheric lines calculated using the equations of transfer as functions of the same temperature at the loop top enabled estimation of this temperature variation from the observed change of chromospheric lines and their comparison with the prediction of coronal models of flares. Comparing these calculations with observations for flares on AD Leo of 12 April 1985, Hawley and Fisher (1992) found quite good agreement of the change of coronal temperatures found from the H_γ line and the model of evolution of a coronal loop, but there was regular divergence in the analysis of the CaII K line, indicating the necessity of utilization of additional heating of flare matter one and a half hours after the beginning of its decay. This suggests that the event was similar to a two-ribbon rather than a compact solar flare.

Hawley and Pettersen (1991) compared the characteristics of two flares on AD Leo: of 28 March 1984 with $\Delta U = 2^m.1$ and of 12 April 1985 with $\Delta U = 4^m.5$. In spite of essential differences in luminosity and duration of these events, they found close similarity in the ratio of line and continuous emissions at different phases of these flares, in the character of development of many spectral details. Thus, they concluded that the structures of flares were substantially similar, and the basic difference was concerned with the sizes and duration of the process of energy release. This conclusion was confirmed in analyzing cooperative observations of AD Leo in March 2000, when eight flares of different intensity were registered: they had strongly different amplitudes and durations but equal blackbody temperatures and close ratios of continuous and line radiation (Hawley et al., 2003).

To interpret the long and strong flare on AU Mic of 15 July 1992 recorded in the extreme ultraviolet from EUVE (Fig. 30), Cully et al. (1994) used the model of quickly extending structure that was earlier used to represent the light curves of optical flares. Fast expansion in the model sharply lowered the density of matter, keeping its temperature rather high for EUV radiation. But later, Katsova et al. (1995) proposed an alternative model: heating of ejected plasma in the vertical current sheet that provided luminosity at sufficiently high density.

Robinson et al. (2001) observed the flare on YZ CMi on 22 December 1994 in optical, microwave, and EUV wavelength ranges and found rather different light curves. At 16:20 UT a sharp maximum of a strong optical flare with $\Delta m \sim 3^m$ and a total duration of about 40 min occurred. Simultaneously, a radio flare started at 3.5 and 6 cm, whereas increased radiation at 13 cm was observed for one and a half hours before and after the optical maximum and reached a maximum level after the decay of higher-frequency radio emission. The flare in EUV was even longer: from 13 to at least 24 UT with a maximum two hours after the optical flare. To interpret these observations, Robinson et al. used the concept of expansion of a magnetic loop, excitation of Alfven waves in it in resonance with convective motions. Dissipation of Alfven waves generates intensive turbulence, which causes heating of the plasma with radiation in the EUV range. Turbulence also generates Langmuir waves, which transform into electromagnetic radiation and accelerate particles. But, in further expansion of the loop it gets out of resonance with convective motions and its heating terminates. When a critical height is achieved, a current sheet is formed in the loop, which corresponds to an optical flare. Instabilities developing in this phase result in the formation of a plasmoid from the loop top. The plasmoid moves away from the star: it slowly cools down and produces long luminosity in EUV. High-frequency microwave radiation comes from the dense part of the loop connected to the stellar surface, whereas low-frequency radiation comes from extending plasmoid.

In calculations of the above semiempirical model for the quiet chromosphere of AD Leo, Mauas and Falchi (1996) constructed the semiempirical hydrostatic models for two moments of the strong flare on AD Leo of 12 April 1985 described by Hawley and Pettersen (1991). Certainly, there are no a priori proofs of the applicability of the hydrostatic models to stellar flares. This can be justified only by numerous constructs of such models for solar flares that provided consistent results for various events.

Mauas and Falchi (1996) analyzed the data for the moments 15 and 20 min after the beginning of this strong flare: they considered fluxes in $UBVR$ bands and in the H_β, H_γ, H_δ, H_8, λ 4227 Å CaI and CaII K lines. Calculations were carried out for independently chosen values of the filling factor of the stellar surface by the flare. The results are presented in Fig. 61 by curves B and C that correspond to filling factors of 5% and 1%. Spectral features arising in the upper layers of the chromosphere are, in general, better represented in calculations within the model of the flare occupying 5% of the stellar surface, whereas the model of flare covering 1% of the stellar surface better represents the features arising at greater depths. In the 1%-model a temperature plateau is distinctly found at a level of about 8200 K that extends from $\log m = -1.1$ to -2.3. This model ensures the best consistency of observed and calculated continuum of the flare. Observations carried out 20 min after the beginning of the flare, when the continuum appreciably weakened, were presented by the model with even smaller area.

Baranovskii et al. (2001b) calculated semiempirical models for six moments of active state of EV Lac: from observations described by Abranin et al. (1998b) the phase of slow decay of the flare of 29 and two fast flares of 31 August 1994, and three moments of decay of the slow flare of 4 September 1995 were selected. The algorithm of Baranovskii et al. (2001a) used for the elaboration of the semiempirical models of the quiet chromosphere was supplemented by an additional unknown, the flare size. The models that were the best in representing the amplitudes of flares ΔU, ΔB, and ΔV, and equivalent widths of emission lines H_β and H_γ of 1994, and amplitudes in the same bands and equivalent widths and line profiles of H_α of 1995 were sought in the calculations. Assuming that the temperature plateaux known in the solar chromosphere and in the quiet chromosphere of EV Lac (Baranovskii et al., 2001a) are formed only in a sufficiently stationary environment, the first Crimean calculations of flare models did not involve the plateau. However, the models did not yield positive results: in adjusting the profile

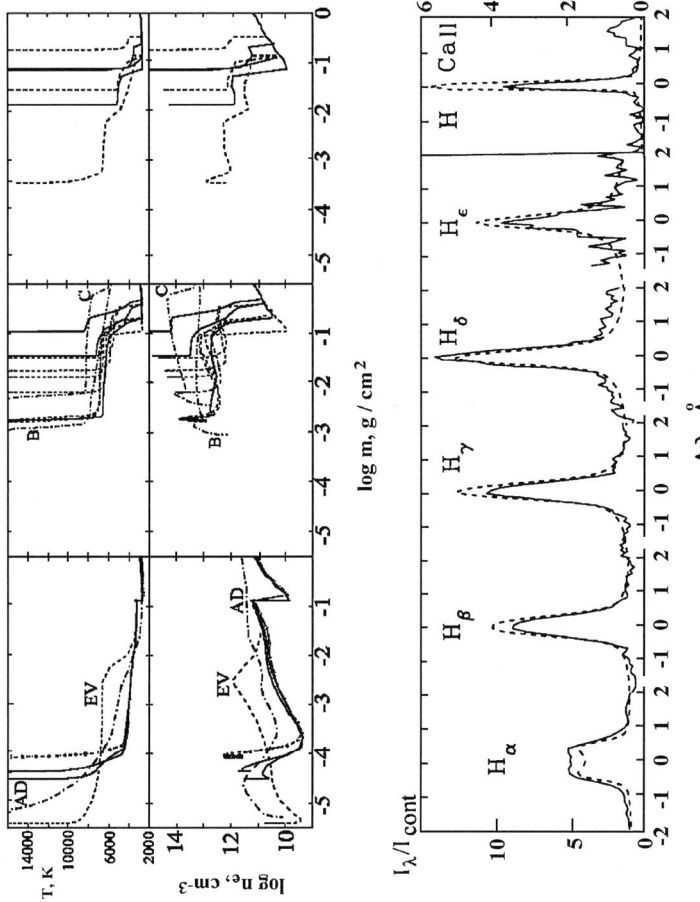

Fig. 61. Semiempirical models of emission sources on EV Lac of 30 August 1994 and representation of emission line profiles by total radiation of active regions, flares, and microflares (Alekseev et al., 2003)

of H_α the intensity of the flare continuum was too small, the temperature rise raised the continuum, but the line was too broad. Moreover, the models allowed too high $\Delta U/\Delta B$ ratios due to a large Balmer jump. Thus, it was concluded that one of the necessary conditions of successful representation of the observations was the inclusion of an extended temperature plateau in the sought flare model. It should be noted that the temperature plateau in the chromosphere appears in the gasdynamic models of solar flares as well (Abbett and Hawley, 1999). The quantitative characteristics of the models obtained by Baranovskii et al. (2001b) are as follows: at increased stellar brightness $\Delta U = 0\overset{m}{.}11$–$0\overset{m}{.}70$ optical depths are $(2$–$6) \times 10^9$ in the center of Ly_α and 200–2000 in H_α, the bottom and upper borders of the plateau are $\log m = -1.0$–-0.7 and -3–-2, respectively, the electron density on them is $2 \times 10^{12-13}$ cm^{-3}, the temperature gradient on the plateau varies from 140 to 400 K at absolute temperatures of 5500–6900 K, and the size of flares is 1.3–4.4% of the stellar surface. Thus, the structures responsible for hydrogen emission of flares on EV Lac are less extended in depth, but have greater electron density than in the quiescent chromosphere. The upper limit of the temperature plateau in flares is at a height of 200–300 km, whereas in the quiescent chromosphere it is equal to 700 km and 1800 km in the quiescent solar chromosphere. On the whole, this picture of hydrostatic flares is consistent with the concept of Grinin and Sobolev (1977) about the localization of a source of optical luminosity of flares in the depth of the stellar atmosphere.

Then, the Crimean algorithm was used for quantitative analysis of the spectrum of the impulsive flare on EV Lac of 30 August 1994 at 23:19 UT and stellar spectra at 15-min intervals before and after the flare recorded at the echelle spectrograph of the Nordic telescope. It was mentioned that the range of wavelengths was recorded from H_α to H CaII. Owing to the high resolution, broad-profile wings were discovered out of the photometrically recorded flare. The results of multicomponent modeling are presented in Fig. 61. The left top diagrams present the models of active regions on EV Lac out of the flare (solid curves) and during the flare (dashes with asterisks). The broken curve (EV) corresponds to the model of homogeneous chromosphere on EV Lac (Fig. 14). The dashed curve (AD) shows the model of AD Leo chromosphere after Mauas and Falchi (1994). The top center diagrams show the models of three flares by solid curves, which in sum present the lines profiles of the flare on EV Lac of 30 August 1994; curves B and C show the models of two moments of the strong flare on AD Leo constructed by Mauas and Falchi (1996), broken lines represent the models of flares on EV Lac from the Crimean observations (Baranovskii et al., 2001b). The right top diagrams show the models of microflares. The models of these structures showed that the depth of flares and microflares overlapped, microflares have slightly lower densities and temperatures. On the other hand, on average, the densities of microflares are two orders of magnitudes higher than those of active regions. They occur at depths that are hundreds of times deeper than active regions. Comparison of the Crimean models and the B and C models of Mauas and

Falchi reveals a substantial similarity in the characteristic sizes of flares, the fact of the existence of a temperature plateau and the depth of occurrence, though these are flares on different stars with various initial data and independent computing programs were used for calculations. However, comparison of the observed and calculated profiles in the lower panel of Fig. 61 with Fig. 14 shows that the stellar spectrum during a flare is represented less successfully than for the quiescent state, probably because the hydrostatic model is insufficient for describing impulsive flares.

Jevremovic et al. (1998b) tried to construct the semiempirical hydrostatic models of three weak flares on the dM5.5e star Gl 866 using the similarity of a designed and observed Balmer decrement as the consistency criterion. They used a spectra in the range of 3400–5400 Å with a spectral resolution of 3 Å obtained at the South African Astronomical Observatory, the model of the photosphere for $T_{\text{eff}} = 2900\,\text{K}$ with the solar content of elements and the condition $dT/d\log m = \text{const}$ in the chromosphere. The model of the quiet chromosphere was successfully constructed with the help of the accepted algorithm. Then, the response of the calculated chromosphere to the bombardment by electron beams with energies from 20 to 200 keV was calculated, which resulted in the change of its height, electron density, and the degree of hydrogen ionization. However, the observed Balmer decrements could not be reproduced in such a way.

Later, Garcia-Alvarez et al. (2002) used this approach to analyze the flare on AT Mic of 21 August 1985 with $\Delta U \sim 4^m$ and a total energy of about 4×10^{33} erg within 3600–4500 Å. They estimated the flare-plasma temperature as 20 000–40 000 K and the density as 10^{13}–$10^{14}\,\text{cm}^{-3}$ and its geometric thickness as 200–500 km.

Though there are various phenomenological models of stellar flares, till now no generally accepted understanding of the process of primary energy release in flares – the mechanism of transformation of magnetic energy into thermal and kinetic ones – has been gained. Over recent years, new directions in the research of this problem have appeared: the concept of avalanches (Lu and Hamilton, 1991; Charbonneau et al., 2001) and the theory of dynamic conductivity of magnetized plasma (Pustil'nik, 1997, 1999). Both deal with a power form of the energy spectrum of flares. However, they have not been elaborated to the extent that would enable a comparison with the observations of actual flares. On the other hand, a fractal structure of plasma in the magnetic field, for which the body of experimental data in solar studies permanently increases (Mogilevsky, 2001), should lead to the revision of some conclusions on the magnetohydrodynamics of a continuous medium.

3
Long-Term Variations in Activity of Flare Stars

3.1 Activity Cycles

Cyclicity in solar spottedness was discovered in the mid-19th century by the German amateur astronomer Heinrich Schwabe. Today, the 11-year cycle of solar activity, along with sunspots and solar flares, is the most well-known phenomenon of solar activity. The body of data collected thus far on the cyclicity of solar activity is huge. It was found that all other characteristics of solar activity change in parallel with the spottedness of the Sun, which is characterized by Wolf's numbers, including the number of spots and their groups, and by the average latitude of spots. These characteristics are the size and number of active regions of the chromosphere, the frequency and intensity of flares, the structure of the solar corona and the intensity of its X-ray emissions, the characteristics of the solar wind, including the parameters of the interplanetary magnetic field. All manifestations of solar–terrestrial relationships and most geophysical phenomena are synchronized with the solar-activity cycles. Based on the close relation between the level of solar activity and the thickness of annual tree rings and the content of carbon isotope ^{14}C in them, the so-called dendrochronological scale was constructed, which helps in tracking the variations of solar activity over eight millennia. There are geological structures evidencing the existence of the cyclicity many millions of years ago.

An average solar cycle is 11.2 years, whereas the duration of individual cycles varies from 7 to 17 years. The degree of variation of all quantitative characteristics of activity varies essentially from one cycle to another. The change in sign of the total magnetic field and those of magnetic fields in the pairs of leading and tail spots on each solar hemisphere during neighboring cycles suggests 22-year solar magnetic cycles.

Two other important features of solar activity connected with spottedness are essential for the subsequent discussion. First, in addition to the 11-year cycle, there is the so-called secular cycle that lasts for 80–90 years. It is suspected that a longer cycle of several centuries exists. Secondly, in 1645–1716 solar spottedness was tens of times lower than usual: spots could hardly be detected at maximum phases and only few aurorae, one of the major geophysical effects that allows determining the solar-cycle phase, were seen from the Earth. It was the so-called Maunder minimum of solar activity. Apparently, this happens once every 2–3 centuries and in total such periods cover 1/3 of the time.

All manifestations of cyclicity of solar activity are associated with various changes of the subphotospheric magnetic field of the Sun, and the solar 11-year cycle was a starting point for the elaboration of the solar dynamo model. Analogously the cyclicity of stellar activity is one of the basic directions in studying the generation mechanisms of stellar magnetic fields. As the data on the cycles of stellar activity were accumulated, attempts were undertaken to select one of the models of stellar dynamo. The studies of the cyclicity in stellar activity use known characteristics of solar activity, but in its turn they support solving this important problem of solar physics.

Further, we consider available data on stellar cycles at different levels of stellar atmospheres.

The first experimental data on possible activity cycles of red dwarfs were obtained in interpreting their broadband photometry after ten years of intense photoelectric observations.

Based on the Crimean observations of BY Dra brightness in the B and V bands, Chugainov (1973) suspected the 8–9-year cycle of spot formation with minimum brightness in 1965. Vogt (1975) doubted this period. The data of Oskanian et al. (1977) on the 20-year interval did not contradict Chugainov's conclusion. From photoelectric observations of the star from the Stephanion Observatory, Mavridis et al. (1982) concluded that in the late 1970s the star entered a new minimum stage with a depth of not less than $0^{m}.3$ and duration of not less than 14 years. Observations by Cutispoto (Rodonò, 1987) and Pettersen et al. (1992b) during the next decade confirmed the 14-year cycle. The data of Panov et al. (1995) evidenced the end of this cycle that, however, was completely dissimilar to the minimum of 1965–67. Further, Mavridis et al. (1982) suspected slow oscillations in quiet brightness of EV Lac in the B band with an amplitude of $0^{m}.3$ and characteristic time of about 5 years. Mahmoud (1993a) confirmed this period on the basis of expanded observations. From the annual average flare radiation in the U band Andrews and Marang (1989) estimated the cyclicity of the dMe star FL Aqr as 10–18 years. On the basis of long UBV observation series, Aslan et al. (1992) found an 8–14-year activity cycle of DH Leo. From a 10-year series of photometric observations of the rapidly rotating K2 dwarf LQ Hya Jetsu (1993) and Cutispoto (1993) estimated the duration of its activity cycle as 6.2 years. On the basis of observations of EK Dra in 1983–94 in the V band, Dorren et al. (1995) suspected the photometric variability cycle of 12–14 years. From UBV observations by two automated telescopes in 1989–98 Messina et al. (1999a) found the 3.9-year cyclicity period of the average brightness change for the K0 V star DX

3.1 Activity Cycles

Leo of the Pleiades moving group. The phase of this cycle correlates with the duration of the photometrically determined axial rotation period. Based on 11-year VRI observations of more than 40 M dwarfs, Weis (1994) revealed 2.7- and 2.9-year photometric periods for Gl 213 and Gl 876, respectively, for which other manifestations of activity were not known. From the 22-year observational series of VY Ari Strassmeier et al. (1997) suspected a 14-year photometric cycle and from the 14-year observations of LQ Hya, a 7-year photometric cycle, which did not contradict the above results of Jetsu and Cutispoto. Continuing these studies, Oláh et al. (2000, 2001) analyzed long-term photometric observations of 10 active stars to find multiperiodicity. For three program dwarfs the following cycles were obtained: LQ Hya – 11.4 and 6.8 years, V 833 Tau – 6.5 and 2.4 years, and BY Dra – 13.7 years. Berdyugina et al. (2002) found 3 periods of spottedness on LQ Hya: a 5.2-year period of switching of active longitudes, a 7.7-year period of change of amplitudes of brightness oscillations, and a 15-year period of variations of average stellar brightness during which the period of stellar rotation, i.e., the differential rotation period, was displayed. Kövári et al. (2004) obtained slightly different values – 13.8, 6.9, and 3.7 years – from a 21-year observational series.

Messina et al. (1999a) and Messina and Guinan (2002) analyzed 10-year observations of six young G-K dwarfs and established photometrical periods: 6.7 years for BE Cet, 5.9 years for κ^1 Cet, 13.1 years for π^1 UMa, 9.2 years for EK Dra, 5.5 years for HN Peg, and 3.2 years for DX Leo. In addition, a secondary period of 2.1 years was found for π^1 UMa. A confident correlation of the brightness amplitude during the cycle and the Rossby numbers was revealed.

One should mention unexpected results obtained by Eaton et al. (1996) in the numerical experiment on the cyclic activity of RS CVn-type stars. They showed that in the case of differential stellar rotation the spots of moderate size and with a characteristic lifetime of about one year and randomly distributed over the stellar surface could yield a light curve, which can be accepted as evidence of cyclicity.

Other parameters considered in searching for activity cycles were the frequency and energy of stellar flares.

In studying EV Lac in the B band, Mavridis et al. (1982) suspected the changes of average frequency and total energy of flares, parallel to variations of the average stellar brightness. However, no periodicity in their distribution was revealed in a more extensive sample of flares on this star in the U band (Alekseev et al., 2000). Pettersen et al. analyzed 241 flares on AD Leo and suspected an 8-year period of flare frequency (Pettersen et al., 1986a), but in considering flares on AD Leo over two decades they did not find variations of the average annual flare amplitudes and energies by more than a factor of 2 (Pettersen et al., 1990a). Pooling together the observational results for UV Cet flares obtained in 1966–88 in Chile, the USA, and Bulgaria, Pettersen et al. (1990b) selected 808 events with an amplitude over 1.4 and considered their time distribution. Following the χ^2 criterion, they rejected the assumption of the constancy of flare frequencies and suggested a 10–15 year cycle. From the observations in the B band, Mahmoud (1993a) suspected the 6.6-year cycle of UV Cet activity. Based on the 20-year observation series of BY Dra in the B band, Mavridis et al. (1995) noted the maximum of average flare energy in 1973–75 and subsequent increase of this value in 1987. As stated above, Alekseev and Gershberg (1997a) found periodic variations of the index of the power energy spectrum of flares on EV Lac with a characteristic time of about 7.5 years without appreciable changes in the level of flare activity.

Mirzoyan and Ohanian (1977) found evidence of variability of the activity level of flare stars in the Pleiades, which they attributed to cyclic activity.

The cycles of stellar activity at the photospheric level can be studied using glass libraries. The Harvard collection of negatives keeps recorded images of the sky since the late 19th century.

Using the Harvard Collection, Phillips and Hartmann (1978) investigated the behavior of average annual brightness of BY Dra type stars: BY Dra itself, CC Eri, YZ CMi, and AU Mic. The images were measured on an iris photometer using more than a hundred negatives from the early 20th century to the 1960s. As a result, a distinct photometric wave was found on BY Dra with an amplitude of $0^{m}\!.3$ and duration of 50–60 years. Considering the binarity of BY Dra, the amplitude of the found maximum in the early 1930s was determined as $0^{m}\!.5$. Similar behavior was found for CC Eri, whereas two other stars did not show brightness cyclicity. Then, Hartmann et al. (1981) studied 225 negatives of the Harvard Collection to determine the long-term brightness dynamics of the dK5e star BD + 26°730 (= Gl 171.2 = V 833 Tau) and for the late 1930s rather confidently found cyclic changes with an amplitude minimum of about $0^{m}\!.5$ with a cycle a duration of about 60 years. In addition, they suspected overlapping cycles of lower amplitude with a duration of about 10 years. Subsequent photoelectric observations by Oláh and Pettersen (1991) yielded results that did not contradict the photographic light curve by Hartmann et al.

From a 20-year series of photographic observations of the flare star HII 2411 in the Hyades Szecsnyi-Nagy (1986) suspected that its activity cycle was 10–15 years.

Bondar (1995, 1996, 2000) undertook the most extensive research: using the photograph collections of Moscow University, and Odessa and Sonneberg observatories, she considered the behavior of average annual brightness levels of 40 dKe-dMe stars. In eye estimates and iris photometer measurements of about 5900 negatives obtained in 1896–1992, she found a variation of the average annual brightness with amplitudes from $0^{m}\!.3$ to $1^{m}\!.0$ and characteristic lifetimes from 3 to 60 years on 21 stars. On eight of them cyclic spottedness could be ascertained confidently, on the others, only presumably. Amplitudes higher than $0^{m}\!.5$ were found in four red dwarfs: V 833 Tau, PZ Mon, EI Cnc, and BY Dra (Fig. 62). The light curves constructed by Bondar for BY Dra and V 833 Tau are in good agreement with the results obtained from the Harvard glass library, and the light curve of PZ Mon, with the data from the

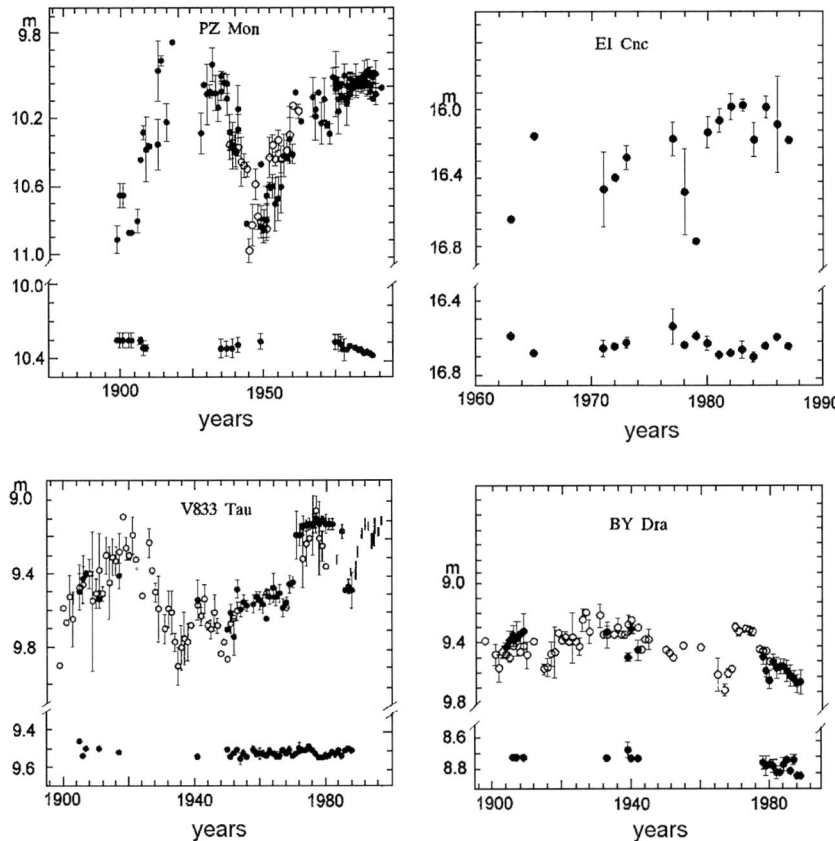

Fig. 62. Long-term brightness variations of red dwarfs: *dark circles* after Bondar (1996), *light circles* on PZ Mon diagram after Wachmann (1968), *light circles* on V 833 Tau diagram after Hartmann et al. (1981), *light circles* on BY Dra diagram after Phillips and Hartmann (1978); *vertical strokes* on PZ Mon diagram from photoelectric measurements by Alekseev and Bondar (1998) The light curves of comparison stars are given at the bottom of each diagram (Bondar, 2000)

Heidelberg collection. The variability amplitudes found by Bondar exceeded the values obtained earlier from shorter intervals and used for constructing of spottedness models of such stars. In Fig. 63, the cyclicity parameters estimated by Bondar are compared with other stellar characteristics. One can see that long activity cycles are typical of stars with a rotational period less than five days (Fig. 63a). Later, this result was confirmed by Messina and Guinan (2002). From less homogeneous data a similar result was obtained earlier by Vogt (1983) and from the chromospheric data for a sample of earlier stars it was suspected by Saar and Baliunas (1992a). Figures 63b and 63c show that the amplitudes of average annual brightnesses above $0\overset{m}{.}5$ are characteristic

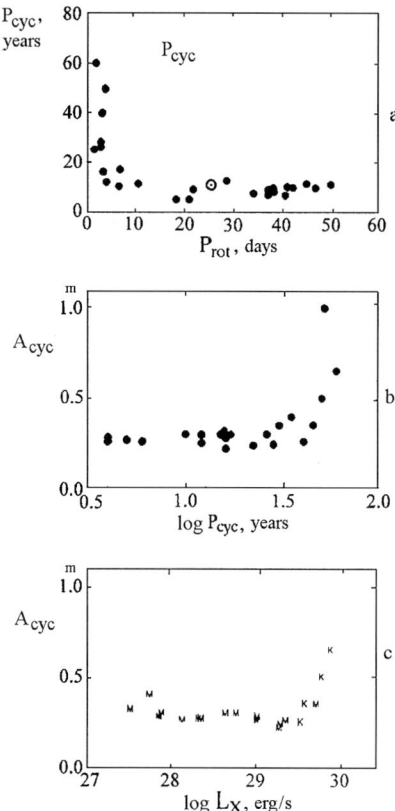

Fig. 63. Parameters of long-term brightness variations and global properties of red dwarf stars: (**a**) axial rotation periods and duration of activity cycles; (**b**) duration of activity cycles and the amplitudes of brightness changes; (**c**) luminosity of K and M dwarfs in soft X-rays and the amplitudes of optical brightness changes (Bondar, 2000)

of stars with cycles longer than 30 years and with luminosity in soft X-rays above 3×10^{29} erg/s. High values of L_X are mainly found for K dwarf stars.

The most reliable and numerous initial data for the cycles of stellar activity are the results of long-term monitoring of CaII H and K emission in the spectra of single late main-sequence stars started in the 1960s by O. Wilson at the 100″ telescope of the Mount Wilson Observatory. In the first paper from the series of works published within this program Wilson (1978) presented the measurement results for calcium emission in the spectra of 91 F5–M2 dwarfs over 9–11 years, during which a measurement accuracy of up to 1–3% was achieved. For 13 program stars the cyclicity of changes of calcium emission was confidently found, for 12 it was established presumably. The duration of found cycles varied from 7 to 14 years. Further, 20% of the studied stars demonstrated the constancy of calcium emission, the others, significant nonsmooth variations. Then, Vaughan (1980) found regular differences in the variability of calcium emission among the stars considered by Wilson. On older stars with rather weak emission the changes were smooth enough, similar to the solar cycle, and the emission bursts were short-lived, as on the Sun, at maximum phases of these cycles. On younger stars with rather strong emission the variation proceeded rather chaotically and, as a rule, no cyclicity was revealed. Then, Vaughan and Preston (1980) added four or five stars with cyclic variations of calcium emission to the Wilson list. On the basis of these observations Vaughan et al. (1981) concluded that 10–12-year cycles were typical of stars whose axial rotation periods exceeded 20 days, and that the duration of these cycles did not correlate with rotational velocity. However, later, Dorren and Guinan (1994) in observations of ultraviolet lines of the transition region found 12-year cycle on "the young Sun", HD 129333 with an axial rotation period of 2.7 days.

Between the groups of young and old stars on the planes $\log S$, $B - V$, where S is a value proportional to the sum of equivalent widths of CaII H and K lines, a certain deficiency of objects was found, which was called the Vaughan–Preston gap. Durney et al. (1981) associated it with fast reduction of the dimensionless dynamo number N_D at a certain stage of stellar evolution. The number is the ratio of generating and dissipative members in the magnetohydrodynamic equation. It determines the efficiency of the dynamo mechanism and is functionally connected with the Rossby number: $N_D \sim \text{Ro}^{-1/2}$. They showed that, according to theoretical expectations of Parker (1971) at large N_D small-scale rapidly variable magnetic fields were generated, whereas at low N_D global magnetic fields of solar type with slow cyclic variations were excited.

After the late 1970s the studies within the Wilson program were actively continued by O. Wilson disciples at the 60″ telescope of the Mount Wilson Observatory and regularly yielded new results with an ever-increasing time interval. Simultaneously, the criteria for selecting cyclic variations of chromospheric emission were specified and the number of stars with such variations was determined more precisely.

When activity cycles were revealed on 27 program stars, Vaughan (1983) compared the distribution of their durations with those in solar cycles since 1740 and found a similarity of these distributions, but some stellar cycles were shorter than the shortest solar cycles.

The analysis of the data obtained over 18 years using the power-spectrum formalism, in general, confirmed the difference in variability of calcium emission on active and nonactive stars found by Vaughan (1980) But the difference was not very large: activity periods of 2.6, 3.8, and 12 years were found for rather active dwarfs. In the sample of stars with activity periods found from the power spectrum, no correlations of cycle durations with the Rossby number or the amplitude of changes in calcium emission were revealed. The periods shorter than five years were found only on F6-G1 stars, about seven years – only on K dwarfs, and periods of 10 years and more were found on stars of all spectral types (Baliunas and Vaughan, 1985)

After successful establishment of the correlation between the intensity of calcium emission and the Rossby number (Noyes et al., 1984a), Noyes et al. (1984b) on 13 slowly rotating late stars, including the Sun, found the correlation between the durations of activity cycles and Rossby numbers as

$$P_{\text{cyc}} \sim \text{Ro}^{1.3\pm0.5} . \tag{64}$$

However, Maceroni et al. (1990) considered the distribution of durations of the cycles of 60 Wilson program stars and did not confirm the relation (64), but found an obviously asymmetrical distribution of cycle durations with maxima of about 6 and 11 years. As stated above, separate consideration of various spectral types of stars revealed regular growth of cycles from F to G and further to K stars. Thus, the suspected bimodality of the general distribution can be a result of summing of different distributions with noncoincident maxima.

On the basis of a 20-year observational series, Baliunas and Jastrow (1990) analyzed the distribution of magnetic activity on 74 G1–G8 dwarfs with mass and age close to those of the Sun. The histogram of S had a precise two-peaked structure: a broad component, whose gravity center practically coincided with the average value on the Sun during the solar cycle, and distinctly separated the narrow component, whose gravity center was appreciably less than the minimum value of S on the Sun during the cycle and very close to the value of S in solar regions with zero magnetic activity. The width of the broad component of the histogram slightly exceeds the range of solar values of S during the cycle. The stars of the broad component displayed activity cycles, while the stars of the narrow component did not show cyclicity. It is natural to consider the stars in the narrow component of the histogram as being at the stage of the Maunder minimum, while the clear split of the components of the histogram means that the transition from the stage of activity observed now on the Sun with well-expressed cyclicity to the stage of Maunder minimum occurs very quickly. This fast transition can be observed directly on HD 3651, where, after the cycle maximum of 1977–78, S decreased below the values recorded earlier, and in the further change of this value the traces of cyclicity disappeared. The ratio of the areas of the histogram components is close to 3:1, which corresponds to the ratio of times spent by the Sun in two stages of different activity. Soderblom and Clements (1987) arrived at a conclusion that the Maunder minima occurred very infrequently on young stars.

For a more stringent analysis of S measurements, Horne and Baliunas (1986) applied a search algorithm for periodicities with respect to nonuniform time series with the estimate of probability of false periodicity, formalized the selection of noise peaks on the power spectra from a useful signal, and estimated a number of independent frequencies, signal-to-noise ratios, and a number of measurements required to determine one and two periodicities in the time series. Gilliland and Baliunas (1987) showed that the basic source of noise in the determination of stellar cycles was rotary modulation of active regions stochastically distributed over the stellar surface.

In 1991, the Wilson program included 99 stars: 36 F, 38 G, and 25 K. Within a strict approach to the establishment of periodicity, according to Horne and Baliunas (1986), for 10–15% of sample stars periodic activity was not revealed, for 40% chaotic variations or rather probable false periodicity were found, for 15% – long-term trends or periods longer than 25 years. Only for 30 stars – 5 F, 10 G, and 15 K dwarfs – were activity cycles confidently determined. As a rule, these stars, as the Sun, are characterized by a fast rise to maximum and a slower decline to minimum. Saar and Baliunas (1992a) compared the periods of activity of the stars with $B-V$ color indices, depths of convective zones, activity indicators S and R_{HK}, axial rotation periods, and Rossby numbers. They did not find a precise correlation, though they noted a significant range of values P_{cyc} at low P_{rot} and narrowing of this range at $P_{\mathrm{rot}} > 30$ days. Considering the amplitudes of cyclic variations, Saar and Baliunas found maximum values of A_{cyc} near the K2 spectral type and a systematic decrease of the maximum amplitude with growing R_{HK}.

For the active G dwarf κ Cet with a chromospheric cycle of about 5.6 years, Saar and Baliunas (1992b) compared the measurements of magnetic field and chromospheric emission over four seasons and revealed a weak correlation between these values, similar to that on the Sun.

Preliminary analysis of the results obtained over 25 years within the O. Wilson program for variations of calcium emission at time intervals more than a year were published by the team of 27 researchers, including the recently deceased initiator (Baliunas et al., 1995). Examples of the long series for S are presented in Fig. 64. The full sample included 111 stars; for 52 of them, including the Sun, the cycles were found, for 31 stars – constant emission or linear trends, and for 29 – acyclic variability. The found periods varied within 2.5 to 25 years, but all "good" and "excellent" periods thus classified based on the probability of false periodicity exceeded seven years. Solar cycles shorter than seven years did not occur for 250 years. Probably, the point is that short periods are less steady and they do not persist even over 25 years. From 25-year statistical data, Donahue (Saar et al., 1994a) found that the average logarithm of stellar age for stars with irregular changes of S was 9.03, for stars with multiperiodic changes – 9.22, for stars with cyclic oscillations – 9.46, and for stars with constant values – 9.86. On stars with cyclic changes calcium emission $\log t < 9.5$ and on stars with constant emission $\log t > 9.6$. Cyclic activity appears on G and K dwarfs at $\log t \sim 8.8$ and on F stars at $\log t \sim 9.3$. According to the results of 30-year observations, 60% of program stars had confident or assumed activity cycles, 25% had acyclic variability, and 15% had constant calcium fluxes (Baliunas et al., 1998).

Soon et al. (1994) considered together the 250-year observational series of the Sun and 25-year observations of the Wilson program stars and found the general regularity: systematic decrease of the amplitude of activity $\Delta R'_{HK}$ with increasing ratio $P_{\mathrm{cyc}}/P_{\mathrm{rot}}$. From the same data, Baliunas and Soon (1995)

390   3 Long-Term Variations in Activity of Flare Stars

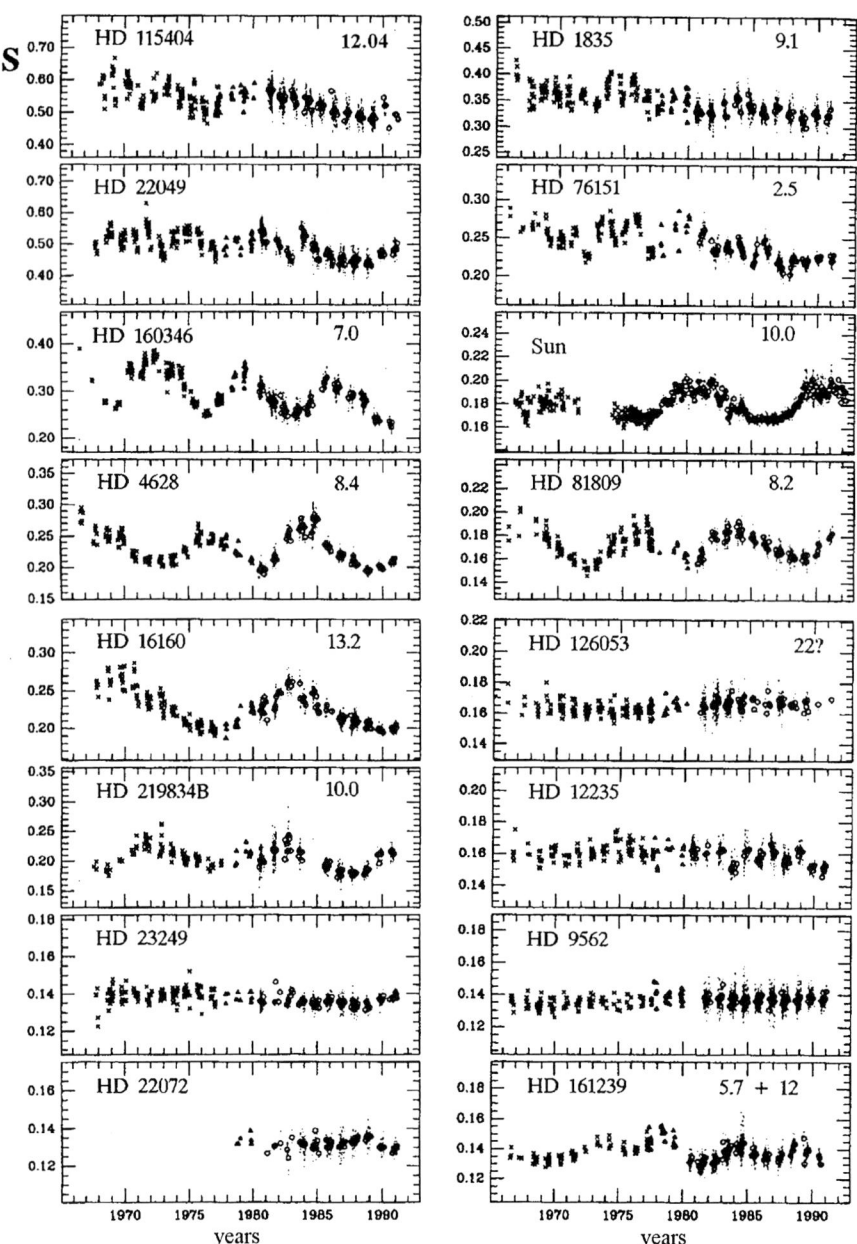

Fig. 64. Long-term variations of calcium emission from 25-year observations within the framework of O. Wilson program. Estimates of cycle durations in years are specified in top right corners of the diagrams (Baliunas et al., 1995)

found an inverse correlation between the cycle duration P_{cyc} and the average activity level $\langle R'_{HK} \rangle$ and a direct correlation between the amplitude S and photometric amplitude during the activity cycle. Baliunas et al. (1996) divided the Wilson program stars, for which cyclic activity was ascertained, into groups of more and less active stars using $\langle R'_{HK} \rangle$ and P_{rot} and found that for the second group of stars, which included the Sun, the following expression was valid:

$$\Delta R'_{HK} / \langle R'_{HK} \rangle \sim (P_{\text{cyc}} / P_{\text{rot}})^{-1.35 + 0.35 / -0.65} . \tag{65}$$

Splitting into the high- and low-activity groups was kept when plotting the stars on the plane (X-ray luminosity, dynamo number). Later, Brandenburg et al. (1998) and Saar and Brandenburg (1999) compared the ratios $P_{\text{cyc}}/P_{\text{rot}}$, Rossby numbers and the level of chromospheric activity $\langle R'_{HK} \rangle$ on 21 stars with well-determined values of P_{cyc} and different levels of activity, including the Sun. They found that on the planes $(P_{\text{cyc}}/P_{\text{rot}}, \text{Ro})$ and $(P_{\text{cyc}}/P_{\text{rot}}, \langle R'_{HK} \rangle)$ the stars were also divided into groups of young active and old nonactive stars. For each group, Brandenburg et al. (1998) found clear power dependences between compared values and a power dependence between $\langle R'_{HK} \rangle$ and Ro that were common for all considered stars. The obtained relations were interpreted as initial growth of the ratio $P_{\text{cyc}}/P_{\text{rot}}$ with stellar age, then a sharp reduction of P_{cyc} by about six times at the age of 2–3 billion years and the subsequent new increase of the ratio following a $t^{0.35}$ law. A sharp reduction of the cycle duration corresponds to the division into young and old stars on the basis of the level of calcium emission mentioned by Vaughan (1980). Some older stars have two periods, and Gleisberg's secular solar cycle satisfies this law.

Frick et al. (1997) used the method of wavelet analysis to survey the Wilson program results. Strictly speaking, the conventional Fourier method is not fairly adequate to the considered cyclic activity of stars with an irregular change of variables and can result in false periodicity. Frick et al. reduced the method of wavelet analysis to the form in which the finiteness of studied time series and missing data did not much influence the results. They applied the constructed algorithm to 25-years observational series of four stars: HD 10 476 with normal cyclicity, HD 201 091 with changing period, the Maunder minimum star HD 10 700, and HD 3651 entering the minimum. The calculations done with the updated wavelet analysis algorithm compared with earlier calculations by Baliunas et al. (1995) confirmed the basic conclusions, but revealed new important details: much lower noise in the obtained periodograms in all cases, a confident change of activity period on HD 201 091 over 25 years from 6.6 to 8 years, and less certain smooth oscillations of the period of HD 10 470 within 10.2–9.7 years. Nothing was added to the characteristic of the activity of HD 10 700; the period of HD 3651 of about 14 years was confirmed, but the reliability of the value was rather low.

In the mid-1990s, spectral observations within an expanded O. Wilson program, the HK project, were recommenced at the upgraded 100″ telescope.

In parallel, an extensive photometric program was started with automated telescopes providing an accuracy of better than $0\overset{m}{.}001$ (Baliunas et al., 1998).

During the solar cycle various indices of the activity level can reach extreme values at different times. Gray et al. (1996) studied this aspect using the example of several G–K dwarfs. The level of calcium emission was considered as an indicator of magnetic activity, photometric data as indicators of temperatures and luminosity, the ratio of depths of two close absorption lines of vanadium and iron as a temperature indicator, and line bisectors as an indicator of stellar granulation. The 10-year observations of G8 ξ Boo A proved that magnetic activity preceded variations of all other considered parameters: photometric brightness – by 1.4 ± 0.4 years, color index – by 1.5 ± 0.5 years, ratios of line depths – by 1.8 ± 0.3 years, and line bisectors – by 2.1 ± 0.3 years. Similar studies were run for several other stars. In all cases, the lead of magnetic activity was revealed and the dependence of the delay of temperature changes on effective temperature was suspected: from three years for the G0 dwarf β Com to 0.3 years for the K2 star ε Eri.

Between 1984 and 1995, Radick et al. (1998) carried out precision photometric observations of 34 F5–K7 stars within the HK project, whose level of chromospheric activity varied from five-fold to a half of the activity level of the Sun. The analysis of the obtained data showed that the amplitudes of variability of photospheric and chromospheric radiation – at short and long time scales – of the sample stars and the Sun were connected by the power ratios with the average level of chromospheric activity $\langle R'_{HK} \rangle$. Young and more active stars were found to weaken with increasing chromospheric emission, i.e., their optical brightness changed due to dark spots, whereas low-activity stars, such as the Sun, brightened, i.e., their optical brightness varied due to bright flocculae.

It is obvious that the CaII H and K lines are the most convenient but not the only chromospheric lines, which can be used to study activity cycles. Thus, Larson et al. (1993) for 12 years conducted spectral observations of the K5 dwarf 61 Cyg A in the near IR range and from variations of the equivalent width of the CaII λ 8662 Å line found a distinct activity period of 7.22 years.

Saar (1998) thoroughly studied the stars in the state of flat activity, i.e., whose S was invariant for many years. He showed that not all of them could be considered as very old objects, in which the dynamo mechanism had become ineffective, for example, because of too slow rotation. But there were stars even younger than the Sun among them, which should be considered as Maunder minimum stars. Saar found that chromospheric CaII and CII lines of these stars corresponded to the basal level presumably identified with acoustic heating of the chromosphere and amplified toward earlier stars, while radiation fluxes from the transition regions and coronae were lower than on the stars with cyclic activity, grew toward later spectral types, and were independent of rotation. Apparently, formation of these fluxes was a weak nonacoustic process, which did not display variability and was more likely dependent on the depth of the convective zone. This process could be the turbulent (distributed) dynamo in the convective zone, which is apparently effective to a certain extent on all stars and is noticeable under the weakening of a solar-type cyclic dynamo.

As stated above, the amplitude of oscillations of X-ray emissions of the Sun during the solar cycle exceeds by one–two orders of magnitude the amplitude of calcium-emission oscillations, but the data on activity cycles of stars in X-rays are much poorer than the data on such cycles of chromospheric emission. This is explained by a much lower variability of active stars in X-rays as compared to the Sun. The second and the most important explanation is that no X-ray studies analogous to the O. Wilson program have been undertaken, since space vehicles operated for 2–3 years only and even the "long-liver" ROSAT did not provide required long-term observation series. Nevertheless, attempts to reveal cyclicity from X-ray observations were undertaken.

Haisch et al. (1990b) considered ultraviolet and X-ray observations of Proxima Cen during four epochs – in March and August 1979, in March 1984, and March 1985 – and found a synchronous decline and the subsequent rise of CIV, MgII, and soft X-ray emission, which they considered as the manifestation of the activity cycle.

The central region of the Pleiades was analyzed by Schmitt et al. (1993b), based on the ROSAT sky survey. They revealed 24 X-ray sources, 20 of them were already discovered by the Einstein Observatory. But Schmitt et al. found changes of X-ray luminosity up to one order of magnitude between the observations separated by a time interval of 10 years, which could not be explained by flares or rotation. Schmitt et al. (1993) associated them with activity cycles. However, later, Gagné et al. (1994a, 1995) considered the variability of X-ray sources in the Pleiades at intervals of 1 and 16 days, 12 months and 10 years. They revealed variability of 22 of 44 bright sources, of which 12 were flares: 33% of the found variable sources were variable on the time scale of 16 days, 64% within 12 months, and 55% within 10 years. The closeness of the two last values put in question the conclusion of Schmitt et al. (1993b) on cyclic activity as a cause of long-term changes of the level of X-ray emission.

Hempelmann et al. (1996) considered the properties of X-ray emission from single F–K dwarfs divided into 3 groups: stars with constant, periodically changing, and chaotically varying calcium emission. They found that average X-ray luminosities in these groups regularly differed: $\langle \log F_X \rangle = 4, 5,$ and 6 in the first, second, and third group, respectively, i.e., the stars with irregular changes of F_{HK} were the most active in X-rays. To determine the cyclicity of X-ray emission, they calculated the phases of chromospheric cycles at the moment of X-ray observations and calculated the differences between the observed and expected values of F_X, which were determined from the dependence $F_X(\text{Ro})$. These differences had a slightly asymmetric curve, similar to the change of Wolf's numbers during the solar cycle, which made it possible to assume the existence of coronal cycles synchronous with chromospheric ones.

Later, Hempelmann et al. (2003) studied two long-term series of ROSAT HRI X-ray observations and HK monitoring of the components of the 61 Cyg system carried out in 1993–98. Though the interval of X-ray observations was noticeably shorter than known periods of chromospheric activity, 7 and 12 years, these observations covered the epochs of maximum activity of each component. Data analysis led to the conclusion on the existence of a close correlation between chromospheric and coronal activity of each component and coronal cycles for both stars. The amplitudes of coronal variations were much higher than the amplitudes of HK fluxes. These cyclic variations were decisive for the total variability of X-ray radiation of the system.

Marino et al. (1999) undertook a systematic search for the cycles of stellar activity in X-rays, comparing the observations of 29 stars in 1978–81 with the Einstein Observatory and in 1990–1994 with PSPC ROSAT. They found that if long-term variations similar to solar cycles existed, their amplitudes were much lower than in variations at short time intervals, which dominated in ROSAT data and that were identified as stellar flares. This result confirmed the earlier conclusion by Schmitt et al. (1995) based on a smaller sample of K–M stars and the conclusion by Stern et al. (1995) obtained for the Hyades stars. This may be due to the fact that on stars of the age of the Hyades and younger and on all red and brown dwarfs the generation of small-scale turbulent magnetic fields dominates over the large-scale dynamo responsible for magnetic cycles on the Sun. It should be noted that Kitchatinov et al. (2000) and Donati et al. (2003b) interpreted long-term magnetometric observations of the young K dwarf LQ Hya through Zeeman–Doppler mapping within the framework of the distributed dynamo model.

3.2 Evolutionary Changes in the Activity of Stars

Unlike all previous chapters, this chapter does not start with the description of the solar situation, since in this case the Sun is the driven and not the driving object: it gives us only one point on the evolutionary tracks of the activity, which should be constructed based on the characteristics of the activity of stars of different age and mass.

Since stars with identical effective temperatures and luminosities display a much varying activity level, i.e., proximity on the Hertzsprung–Russell diagram does not guarantee closeness of activity levels, one should find another essential parameter defining the activity. Originally, UV Cet-type flare stars were attributed to young stars, since they had certain similarities with T Tau-type stars (Ambartsumian, 1954). The concept of the youth of T Tau-type stars was advanced by Ambartsumian shortly before (1952) and got wide recognition. The idea of the youth of flare stars obtained independent support, when the analysis of kinematic characteristics of red dwarfs showed that dMe stars were much younger than dM stars (Herbig, 1962). Then, Kraft and Greenstein (1969) found that there were many more K5–M3 dwarfs with calcium emission in the Pleiades than in the Hyades. Robinson and Kraft (1974) found spotted K3–M0 stars in the Pleiades and did not find such objects in the Hyades. But, as the number of known UV Cet-type stars increased, in particular, in clusters of various age and among binary systems with flare components, rather old objects were found among them, even subdwarfs of up to 10 billion years of age. Old UV Cet-type stars were found in studying kinematic characteristics of such objects (Shakhovskaya, 1975; Poveda et al., 1995). This fact disproved short-lived relics of star formation and youth in itself as the main cause of the activity. The sources of modern evolutionary concepts of this activity go back to the studies of stellar rotation, the parameter, which practically does not influence the position of a star on the Hertzsprung–Russell diagram, but determines its magnetism and all related processes in the photosphere and atmosphere, to the studies of evolution of rotation and chromospheric emission, as the most accessible manifestations of stellar magnetism.

In 1955, Parker (1955b) proposed the idea of a stellar dynamo: the scheme of amplification of the stellar magnetic field due to the interaction of convective motions and rotation. Soon, Schatzman (1959, 1962) explained the strong distinction of rotational velocities of stars of early and late spectral types by magnetic deceleration of stars with convective zones, assuming that this process was the most effective when stars approached the main sequence.

In the mid-1950s, O. Wilson started systematic observations of calcium emission to reveal its relation to other parameters. In the early 1960s, he summarized preliminary results of studying G0–K2 dwarfs among field stars and in four stellar clusters. He showed that emission stars occurred in clusters more often than among field stars and that the level of emission from the stars of the young cluster Pleiades was higher than in the older clusters Praesepe and the Hyades (Wilson, 1963). Being completed with the photomtry of more than 100 main-sequence stars (Wilson and Skumanich, 1964) and the spectroscopy of more than 300 stars (Wilson, 1966), this study initiated an evolutionary approach to the phenomenon of calcium emission and chromospheric activity in general. Soon, Wilson and Wooley (1970) compared the intensity of calcium emission in the spectra of 325 late main-sequence stars close to the Sun with the parameters of their galactic orbits and found that stars with strong emission had orbits close to circular and with small inclination angles to the galaxy plane, i.e., the orbits of young objects, whereas stars with weak emission had orbits typical of old objects. Thus, the kinematic characteristics of a large number of stars in the solar vicinity provided independent confirmation of the slow decay of calcim emission.

From the spectrograms with a dispersion of 6Å/mm, Kraft (1967) found that F–G stars with calcium emission had systematically higher rotational velocities than stars without the emission, and solar-type stars in the Pleiades rotated faster than the Hyades stars. On this basis, he suggested secular deceleration of rotation of main-sequence stars with convective envelopes and linked this effect to the Schatzman concept, assuming, however, that magnetic braking continued during the main-sequence life of the stars. According to Kraft, the velocity of a star whose mass was equal to 1.2 solar masses decreased by a factor of 2 over 4×10^8 years. Soon, the close relation between chromospheric emission and local magnetic fields was established in solar studies (Frazier, 1970), which made it possible to advance an hypothesis about secular decay of the stellar magnetic fields eventually caused by the deceleration of stellar rotation. The concept of deceleration of stellar rotation and simultaneous decay of the magnetic activity on evolutionary time scales has become conventional and is enriched by the studies of evolution of various manifestations of this activity. Results of the researches are reviewed below.

3.2.1 Evolution of Stellar Activity

Magnetohydrodynamic calculations of the rise of magnetic flux tubes to the stellar surface carried out by Schüssler and Solanki (1992) for detecting high-latitude spots on rapidly rotating RS CVn-type stars showed that fast rotation displaced the rise of such tubes to high latitudes. This idea was advanced by Granzer et al. (2000) for stellar evolution: they calculated the dynamics of magnetic flux tubes for stars with masses of 0.4–1.7 solar masses and with angular rotational velocity within 0.25–63 solar values, which covered the interval from the classical T Tau-type stars to the stars of the α Per cluster. Granzer et al. found that the latitudes of spots should grow quickly with the increase of rotational velocities, slightly decrease for stars with large masses, and quickly decrease with age. Experimental data are still too fragmentary to check the calculations.

The most abundant data on the evolutionary change of the level of stellar activity were accumulated in the studies of stellar chromospheres.

After qualitative conclusions by Wilson and Kraft on the relation of the chromospheric calcium emission to the age and rotation of stars, based on the observations of stars in the Pleiades and the Hyades, the UMa moving group, and the Sun, Skumanich (1972) proposed the $t^{-1/2}$ law describing the weakening of calcium emission and deceleration of stellar rotation. (According to Smith (1979), this relation is valid after the transition to photoelectric recording of spectral-line profiles, which noticeably increased the estimates of rotary line broadening.) Braking of rotation was attributed to permanent loss of angular momentum in the coronal wind (Durney, 1972), which is the product of activity. The decay of chromospheric activity was connected with the weakening of the magnetic dynamo initiated by rotation (Parker, 1970).

Continuing the studies of calcium emissions started by Wilson, Vaughan and Preston (1980) and Vaughan (1980) discovered a certain "Vaughan–Preston gap", dividing young stars with strong emission and less-active old (older than 10^9 years) stars. The ratio of more and less active stars complied with the hypothesis about secular decay of this emission at a constant rate of star formation. From the sample of 486 stars in the solar vicinity Soderblom (1985) calculated R'_{HK} and found that these values satisfied $t^{-1/2}$. Later, Soderblom and Clements (1987) found that this relation started to be valid for an age close to that of the UMa moving group (3×10^8 years) or the Hyades moving group (6×10^8 years).

From the subsample of G0–K5 dwarfs of the sample of 111 F2–M2 stars investigated by the O. Wilson team, Baliunas et al. (1995) picked out young stars with a high average level of activity, fast rotation, an absence of the Maunder minima and rare smooth cycles, stars of intermediate age of 1–2 billion years with an average level of activity and rotation and more frequent cycles, and old stars, such as the Sun and older, with slower rotation, lower level of activity, smooth cycles, and the Maunder minima.

Based on the depths of absorption details of H_1 and K_1 of calcium lines, Barry et al. (1981) considered regular changes of the level of the chromospheric activity of solar-type stars in six clusters with ages from 10^7 to 5×10^9 years. They concluded that these spectral characteristics were dependent on the intensity of emissions and could be used for age calibration, which for stars younger than the Hyades was determined to an accuracy of up to a factor of 2 and for stars older than 4×10^9 years to an accuracy of up to 25%.

Hartmann et al. (1984b) analyzed the "Vaughan–Preston gap". They concluded that the minimum of the measured S caused by the photospheric contribution and independent of the chromospheric activity, on the one hand, and emission saturation on the youngest stars found in observations of the Pleiades, where $t^{-1/2}$ definitely was not satisfied, on the other hand, resulted in the concentration of the values of S at the band borders, which produced an impression of a "gap". Further, they showed that the observed distribution of chromospheric emission could be presented on the assumption of constant birth of stars provided that on young stars chromospheric emission decayed exponentially and then $t^{-1/2}$ took effect.

In 1964, Herbig (1965) discovered a secular decrease of the content of lithium in main-sequence stars. Skumanich (1972) compared this effect with the weakening of calcium emission and braking of rotation, but did not find confident synchronism. Then, from the spectra of about a hundred F5–G5 dwarfs, Duncan (1981) estimated the intensity of calcium emission and the content of lithium. He found that on the general background of parallel secular weakening of these values there were stars with weak emission but with high content of lithium. To interpret the objects, Duncan assumed that at the age of 1–2 billion years a sharp decay of emission occurred, whereas the jump was absent in the burning out of lithium. Walter (1982) showed that at power representation of the ratio L_X/L_{bol} as a function of the rotational period there was a break near the period of 12 days, this period corresponded to the Vaughan–Preston gap and the age of single G dwarfs of one billion years. On the other hand, the strongest lithium line in the spectrum of V 1005 Ori is not related to the extremely high activity. Probably, this is due to the fact that in the atmospheres of M dwarfs the lithium line λ 6708 Å becomes dependent on the activity level, as the H_α line, and ceases to be an independent indicator of age (Houdebine and Doyle, 1994b).

Using the observations with the echelle spectrograph of the 3-m Lick telescope and his own calibration of the content of lithium as the age characteristic, Soderblom (1983) found that for solar-type stars the relation $\langle v \sin i \rangle \sim t^{-1/2}$ was valid. However, it was not satisfied for the stars younger than the Pleiades and appreciably overstated their velocities as compared to the observed values. The intensity of calcium emission was proportional to the rotational velocity, but the value $v \sin i$ did not display an analog to the Vaughan–Preston gap.

Observational evidence of the fact that rotation rather than the age in itself determines the level of stellar activity was obtained in studying the components of stellar pairs: as stated above (Middlekoop, 1982; Walter, 1982; Rutten, 1986, 1987; Maggio et al. 1987), the dependences of the fluxes F_{HK} and F_X on rotational velocity were common for the components of binary systems and single stars. This commonness is violated only in the systems with very short orbital periods – in semidetached and contact systems. Therefore, all research results for activity as a function of rotation listed above have an exact evolutionary sense.

From a dozen stars observed by IUE, Blanco et al. (1982) found a linear dependence between $\log F_{MgII}$ and the axial rotational period, common for the considered F0–K5 dwarfs. Then, Catalano and Marilli (1983) compared the luminosity of calcium emission and rotation of twenty F8-K7 stars and found that this luminosity decreases with growing axial rotation period as $dex(-P/27^d)$. Comparison of luminosities of the Pleiades and the Hyades stars with their masses revealed a practically identical dependence $L_K \sim M^{5.1}$ for both clusters, but with a shift toward lower luminosities in older Hyades stars. If one excludes young G stars of the Pleiades, for the stars within the range of 10^8–5×10^9 years the relation is valid

$$L_K(M/M_\odot, t) = L_K(1,0)(M/M_\odot)^{5.1} \times 10^{-1.5 \times 10^{-5} \times t^{1/2}}, \qquad (66)$$

with a correlation coefficient of 0.98 and at $L_K(1,0) = 1.90 \times 10^{29}$. Further, Catalano and Marilli found that the relation $L_K \sim t^{-1/2}$ represented observations poorer than the exponent. But exponential decay of the emission yields the rotary deceleration rate as $\Omega \sim t^{-1/2}$ for stars older than 3×10^8 years. Finally, they found that the content of lithium on the Sun and the appropriate stellar values determined by Duncan and Soderblom well suited the general exponential dependence for the rate of lithium burning out, and the exponent of the dependence included $t^{1/2}$.

Using a more complete sample of single stars, Marilli et al. (1986) revised the results of Catalano and Marilli (1983). They found that the coefficient of linear relation between the logarithm of luminosity of calcium emission and the axial rotation period depended on $B-V$, i.e., the depth of the convective zone. Rutten (1986) arrived at a similar conclusion. Taking into account this dependence, Marilli et al. concluded that there was a relation between L_{HK} and the Rossby number.

Using the content of lithium after Duncan (1981) and the depths of H_1 and K_1 absorptions after Barry et al. (1984) as time scales, Cabestany and Vazquez (1983) considered the time change of magnesium emission and found that luminosity in the k MgII line could be presented by the relation

$$L_k(MgII) = 2.8 \times 10^{29-1.82\times 10^{-5}\times t^{1/2}}, \qquad (67)$$

which is rather similar to (66) for the CaII K line found by Catalano and Marilli (1983).

Barry et al. (1984) constructed, for the low-resolution spectra, an analog of the Mount Wilson value S and determined R_{HK} from the spectra of F-G dwarfs in seven open stellar clusters of various age and the Sun. The obtained values did not match the Skumanich relation $t^{-1/2}$, but could be presented by one or two exponents: in the first case, the exponent included $t^{1/2}$, in the second, the first addend described the long-term evolution of chromospheric emission in the main sequence, and the second addend – fast evolution of emission from the youngest stars. Having added the data on the middle-aged cluster NGC 752, Barry et al. (1987) specified the chromospheric scale of the age of solar-type stars within 10^7–6×10^9 years and confirmed that the above two analytical representations equally well described the secular decay of chromospheric emission without a break as the Vaughan–Preston gap.

According to Noyes et al. (1984a), Fig. 9 shows the dependence of R'_{HK}, the ratio of fluxes in CaII H and K lines corrected for the contribution of a radiative atmosphere to the bolometric luminosity on the Rossby number, the key parameter of the dynamo theory. As stated above, this dependence appeared to be so close that it was used to estimate the axial rotation period from the observed luminosity of calcium emission. Similar results were obtained in comparing the values of R_{hk} for G–K stars with the axial rotation periods from 5 to 50 days and the Rossby numbers, but F and the earliest G dwarfs had some differences (Hartmann et al., 1984a). Doyle (1987) considered the most active emission dwarfs with the rotational periods from 0.8 to 6 days and confirmed the correlation of R_{hk} and Rossby numbers, which again was so close that it allowed estimation of axial rotation periods. At $P_{\rm rot}/\tau_C \sim 1/10$, there is saturation analogous to that found by Vilhu (1984) for the radiation of the transition region. For late spectral types this border corresponds to four days. The times of rise of convective elements involved in Rossby numbers were thoroughly studied by Stępień (1994) for the whole range of F–M stars: he included experimental corrections to theoretically calculated values and essentially decreased the scattering of points on "activity level–Rossby number" diagrams for dwarf stars.

Using the sample of 115 chromospherically active early F and G dwarfs, Barry (1988) estimated possible errors of chromospheric ages. He confronted rotational velocities and the chromospheric age of sample stars and found that the rotational periods $P_{\rm rot}$ increased as $t^{0.37}$. The appropriate change of angular velocity is shown in Fig. 65. In this case, braking is proportional to angular momentum and inversely proportional to age and rotation. Rotation depends only on the stellar mass and age. The entire activity decay is determined by the increase of the Rossby number. For a wider range of stellar masses from 0.5 to 1.1 $M_\odot$ Catalano et al. (1989) found that $P_{\rm rot} \sim t^{-1/2}$ with the factor growing toward low masses. According to Barry, the Vaughan–Preston gap is caused by nonuniformity of star formation.

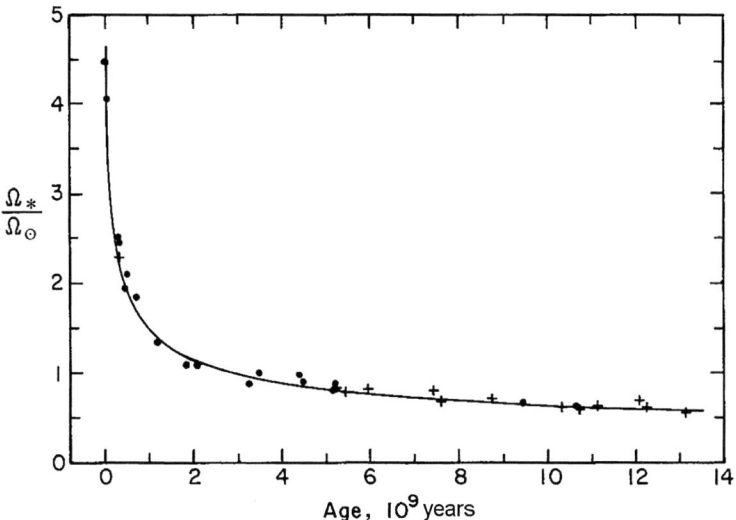

Fig. 65. Evolution of angular velocity of F–G dwarfs (after Barry, 1988)

As stated above, Herbig (1985) measured the intensity of the H_α emission line in the spectra of 40 F8–G3 dwarfs and, using the age scale of Duncan (1981), found secular decay of H_α emission on the background of great variability of data, which can be presented as $\sim t^{-0.4}$, or using the exponent with an e-fold decay over $(3$–$5) \times 10^9$ years. Mekkaden (1985) compared the intensity of H_α emission measured by Herbig for F8–G3 dwarfs with their rotation and found a close correlation of R_{H_α} with axial rotational periods and a less close relation to Rossby numbers.

Simon et al. (1985) thoroughly analyzed the UV spectra of the Sun and 31 F7–G2 dwarfs, for which lithium or other age estimates were available, and found that the sum of surface fluxes in four strongest UV lines of the chromosphere and CaII H and K lines weakened with time as $t^{-0.51\pm0.05}$, whereas the sum of surface fluxes in five strongest transition-region lines weakens as $t^{-1.00\pm0.09}$. But in both cases observations are better presented by an exponential decay with the time of an e-fold decay of 2.6 and 1.4 billion years for the chromosphere and the transition region, respectively. Further, following Noyes et al. (1984a), Simon et al. constructed the dependences of flux ratios normalized to bolometric luminosity on Rossby numbers in all considered UV lines. They found that the decay of CaII and MgII emissions was slower than that of CIV and SiIV and X-ray emission was faster than the emission of the transition region. They presented arguments supporting the hypothesis of Hartmann et al. (1984b) and Barry et al. (1984) that stellar activity decayed at a different rate in the phase of T Tau and in the main sequence. They proposed various physical mechanisms of the phenomenon. In the first case, decay occurred due to a decrease of the depth of the convective zone, inherited from the evolution during the Hayashi phase, whereas lowered activity of main-sequence stars starting from a certain plateau inherent in the youngest stars of the sequence was caused by deceleration of rotation. These two processes can be combined assuming the decisive role of Rossby numbers: reduced activity in the first case is connected with the growth of the Rossby number due to the reduction of the denominator, whereas in the second case the former grows together with the numerator. Changes in integrated intensity of chromospheric emission can occur due to changes of the brightness of the chromospheric network, which on the Sun determines this integrated intensity, or due to the changes of the filling factor, as during the solar-activity cycle. Based on detailed comparison with solar data, Simon et al. (1985) concluded that on young stars excessive emission could be caused by flares covering up to 4% of the surface and flocculae occupying up to 30% of the surface, whereas on old stars the filling factors were 5–10 times less, and in all cases radiation in the transition-region lines occurred basically in flares, while in chromospheric lines it was from flocculae. Finally, they showed that the ratios of R_{MgII}, R_{CIV}, and R_X found by Ayres et al. (1981a) corresponded to a regular change of these values at a monotonic change of the Rossby number. Being based on the observations of solar-type stars, these correlations appeared to be fair for the whole range of F–M dwarfs.

According to Stauffer and Hartmann (1986), an e-fold decay of H_α emission takes approximately one billion years. From independent observations of a number of moving stellar groups, Eggen (1990) estimated the rate of decay of equivalent widths of H_α emission as $t^{-0.5}$ and MgII k and h emissions as $t^{-0.3}$. Later, Stauffer and Hartmann (1987) examined the distribution of rotational velocities of low-mass stars in the Pleiades. Detection of fast rotators among K and M dwarfs and minimum rotational velocities of G stars led to the conclusion that low-mass stars before achieving the main sequence, appreciably accelerated and then quickly decelerated rotation. Thus, they assumed that the wide range of angular momenta found on K and M stars of the Pleiades could be caused by the considerable range of the initial angular momenta in combination with braking in the main sequence, regardless of rotational velocities, rather than a combination of braking in the main sequence with the considerable age range of these stars. However, by the age of the Hyades the effect of the initial distribution of angular momentum practically disappears.

As compared to the Pleiades, in the Hyades the beginning of H_α emission and fast axial rotation occur at later stars (Stauffer et al., 1991). According to Leggett et al. (1994) and Stauffer et al. (1995), all cluster members in the Pleiades and the Hyades with a mass less than 0.3 solar masses displayed H_α emission, and in the Pleiades R_{H_α} achieves a maximum at about 0.3 $M_\odot$.

Simon (1990) analyzed the emission of MgII lines in the spectra of G–M stars of several young clusters observed with IUE. He established that as the stars achieved the main sequence, there occurred very fast braking of rotation and only then did the Skumanich ratio $t^{-1/2}$ come into force. But for low-mass stars the time of initial loss of angular momentum and activity decay was much longer than for solar-type stars. This conclusion was confirmed by Stauffer et al. (1991). Having found the closeness of radio luminosity of fast rotators in the Pleiades and T Tau-type stars, Lim and White (1995) concluded that the stars reached the main sequence in the mode of saturated activity.

Soderblom et al. (1991) measured the calcium emission from solar-type stars in binary systems with known age and, upon combining the results with the data on open clusters, inferred that there was a functional rather than statistic relation between age and chromospheric emission. But they could not resolve the dilemma whether the decay of chromospheric emission was described by the power function of time or by variations of the star-formation rate over the last billion years.

Figure 66 presents the decay curve of calcium emission constructed by Baliunas et al. (1998) from observations of stellar clusters and moving groups

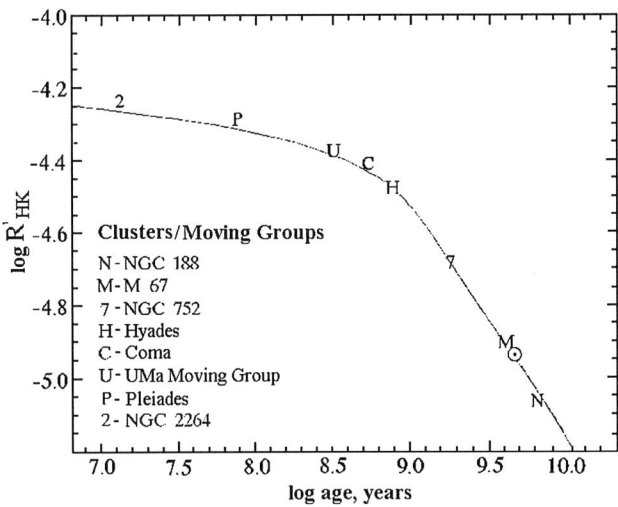

Fig. 66. The decay of calcium emission from observations of stars in clusters and moving groups of different ages (Baliunas et al., 1998)

3.2 Evolutionary Changes in the Activity of Stars

of different age. The study revealed cyclic emission variability in the Groombridge 1830 subdwarf, whose age is about 10 billion years.

Above, we mentioned the results of Simon et al. (1985) proving that the rate of evolutionary decay of emission of the transition region was much greater than that in the chromosphere. A similar conclusion was made by Marilli and Catalano (1984) who revised the preliminary results and found the following ratios:

$$L_{HK} = 1.1 \times 10^{29 - P_{\rm rot}/27\overset{d}{.}0} \, ,$$
$$L_{CIV} = 5.5 \times 10^{27 - P_{\rm rot}/22\overset{d}{.}8}, \quad \text{and} \qquad (68)$$
$$L_X = 5.7 \times 10^{29 - P_{\rm rot}/10\overset{d}{.}4} \, ,$$

governing absolute luminosities of the chromosphere, the transition region and corona as a function of the rotational period. Simon and Fekel (1987) compared the values of R_{CIV} and axial rotation periods and confirmed the importance of the Rossby number for unifying the relations of activity and rotation of dwarf stars. From the very precise dependence $R_{CIV}(\mathrm{Ro})$ Simon et al. (1985) estimated the deceleration of rotational velocity of a star with a mass of 1.1 $M_\odot$ with increasing rotational period from four days for the age of 10^8 to 22 days for the age of 4×10^9.

Using HST, Ayres et al. (1996) and Ayres (1999) observed ten early G dwarfs in three open clusters of different age: in α Per, the Pleiades and the Hyades. Combining these results with those of Ayres et al. (1995) for F9–G2 field dwarfs, they found an essential decline in CIV emission from the youngest to the oldest stars well correlated with $v_{\rm rot}$. Saturation of this and soft X-ray emission occurred at about $v_{\rm rot} \sim 35\,\mathrm{km/s}$. The correlation of intensity of these carbon lines and soft X-rays is of the form $R_X \sim R_{CIV}^{1.7}$, and the Sun at different phases of its activity cycle matched the relation.

The character of decay of coronal activity of stars was determined by comparing the activity of individual stars of different age and by analyzing the objects in various stellar clusters.

Johnson (1983b) found that, as a rule, M dwarfs with maximum X-ray luminosity, according to their kinematic characteristics, could be ranked among young objects, but some of them belonged to the old-disk population. Thus, he suggested that these M dwarfs recently appeared in fast clouds of interstellar gas.

Dobson and Radick (1989) compared soft X-ray emission and rotation of 157 single main-sequence late stars in the solar vicinity and in different clusters and found that the best correlation was between normalized X-ray luminosity R_X and Rossby numbers; this correlation applies to the stars of all spectral types from early F and covers three orders of magnitude of Ro and four orders of magnitude of R_X.

On the basis of kinematic properties, Hawley and Feigelson (1994) selected five nearest and oldest M dwarfs of the halo population without H_α emission and observed them in X-rays. Four stars displayed the luminosity $L_X = (1\text{--}5) \times 10^{26}$ erg/s, the fifth $L_X < 6 \times 10^{25}$ erg/s. Thus, the activity can be maintained within the Hubble time. It should be noted that, according to Peres et al. (2000), the luminosity of the solar corona measured by ROSAT PSPC would make 3×10^{26} erg/s at the activity minimum and 5×10^{27} erg/s at the activity maximum.

Micela et al. (1997) analyzed the observations of 12 nearby K4–M6 halo and old-disk stars using ROSAT PSPC and WFC. They found that for halo stars the values of L_X were systematically lower than for the old-disk stars: $(1\text{--}5) \times 10^{26}$ and $(1\text{--}9) \times 10^{27}$ erg/s, respectively, and extremely low $L_X \sim 3 \times 10^{25}$ erg/s of Barnard's star (GJ 699), which showed no signs of activity in optical range. The obtained values of L_X of the halo stars correspond to the lowest luminosities of nearby dK and dM stars of a spatially full sample by Schmitt et al. (1995) whereas L_X of the old-disk stars is within average luminosities.

Hempelmann et al. (1995) selected about 100 single stars of late spectral types with directly measured rotational periods and different age – both field stars and members of the Pleiades and the Hyades – and analyzed RASS data for them. They found that field stars and members of young clusters satisfied the same correlations $L_X/L_{\text{bol}}(P_{\text{rot}})$ and $L_X/L_{\text{bol}}(\text{Ro})$ that were valid over three orders of magnitude, which evidenced the decisive role of rotation for the level of X-ray activity, but not the age in itself. Another argument supporting this statement was recently obtained by James et al. (2000) who compared the features of X-ray emission of rapidly rotating M dwarfs – single stars and components of binary systems with $v_{\text{rot}} > 6$ km/s. They found that for the former, including objects with $0.2 < P_{\text{rot}} < 10.1$ days, $\langle \log(L_X/L_{\text{bol}}) \rangle = -3.21 \pm 0.04$, whereas for the latter, including objects with $0.8 < P_{\text{rot}} < 10.4$ days, $\langle \log(L_X/L_{\text{bol}}) \rangle = -3.19 \pm 0.10$. On looking for statistical dependences of L_X, F_X and L_X/L_{bol} on some power of the period or the Rossby number, Hempelmann et al. found that exponents in all cases were close to -1. But for Ro>1/3 L_X and F_X decreased with growing Ro much faster than at lower Rossby numbers.

Combining HST and ROSAT data, Ayres (1999) found in early G field dwarfs and those from α Per, the Pleiades, and the Hyades clusters a distinct correlation $R_X \sim R_{CIV}^{2.0}$ with a certain saturation of the youngest α Per stars. Comparison of these photometric values with rotational velocities revealed precise correlations $R_X \sim v_{\text{rot}}^{3.0 \pm 0.6}$ and $R_{CIV} \sim v_{\text{rot}}^{1.5 \pm 0.3}$ with saturation at about 35 km/s and 100 km/s, respectively. During the magnetic activity cycle the Sun drifts along the correlation (R_X, R_{CIV}) constructed using the data on the considered stars. Another important conclusion is that young stars have a noticeable dispersion of high-temperature emission, whereas in the Hyades the scattering is insignificant. Within the model of simple magnetic loops (Rosner et al., 1978) the nonlinear correlation of R_{CIV} and v_{rot} should be connected with the pressure rise at the feet of such loops as compared to the appropriate solar values, and the nonlinearity of the correlation of R_X and R_{CIV} indicates the rise of average temperatures at the loop tops. Thus, Ayres et al. concluded that on old stars, such as the Sun, loops of X-ray bright points and of active regions dominated, whereas on the youngest and most active stars large-scale structures or postflare loops prevailed.

The data on evolution of magnetic activity of stars obtained in X-ray observations of stellar clusters containing stars of different masses but the same age, identical chemical composition, and remote by the same distance, show that the decay of X-ray activity with age is typical of all stars of late spectral types and this process depends on the stellar spectrum.

Stern et al. (1981) found that the higher X-ray luminosity of the Hyades as compared to field stars corresponds to the dependence $L_X \sim (v \sin i)^2$, while the Pleiades demonstrate a more complex dependence and greater scattering.

Caillault and Helfand (1985) found that L_X in several clusters decreased not as $t^{-1/2}$, but first as $\log t = 9$ very slowly and then much faster than according to Skumanich. Since L_X of the Sun, if it were completely covered by active regions, would be 10 times lower than that of the brightest G stars in the Pleiades, Caillault and Helfand assumed that the coronae of young stars had bigger loops than the solar corona, which was later confirmed by Ayres (1999). The fact that L_X on K0–M3 stars does not correlate with $v \sin i$ found by Pallavicini et al. (1982), but fits the exponential relation of Walter (1982) suggests the existence of an additional factor that determines L_X of young stars along with rotation.

Stern (1984) and Maggio et al. (1987) compared the X-ray luminosity of stars in Orion, the Pleiades, the Hyades, and the Sun and found a systematic decrease of maximum values of L_X with age. Using the Einstein Observatory data, Micela et al. (1985) detected that the X-ray luminosity of G stars in the Pleiades was much higher than in the Hyades. Essentially, the greater number of X-ray bright G dwarfs as compared to K dwarfs, the absence of M dwarfs, and the availability of rapidly rotating K stars in the Pleiades led to the suspicion that in this young cluster the rotation of K dwarfs accelerated, whereas G stars already entered the phase of magnetic braking.

According to Caillault (1996), by the end of 1995 ROSAT provided data on 13 stellar clusters with ages varying from 20 to 600 million years separated by 45 to 400 pc, the number of identified X-ray sources varied from 15 to 185. Figure 67 shows normalized distribution functions of the X-ray luminosity of G dwarfs in seven clusters. One can see a distinct change of the total decay of X-ray emission with age, though significant divergence of these functions for the Hyades and Praesepe – clusters of the same age – requires additional explanations. One cause is the different contribution of binary systems to the total X-ray luminosity or different metallicity of stars in the clusters (Barrado y Navascues et al., 1997). On the whole, X-ray data obtained for the clusters make it possible to conclude that the distribution function of L_X depends on the age of clusters and the mass of the considered stars, which is caused by

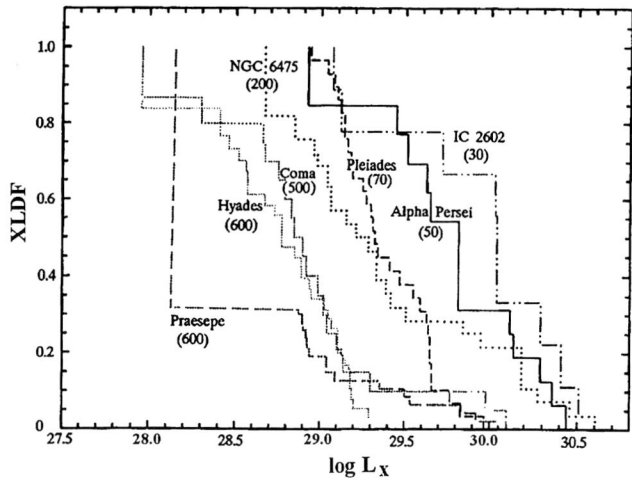

Fig. 67. Normalized distribution functions of X-ray luminosity of G dwarfs in seven stellar clusters of different age after Caillault (1996); numbers in brackets indicate the age in million years

the dependence of the deceleration scale on stellar mass, which in turn results in the age dependence of the distributions of rotational velocities of stars in a cluster. Comparison of coronal activity and rotational velocities reveals the saturation of the ratio L_X/L_{bol} at the level of 10^{-3} for rapidly rotating stars.

Patten and Simon (1993, 1996) analyzed ROSAT PSPC observations of the very young – about 30 million years – cluster IC 2391, where at the center 76 X-ray sources with $L_X(0.2-2.0\,\text{keV}) > 2 \times 10^{28}$ erg/s were found, 19 of them were identified as cluster members. Using ground photometric and spectral observations of new members of the cluster, they concluded that for solar-type stars rotational periods differed by more than 20 times, thus stars of very different activity level reached the main sequence. In comparison with older clusters, the Pleiades and the Hyades, it was found that in the general decrease of $\langle v_{\text{rot}} \rangle$ and $\langle L_X \rangle$ with age, scattering of L_X grew because the stars left the saturation mode for the phase, where their X-ray luminosity was determined by rotation.

In a later review of the results of a study of 15 clusters younger than 1 billion years, Randich (1997) considered distribution functions of X-ray luminosities of stars with different masses. The distributions are determined by the number of fast rotators in the cluster, whose Rossby numbers were less than 0.16 and X-ray luminosity was at the saturation level, and by the distribution of rotational velocities among slow rotators. Since the characteristic times of braking of fast and slow rotators differ and depend on stellar masses, in the age range from 3×10^7 to 6×10^8 years the values of L_X are not described by the simple Skumanich or other relation common for different mass stars.

In ROSAT PSPC studies of old and intermediate age open clusters, Belloni (1997) and Belloni and Tagliaferri (1997) confirmed the conclusion obtained for other old clusters: at an age of more than 10^9 years, high X-ray luminosity was within 10^{29}–10^{30} erg/s only in binary systems.

Stellar wind is one of the corona properties that today cannot be measured in other dwarf stars but for the Sun. Though modern properties of solar wind are investigated in detail, its history is practically unknown. Some indirect evidences are accessible from the observations of very young clusters α Per and IC 2391, which are characterized by the wide distribution of rotational periods of G dwarfs, whereas in older clusters, as the Hyades, this distribution is much narrower. Thus, it was suggested that at a rotational velocity below some critical value the braking mechanism did not depend on rotational velocity, whereas at velocities higher than the critical level it became a steep function of the velocity.

On the basis of the theory of magnetic braking of stars by thermal wind advanced by Mestel (1984), Stępień (1989) calculated the evolution of stellar rotation in the range of color indices $(B - V)$ from $0\overset{m}{.}5$ up to $1\overset{m}{.}3$ for the Hyades, assuming initial rotational periods of all stars as 0.7, 1.3 and 3 days, and for slowly rotating single stars in the solar vicinity with age from 5 to 13 billion years, assuming initial periods of their rotation of two and five days. In the first case, the calculated periods to the age of the Hyades reached 14 days, in the second – 48 days. These values and distributions $P_{\mathrm{rot}}(B-V)$ well represented the observations and were used by Stępień to calculate expected fluxes ΔF_{CaII} and F_X of stars in the Pleiades, young cluster NGC 2264, and the Hyades. The calculations yielded observed fluxes of calcium emission to an accuracy up to ± 0.05 dex and X-ray emission up to ± 0.5 dex. Then, Stępień considered evolution of solar-type stars assuming the initial rotational period of 1, 3, and 5 days (see Fig. 68). According to these diagrams, at the age of over 10^8 years ΔF_{CaII} decreases as $t^{-0.5}$ and F_X as $t^{-1.7}$. The agreement of the calculations with observations confirms the validity of the Mestel theory (1984).

Here, it is appropriate to remember the conclusion made by Wood and Linsky (1998) in studying "a hydrogen wall" in circumstellar space: during the decay of a surface X-ray flux pressure in the stellar wind increases and one should expect a growing rate of secular loss of stellar mass. On the other hand, at reducing coronal temperature below a critical value a solar-type wind cannot be maintained on a star.

The data on evolutionary changes of the level of flare activity of red dwarf stars are relatively few. In essence, they suggest the revision of the initial opinion about flare stars as extremely young objects. Affiliation of several

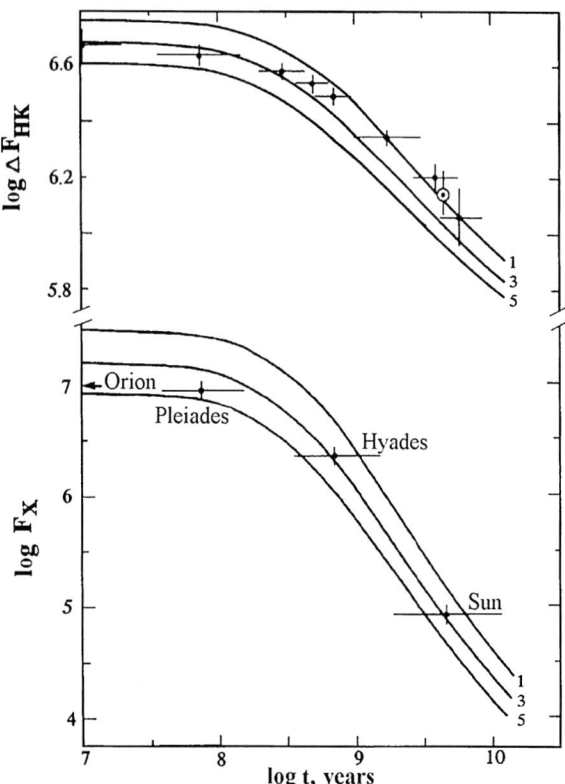

Fig. 68. Secular decay of calcium and X-ray emission from observations of cluster stars and the Sun and from calculations of the evolution of stellar rotation by Stępień (1989)

flare stars to old objects was suspected by van de Kamp (1969), Chugainov (1972b), Lee and Hoxie (1972), and Kunkel (1972). Shakhovskaya (1975) considered the kinematic characteristics of 646 red dwarfs in the solar vicinity. She concluded that flare activity detected both from directly recorded flares and from hydrogen emission was observed among the objects of the young-disk population, whose age is less than 5×10^8 years, and among those of the old-disk population, whose age varies from 5×10^8 to 5×10^9 years, and even among halo objects, whose age exceeds 5×10^9 years. But on young stars, this activity was the strongest and it weakened toward older objects. From photographic observations of flare stars in four stellar clusters, Parsamian (1976) found a systematic decrease of the maximum luminosity of such stars with increasing age of clusters.

Figure 40 illustrates evolutionary changes of the spectral index of the power-energy spectrum of flares: with the age this spectrum flattens and, hence, the contribution of rare but strong flares to the total flare radiation increases.

Skumanich (1986) imparted an evolutionary sense to the above correlation of total flare radiation and quiet X-ray emission from the corona. Based on the statistics on solar white-light flares, which are the most similar to stellar flares, and the luminosity of quiet X-ray emission from the Sun, he showed that the Sun also satisfied this correlation. Further, he found that dM dwarfs, on which optical flares and X-ray emission were recorded, satisfied the relation. Considering dM stars as evolved dMe dwarfs, Skumanich arrived at a conclusion that during stellar evolution the frequency of flares was preserved, while their power decreased, as did the X-ray luminosity of corona.

Thus, evolution of the activity of lower main-sequence stars depends on a set of processes. For the youngest stars, one can find the final phase of rotary acceleration related to the end of compression when a star reaches the main sequence, and stars arrive at the sequence with significant dispersion of angular momenta. When this phase is over, rotational velocities of most stars are rather high, and there is activity saturation with maximum values of R_{HK}, R_{hk}, R_{H_α}, R_{CIV}, R_X, and R_R. When stellar compression is over, magnetic braking of rotation starts, which first operates very efficiently, and when the rotational velocity decreases to some critical values, a smoother monotonic activity decay begins. It is obvious that all critical points in the specified sequence of events are determined by both stellar mass and particularly the considered manifestation of its activity, which results in significant variety of activity indicators of various stars even inside one cluster. It should be noted that, according to D'Antona and Mazzitelli (1985), the time it takes a star of 0.6 solar masses to reach the main sequence is 1.2×10^8 years, whereas for stars of 0.1 $M_\odot$ it is 1.5×10^9 years and 0.08 $M_\odot$ it is 1.3×10^{10} years.

3.2.2 Evolution of Solar Activity

Above, we mentioned some properties of G dwarfs, which should be included in the general picture of evolution of the solar activity: the initial angular momentum of the Sun in the wide range of probable values; expected high-latitude spots on the quickly rotating young Sun; the change of speed of secular braking shown in Fig. 65; Marilli–Catalano relationship (68) and the ratios of R_{MgII}, R_{CIV}, and R_X and the ensuing conclusion on different rates of activity decay at various levels of the atmosphere, the general character of the decay of coronal activity shown in Fig. 66; ratios of R_X, R_{CIV}, and v_{rot} and the ensuing conclusion about evolutionary changes of parameters of coronal loops. Solar-type stars achieve a universal dependence of activity on rotation at the age of 200–300 million years. Below, all these general conclusions are specified and elaborated upon.

There are two approaches to studying the evolution of solar activity: statistical consideration of various characteristics of the activity of G dwarfs of different age and detailed research of actual stars similar to the Sun using as many parameters as possible, except for the age. For the second problem, Gaidos et al. (2000) selected 38 G–K dwarfs as the analogs of the young Sun within 25 pc from the Sun; the less active among them turned out to be slow rotators of the greatest ages. Both approaches yielded certain results.

A stumbling block of former evolutionary schemes of the solar system was the distribution of angular momenta between the Sun and planets: though their mass is negligible, planets are responsible for the overwhelming part of the total angular momentum of the system. The concept of magnetic braking cuts the Gordian knot: soon after the achievement of the main sequence a star quickly dumps the major part of the initial angular momentum. In other words, magnetic braking is not only universal, but also an effective mechanism of fast dumping of the initial angular momentum resulting in further braking depending on stellar mass and age. These reasons enable discussions of the early stages of the development of the Sun with faster rotation and higher magnetic activity and predictions for its future. According to Soderblom (1983), the rotational velocity of the Sun corresponds to the average rotational velocity of G dwarfs of solar age, which is justified by the statistical approach to its evolution.

Feigelson et al. (1991) estimated the expected properties of solar activity at the earliest stages of its development, assuming that at the age of 10^6 years the Sun was a T Tau-type star with weak lines (WTT), for which, unlike for classical T Tau-type stars (CTT), the surrounding circumstellar medium was not a decisive factor in the observed activity. Spots occupy from 5 to 40% of the surface of WTT stars, the level of optical and ultraviolet activity of their chromospheres exceeds the solar one by approximately 50 times, the intensity of X-ray and radio emissions – by 3–5 orders of magnitude, and these fluxes vary on time scales of a day, hours and minutes and with the greatest probability are caused by strong flare activity of a magnetic nature. Within this approach to the young Sun, at the age of 0.3 to 1.5 million years it should have a constant temperature of 4200 ± 500 K and spectral type K5 $\pm$2 IV. At the age of one million years its luminosity was 1.7 of the current luminosity, $L_X \sim 10^{30}$ erg/s, and radius a 2.5 of the current radius. It was compressed to the current size at the age of 10–20 million years. At an axial rotational velocity of about 25 km/s the period $P_{\rm rot}$ was about 5 days. Spots covered up to 25% of the solar surface and their temperature was 3300 ± 500 K.

Stępień and Geyer (1996) carried out photometric observations of 16 solar-type stars and on 9 of them detected appreciable brightness oscillations with amplitudes of several hundredths of the stellar magnitude, for most of them $P_{\rm rot} < 10$ days. Probably, as in the Hyades, small amplitudes of brightness oscillation are common characteristics of active solar-type field stars with a rotational periods of about one week and less, some variables display strong modulation of variability amplitude within one year.

The ideas on the character of evolution of solar activity on the basis of the HK project were stated by Baliunas (1991). From the full sample of project

3.2 Evolutionary Changes in the Activity of Stars

Table 17. Parameters of magnetic activity of solar-type stars of different age (after Baliunas, 1991)

	Young Stars	Old Stars	The Sun
Number of stars	7	12	
Age (billion years)	~1	several	4.6
$\langle S \rangle$	0.314	0.165	0.171
$\langle P_{\rm rot} \rangle$ (days)	9.1	27	25
Character $S(t)$	periodic or irregular without minima	periodic with minima taking 1/4 of time	11-year cycle with minima taking up 1/3 of time
Correlation of brightness and magnetic activity	reverse	direct	direct
Amplitude	<several per cent	<0.4%	~0.1%

stars she selected single objects with close to solar $B-V$ color indices ($\sim 0\overset{m}{.}66$) and the range of S in this subsample was explained by the age difference. Table 17 presents statistical data for the stars of this subsample with axial rotation periods estimated from the observations of calcium emission. The second column of the table contains data for young stars, whose age is about one billion years, only the objects with variable S, both periodic and irregular, are cited. The objects with ages of several billions years (third column) display periodic changes of S or its constancy. Figure 69 illustrates S of two stars from the second column and two stars from the third column. The second from the bottom row presents the correlations of stellar brightness with the level of magnetic activity: for young stars there is an anticorrelation, i.e., large dark spots dominate in brightness variations, whereas on the Sun and old stars, brightening magnetic areas dominate (Radick et al., 1998). Figure 70 shows the profiles of MgII h and k lines in the spectra of G stars of various ages (Dorren and Guinan, 1994).

Thus, the young Sun had several times higher S, which had considerably greater nonperiodic or periodic variations with a shorter period, and did not experience the Maunder-type minima. One can expect that in the future the current periodicity will be preserved until the Sun leaves the main sequence, because this pattern was observed for stars up to 10 billion years old, but one cannot specify the expected frequency of the Maunder-type minima.

To trace the expected changes of solar activity, Dorren et al. (1994) selected nine single G0–G5 dwarfs with masses of 0.9 to 1.1 solar masses, with known axial rotation periods of 2.7–45 days, and with ages of $7 \times 10^7 - 9 \times 10^9$ years (Fig. 71). Comparison of X-ray luminosity, luminosity of the CIV lines of the transition region and chromospheric MgII lines revealed the power dependence of these emissions on the rotational period with exponents of -2.5, -1.6, and

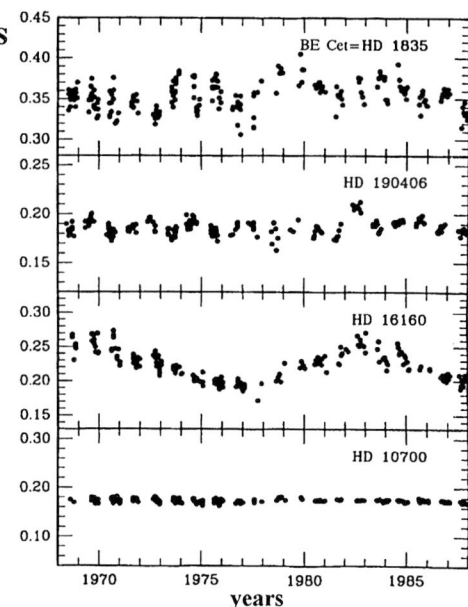

Fig. 69. Four characteristic time changes of S on solar-type stars (after Baliunas, 1991)

-0.76, respectively. In comparing the amplitudes of brightness oscillations in the V band due to stellar spottedness and the rotational periods a decrease of these amplitudes from $0\overset{m}{.}09$ at $P_{\rm rot} = 1.5$ days to $0\overset{m}{.}008$ at $P_{\rm rot} = 14$ days was revealed. For greater rotational periods, brightness oscillations were below the detection limit. (Later, Güdel and Gaidos (2001) suspected a similar dependence of $L_R/L_{\rm bol}$ on $P_{\rm rot}$ for G dwarfs.) Comparison of $P_{\rm rot}$ with the age yielded the relation $P_{\rm rot} = 0.21 \times t^{0.57}$, where $P_{\rm rot}$ was in days and age, in millions years.

Güdel et al. (1997b) analyzed ROSAT and ASCA spectra of 11 single G0–G5 dwarfs and the subgiant β Hyi within the age range from 70 million to 9 billion years and with L_X varying from 1 to 500 X-ray solar luminosities and relevant data for the Sun. They found that with increasing age the energy distribution changed systematically in the X-ray spectrum: both the temperature and emission measure of the high-temperature component of stellar corona decreased (Fig. 72). Thus, on the young star EK Dra the temperature of the hot component of the corona reached 20–30 MK, which is equal to the temperature of solar flares. On the whole, the change of this temperature with the age can be presented as $T_2 \sim t^{-0.3}$ or $P_{\rm rot}^{-0.55}$, and the dependence of X-ray luminosity on this temperature as $L_X \sim T_2^4$. EM_2 also decreases rapidly with age and near 500 million years this corona component becomes insignificant. Radio observations revealed a similar correlation $L_R \sim T_2^4$ and fit into this evolutionary picture. This suggests that the existence of energetic

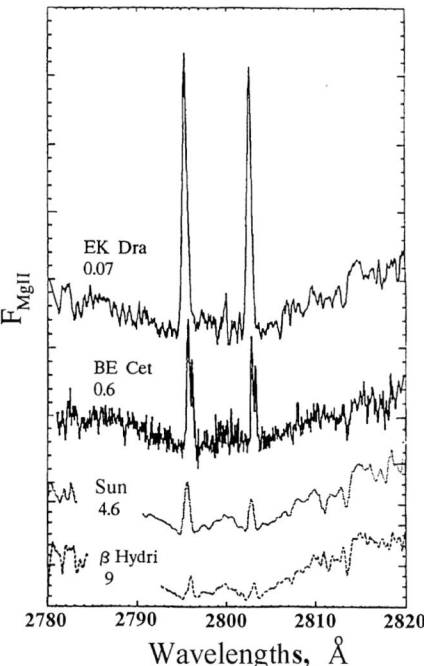

Fig. 70. Four characteristic MgII h and k line profiles in the spectra of G dwarfs of different age; the age in billion years is given below the names of stars (Dorren and Guinan, 1994)

electrons is connected with the hot component of the corona. But saturation of L_R occurs at rotational periods less than two days, i.e., when rotation is faster than the saturation of L_X. Developing this scheme, Güdel et al. (1998) described the following evolutionary picture. On young stars, a dynamo generates high surface magnetic activity with large volume filling factor of the corona with magnetic loops. In this case, fast strong flares spend a considerable part of the energy for the acceleration of fast particles and plasma heating in loops to 20–30 MK, which are recorded as microwave gyrosynchrotron radiation and a high-temperature thermal component in X-rays. On fast rotators X-ray coronae are saturated, which may suggest filling of all loops with hot and dense plasma, whose temperature is maintained by frequent flares. With age, the filling factor of the corona with loops decreases, strong flares become less common, and after 500 million years such flares become isolated events, therefore quasipermanent microwave radiation disappears. It should be noted that similar evolution of flare activity follows from Fig. 40.

Let us consider in detail actual stars to illustrate the evolution of solar activity.

414 3 Long-Term Variations in Activity of Flare Stars

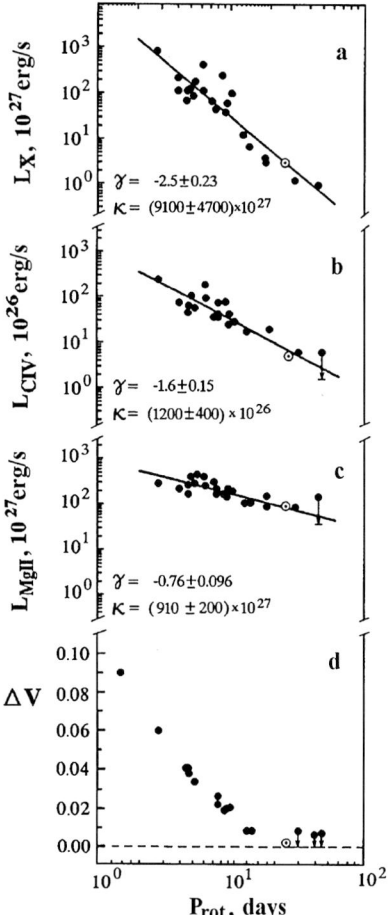

Fig. 71. X-ray luminosity, luminosities of CIV lines in the transition region and chromospheric lines Mg II, and amplitudes of apparent brightness of G dwarfs vs. rotational period; dependences $L(P_{\rm rot})$ were approximated by the power functions of $L = kP_{\rm rot}^{\gamma}$ type, the values of k and γ are presented in each diagram (Dorren et al., 1994)

Dorren and Guinan (1994) analyzed the activity of the G0 dwarf HD 129333 (= EK Dra), as an analog of a very young Sun that had passed the T Tau stage and recently achieved the main sequence. The star is close to the Sun and belongs to the Pleiades moving group. It is about 70 million years old, the planet-formation process being completed. The effective temperature of EK Dra is 150 K higher than the solar one, its mass is $\sim 1.05 M_\odot$ and radius $\sim 0.92 R_\odot$. Photometry of the star showed its spottedness up to 6%, the largest among G stars, and $\Delta T \sim 500$ K. Scheible and Guinan (1994) presented the light curve of the star with the help of two spots with radii

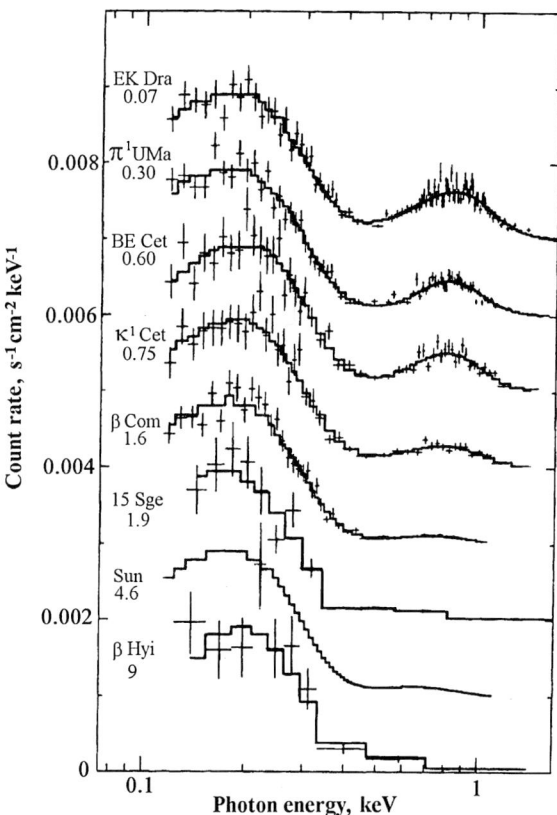

Fig. 72. ROSAT and ASCA X-ray spectra of G stars of different age; the age is given in billion years (Güdel et al., 1997b)

of 20°, while the Doppler mapping revealed cool spots at the equator, at middle latitudes and at the pole, though the polar spot was seen in only 9 of 12 spectral lines (Strassmeier and Rice, 1998). The emission of H_α is maximum on the hemisphere with the greatest spottedness. The luminosity of EK Dra in soft X-rays exceeds the solar luminosity by a factor of 300, in the ultraviolet lines of the transition region by 20–100 times, in ultraviolet chromospheric lines by 3–20 times, CaII and MgII emission by 3–5 times, which is the strongest among G dwarfs that do not belong to binary systems. At $v \sin i \sim 18-20 \, \text{km/s}$ the rotation period is 2.7–2.8 days, the shortest among G stars, it modulates the calcium emission with an amplitude of 5%. The activity level of EK Dra satisfies the Skumanich relation, though it is already close to saturation. Saar and Bookbinder (1998a) found that low-amplitude flares contributed up to 8% to the flux in CIV and SiIV lines. X-ray emission of the star was presented by the two-temperature coronal model with $T_1 = 1.3 \times 10^6$ and $T_2 = 9.6 \times 10^6$ K, but later Güdel et al. (1997b) estimated the temperature of the hot component as 20–30 MK. Radio emission at 3.6 cm was 3000 times higher than the appropriate solar value. According to Dorren and Guinan (1994) and Dorren et al. (1995), variations of average brightness and fluxes in ultraviolet lines suggest an activity cycle of about 12 years. But, unlike the old Sun, young EK Dra weakens at the activity maximum phase and brightens at its minimum. The amplitude of changes of CIV lines in the spectrum of EK Dra during the cycle is much higher than that of CII and MgII lines.

Güdel et al. (1995a) investigated EK Dra in the microwave and X-ray ranges. In both ranges the star was brighter than single-flare M dwarfs and G dwarfs in the Pleiades. For the brightness temperature of microwave radiation of about 10^8 K, the most suitable mechanism of this radiation is optically thin gyrosynchrotron emission of relativistic electrons. According to RASS data, $L_X = 8 \times 10^{29}$ erg/s, in the two-temperature coronal model its components had temperatures of 1.9 and 10 MK and emission measures of 1.2×10^{52} and 2.5×10^{52} cm^{-3}, and cooler component displayed much greater rotational modulation than the hotter one. Güdel and Schmitt (1996) from the modulation of X-ray radiation of EK Dra, estimated an electron density of stellar corona as 3×10^{10} cm^{-3}. From this modulation, the electron density of the cooler component was estimated as 4×10^{10} cm^{-3} and its size as not more than 0.1–0.2 R_*. Thus, high X-ray luminosity of the star is caused by high density, rather than by a great volume of luminous matter, and for retaining such structures a magnetic field of about 240 G is required.

Using four ASCA instruments and two EUVE (SW and MW) spectra of EK Dra Güdel et al. (1997a) analyzed the structure of DEM in the range of temperatures from 0.1 up to 100 MK. Five different analysis methods led to the coordinated conclusion that DEM had broad maxima in the ranges of 5–8 and 15–40 MK, a narrow deep minimum at about 8–20 MK, and a small amount of plasma at 1–3 MK. This coldest part of the EK Dra corona corresponds to the solar corona, though its EM is higher by an order of magnitude than the solar value. Apparently, the general view of DEM is determined by the structure of the cooling function of the optically thin plasma with regard to its thermodynamic stability, and the hottest corona component arises in flares.

Messina et al. (1999b) compared the young Sun and the single G5 dwarf HD 134319, a probable member of the Hyades moving group with strong chromospheric emission. According to their photometric observations in 1991–95, the star has an axial rotational period of 4.45 days and stable spottedness with spots uniformly distributed along the equator plus two active longitudes of opposite longitude with a total area of dark spots from 7 up to 24%, depending on the accepted inclination angle of the rotational axis of the star and the algorithm selected for the inverse photometric problem.

Marino et al. (2003b) observed the young fast rotator VXR45 in IC 2391 on XMM-Newton/pnEPIC. This G9 dwarf has $v \sin i > 200$ km/s and $P_{\text{rot}} = 0.223$ days, i.e., it is a supersaturated star and its X-ray radiation is slightly lower than the maximum observed at rotational velocities of several tens of kilometers per second. From the X-ray light curve they confidently established rotational modulation during two photometric periods. As opposed to the case of EK Dra observed by Güdel et al. (1995a), this modulation did not impact the hardness of radiation. From the spectrum within 0.3–7.8 keV they constructed the two-component coronal model: $T_1 = 7$, $T_2 = 14$ MK, and $EM_1/EM_2 = 1.4$ at an FIP effect of about 0.27. The revealed rotational modulation evidences structural inhomogeneities of the corona and disproves the hypothesis on supersaturated stars as the objects completely covered by active regions. Jardine (2004) proposed the model in which rotational modulation was directly related to supersaturation.

The early stages of solar activity are successfully analyzed by the example of numerous solar-type stars with a high activity level and confidently determined age. However, studying the future of solar activity is much more difficult because the activity level is much lower and old stars of known age are, as a rule, in remote globular clusters. To get an idea of the far future of solar activity, i.e., to determine its absolute minimum, Dravins et al. (1993a–c) studied the single G2 IV star β Hyi, which recently left the main sequence. This nearest subgiant is at a distance of 6.5 pc from the Sun, its age determined from the position on the evolutionary track is 9.5 billion years, i.e., the star is twice as old as the Sun and brighter by 1^m than the Sun by absolute value. With the help of high-resolution spectroscopy and three-dimensional hydrodynamic calculations the parameters of the β Hyi photosphere were determined: the temperature differs from the solar temperature by no more than 100 K, the acceleration of gravity is four times lower than on the Sun, and the stellar radius is 1.6 $R_\odot$. The photosphere contains several times larger and more contrast granules, the velocities of matter are higher by a factor of 1.5–2. Lower gas pressure can admit approximately twice weaker local magnetic fields. At $v \sin i = 2 \pm 1$ km/s the axial rotation period is about 45 days. A general small deficiency of metals [Fe/H] ~ -0.2, corresponding to old stars, the content of lithium is higher by an order of magnitude than the solar content, which can be associated with deepening of the convective zone when a star leaves the main sequence. This mechanism of lithium enrichment of the atmospheres of subgiants has not been completely clarified, but the excess of this element observed on other subgiants does not raise doubts about the considerable age of β Hyi.

Calcium emission in the spectrum of β Hyi is half that in the solar spectrum, but it displays the same profile asymmetry. About a hundred IUE high-dispersion spectrograms of MgII emission obtained over 12 years show smooth and regular changes of emission without signs of fast appearance and growth of individual active regions. The character of calcium emission was determined in low-activity stars within the HK project. The recorded minimum and maximum of magnesium emission allows one to estimate the activity cycle of β Hyi as 15–18 years. The measured amplitude of MgII emission of about 30%

corresponds to variations of the Mount–Wilson index S of 1%, which is the lowest recorded amplitude of the stellar activity cycle. However, the measured level of chromospheric activity and established cyclicity evidence that the observed chromosphere does not comply with the basal level but is due to the magnetic activity of the star.

The surface fluxes of β Hyi measured by IUE in ultraviolet high-excitation lines are comparable with the appropriate solar values in the minimum phase, but different algorithms of data processing yielded essentially different results.

Observations in soft X-rays with different EXOSAT filters revealed that the flux and spectrum of β Hyi radiation differed from those of the Sun. The analysis of similar solar observations within the isothermal coronal model yields $\log T = 6.5$ and $\log EM = 49.6$, but the data on β Hyi provided two solutions: $\log T = 5.7$ with $\log EM = 49.1$ and $\log T = 6.6$ with $\log EM = 49.6$. The "hot" model is rather close to the solar corona, but the "cold" variant falls in another steady thermodynamic minimum of optically thin high-temperature plasma. The temperature of "cold" β Hyi corona approaches or has already reached the critical level, below which a stellar wind similar to the solar one cannot exist and the termination of such an outflow should lead to termination of magnetic braking of stellar rotation.

4
Conclusion

Semicentennial intensive studies of UV Cet-type flare stars coincided with the period of formation and fast development of multiwavelength astrophysics. By now, observational data on the activity of such stars were accumulated practically in the whole range of electromagnetic radiation – from medium X-rays to decameter radio waves. As a result, the identity of the physical nature of the activity of such stars and of the Sun has been ascertained. It has been revealed that such activity is inherent in all lower main-sequence stars at a certain – frequently rather long – stage of their development, that the energy of this activity is eventually derived from stellar rotation, and the "driving belts" are hydrodynamic and magnetohydrodynamic processes. This book is devoted to the phenomena of magnetic stellar activity, but until now the term has not been defined. According to Dupree (1981), stellar activity is something that cannot be determined by basic optical parameters – stellar temperature and luminosity, i.e., the position of a star on the Hertzsprung–Russell diagram. Linsky (1988) defines stellar activity as a set of phenomena that occur when basic assumptions of the classical theory of the stationary stellar atmosphere – energy transfer from the center by radiation and convection, the resulting convective motions do not impact temperature; hydrostatic atmosphere; too weak magnetic fields to control the fluxes of matter do not change the energy balance of stellar atmospheres – are violated.

We showed that for the considered activity, which involves strong local magnetic fields, nonuniform nonradiative heating of the stellar surface, regular motions and nonstationary phenomena with characteristic times from milliseconds to several decades, the decisive parameter is stellar rotation, and it practically does not influence its position on the Hertzsprung–Russell diagram. Though the rotation of upper main-sequence stars is maximum, there are no manifestations of the discussed activity on them. Hence, there should be an important intermediate link including this energy source in the chain of considered activity. This is subphotospheric convection. Indeed, this activity occurs on stars with a mass of less than 1.5 solar masses, at this threshold stars acquire a convective envelope. How abrupt is the threshold at which the activity is triggered? The full range of masses of the main sequence stars is more than three orders of magnitude, but cardinal changes occur on the interval from two to one solar masses: stars of two solar masses have a characteristic rotational velocity of 120 km/s and display no magnetic activity, whereas the rotational velocity of the Sun is 2 km/s and it is characterized by intense magnetic activity. The decisive role of rotation is observed both for young single medium- and low-mass stars, and for similar components of close binary systems of different age, in which high rotational velocity is maintained for long periods due to the energy of their orbital motion.

The experimentally established close connection of sunspots and faculae, chromospheric network and flocculae with local magnetic fields, development of transient flares in the regions of a complicated magnetic configuration, strong coronal radiation in the region of closed magnetic loops and coronal holes in the regions of open magnetic structures – all suggests the magnetic nature of the solar activity. The fundamental role of magnetic fields is stipulated by the fact that their nonpotential components can contain considerable reserves of energy, which can rapidly convert into other forms accessible to direct observations, and that hydromagnetic waves governing the atmospheric energetics and its nonthermal heating are excited in magnetic flux tubes and efficiently transferred along them. The loop configuration of magnetic fields determines the geometry of plasma structures in chromospheres and coronae and the time character of the development of flares in them. During the life of a single main-sequence star the loss of angular momentum due to magnetic braking noticeably slows its rotation and lowers the level of magnetic activity.

Therefore, the exact definition formulated by Parker 20 years ago "solar activity is first of all the result of displacement of magnetic flux tubes by convective gas motions under the solar surface and the subsequent shift of rarefied gas by magnetic-field discontinuities and instabilities above the solar surface" is completely valid for stars with the solar-type activity considered in this monograph.

We thoroughly considered relevant observational facts and briefly outlined theoretical models constructed on their basis. The facts are diverse, which is natural, since the phenomena of magnetic solar-type activity are recorded on stars with masses varying from 1.2 to 0.06 solar masses, with the axial rotation periods from 40 days to 4 h, and age from millions to ten billion years, whereas quantitative characteristics of the phenomena – surface spottedness, energy of permanent chromospheres, coronae and sporadic flares – vary within two to five orders of magnitude. Omitting the details of numerous theoretical studies of stellar magnetism, here we will discuss two basic groups of principal problems of the considered activity of stars concerned with their subphotospheric and atmospheric magnetic fields, following Parker's division. The first of them mainly determine various time characteristics of the stellar activity, while the second are connected with energy, since real energy release attributed to magnetic activity is observed in stellar atmospheres where dissipation of the active phenomena occurs from the photosphere to the corona.

In considering these major problems, some of the above observational data should be discussed anew.

4.1 Activity and Subphotospheric Magnetic Fields

Studies of the 11-year solar cycle gave rise to observational and theoretical research of the solar dynamo, i.e., the interaction of convection and rotation resulting in the emergence of magnetic structures in the convective envelope and atmosphere, and stimulated study of local magnetic structures on the stellar surface. However, the variety of phenomena found on stars was incomparably vaster than on the Sun. To interpret them, various dynamo models, models of generation of stellar magnetic fields, were advanced: the kinetic or linear model, which specifies the velocity field and disregards the back coupling of magnetic field and plasma motion; the hydrodynamic or nonlinear model that takes into account the back coupling and solves the complete system of magnetohydrodynamic equations; the cyclic envelope model of solar type, in which the interface between the convective zone and radiative core is essential; and the model of a distributed or turbulent dynamo, where boundary conditions are insignificant. In the process of accumulation of data on stellar activity cycles researchers tried to give preference to one or another model of the stellar dynamo or to correlate them with various structures of stellar interiors and/or evolution phases. But solutions of magnetohydrodynamic equations in different approximations sometimes resulted in opposite conclusions. Thus, Belvedere et al. (1982) constructed the dynamo theory, in which the duration of activity cycles should grow from F5 to M0 stars, whereas the theory of Robinson and Durney (1982) decreased it from G0 to M5 stars. Since the first work Parker (1955a) on the mechanism of the solar dynamo required generation of a magnetic field, the spiral character of plasma motion, the α-effect, has been attributed to convection, but recently Brandenburg (1998) put forward an alternative idea that this effect could be due to magnetic buoyancy and instability, whereas Getling (2001) completely rejected the concept of emerging magnetic tubes and explained the generation of magnetic fields only by convective motions.

Applying the nonlinear dynamo of a rotating spherical envelope with three-dimensional convection to the Sun, whose full and differential rotation and surface convection parameters are well known, Gilman (1983) could not reproduce the observed duration of the solar cycle and observed migrating toroidal fields, but showed that in the nonlinear $\alpha - \omega$ dynamo the cyclic behavior was possible in the limited ranges of medium electroconductivity and rotational velocities: at very low velocities the α-effect was very weak, at very high velocities the rotation suppressed convection. In the nonlinear dynamo, the induced magnetic field should suppress the differential rotation, and this feedback is strong enough: when a certain critical level is exceeded by a factor of 3–4, magnetic cycles should disappear.

4 Conclusion

Saar and Baliunas (1992a,b) studied more complex relations between P_{cyc} and stellar parameters within the framework of the $\alpha - \omega$ dynamo theory. On comparing the normalized frequencies of cyclic variations and dimensionless dynamo numbers, they found correlations between these values in two subsamples containing active and nonactive stars: most stars considered by Noyes et al. (1984b) and the Sun were in the subsample of nonactive older stars, whereas in the subsample of young active stars the ratios of R'_{HK} were twice as high, the dynamo number was greater, and the frequency of cyclic variations was lower. Considering the amplitudes of cyclic variations, Saar and Baliunas found that A_{cyc} grew together with P_{rot} – see similar results by Bondar (2000) in Fig. 63 – and with the frequency of cyclic variations.

Baliunas et al. (1996) found for the low-activity stars, including the Sun, the correlation (65), which evidenced the role of a nonlinear dynamo in the generation of magnetic fields. Brandenburg et al. (1998) assumed that the cause of the jump of P_{cyc} on 2–3 billion year old stars could be concerned with the rearrangement of internal magnetic fields, excitation of high dynamo modes or a change of the dominating factor of instability.

In studying stellar magnetism, the features of axial rotation, in particular surface differential rotation, were considered together with cyclic activity. Donahue et al. (1996) analyzed the rotational periods of 36 stars of the Wilson program, on which the rotational periods were confidently determined using the algorithm of Horne and Baliunas (1986) at least over five seasons. They found the following correlation between the average periods $\langle P_{rot} \rangle$ and their complete ranges ΔP

$$\Delta P \sim \langle P_{rot} \rangle^{1.3 \pm 0.1} . \tag{69}$$

The correlation coefficient was 0.90. It matched the solar situation well. The nonlinear relation (69) contradicted the assumption on proportionality of radial gradient of angular rotational velocity to the angular velocity used in theoretical mean-field dynamo models. It did not match the $\alpha - \omega$ dynamo model, in which surface differential rotation should grow together with angular rotational velocities, whereas, according to (69), it should grow with increasing $\langle P_{rot} \rangle$, i.e., with stellar age.

For 22 slowly rotating stars with well-determined periods of chromospheric activity, Ossendrijver (1997) found the correlation

$$P_{cyc} \sim Ro^{2.0 \pm 0.3} . \tag{70}$$

Within the framework of the mean-field linear dynamo this relation corresponds to decelerating differential rotation and fast growth of the α-effect with increasing stellar rotational velocity.

Using high-resolution spectra of 32 F–K stars Reiners and Schmitt (2003) determined differential rotation and found that it was more common in slowly rotating stars, though deviations from solid-body rotation took place in some fast rotators as well. They got evidence against significant differential rotation of the most active stars.

4.1 Activity and Subphotospheric Magnetic Fields

On the basis of a long-term series of high-precision photometric observations of six young solar-type stars, Messina and Guinan (2003) found that all the considered stars demonstrated variable duration of rotational period, i.e., were characterized by differential rotation. These variations were also periodical with the phase of the spottedness cycle of BE Cet and DX Leo. Apparently, a similar situation is observed for π^1 UMa, EK Dra, and HN Peg. For BE Cet, π^1 UMa, and EK Dra, the rotational period decreases in the course of the cycle, then abruptly returns to the initial value at the beginning of the next cycle, but DX Leo, κ Cet, HN Peg, and LQ Hya demonstrate an inverse change of the period duration. The dependence of the amplitude of variation of the rotational period is close to (69). The cycle duration does not correlate with the dynamo number but is in positive correlation with the amplitude of differential rotation.

The above survey shows that theoretical studies are still far from being completed. Current general notions of subphotospheric magnetic fields in stars with solar-type activity are briefly as follows.

Durney et al. (1981) connected the dispersion of chromospheric emission of stars of different age with the evolutionary reduction of the dynamo number, the ratio of generating and dissipative terms in the magnetohydrodynamic equation determining the efficiency of the magnetic-field generation. Knobloch et al. (1981) arrived at a conclusion that the Vaughan–Preston gap was caused by a nonmonotic dependence of the strength of the magnetic field generated by the stellar dynamo on the Rossby number: at some critical value of the number, depending on stellar mass, one should expect a fast decrease of the strength. Durney and Robinson (1982) estimated the expected magnetic-field strength and sizes of the regions occupied by such fields on the lower main-sequence stars assuming that field strength was determined by the condition of equal time of emerging of magnetic flux tube and the time of an e-fold amplification of the field through the dynamo mechanism. As a result, the following conclusions were drawn: at a given axial rotation period the magnetic field should grow from several hundred gauss on G0 stars to many kilogauss on M5 stars; the part of the surface covered by magnetic fields should quickly grow with the decrease of the axial rotation period, and for periods close to solar period, M5 stars should be practically completely covered by magnetic fields. Based on the model by Durney and Robinson, Montesinos et al. (1987) found relations connecting the parameters of the internal structure of stars with the strength of the magnetic field generated in it and the surface filling factor with magnetic structures. Applying the relations to the sample of more than 30 G0–K3 main-sequence stars, they estimated the expected strength of magnetic fields and surface filling factors with them. The resulting filling factors displayed a clear dependence on Rossby numbers, and the measured total fluxes in CaII and MgII lines, on the calculated filling factors. Further, analyzing Mount–Wilson monitoring measurements of calcium emission in the spectra of single F–G dwarfs, Hempelmann et al. (1996) divided all the examined stars into 3 groups with regard to the emission character:

constant, regular and irregular variables. They found that on regular variables the Rossby number, as a rule, was above unity and on irregular variables it was less than unity. Then, Hempelmann et al. found that stars belonging to different groups had different ΔF_{HK} and F_X: coefficients of regression between the logarithms of these values were 0.99 ± 0.13 on stars with periodic changes of F_{HK} and 1.7 ± 0.4 on stars with chaotic changes of fluxes. These features can be related to the transition from the nonlinear dynamo operating on stars with irregular changes of F_{HK} to a linear dynamo in stars with periodic variations of fluxes.

Recently, Bruevich et al. (2001) updated the concept of Durney et al. (1981) within the framework of the theory of dynamic systems. They found that young stars with high dynamo numbers should be compared with dynamic systems with a high degree of freedom, which are characterized by the modes of chaotic behavior and strange attractors. In the course of evolutionary braking of rotation the dynamo number decreases, the dynamic system enters the limit cycle mode with auto-oscillations, which is manifested in the case of cyclicities. In other words, the evolutionary transition from chaotic to cyclic activity is concerned with the decrease of the effective dimension of the dynamic system.

But the concept of evolution of dynamic systems through a cyclic dynamo is certainly insufficient for describing the whole variety of stars with solar-type activity, in particular, the youngest and lowest-mass stars.

For the youngest and, hence, the fastest rotators, the effect of saturation when the activity level ceased to depend on rotational velocity was found (Simon, 2001). Apparently, this is an approximation to full coverage of stellar surface by magnetic fields and physical restrictions of the dynamo mechanism are triggered, for example, in suppressing differential rotation necessary for a solar-type dynamo, by fast rotation of the whole star. Only the first factor f $\rightarrow$ 1 is insufficient, since in this case the level of activity can hardly surpass the solar value by two orders of magnitude, whereas a much greater increase of this level has been observed. On the other hand, the growth of the degree of spottedness on active K–M dwarfs is an observationally established fact, and this cause of saturation should not be rejected.

Donati et al. (2003a,b) estimated differential rotation on the surfaces of AB Dor and LQ Hya, revealed differences in the estimates based on cool spots and magnetic structures and changes of its amplitude over several years. They assumed that their results evidenced the effect of dynamo distributed over the whole convective zone, rather than concentrated on the bottom of the zone, as on the Sun.

The simple relation of the rotational velocity with the activity level on late M dwarfs is violated as well: there are late M dwarfs – fast rotators without H_α emission, which contradicts known correlations on G stars, and active M dwarfs within the wide range of rotational velocities, whereas L dwarfs rotate quickly, but are nonactive (Basri, 2001). In other words, the activity weakens in achieving completely convective structures and disappears

4.1 Activity and Subphotospheric Magnetic Fields

on objects later than M9. Apparently, on low-mass stars the main role is played by a distributed or turbulent small-scale dynamo (Rosner, 1980; Durney et al., 1993). Remember the study of a very low-mass star VB 8 in the extreme ultraviolet, whose activity, according to Drake et al. (1996), is maintained by a turbulent dynamo. Its main difference from the large-scale envelope dynamo is a weak connection with rotation, slower transfer of the angular momentum. This circumstance can cause fast rotation of M dwarfs in the Pleiades and the Hyades and lack of X-ray variability in late stars in the Hyades, which is typical of the Pleiades (Stauffer et al., 1998; Gagné et al., 1995; Stern et al., 1995; Durney et al., 1993).

Messina et al. (2003) found correlations between the maximum amplitude of rotational modulation of stellar brightness $A_{\max}$, which they considered as a value related to the total magnetic flux fB and rotational period, and the Rossby number. These correlations distinctly show the breaks near a period of 1.1 days, which were interpreted as a evidence of the transfer from one dynamo mechanism to another.

According to Mohanty et al. (2002), due to the low temperature, high total density, and therefore very low degree of ionization of matter on the latest M dwarfs and on L dwarfs, one of the main features of cosmic plasma is violated – its freezing into magnetic fields. This results in a qualitatively new magnetohydrodynamic situation: noticeable nonpotential magnetic fields responsible for the stellar activity cannot exist in these layers of the stellar atmosphere.

Walter (1990) found that the dependence of soft X-rays on rotation appeared at $B-V > 0\overset{m}{.}45$ as a threshold of emergence of the solar-type dynamo. Probably, the mechanism of the turbulent dynamo is effective not only on the objects where cyclic magnetic fields of solar type are absent. It is probable that the turbulent dynamo is responsible for the activity in the equatorial zone of the Sun with a low phase activity level independent of the cycle, for the ubiquitous "magnetic carpet" on the Sun, responsible for the diffuse X-ray radiation and the residual activity of the Sun during the Maunder minimum. Stars with flat activity can also be candidates for turbulent dynamo objects: their coronae are weak and there is no dependence of coronal emission and radiation of transition regions on rotation (Saar, 1998, 2001).

Schmitt et al. (1998) considered the simultaneous action of two different mechanisms of magnetic-field generation.

According to Kitchatinov et al. (2001), the interaction of magnetic fields in the radiative core and convective envelope can lead to the effect of active longitudes.

However, recently Solov'ev and Kiritchek (2004) proposed the theory of the solar diffuse magnetic cycle that did not include the concept of a dynamo but successfully represented many solar-activity features.

All the above considerations should be included in the general theory of magnetic fields of medium- and low-mass stars or disproved by new observations.

An important step toward formulating the theory was recently made by Barnes (2003a,b). He analyzed a large sample of stars with known rotational periods within $0\overset{m}{.}5 < B - V < 1\overset{m}{.}5$ and found two sequences on the diagram $(P_{\rm rot}, B - V)$, which he compared with different structures of stellar magnetic fields. Clear localization of these sequences on the diagram depending on the age of stars within the range of 30 to 4500 million years made it possible to advance the concept of gyrochronolgy and estimate the characteristic time of the evolutionary transfer of the global magnetic field from one structure to another in stars of different masses.

4.2 Activity and Magnetism of Stellar Atmospheres

Flux tubes of photospheric magnetic fields go up to the solar atmosphere and determine its spatial structure and physical state. Close connection of the magnetic field and various manifestations of solar activity – dark spots and bright faculae in the photosphere, bright flocculae in the chromosphere, coronal condensations and sporadic flares – is found directly in comparing magnetograms with the localization of the structures on the solar disk. This relation is expressed in precise correlations of local values of fB with fluxes ΔF_{HK} and F_{CIV}. A similar situation is typical of active red dwarfs – see (22) and (23) – though an analog of solar magnetograms for stars is unavailable as yet, thus direct comparisons are impossible in this case. However, for stars, there are less detailed but equally convincing evidences of the relation between photospheric magnetic fields and the level of atmospheric activity. Thus, the comparison of average magnetic fluxes from stellar photospheres with X-ray luminosity of coronae $L_X/L_{\rm bol}$ presented in Fig. 73 illustrates the clear correlation of the values. Saar (2001) found the quantitative relation (22) for stars whose ratios were similar to the appropriate local solar ones. Jordan and Montesinos (1991) found the correlation between coronal temperature and the Rossby number, which also suggested the relation of the dynamo mechanism deep in the convective zone to physical processes in the upper atmosphere. The correlation of the maximum amplitudes of rotational modulation of brightness and the ratios $L_X/L_{\rm bol}$ found by Messina et al. (2003) has a similar meaning.

Let us consider two more actual examples.

The wide campaign of comprehensive investigation of the active G8 star ξ Boo A carried out in 1986 included magnetometric observations of the λ 6173 Å line and spectral observations in the region of the helium D_3 line, multicolor broadband polarimetric observations, spectral observations in the ultraviolet and in CaII H and K lines (Saar et al., 1988) . Data analysis revealed clear synchronous changes of the magnetic flux, the intensity of ultraviolet emission in CIV and CII lines and CaII emission, which directly evidenced the relation of the magnetic flux in the photospheric field to the emission from the outer atmosphere. Maximum absorption in the helium D_3

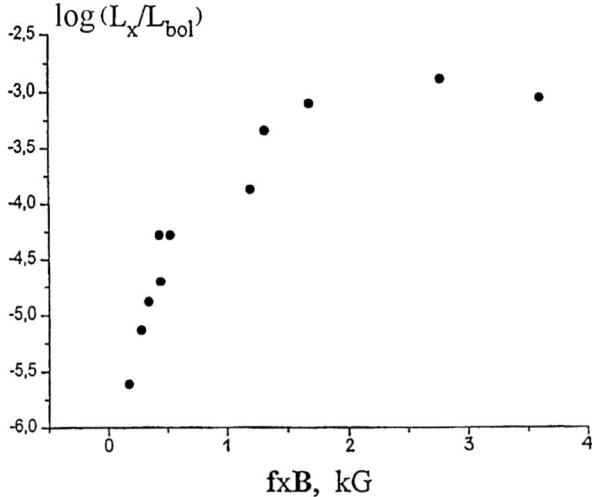

Fig. 73. Comparison of average magnetic fluxes fB on stellar surface and relative luminosity of coronal X-ray emission (Katsova, 1999)

line and maximum flooding of the D cores of sodium lines were near the maximum of the magnetic flux. On the basis of comparison of the obtained data, the existence of four longitudinal sectors with enhanced magnetic fluxes was suggested.

By analogy with the Sun, one could expect that monitoring of stars with asymmetric distribution of active regions will reveal the growth and decline of ultraviolet emission, X-ray emission, and magnetic flux at certain phases of the rotational period. This prediction was confirmed by the observations of ϵ Eri (Saar et al., 1986b) and ξ Boo A (Saar et al., 1986b, 1988).

The first two parts of the book consistently reported the data on the quiescent state of flare stars and flares on them. Experimental data obtained for the objects with different scales of time changes were naturally grouped into these categories, and their interpretation required different theoretical constructs: models of atmospheres of active stars and models of flares. Independent development of these models led to the notions on active processes – flares – in the passive medium of stellar atmospheres. However, this dichotomy should not be absolutized, since there is a deep physical commonness of these phenomena, which consists in possible relations of the heating mechanism of quiet atmosphere of an active star and the energy source of flares. In solar studies, these problems were unified due to the experimental discovery of microflares in hard X-rays and bright ultraviolet points: both phenomena, as well as flares, are caused by fluxes of fast electrons interacting with a dense chromosphere in the feet of coronal loops. Their energy is apparently sufficient for heating of the solar atmosphere. This concept is supported by the data on the heating of the solar wind not only in the base, but also at great distance from the solar surface due to surges, transients, and coronal eruptions. For active stars, the above statistical correlations of X-ray luminosity of quiet coronae and the average power of optical flares and the very hot component of stellar coronae, whose temperature is close to the temperature of solar flares, can be considered as unifying facts.

Using this unifying approach, we shall reconsider the problem of the deficiency of photospheric radiation of spotted stars assuming that the general photometric effect of stellar spottedness is, in a way, connected with the integral magnetic flux. In Fig. 74 the bolometric deficits of photospheric radiation of two dozen spotted stars are compared with the estimates of radiative losses by the chromospheres and coronae during the epochs close to optical observations (Alekseev et al., 2001). The diagrams show certain correlations of compared values, which eventually, as in Fig. 73, correspond to correlations of photospheric and atmospheric magnetic fluxes. On the basis of quantitative analysis of the found correlations, Alekseev et al. (2001) concluded that the bolometric deficit of photospheric radiation of the most active spotted stars definitely exceeded their radiative losses for permanent radiation from all atmospheric layers and for the radiation of sporadic flares. In this connection, Katsova and Livshits formulated the idea about the transfer of deficient energy of photospheric radiation into a global rearrangement of the structures in the upper atmospheres of such stars analogous to local reorganization of atmospheres during flares.

On the Sun and on other less-active dwarfs the deficit of photospheric radiation can be converted into the spread of this energy over the near photospheric regions, formation of facular areas and related chromospheric flocculae. Indeed, exact measurements of the solar constant showed that the deficit of photospheric radiation of the Sun was substantially balanced by additional radiation of the atmosphere: the measurements detected the decrease of this constant during the passage of a large group of spots, but during the maximum of solar activity at the greatest spottedness this constant grew at the expense of numerous active regions. Precision photometric measurements of some stars of the Wilson program showed that the total brightness of the most active stars slightly decreased during the epoch of maximum chromospheric activity, i.e., the brightness variations were determined by large spotted regions with decreased temperature. As to the less-active stars, during the maxima of chromospheric activity their brightness increases, i.e., as on the Sun, the additional radiation of faculae prevails (Radick et al., 1998; Alekseev et al., 2001). It should be noted that domination of faculae above spots during the solar maximum suggests the increasing contribution of large magnetic structures to their total spectrum. Recently, Abramenko (2002) found a local manifestation of this process in considering some active solar regions: the closer the region to the realization of flares, the flatter is the structural function of photospheric magnetic fields, i.e., the contribution of large-scale components increases.

A number of characteristics systematically distinguishes the atmospheres of the most active red dwarfs from the solar atmosphere. Broadly speaking,

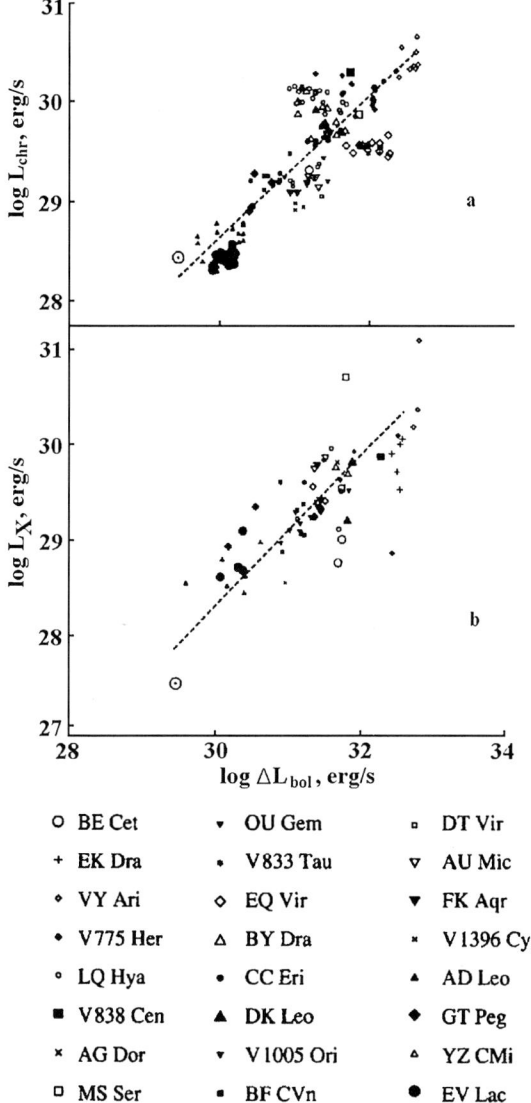

Fig. 74. Comparison of the deficits of bolometric luminosity of photospheric radiation with radiative losses of chromospheres (**a**) and coronae (**b**) (Alekseev et al., 2001)

on the Sun, spots occupy fractions of a per cent of the entire surface and the deficit of photospheric radiation can spread over large active regions surrounding spots. But on the most active stars where spots occupy up to tens of per cent of the surface, there is no place for extended active regions, and the deficit energy of photospheric radiation is channeled to the upper layers of the atmosphere. In Sects. 1.3 and 1.4 we considered qualitative distinctions in the degree of excitation of the atmospheres of the most active stars and the Sun: the stars are characterized by higher electron density of the atmospheres, higher total radiative losses, a great part of which is caused by the corona, availability of a hot component in their coronae, which can arise in flare processes, and higher flare activity. Rather high absolute values of bolometric deficits of radiation energy for the most active stars – up to 5×10^{32} erg/s – definitely exceed radiative losses of atmospheres and flares, which do not reach $10^{-2} L_{\mathrm{bol}}$ even on the most active stars (Pettersen, 1988). In the Katsova–Livshits concept, the governing parameter of stellar activity is the degree of stellar spottedness, which varies both during the evolution of an individual star due to secular braking of its rotation and decay of its magnetic activity and in the spectral sequences of stars due to systematic changes of the structure of their convective zones along the main sequence.

Phenomenological notions on the decisive role of the deficit of photospheric radiation in the energy of atmospheres of active stars brings us back to the idea proposed by Mullan (1975b) on Alfven waves as a source of heating of the atmospheres of flare stars and the energy of flares. (But already in 1968 de Jager stated that the deficit of radiation of a large sunspot can save energy sufficient for a large solar flare up to 10^{32} erg.) Mogilevsky (1980, 1986) applied the concept of self-organizing plasma to solar-activity events: the plasma in the convective zone is self-organized in magnetohydrodynamic solitons, which directly connect subphotospheric magnetism with the activity manifestations in the stellar atmosphere and have a higher propagation velocity and negligible dissipation than magnetohydrodynamic waves. Not giving preference to any specific form of energy flux coming from subphotospheric layers, one can cite some arguments for this scheme of energy supply for the whole atmospheric activity: "All forms of observed radiative losses – quiet atmospheres and flares – can have a common cause, and this cause should be sought inside a star" (Pettersen, 1988). Passage of the flux of nonradiative energy through a rather stratified stellar atmosphere in different configurations of magnetic flux tubes can include more than one change of the actual energy carrier: various waves, including slow magnetohydrodynamic waves and magnetohydrodynamic solitons, fluxes of fast particles, reverse – top down – heat conductivity, radiative heating of the upper chromosphere by the Lyman series from the transition region and heating of the temperature minimum region by the Balmer series from the chromosphere. The character of the energy flux should have threshold properties and significantly depend on its power: at low energy density the flux proceeds in a "laminar" way, heating up the atmosphere, whereas at high energy density there should be breaks of the flux, which can be identified with flares. These phase transitions are essential in the present-day models of flares (Pustil'nik, 1997, 1999). A strong energy flux through a nonlinear medium is one of the basic requirements of thermodynamics of open systems needed for self-organization of the medium. The formation of current sheets, which are often considered as direct energy sources of flares and heating of stellar atmospheres, can be one of the results of the self-organization, while the actual channeling of their energy can depend on the amount of free energy contained in them (Mullan, 1989). Probably, the slowly varying deficit of photospheric radiation during a flare can be a source of additional heating of plasma at the phase of flare decay.

Concluding their review "Flares on the Sun and Other Stars" Haisch et al. (1991) stated that, since flare studies yielded ever-decreasing scales and hence higher energy densities in the sources of initial energy release, astrophysics will eventually deal with vacuum instability. In my opinion, for the magnetic activity under consideration such depths of the Universe are an exaggeration, macroscopic nonlinear processes in magnetized plasma resulting in the self-organization of matter at different spatial and time scales will suffice. In the thermodynamics of open systems, a star as a whole is a dissipative system of the greatest scale, in which global magnetic fields are self-organized due to the energy of rotation and convective motions (Gershberg, 1986), while dissipative systems are realized in small-scale structures due to the energy of the carrier of the deficit of photospheric radiation resulting in stellar flares and other local phenomena. Probably, this general synergetic approach will give a key to understanding the various manifestations of the solar-type activity on main-sequence stars.

References

Abada-Simon M., Lecacheux A., Louarn P., Dulk G.A. et al., 1994. Astron. Astrophys., **288**, 219.
Abada-Simon M., Lecacheux A., Aubier M., and Bookbinder J.A., 1997. Astron. Astrophys., **321**, 841.
Abbett W.P. and Hawley S.L., 1999. Astrophys. J., **521**, 906.
Abdul-Aziz H., Abranin E.P., Alekseev I. Yu. et al., 1995. Astron. Astrophys. Suppl. Ser., **114**, 509.
Abell G.O., 1959. Publ. Astron. Soc. Pacific, **71**, 517.
Abramenko V.I., 2002. Astron. Zhur., **79**, 182 (= Astron. Rep., **46**, 161).
Abranin E.P., Alekseev I. Yu., Bazelyan L.L., Brazhenko A.I. et al., 1994. Kinemat. Phys. Celest. Bodies, **10**, 4, 70.
Abranin E.P., Alekseev I. Yu., Avgoloupis S. et al., 1997. In: G. Asteriadis, A. Bantelas, M.E. Contadakis, K. Katsambalos et al. (eds). The Earth and Universe. Aristotle Univ. Thessalonoki. Ziti Editions. p. 33.
Abranin E.P., Bazelyan L.L., Alekseev I. Yu., Gershberg R.E. et al., 1998a. Astrophys. Space Sci., **257**, 131.
Abranin E.P., Alekseev I. Yu., Avgoloupis S. et al., 1998b. Astron. Astrophys. Trans., **17**, 221.
Agrawal P.C., 1988. Astron. Astrophys., **204**, 235.
Agrawal P.C., Rao A.R., and Sreekantan B.V., 1986. Monthly Not. Roy. Astron. Soc., **219**, 225.
Airapetian V.S. and Holman G.D., 1998. Astrophys. J., **501**, 805.
Alef W., Benz A.O., and Güdel M., 1997. Astron. Astrophys., **317**, 707.
Alekseev I. Yu., 2000. Astron. Zhur., **77**, 207 (= Astron. Rep., **44**, 178).
Alekseev I. Yu., 2001. Low Mass Spotted Stars. Astroprint, Odessa.
Alekseev I. Yu. and Bondar N.I., 1998. Astron. Zhur., **75**, 742 (= Astron. Rep., **42**, 655).
Alekseev I. Yu. and Gershberg R.E., 1996a. Astron. Zhur., **73**, 589 (= Astron. Rep., **40**, 538).
Alekseev I. Yu. and Gershberg R.E., 1996b. Astron. Zhur., **73**, 579 (= Astron. Rep. **40**, 528).

Alekseev I. Yu. and Gershberg R.E., 1996c. Astrofizika, **39**, 67.
Alekseev I. Yu. and Gershberg R.E., 1997a. In: G. Asteriadis, A. Bantelas, M.E. Contadakis, K. Katsambalos et al. (eds). The Earth and Universe. Aristotle Univ. Thessalonoki. Ziti Editions. 43.
Alekseev I. Yu. and Gershberg R.E., 1997b. Astron. Zhur., **74**, 240 (= Astron. Rep., **41**, 207).
Alekseev I. Yu. and Kozlova O.V., 2000. Astrofizika, **43**, 339.
Alekseev I. Yu. and Kozlova O.V., 2001. Astrofizika, **44**, 529.
Alekseev I. Yu. and Kozlova O.V., 2002. Astron. Astrophys., **396**, 203.
Alekseev I. Yu. and Kozlova O.V., 2003a. Astron. Astrophys., **403**, 205.
Alekseev I. Yu. and Kozlova O.V., 2003b. Astrofizika, **46**, 41.
Alekseev I. Yu., Gershberg R.E., Ilyin I.V., Shakhovskaya N.I. et al., 1994. Astron. Astrophys., **288**, 502.
Alekseev I. Yu., Chalenko V.E., and Shakhovskoi D.N., 2000. Astron. Zhur., **77**, 777 (= Astron. Rep., **44**, 689).
Alekseev I. Yu., Gershberg R.E. Katsova M.M., and Livshits M.A., 2001. Astron. Zhur., **78**, 558 (= Astron. Rep., **45**, 482).
Alekseev I. Yu., Baranovskii E.A., Gershberg R.E., Il'in I.V. et al., 2003. Astron. Zhur., **80**, 42 (= Astron. Rep., **47**, 312).
Alfven H. and Carlqvist P., 1967. Solar Phys., **1**, 220.
Allard F. and Hauschildt P.H., 1995. Astrophys. J., **445**, 433.
Allard F., Hauschildt P.H., Alexander D.R., and Starrfield S., 1997. Ann. Rev. Astron. Astrophys., **35**, 137.
Amado P.J., 1997. Physical properties of starspots. Thesis. Queen's University of Belfast.
Amado P.J., Doyle J.G., Byrne P.B. et al., 2000. Astron. Astrophys., **359**, 159.
Ambartsumian V.A., 1952. Trans. IAU, **8**, 665.
Ambartsumian V.A., 1954. Bull. Byurakan Astrophys. Obs., 13.
Ambartsumian V.A., 1969. Stars, Nebulae, Galaxies. Publ. House Acad. Sci. Armenian SSR. Yerevan. p. 283.
Ambartsumian V.A., 1978. Astrofizika, **14**, 367.
Ambruster C.W., 1995. Private communication.
Ambruster C. and Wood K.S., 1984. In: S.L. Baliunas and L. Hartmann (eds). Cool Stars, Stellar Systems, and the Sun. Lecture Notes in Physics, **193**, 191.
Ambruster C.W. and Wood K.S., 1986. Astrophys. J., **311**, 258.
Ambruster C., Snyder W.A., and Wood K.S., 1984. Astrophys. J., **284**, 270.
Ambruster C.W., Sciortino S., and Golub L., 1987. Astrophys. J. Suppl. Ser., **65**, 273.
Ambruster C.W., Pettersen B.R., Hawley S.L. et al., 1989a. In: B.M. Haisch and M. Rodonò (eds). Solar and Stellar Flares. Catania Astrophysical Observatory Special Publication. p. 27.
Ambruster C.W., Pettersen B.R., and Sundland S.R., 1989b. Astron. Astrophys., **208**, 198.

Ambruster C. and Fekel F., 1990. Bull. Am. Astron. Soc., **22**, 857.
Ambruster C.W., Brown A., and Fekel F.C., 1994a. In: J.P. Caillault (ed). Cool Stars, Stellar Systems, and the Sun. Astron. Soc. Pacific Conf. Ser, **64**, 348.
Ambruster C.W., Brown A., Pettersen B., and Gershberg R.E., 1994b. Bull. Am. Astron. Soc., **26**, 866.
Ambruster C.W., Brown A., Fekel F.C. et al., 1998. In: R.A. Donahue and J.A. Bookbinder (eds). Cool Stars, Stellar Systems, and the Sun. Astron. Soc. Pacific Conf. Ser., **154**, CD 1205.
Andersen B.N., 1976. Info. Bull. Var. Stars No 1084.
Andersen B.N., Kjeldseth-Moe O., and Pettersen B.R., 1986. In: E.J. Rolfe (ed). New Insight in Astrophysics. ESA SP-263, 87.
Anderson C.M., 1979. Publ. Astron. Soc. Pacific, **91**, 202.
Anderson C.M., Hartmann L.W., and Bopp B.W., 1976. Astrophys. J., **204**, L51.
Andretta V. and Giampapa M.S., 1995. Astrophys. J., **439**, 405.
Andretta V., Doyle J.G., and Byrne P.B., 1997. Astron. Astrophys., **322**, 266.
Andrews A.D., 1965. Irish Astron. J., **7**, 20.
Andrews A.D., 1966a. Publ. Astron. Soc. Pacific, **78**, 324.
Andrews A.D., 1966b. Publ. Astron. Soc. Pacific, **78**, 542.
Andrews A.D., 1982. Info. Bull. Var. Stars, No 2254.
Andrews A.D., 1989a. Astron. Astrophys., **210**, 303.
Andrews A.D., 1989b. Astron. Astrophys., **214**, 220.
Andrews A.D., 1990a. Astron. Astrophys., **227**, 456.
Andrews A.D., 1990b. Astron. Astrophys., **229**, 504.
Andrews A.D., 1990c. Astron. Astrophys., **239**, 235.
Andrews A.D., 1991. Astron. Astrophys., **245**, 219.
Andrews A.D., Chugainov P.F., Gershberg R.E., and Oskanian V.S., 1969. Info. Bull. Var. Stars, No 326.
Andrews A.D. and Marang F., 1989. Armagh Obs. preprint, ser. No 72.
Arakelian M.A., 1959. CR Acad. Sci. Armenian SSR, **29**, 35.
Arsenijevic J., 1985. Astron. Astrophys., **145**, 430.
Arzner K. and Güdel M., 2004. Astrophys. J., **602**, 363.
Aslan Z., Derman E., Akalin A., and Ozdemir T., 1992. Astron. Astrophys., **257**, 580.
Athay R.G. and Thomas R.N., 1956. Astrophys. J., **123**, 299.
Audard M., Güdel M., and Guinan E.F., 1999. Astrophys. J., **513**, L53.
Audard M., Güdel M., Drake J.J., and Kashyap V.L., 2000. Astrophys. J., **541**, 396.
Audard M., Güdel M., Skinner S.L., 2003. Astrophys. J., **589**, 983.
Avgoloupis S., 1986. Astron. Astrophys., **162**, 151.
Ayres T.R., 1999. Astrophys. J., **525**, 240.
Ayres T.R., Linsky J.L., Rodgers A.W., and Kurucz R.L., 1976. Astrophys. J., **210**, 199.

Ayres T.R., Linsky J.L., Garmire G., and Cordova F., 1979. Astrophys. J., **232**, L117.
Ayres T.R., Marstad N.C., and Linsky J.L., 1981a. Astrophys. J., **247**, 545.
Ayres T.R., Linsky J.L., Vaiana G.S. et al., 1981b. Astrophys. J., **250**, 293.
Ayres T.R., Linsky J.L., Simon T. et al., 1983a. Astrophys. J., **274**, 784.
Ayres T.R., Eriksson K., Linsky J.L., and Stencel R.E. 1983b. Astrophys. J., **270**, L17.
Ayres T.R., Stauffer J.R., Simon T. et al., 1994. Astrophys. J., **420**, L33.
Ayres T.R., Fleming T.A., Simon T. et al., 1995. Astrophys. J. Suppl. Ser., **96**, 223.
Ayres T.R., Simon T., Stauffer J.R. et al., 1996. Astrophys. J., **473**, 279.
Ayres T.R., Brown A., Harper G.M. et al., 2003. Astrophys. J., **583**, 963.
Babcock H.W., 1958. Astrophys. J. Suppl. Ser., **3**, 141.
Babel J., Queloz D., North P., and Mayor M., 1995. In: K.G. Strassmeier (ed). Stellar Surface Structure. Poster Proceedings. Inst. Astron. Univ. Wien. p. 15.
Badalyan O.G. and Livshits M.A., 1992. Astron. Zh., **69**, 138 (= Sov. Astron., **36**, 70).
Badalyan O.G. and Obridko V.N., 1984. Astron. Zh., **61**, 968 (= Sov. Astron., **28**, 564).
Baliunas S.L., 1988. In: A.K. Dupree et al. (eds). Formation and Evolution of Low Mass Stars. Kluwer, Boston. p. 319.
Baliunas S.L., 1991. In: C.P. Sonett, M.S. Giampapa, and M.S. Matthews (eds). The Sun in Time. Space Sci. Ser. University of Arizona Press. Tucson. USA. p. 809.
Baliunas S.L. and Jastrow R., 1990. Nature, **348**, 520.
Baliunas S.L. and Raymond J.C., 1984. Astrophys. J., **282**, 728.
Baliunas S.L. and Vaughan A.H., 1985. Ann. Rev. Astron. Astrophys., **23**, 379.
Baliunas S. and Soon W., 1995. Astrophys. J., **450**, 896.
Baliunas S.L. et al., 1995. Astrophys. J., **438**, 269.
Baliunas S.L., Hartmann L., Vaughan A.H. et al., 1981. Astrophys. J., **246**, 473.
Baliunas S.L., Vaughan A.H., Hartmann L. et al., 1983. Astrophys. J., **275**, 752.
Baliunas S.L., Horne J.H., Porter A. et al., 1985. Astrophys. J., **294**, 310.
Baliunas S.L., Raymond J.C., and Loeser J.G., 1986. New Insight in Astrophysics. ESA SP-263, 181.
Baliunas S.L., Nesme-Ribes E., Sokoloff D., and Soon W.H., 1996. Astrophys. J., **460**, 848.
Baliunas S.L., Donahue R.A., Soon W., and Henry G.W., 1998. In: R.A. Donahue and J.A. Bookbinder (eds). Cool Stars, Stellar Systems, and the Sun. Astron. Soc. Pacific Conf. Ser., **154**, 153.
Ball B. and Bromage G., 1995. In: J. Greiner, H.W. Duerbeck, and R.E. Gershberg (eds). Flares and Flashes. Lecture Notes in Physics, **454**, 65.

Ball B. and Bromage G., 1996. In: R. Pallavicini and A.K. Dupree (eds). Cool Stars, Stellar Systems, and the Sun. Astron. Soc. Pacific Conf. Ser., **109**, 585.
Banks T., Kilmartin P.M., and Budding E., 1991. Astrophys. Space Sci., **183**, 309.
Baranovskii E.A., Gershberg R.E., and Shakhovskoi D.N., 2001a. Astron. Zh., **78**, 78 (= Astron. Rep., **45**, 67).
Baranovskii E.A., Gershberg R.E., and Shakhovskoi D.N., 2001b. Astron. Zh., **78**, 359 (= Astron. Rep., **45**, 309).
Barbera M., Micela G., Sciortini S. et al., 1993. Astrophys. J., **414**, 846.
Barbera M., Bocchino F., Damiani F. et al., 2002. Astron. Astrophys., **387**, 463.
Barden S.C., Ramsey L.W., Fried R.E. et al., 1986. In: M. Zeilik and D.M. Gibson (eds). Cool Stars, Stellar Systems, and the Sun. Lecture Notes in Physics, **254**, 241.
Barnes J.R., Collier Cameron A., Unruh Y.C. et al., 1998. Monthly Not. Roy. Astron. Soc., **299**, 904.
Barnes S., 2003a. Astrophys. J., **586**, 464.
Barnes S., 2003b. Astrophys. J., **586**, L145.
Barrado y Navascues D., Stauffer J.R., and Randich S., 1997. In: G. Micela, R. Pallavicini, and S. Sciortino (eds). Cool Stars in Clusters and Associations: Magnetic Activity and Age Indicators. Mem. Soc. Astron. Ital., **68**, 985.
Barry D.C., 1988. Astrophys. J., **334**, 436.
Barry D.C., Cromwell R.H., Hege K., and Schoolman S.A., 1981. Astrophys. J., **247**, 210.
Barry D.C., Hege K., and Cromwell R.H., 1984. Astrophys. J., **277**, L65.
Barry D.C., Cromwell R.H., and Hege E.K., 1987. Astrophys. J., **315**, 264.
Barstow M.A., Bromage G.E., Pankiewicz G.S. et al., 1991. Nature, **343**, 635.
Basri G., 2001. In: R.J. Garcia López, R. Rebolo, and M.R. Zapatero Osorio (eds). Cool Stars, Stellar Systems, and the Sun. Astron. Soc. Pacific Conf. Ser., **223**, 261.
Basri G. and Marcy G.W., 1994. Astrophys. J., **431**, 844.
Basri G. and Marcy G.W., 1995. Astron. J., **109**, 762.
Basri G., Marcy G.W., and Valenti J.A., 1992. Astrophys. J., **390**, 622.
Basri G., Marcy G., Oppenheimer B. et al., 1996. In: R. Pallavicini and A.K. Dupree (eds). Cool Stars, Stellar Systems, and the Sun. Astron. Soc. Pacific Conf. Ser., **109**, 587.
Basri G.S. and Linsky J.L., 1979. Astrophys. J., **234**, 1023.
Bastian T.S., 1990. Solar Phys., **130**, 265.
Bastian T.S. and Bookbinder J.A., 1987. Nature, **326**, 678.
Bastian T.S., Bookbinder J.A., Dulk G.A., and Davis M., 1990. Astrophys. J., **353**, 265.
Batyrshinova V.M. and Ibragimov M.A., 2001. P. Astron. Zh., **27**, 36 (= Astron. Lett. **27**, 29).

Belloni T., 1997. In: G. Micela, R. Pallavicini, and Sciortino (eds). Cool Stars in Clusters and Associations: Magnetic Activity and Age Indicators. Mem. Soc. Astron. Ital., **68**, 993.
Belloni T. and Tagliaferri G., 1997. Astron. Astrophys., **326**, 608.
Belvedere G., Chuideri C., and Paterno L., 1982. Astron. Astrophys., **105**, 133.
Benedict G.F., Nelan E., McArthur B. et al., 1993. Publ. Astron. Soc. Pacific, **105**, 487.
Benz A.O., 1995. In: J. Greiner, H.W. Duerbeck, and R.E. Gershberg (eds). Flares and Flashes. Lecture Notes in Physics, **454**, 23.
Benz A.O. and Alef W., 1991. Astron. Astrophys., **252**, L19.
Benz A.O. and Güdel M., 1994. Astron. Astrophys., **285**, 621.
Benz A.O., Alef W., and Güdel M., 1995. Astron. Astrophys., **298**, 187.
Benz A.O., Conway J., and Güdel M., 1998. Astron. Astrophys., **331**, 596.
Berdyugin A.V., Gershberg R.E., Ilyin I.V. et al., 1995. Izv. Crimean Astrophys. Obs., **89**, 81.
Berdyugina S.V., Ilyin I.V., and Tuominen I., 2001. In: R.J. Garcia López, R. Rebolo, and M.R. Zapatero Osorio (eds). Cool Stars, Stellar Systems, and the Sun. Astron. Soc. Pacific Conf. Ser., **223**, CD 1207.
Berdyugina S.V., Pelt J., and Tuominen I., 2002. Astron. Astrophys., **394**, 505.
Berrios-Salas M.L., Guzman-Vega M., Fernandez-Labra J., and Maldini-Sanches M., 1989. Astrophys. Space Sci, **162**, 205.
Beskin G.M., Neizvestny S.I., Plakhotnichenko V.L. et al., 1988. Izv. Crimean Astrophys. Obs., **79**, 71.
Beskin G.M., Mitronova S.N., and Panferova I.P., 1995. In: J. Greiner, H.W. Duerbeck, and R.E. Gershberg (eds). Flares and Flashes. Lecture Notes in Physics, **454**, 85.
Bidelman W.P., 1954. Astrophys. J. Suppl. Ser., **1**, 175.
Blanco C., Catalano S., Marilli E., and Rodonò M., 1974. Astron. Astrophys., **33**, 257.
Blanco C., Bruca L., Catalano S., and Marilli E., 1982. Astron. Astrophys., **115**, 280.
Boesgaard A.M., 1974. Astrophys. J., **188**, 567.
Boesgaard A.M., Chesley D., and Preston G.W., 1975. Publ. Astron. Soc. Pacific, **87**, 353.
Boesgaard A.M. and Simon T., 1984. Astrophys. J., **277**, 241.
Bohn H.U., 1984. Astron. Astrophys., **136**, 338.
Bondar N.I., 1995. Astron. Astrophys. Suppl. Ser., **111**, 259.
Bondar N.I., 1996. Izv. Crimean Astrophys. Obs., **93**, 111.
Bondar N.I., 2000. Studies of manifestations of non-stationarity of stars of different ages by photometric and spectrophotometric methods. Thesis. Odessa.
Bookbinder J., 1985. Ph.D. Thesis, Harvard Univ.

Bookbinder J.A., Walter F.M., and Brown A., 1992. In: M.S. Giampapa and J.A. Bookbinder (eds). Cool Stars, Stellar Systems, and the Sun. Astron. Soc. Pacific Conf. Ser., **26**, 27.
Bopp B.W., 1974a. Publ. Astron. Soc. Pacific, **86**, 281.
Bopp B.W., 1974b. Monthly Not. Roy. Astron. Soc., **166**, 79.
Bopp B.W., 1974c. Monthly Not. Roy. Astron. Soc., **168**, 255.
Bopp B.W., 1974d. Astrophys. J., **193**, 389.
Bopp B.W., 1987. Astrophys. J., **317**, 781.
Bopp B.W. and Espenak F., 1977. Astron. J., **82**, 916.
Bopp B.W. and Evans D.S., 1973. Monthly Not. Roy. Astron. Soc., **164**, 343.
Bopp B.W. and Fekel F., 1977a. Astron. J., **82**, 490.
Bopp B.W. and Fekel F., 1977b. Publ. Astron. Soc. Pacific, **89**, 65.
Bopp B.W. and Ferland G., 1977. Publ. Astron. Soc. Pacific, **89**, 69.
Bopp B.W. and Moffett T.J., 1973. Astrophys. J., **185**, 239.
Bopp B.W. and Noah P.V., 1980. Publ. Astron. Soc. Pacific, **92**, 717.
Bopp B.W., Noah P.V., Klimke A., and Africano J., 1981. Astrophys. J., **249**, 210.
Bopp B.W., Africano J.L., Stencel R.E. et al., 1983. Astrophys. J., **275**, 691.
Borra E.F., Edwards G., and Mayor M., 1984. Astrophys. J., **284**, 211.
Bouchy F. and Carrier F., 2002. Astron. Astrophys., **390**, 205.
Brandenburg A., 1998. In: R.A. Donahue and J.A. Bookbinder (eds). Cool Stars, Stellar Systems, and the Sun. Astron. Soc. Pacific Conf. Ser., **154**, 173.
Brandenburg A., Saar S.H., and Turpin C., 1998. Astrophys. J., **498**, L51.
Bromage G.E., 1992. In: M.S. Giampapa and J.A. Bookbinder (eds). Cool Stars, Stellar Systems, and the Sun. Astron. Soc. Pacific Conf. Ser., **26**, 61.
Bromage G.E., Patchett B.E., Phillips K.J.H. et al., 1983. In: P.B. Byrne and M. Rodonò (eds). Activity in Red Dwarf Stars. Reidel, Dordrecht. p. 245.
Bromage G.E., Phillips K.J.H., Dufton P.L., and Kingston A.E., 1986. Monthly Not. Roy. Astron. Soc., **220**, 1021.
Bromage G.E., Kellett B.J., Jeffries R.D. et al., 1992. In: M.S. Giampapa and J.A. Bookbinder (eds). Cool Stars, Stellar Systems, and the Sun. Astron. Soc. Pacific Conf. Ser., **26**, 80.
Brosius J.W., Robinson R.D., and Maran S.P., 1995. Astrophys. J., **441**, 385.
Brown A., 1994. In: J.P. Caillault (ed). Cool Stars, Stellar Systems, and the Sun. Astron. Soc. Pacific Conf. Ser., **64**, 23.
Brown A., 1996. In: S. Bowyer and R.F. Malina (eds). Astrophysics in the Extreme Ultraviolet. Kluwer Acad. Publ. Dordrecht. p. 89.
Brown A., Osten R.A., Ayres T.R., and Harper G., 2002. In: F. Favata and J.J. Drake (eds). Stellar Coronae in the Chandra and XMM-Newton Era. Proc. ESTEC Symp., Noordwijk, the Netherlands, June 25–29, 2001. Astron. Soc. Pacific Conf. Ser., **277**, 223.
Brown D.N. and Landstreet J.D., 1981. Astrophys. J. 246, 899.
Bruevich V.V., Burnashev V.I., Grinin V.P. et al., 1980. Izv. Crimean Astrophys. Obs., **61**, 90.

Bruevich E.A., Katsova M.M., and Livshits M.A., 1990. Astron. Zh., **67**, 115 (= Sov. Astron., **34**, 60).
Bruevich E.A., Katsova M.M., and Sokoloff D.D., 2001. Astron. Zh., **78**, 827 (= Astron. Rep., **45**, 718).
Bruls J.H.M.J., Schüssler M., and Solanki S.K., 1999. In: C.J. Butler and J.G. Doyle (eds). Solar and Stellar Activity: Similarities and Differences. Astron. Soc. Pacific Conf. Ser., **158**, 182.
Bruning D.H., Chenoweth R.E., and Marcy G.W., 1987. In: J.L. Linsky and R.E. Stencel (eds). Cool Stars, Stellar Systems, and the Sun. Lecture Notes in Physics, **291**, 36.
Buchholz B., Ulmschneider P., and Cuntz M., 1998. Astrophys. J., **494**, 700.
Budding E., 1977. Astrophys. Space Sci., **48**, 207.
Budding E. and Zeilik M., 1987. Astrophys. J. 319, 827.
Burnasheva B.A., Gershberg R.E., Zvereva A.M. et al., 1989. Astron. Zh., **66**, 328 (= Sov. Astron., **33**, 165).
Busko I.C. and Torres C.A.O., 1978. Astron. Astrophys., **64**, 153.
Busko I.C., Quast G.R., Torres C.A.O., 1977. Astron. Astrophys., **60**, L27.
Butler C.J., 1991. In: B.P. Pettersen (ed). Stellar Flares. Mem. Soc. Astron. Ital., **62**, 243.
Butler C.J., 1992. Astron. Astrophys., **272**, 507.
Butler C.J., 1996. In: K.G. Strassmeier and J.L. Linsky (eds). Stellar Surface Structure. Kluwer, Dordrecht. p. 423.
Butler C.J. and Rodonò M., 1985. Irish Astron. J., **17**, 131.
Butler C.J., Byrne P.B., Andrews A.D., and Doyle J.G., 1981. Monthly Not. Roy. Astron. Soc., **197**, 815.
Butler C.J., Andrews A.D., Doyle J.G. et al., 1983. In: P.B. Byrne and M. Rodonò (eds). Activity in Red Dwarf Stars. Reidel, Dordrecht. p. 249.
Butler C.J., Doyle J.G., Andrews A.D. et al., 1984. ESA SP-218, 234.
Butler C.J., Rodonò M., Foing B.H., and Haisch B.M., 1986. Nature, **321**, 679.
Butler C.J., Doyle J.G., Andrews A.D. et al., 1987. Astron. Astrophys., **174**, 139.
Butler C.J., Rodonò M., and Foing B.H., 1988. Astron. Astrophys., **206**, L1.
Butler C.J., Doyle J.G., Budding E., and Foing B., 1994. Armagh Observatory preprint No 181.
Butler C.J., Doyle J.G., and Budding E., 1996. In: R. Pallavicini and A.K. Dupree (eds). Cool Stars, Stellar Systems, and the Sun. Astron. Soc. Pacific Conf. Ser., **109**, 589.
Byrne P.B., 1983. In: P.B. Byrne and M. Rodonò (eds). Activity in Red Dwarf Stars. Reidel, Dordrecht. p. 157.
Byrne P.B., 1986. In: M. Zeilik and D.M. Gibson (eds). Cool Stars, Stellar Systems, and the Sun. Lecture Notes in Physics, **254**, 320.
Byrne P.B., 1989. Solar Phys., **121**, 61.

Byrne P.B., 1992. In: C.S. Jeffrey and R.E.M. Griffin (eds). Stellar Chromospheres, Coronae, and Winds. Publ. Institute of Astronomy Univ. Cambridge.

Byrne P.B., 1993a. In: J.F. Linsky and S. Serio (eds). Physics of Solar and Stellar Coronae. Astrophysics and Science Library, **184**, 431, 489.

Byrne P.B., 1993b. Astron. Astrophys., **272**, 495.

Byrne P.B., 1993c. Astron. Astrophys., **278**, 520.

Byrne P.B., 1996. In: K.G. Strassmeier and J.L. Linsky (eds). Stellar Surface Structure. Kluwer, Dordrecht. p. 299.

Byrne P.B., 1997. In: G. Asteriadis, A. Bantelas, M.E. Contadakis et al. (eds). The Earth and the Universe. Thessaloniki. Ziti Editions. p. 59.

Byrne P.B. and Doyle J.G., 1989. Astron. Astrophys., **208**, 159.

Byrne P.B. and Doyle J.G., 1990. Astron. Astrophys. **238**, 221.

Byrne P.B. and Gary D.E., 1989. In: B.M. Haisch and M. Rodonò (eds). Solar and Stellar Flares. Catania Astrophysical Observatory Special Publication. p. 63.

Byrne P.B. and Mathioudakis M., 1993. In: J.F. Linsky and S. Serio (eds). Physics of Solar and Stellar Coronae. Astrophysics and Science Library, **184**, 435.

Byrne P.B. and McFarland J., 1980. Monthly Not. Roy. Astron. Soc., **193**, 525.

Byrne P.B. and McKay D., 1989. Astron. Astrophys., **223**, 241.

Byrne P.B. and McKay D., 1990. Astron. Astrophys., **227**, 490.

Byrne P.B. and Mullan D.J., 1992. Surface Inhomogeneities on Late-Type Stars. Lecture Notes in Physics, 397. 356p.

Byrne P.M. and Lanzafame A.C., 1994. In: J.P. Caillault (ed). Cool Stars, Stellar Systems, and the Sun. Astron. Soc. Pacific Conf. Ser., **64**, 378.

Byrne P.B., Doyle J.G., and Menzies J.W., 1985. Monthly Not. Roy. Astron. Soc., **214**, 119.

Byrne P.B., Doyle J.G., Brown A. et al., 1987. Astron. Astrophys., **180**, 172.

Byrne P.B., Butler C.J., and Lyons M.A., 1990. Astron. Astrophys., **236**, 455.

Byrne P.B., Agnew D.J., Cutispoto G. et al., 1992a. In: P.B. Byrne and D.J. Mullan (eds). Surface Inhomogeneities on Late-Type Stars. Lecture Notes in Physics, **397**, 255.

Byrne P.B., Doyle J.G., and Mattioudakis M., 1992b. In: M.S. Giampapa and J.A. Bookbinder (eds). Cool Stars, Stellar Systems, and the Sun. Astron. Soc. Pacific Conf. Ser., **26**, 438.

Byrne P.B., Mathioudakis M., Young A., and Skumanich A., 1994. In: J.P. Caillault (ed). Cool Stars, Stellar Systems, and the Sun. Astron. Soc. Pacific Conf. Ser., **64**, 375.

Byrne P.B., Eibe M.T., and Rolleston W.R.J., 1996. Astron. Astrophys., **311**, 651.

Byrne P.B., Sarro L.M., and Lanzafame A.C., 1998. In: R.A. Donahue and J.A. Bookbinder (eds). Cool Stars, Stellar Systems, and the Sun. Astron. Soc. Pacific Conf. Ser., **154**, CD 1392.

Cabestany J. and Vazquez M., 1983. Astrophys. Space Sci., **97**, 151.
Caillault J.-P., 1982. Astron. J., **87**, 558.
Caillault J.-P., 1996. In: R. Pallavicini and A.K. Dupree (eds). Cool Stars, Stellar Systems, and the Sun. Astron. Soc. Pacific Conf. Ser., **109**, 325.
Caillault J.-P. and Drake S., 1991. In: I. Tuominen, D. Moss, and G. Rüdiger (eds). The Sun and Cool Stars: Activity, Magnetism, Dynamos. Lecture Notes in Physics, **380**, 494.
Caillault J.-P. and Helfand D.J., 1985. Astrophys. J., **289**, 279.
Caillault J.-P., Drake S., and Florkowski D., 1988. Astron. J., **95**, 887.
Caillault J.-P., Gagné M., and Stauffer J., 1992. In: M.S. Giampapa and J.A. Bookbinder (eds). Cool Stars, Stellar Systems, and the Sun. Astron. Soc. Pacific Conf. Ser., **26**, 97.
Campbell B. and Cayrel R., 1984. Astrophys. J., **283**, L17.
Cappelli A., Cerruti-Sola M., Cheng C.-C., and Pallavicini R., 1989. Astron. Astrophys., **213**, 226.
Carpenter K.G. and Wing R.F., 1979. Bull. Am. Astron. Soc., **11**, 419.
Carrier F. and Bourban G., 2003. Astron. Astrophys., **406**, L23.
Carter B.D., O'Mara B.J., and Ross J.E., 1988. Monthly Not. Roy. Astron. Soc., **231**, 49.
Cash W., Charles P., Bowyer S. et al., 1979. Astrophys. J., **231**, L137.
Cash W., Charles P., and Johnson H.M., 1980. Astrophys. J., **239**, L23.
Catalano S. and Marilli E., 1983. Astron. Astrophys., **121**, 190.
Catalano S., Marilli E., and Trigilio C., 1989. Mem. Soc. Astron. Ital., **60**, 119.
Cerruti-Sola M., Cheng C.-C., and Pallavicini R., 1992. Astron. Astrophys., **256**, 185.
Charbonneau P., McIntosh S.W., Liu H.-L., and Bogdan T., 2001. Solar Phys. **203**, 321.
Cheng C.-C. and Pallavicini R., 1991. Astrophys. J., **381**, 234.
Christian D.J., 2001. In: R.J. Garcia López, R. Rebolo, and M.R. Zapatero Osorio (eds). Cool Stars, Stellar Systems, and the Sun. Astron. Soc. Pacific Conf. Ser., **223**, CD 1127.
Christian D.J. and Mathioudakis M., 2002. Astron. J., **123**, 2796.
Christian D.J. and Vennes S., 1999. Astron. J., **117**, 1852.
Christian D.J., Drake J.J., and Mathioudakis M., 1998. Astron. J., **115**, 316.
Christian D.J., Craig N., Cahill W. et al., 1999. Astron. J., **117**, 2466.
Christian D.J., Mathioudakis M., Jevremovic D. et al., 2003. Astrophys. J., **593**, L105.
Chugainov P.F., 1961. Izv. Crimean Astrophys. Obs., **26**, 171.
Chugainov P.F., 1962. Izv. Crimean Astrophys. Obs., **28**, 150.
Chugainov P.F., 1965. Izv. Crimean Astrophys. Obs., **33**, 215.
Chugainov P.F., 1966. Info. Bull. Var. Stars, No 122.
Chugainov P.F., 1968. Izv. Crimean Astrophys. Obs., **38**, 200.
Chugainov P.F., 1969a. Izv. Crimean Astrophys. Obs., **40**, 33.

Chugainov P.F., 1969b. In: L. Detre (ed). Non-Periodic Phenomena in Variable Stars. Proc. IAU Coll., No 4. Academic Press, Budapest. p. 127.
Chugainov P.F., 1972a. Izv. Crimean Astrophys. Obs., **44**, 3.
Chugainov P.F., 1972b. Izv. Crimean Astrophys. Obs., **46**, 14.
Chugainov P.F., 1973. Izv. Crimean Astrophys. Obs., **48**, 3.
Chugainov P.F., 1974. Izv. Crimean Astrophys. Obs., **50**, 93.
Chugainov P.F., 1976. Izv. Crimean Astrophys. Obs., **55**, 94.
Chugainov P.F., 1979. Transactions of the IAU. XVII A, Part 2, 139.
Chugainov P.F., 1983. Izv. Crimean Astrophys. Obs., **67**, 42.
Chugainov P.F. and Lovkaya M.N., 1988. Izv. Crimean Astrophys. Obs., **79**, 63.
Chugainov P.F. and Lovkaya M.N., 1989. Astrofizika, **30**, 243.
Chugainov P.F. and Lovkaya M.N., 1992. Izv. Crimean Astrophys. Obs., **84**, 73.
Chugainov P.F., Havlen R.J., Westerlund B.E., and White R.E., 1969. Info. Bull. Var. Stars, No 343.
Ciaravella A., Peres G., and Serio S., 1993. Solar Phys., **145**, 45.
Ciaravella A., Magghio A., and Peres G., 1997. Astron. Astrophys., **320**, 945.
Collier Cameron A., 1999. In: C.J. Butler and J.G. Doyle (eds). Solar and Stellar Activity: Similarities and Differences. Astron. Soc. Pacific Conf. Ser., **158**, 146.
Collier Cameron A. and Woods J.A., 1992. Monthly Not. Roy. Astron. Soc., **258**, 360.
Collura A., Pasquini L., and Schmitt J.H.M.M., 1988. Astron. Astrophys., **205**, 197.
Coluzzi R., De Biase G.A., Ferraro I., and Rodonò M., 1978. Mem. Soc. Astron. Ital., **49**, 709.
Connors A., Serlemitsos P.J., and Swank J.H., 1986. Astrophys. J., **303**, 769.
Contadakis M.E., 1995. Astron. Astrophys., **300**, 819.
Contadakis M.E., 1996. In: R. Pallavicini and A.K. Dupree (eds). Cool Stars, Stellar Systems, and the Sun. Astron. Soc. Pacific Conf. Ser., **109**, 597.
Contadakis M.E., 1997. In: G. Asteriadis, A. Bantelas, M.E. Contadakis et al. (eds). The Earth and the Universe.Thessaloniki. Ziti Editions. p. 67.
Covino S., Panzera M.R., Tagliaferri G., and Pallavicini R., 2001. Astron. Astrophys., **371**, 973.
Cox J.J. and Gibson D.M., 1984. Bull. Am. Astron. Soc., **16**, 900.
Craig N., Abbott M., Finley D. et al., 1997. Astrophys. J. Suppl. Ser., **113**, 131.
Cram L.E., 1982. Astrophys. J., **253**, 768.
Cram L.E. and Giampapa M.S., 1987. Astrophys. J., **323**, 316.
Cram L.E. and Mullan D.J., 1979. Astrophys. J., **234**, 579.
Cram L.E. and Mullan D.J., 1985. Astrophys. J., **294**, 626.
Cram L.E. and Woods D.T., 1982. Astrophys. J., **257**, 269.
Crannell C.J., McClintock J.E., and Moffett T., 1974. Nature, **252**, 659.
Cristaldi S. and Rodonò M., 1970. Astron. Astrophys. Suppl. Ser., **2**, 223.

Cristaldi S. and Rodonò M., 1973. Astron. Astrophys. Suppl. Ser., **10**, 47.
Cristaldi S. and Rodonò M., 1975. In: V.E. Sherwood and L. Plaut (eds). Variable Stars and Stellar Evolution. Reidel, Dordrecht. p. 75.
Cristaldi S., Godoli G., Narbone M., and Rodonò M., 1969. In: L. Detre (ed). Non-Periodic Phenomena in Variable Stars. Proc. IAU Coll., No 4. Academic Press, Budapest. p. 149.
Cristaldi S., Gershberg R.E., and Rodonò M., 1980. Astron. Astrophys., **89**, 123.
Cully S.L., Siegmund O.H.W., Vedder P.W., and Vallerga J.V., 1993. Astrophys. J., **414**, L49.
Cully S.L., Fisher G.H., Abbott M.J., and Siegmund O.H.W., 1994. Astrophys. J., **435**, 449.
Cully S.L., Fisher G.H., Hawley S.L., and Simon T., 1997. Astrophys. J., **491**, 910.
Cuntz M., Ulmschneider P., Rammacher W. et al., 1999. Astrophys. J., **522**, 1053.
Cutispoto G., 1993. Astron. Astrophys. Suppl. Ser., **102**, 655.
Cutispoto G. and Giampapa M.S., 1988. Publ. Astron. Soc. Pacific, **100**, 1452.
Cutispoto G., Tagliaferri G., de Medeiros J.R. et al., 2003. Astron. Astrophys., **397**, 987.
Danks A.C. and Lambert D.L., 1985. Astron. Astrophys., **148**, 293.
D'Antona F. and Mazzitelli I., 1985. Astrophys. J., **296**, 502.
Davidson J.K. and Neff J.S., 1977. Astrophys. J., **214**, 140.
Davis R.J., Lovell B., Palmer H.P., and Spencer R.E., 1978. Nature, **273**, 644.
de Jager C., 1968. In: K.O. Kiepenheuer (ed). Structure and Development of Active Regions. Reidel, Dordrecht. p. 481.
de Jager C., 1976. Mem. Soc. Royale des Sciences de Liege, **6**, Ser. 9, 369.
de Jager C. and Nieuwenhuijzen H., 1987. Astron. Astrophys., **177**, 217.
de Jager C., Heise J., Avgoloupis S. et al., 1986. Astron. Astrophys., **156**, 95.
de Jager C., Heise J., van Genderen A.M. et al., 1989. Astron. Astrophys., **211**, 157.
de la Reza R., Torres C.A.O., Busko I.C., 1981. Monthly Not. Roy. Astron. Soc., **194**, 829.
Del Zanna G., Landini M., Migliorini S., and Monsignori Fossi B.C., 1996. In: R. Pallavicini and A.K. Dupree (eds). Cool Stars, Stellar Systems, and the Sun. Astron. Soc. Pacific Conf. Ser., **109**, 261.
Del Zanna G., Landini M., and Mason H.E., 2002. Astron. Astrophys., **385**, 968.
Delfosse X., Forveille T., Perrier C., and Mayor M., 1998. Astron. Astrophys., **331**, 581.
Dempsey R.C., Linsky J.L., Fleming T.A., and Schmitt J.H.M.M., 1997. Astrophys. J., **478**, 358.
Dennis B.R., 1985. Solar Phys., **100**, 465.
Dennis B.R. and Schwartz R.A., 1989. Solar Phys., **121**, 75.
Dobson A.K. and Radick R.R., 1989. Astrophys. J., **344**, 907.

Donahue R.A. and Baliunas S.L., 1992. Astrophys. J., **393**, L63.
Donahue R.A. and Baliunas S.L., 1994. In: J.P. Caillault (ed). Cool Stars, Stellar Systems, and the Sun. Astron. Soc. Pacific Conf. Ser., **64**, 396.
Donahue R.A., Saar S.H., and Baliunas S.L., 1996. Astrophys. J., **466**, 384.
Donahue R.A., Dobson A.K., and Baliunas S.L., 1997. Solar Phys. **171**, 191.
Donati J.-F., 1999. Monthly Not. Roy. Astron. Soc., **302**, 457.
Donati J.-F., Semel M., Carter B.D. et al., 1997. Monthly Not. Roy. Astron. Soc., **291**, 658.
Donati J.-F., Collier Cameron A., and Petit P., 2003a. Monthly Not. Roy. Astron. Soc., **345**, 1187.
Donati J.-F., Collier Cameron A., Semel M. et al., 2003b. Monthly Not. Roy. Astron. Soc., **345**, 1145.
Dorren J.D., 1987. Astrophys. J., **320**, 756.
Dorren J.D. and Guinan E.F., 1982a. Astrophys. J., **252**, 296.
Dorren J.D. and Guinan E.F., 1982b. In: M.S. Giampapa and L. Golub (eds). Cool Stars, Stellar Systems, and the Sun. SAO Special Report, No 392/2, 49.
Dorren J.D. and Guinan E.F., 1994. Astrophys. J., **428**, 805.
Dorren J.D., Siah M.J., Guinan E.F., and McCook G.P., 1981. Astron. J., **86**, 572.
Dorren J.D., Guinan E.F., and Dewarf L.E., 1994. In: J.P. Caillault (ed). Cool Stars, Stellar Systems, and the Sun. Astron. Soc. Pacific Conf. Ser., **64**, 399.
Dorren J.D., Güdel M., and Guinan E.F., 1995. Astrophys. J., **448**, 431.
Doyle J.G., 1987. Astron. Astrophys., **177**, 201.
Doyle J.G., 1989a. Astron. Astrophys., **214**, 258.
Doyle J.G., 1989b. Astron. Astrophys., **218**, 195.
Doyle J.G., 1996a. Astron. Astrophys., **307**, 162.
Doyle J.G., 1996b. Astron. Astrophys., **307**, L45.
Doyle J.G. and Butler C.J., 1985. Nature, **313**, 378.
Doyle J.G. and Byrne P.B., 1987. In: J.L. Linsky and R.E. Stencel (eds). Cool Stars, Stellar Systems, and the Sun. Lecture Notes in Physics, **291**, 173.
Doyle J.G. and Byrne P.B., 1990. Astron. Astrophys., **238**, 221.
Doyle J.G. and Collier Cameron A., 1990. Monthly Not. Roy. Astron. Soc., **244**, 291.
Doyle J.G. and Cook J.W., 1992. Astrophys. J., **391**, 393.
Doyle J.G. and Mathioudakis M., 1990. Astron. Astrophys., **227**, 130.
Doyle J.G., Butler C.J., Haisch B.M., and Rodonò M., 1986. Monthly Not. Roy. Astron. Soc., **223**, 1.
Doyle J.G., Butler C.J., Callanan P.J. et al., 1988a. Astron. Astrophys., **191**, 79.
Doyle J.G., Butler C.J., Byrne P.B., and van den Oord G.H.J., 1988b. Astron. Astrophys., **193**, 229.
Doyle J.G., van den Oord G.H.J., and Butler C.J., 1989. Astron. Astrophys., **208**, 208.
Doyle J.G., Panagi P., and Byrne P.B., 1990a. Astron. Astrophys., **228**, 443.

Doyle J.G., Mathioudakis M., Panagi P.M., and Butler C.J., 1990b. Astron. Astrophys. Suppl. Ser., **86**, 403.
Doyle J.G., van den Oord G.H.J., Butler C.J., and Kiang T., 1990c. Astron. Astrophys., **232**, 83.
Doyle J.G., Houdebine E.R., Mathioudakis M., and Panagi P.M., 1994. Astron. Astrophys., **285**, 233.
Doyle J.G., Short C.I., Byrne P.B., and Amado P.J., 1998. Astron. Astrophys., **329**, 229.
Drake J.J., 1996. In: R. Pallavicini and A.K. Dupree (eds). Cool Stars, Stellar Systems, and the Sun. Astron. Soc. Pacific Conf. Ser., **109**, 203.
Drake J.J., 1998. Astrophys. J., **496**, L33.
Drake J.J., 1999. Astrophys. J. Suppl. Ser., **122**, 269.
Drake J.J., 2002. In: F. Favata and J.J. Drake (eds). StellarCoronae in the Chandra and XMM-Newton Era. Proc. ESTEC Symp., Noordwijk, the Netherlands, June 25–29, 2001. Astron. Soc. Pacific Conf. Ser., **277**, 75.
Drake J.J. and Kashyap V., 2001. Astrophys. J., **547**, 428.
Drake J.J. and Sarma M.J., 2003. Astrophys. J., **594**, L55.
Drake J.J., Brown A., Bowyer S. et al., 1994a. In: J.P. Caillault (ed). Cool Stars, Stellar Systems, and the Sun. Astron. Soc. Pacific Conf. Ser., **64**, 35.
Drake J.J., Stern R.A., Stringfellow G.S. et al., 1996. Astrophys. J., **469**, 828.
Drake J.J., Laming J.M., and Widing K.G., 1997. Astrophys. J., **478**, 403.
Drake J.J., Peres G., Orlando S. et al., 2000. Astrophys. J., **545**, 1074.
Drake S.A. and Caillault J.-P., 1991. Bull. Am. Astron. Soc., **23**, 942.
Drake S.A. and Ulrich R.K., 1980. Astrophys. J. Suppl. Ser., **42**, 351.
Drake S.A., Singh K.P., White N.E., and Simon T., 1994b. Astrophys. J., **436**, L87.
Dravins D., Lindegren L., Nordlund A., and VandenBerg D.A., 1993a. Astrophys. J., **403**, 385.
Dravins D., Linde P., Fredga K., and Gahm G.F., 1993b. Astrophys. J., **403**, 396.
Dravins D., Linde P., Ayres T.R. et al., 1993c. Astrophys. J., **403**, 403.
Dulk G.A., 1985. Ann. Rev. Astron. Astrophys., **23**, 169.
Dulk G.A., 1987. In: J.L. Linsky and R.E. Stencel (eds). Cool Stars, Stellar Systems, and the Sun. Lecture Notes in Physics, **291**, 72.
Duncan D.K., 1981. Astrophys. J., **248**, 651.
Duncan D.K. and 17 coauthors, 1991. Astrophys. J. Suppl. Ser., **76**, 383.
Dupree A.K, 1981. In: M.S. Giampapa and L. Golub (eds). Cool Stars, Stellar Systems, and the Sun. SAO Special Report No 392, 3.
Durney B.R., 1972. In: C.P. Sonnett, P.J. Coleman, and J.M. Wilcox (eds). Proc. First Solar Wind Conf. Washington, NASA. p. 282.
Durney B.R. and Robinson R.D., 1982. Astrophys. J., **253**, 290.
Durney B.R., Mihalas D., and Robinson R.D., 1981. Publ. Astron. Soc. Pacific, **93**, 537.
Durney B.R., De Young D.S., and Roxburgh I.W., 1993. Solar Phys. **145**, 209.
Eason E.L., Giampapa M.S., Radick R.R. et al., 1992. Astron. J., **104**, 1161.

Eaton J.A., 1992. In: P.B. Byrne and D.J. Mullan (eds). Surface Inhomogeneities on Late-Type Stars. Lecture Notes in Physics, **397**, 15.
Eaton J.A. and Hall D.S., 1979. Astrophys. J., **227**, 907.
Eaton J.A., Henry G.W., and Fekel F.C., 1996. Astrophys. J 462, 888.
Edwards P.J., 1971. Nature, Phys. Sci., **234**, 75.
Efimov Yu.S., 1970. Izv. Crimean Astrophys. Obs., **41-42**, 357.
Efimov Yu.S. and Shakhovskoy N.M., 1972. Izv. Crimean Astrophys. Obs., **45**, 111.
Eggen O.J., 1990. Publ. Astron. Soc. Pacific, **102**, 166.
Eibe M.T., Byrne P.B., Jeffries R.D., and Gunn A.G., 1999. Astron. Astrophys., **341**, 527.
Eker Z., 1994. Astrophys. J., **420**, 373.
Eker Z., 1995. Astrophys. J., **445**, 526.
Eker Z., 1996. Astrophys. J., **473**, 388.
Elgaroy O., 1988. Astron. Astrophys., **204**, 147.
Elgaroy O., Joras P., Engvold O. et al., 1988. Astron. Astrophys., **193**, 211.
Elgaroy O., Engvold O., and Carlsson M., 1990. Astron. Astrophys., **234**, 308.
Engelkemeir D., 1959. Publ. Astron. Soc. Pacific, **71**, 522.
Eritsian M.A., 1978. Bull. Byurakan Obs. **50**, 40.
Evans D.S., 1971. Monthly Not. Roy. Astron. Soc., **154**, 329.
Evans D.S., 1975. In: V.E. Sherwood and L. Plaut (eds). Variable Stars and Stellar Evolution. Reidel, Dordrecht. p. 93.
Evans D.S. and Bopp B.W., 1974. Observatory, **94**, 80.
Falchi A., Falciani R., Smaldone L.A., and Tozzi G.P., 1990. Astrophys. Lett. Comm., **28**, 15.
Favata F., 1998. In: R.A. Donahue and J.A. Bookbinder (eds). Cool Stars, Stellar Systems, and the Sun. Astron. Soc. Pacific Conf. Ser., **154**, 511.
Favata F., Barbera M., Micela G., and Sciortino S., 1995. Astron. Astrophys., **295**, 147.
Favata F., Micela G., and Sciortino S., 1997. Astron. Astrophys., **322**, 131.
Favata F., Micela G., and Reale F., 2000a. Astron. Astrophys., **354**, 1021.
Favata F., Reale F., Micela G. et al., 2000b. Astron. Astrophys., **353**, 987.
Favata F., Reale F., Micela G. et al., 2001. In: R.J. Garcia López, R. Rebolo and M.R. Zapatero Osorio (eds). Cool Stars, Stellar Systems, and the Sun. Astron. Soc. Pacific Conf. Ser., **223**, CD 1133.
Fawzy D., Rammacher W., Ulmschneider P. et al., 2002. Astron. Astrophys., **386**, pp. 971, 983, 994.
Feigelson E.D., Giampapa M.S., and Vrba F.J., 1991. In: C.P. Sonett, M.S. Giampapa and M.S. Matthews (eds). The Sun in Time. Space Sci. Ser., University of Arizona Press. Tucson. p. 658.
Feldman U., Laming J.M., and Doschek G.A., 1995. Astrophys. J., **451**, L79.
Ferland G. and Bopp B.W., 1976. Publ. Astron. Soc. Pacific, **88**, 451.
Fernandez-Figueroa M.J., de Castro E., and Rego M., 1983. Astron. Astrophys., **119**, 243.

Fernandez-Figueroa M.J., Montes D., de Castro E., and Cornide M., 1994. Astrophys. J. Suppl. Ser., **90**, 433.
Ferreira M., 2001. In: R.J. Garcia López, R. Rebolo, and M.R. Zapatero Osorio (eds). Cool Stars, Stellar Systems, and the Sun. Astron. Soc. Pacific Conf. Ser., **223**, CD 1139.
Fisher G.H. and Hawley S.L., 1990. Astrophys. J., **357**, 243.
Fisher P.L. and Gibson D.M., 1982. In: M.S. Giampapa and L. Golub (eds). Second Cambridge Workshop on Cool Stars, Stellar Systems, and the Sun. SAO Special Report No 392 II, 109.
Fleming T.A., 1998. Astrophys. J., **504**, 461.
Fleming T.A. and Giampapa M.S., 1989. Astrophys. J., **346**, 299.
Fleming T.A., Liebert J., Gioia I.M., and Maccacaro T., 1988. Astrophys. J., **331**, 958.
Fleming T.A., Gioia I.M., and Maccacaro T., 1989. Astrophys. J., **340**, 1011.
Fleming T.A., Giampapa M.S., Schmitt J.H.M.M., and Bookbinder J.A., 1993. Astrophys. J., **410**, 387.
Fleming T.A., Schmitt J.H.M.M., and Giampapa M.S., 1995. Astrophys. J., **450**, 401.
Fleming T.A., Giampapa M.S., and Schmitt J.H.M.M., 2000. Astrophys. J., **533**, 372.
Fleming T.A., Giampapa M.S., and Garza D., 2003. Astrophys. J., **594**, 982.
Flesh T.R. and Oliver J.P., 1974. Astrophys. J., **189**, L127.
Foing B.H., Crivellari L., Vladilo G. et al., 1989. Astron. Astrophys. Suppl. Ser., **80**, 189.
Fomalont E.B. and Sanders W.L., 1989. Astron. J., **98**, 279.
Fomin V.P. and Chugainov P.F., 1980. In: L.V. Mirzoyan (ed). Flare Stars, FU Ors, and Herbig-Haro objects. Publ. House Acad. Sci. Armenian SSR. Yerevan. p. 157.
Forbes T.G., 1988. In: O. Havnes, B.R. Pettersen, J.H.M.M. Schmitt, and J.E. Solheim (eds). Activity in Cool StarEnvelopes. Kluwer, Dordrecht. p. 115.
Fosbury R.A.E., 1973. Astron. Astrophys., **27**, 129.
Fosbury R.A.E., 1974. Monthly Not. Roy. Astron. Soc., **169**, 147.
Franciosini E., Randich S., and Pallavicini R., 2000. Astron. Astrophys., **357**, 139.
Franciosini E., Randich S., and Pallavicini R., 2003. Astron. Astrophys., **405**, 551.
Frasca A., Freire Ferrero R., Marilli E., and Catalano S., 2001. In: R.J. Garcia López, R. Rebolo, and M.R. Zapatero Osorio (eds). Cool Stars, Stellar Systems, and the Sun. Astron. Soc. Pacific Conf. Ser., **223**, CD 937.
Frazier E.N., 1970. Solar Phys., **14**, 89.
Frick P., Baliunas S.L., Galyagin D. et al., 1997. Astrophys. J., **483**, 426.
Friedemann C. and Gürtler J., 1975. Astron. Nach. Heft 3 B and 296, 125.
Fuhrmeister B. and Schmitt J.H.M.M., 2003. Astron. Astrophys., **403**, 247.

Fuhrmeister B., Schmitt J.H.M.M., and Wichmann R., 2004. Astron. Astrophys., **417**, 701.
Gagné M. and Caillault J.-P., 1994. In: J.P. Caillault (ed). Cool Stars, Stellar Systems, and the Sun. Astron. Soc. Pacific Conf. Ser., **64**, 408.
Gagné M., Caillault J.-P., Hartmann L.W. et al., 1994a. In: J.P. Caillault (ed). Cool Stars, Stellar Systems, and the Sun. Astron. Soc. Pacific Conf. Ser., **64**, 80.
Gagné M., Caillault J.-P., Sharkey E., and Stauffer J.R., 1994b. In: J.P. Caillault (ed). Cool Stars, Stellar Systems, and the Sun. Astron. Soc. Pacific Conf. Ser., **64**, 83.
Gagné M., Caillault J.-P., and Stauffer J.R., 1995. Astrophys. J., **450**, 217.
Gagné M., Valenti J.A., Johns-Krull Ch. et al., 1998. In: R.A. Donahue and J.A. Bookbinder (eds). Cool Stars, Stellar Systems, and the Sun. Astron. Soc. Pacific Conf. Ser., **154**, CD 1484.
Gagné M., Valenti J.A., Linsky J.L. et al., 1999. Astrophys. J., **515**, 423.
Gaidos E.J., Henry G.W., and Henry S.M., 2000. Astron. J., **120**, 1006.
Gaposhkin S., 1955. Tonantzintla y Tacubaya Bol., No 13, 39.
Garcia López R.J., Crivellari L., Beckman J.E., and Rebolo R., 1992. Astron. Astrophys., **262**, 195.
Garcia López R.J., Rebolo R., Beckman J.E., and McKeith C.D., 1993. Astron. Astrophys., **273**, 482.
Garcia-Alvarez D., Jevremovic D., Doyle J.G., and Butler C.J., 2002. Astron. Astrophys., **383**, 548.
Gary D.E., 1985. In: R.M. Hjellming and D.M. Gibson (eds). Radio Stars. Astrophys. Space Science Lib., **116**, 185.
Gary D.E., 1986. In: M. Zeilik and D.M. Gibson (eds). Cool Stars, Stellar Systems, and the Sun. Lecture Notes in Physics, **254**, 235.
Gary D.E. and Linsky J.L., 1981. Astrophys. J., **250**, 284.
Gary D.E., Linsky J.L., and Dulk G.A., 1982. Astrophys. J., **263**, L79.
Gary D.E., Byrne P.B., and Butler C.J., 1987. In: J.L. Linsky and R.E. Stencel (eds). Cool Stars, Stellar Systems, and the Sun. Lecture Notes in Physics, **291**, 106.
Gayley K.G., 1994. Astrophys. J., **431**, 806.
Gershberg R.E., 1964. Izv. Crimean Astrophys. Obs., **32**, 133.
Gershberg R.E., 1967. Izv. Crimean Astrophys. Obs., **36**, 216.
Gershberg R.E., 1968. Izv. Crimean Astrophys. Obs., **38**, 177.
Gershberg R.E., 1969. In: L. Detre (ed). Non-Periodic Phenomena in Variable Stars. Proc. IAU Coll., No 4. Academic Press, Budapest. p. 111.
Gershberg R.E., 1970a. Flares of Red Dwarf Stars. Nauka. Moscow. 168 p.
Gershberg R.E., 1970b. Astrofizika, **6**, 191.
Gershberg R.E., 1972a. Astrophys. Space Sci., **19**, 75.
Gershberg R.E., 1972b. Izv. Crimean Astrophys. Obs., **45**, 118.
Gershberg R.E., 1974a. Astron. Zh., **51**, 552.
Gershberg R.E., 1974b. Izv. Crimean Astrophys. Obs., **51**, 117.
Gershberg R.E., 1978. Low-Mass Flare Stars. Nauka. Moscow. 128 p.

Gershberg R.E., 1980. Astrofizika, **16**, 375.
Gershberg R.E., 1982. Transactions of the IAU XVIII A, 285.
Gershberg R.E., 1985. Astrofizika, **22**, 531.
Gershberg R.E., 1986. In: C.R. Cowley et al. (eds). Upper MainSequence Stars with Anomalous Abundances. Reidel, Dordrecht. p. 25.
Gershberg R.E., 1989. Mem. Soc. Astron. Ital., **60**, 263.
Gershberg R.E. and Chugaiunov P.F., 1966. Astron. Zh., **43**, 1168 (= Sov. Astron., **10**, 934).
Gershberg R.E. and Chugainov P.F., 1967. Astron. Zh., **44**, 260 (= Sov. Astron., **11**, 205).
Gershberg R.E. and Chugainov P.F., 1969. Izv. Crimean Astrophys. Obs. **40**, 7.
Gershberg R.E. and Petrov P.P., 1986. In: L.V. Mirzoyan (ed). Flare Stars and Related Objects. Publ. House Acad. Sci. Armenian SSR, Yerevan. p. 37.
Gershberg R.E. and Pikelner S.B., 1972. Comments Astrophys. Space Physics, **4**, 113.
Gershberg R.E. and Shakhovskaya N.I., 1971. Astron. Zh., **48**, 934. (= Sov. Astron., **15**, 737).
Gershberg R.E. and Shakhovskaya N.I., 1973. Nature, Phys. Sci., **242**, 85.
Gershberg R.E. and Shakhovskaya N.I., 1974. Izv. Crimean Astrophys. Obs., **49**, 73.
Gershberg R.E. and Shakhovskaya N.I., 1983. Astrophys. Space Sci., **95**, 235.
Gershberg R.E. and Shakhovskaya N.I., 1985. Transactions of the IAU XIX, A, 298.
Gershberg R.E. and Shakhovskaya N.I., 1988. Transactions of the IAU XX, A, 283.
Gershberg R.E. and Shakhovskaya N.I., 1991. Transactions of the IAU XXI, 273.
Gershberg R.E. and Shnol E.E., 1972. Izv. Crimean Astrophys. Obs., **46**, 59.
Gershberg R.E., Neshpor Yu.I., and Chugainov P.F., 1969. Izv. Crimean Astrophys. Obs., **39**, 140.
Gershberg R.E., Mogilevsky E.I., and Obridko V.N., 1987. Kinemat. Phys. Celest. Bodies, **3**, No 5, 3.
Gershberg R.E., Grinin V.P., Il'in I.V. et al., 1991a. Astron. Zh., **68**, 548 (= Sov. Astron., **35**, 269).
Gershberg R.E., Ilyin I.V., and Shakhovskaya N.I., 1991b. Astron. Zh., **68**, 959 (= Sov. Astron., **35**, 479).
Gershberg R.E., Il'in I.V., Rostopchina A.N. et al., 1993. Astron. Zh., **70**, 984 (= Astron. Rep., **37**, 497).
Gershberg R.E., Katsova M.M., Lovkaya M.N. et al., 1999. Astron. Astrophys. Suppl. Ser., **139**, 555.
Getling A.V., 2001. Astron. Zh., **78**, 661 (= Astron. Rep., **45**, 569).
Giampapa M.S., 1980. In: A.K. Dupree (ed). Proc. 1st Cambridge Workshop on Cool Stars, Stellar Systems, and the Sun. SAO SP-389, 119.

Giampapa M.S., 1983. In: J.O. Stenflo (ed). Solar and Stellar Magnetic Fields: Origins and Coronal Effects. Reidel, Dordrecht. p. 187.

Giampapa M.S., 1985. Astrophys. J., **299**, 781.

Giampapa M.S., 1987. In: J.L. Linsky and R.E. Stencel (eds). Cool Stars, Stellar Systems, and the Sun. Lecture Notes in Physics, **291**, 236.

Giampapa M.S., 1992. In: P.B. Byrne and D.J. Mullan (eds). Surface Inhomogeneities on Late-Type Stars. Lecture Notes in Physics, **397**, 90.

Giampapa M.S. and Liebert J., 1986. Astrophys. J., **305**, 784.

Giampapa M.S., Linsky J.L., Schneeberger T.J., and Worden S., 1978. Astrophys. J., **226**, 144.

Giampapa M.S., Worden S.P., Schneeberger T.J., and Cram L.E., 1981a. Astrophys. J., **246**, 502.

Giampapa M.S., Golub L., Rosner R. et al., 1981b. In: M.S. Giampapa and L. Golub (eds). Cool Stars, Stellar Systems, and the Sun. SAO SP-392, 73.

Giampapa M.S., Worden S.P., and Linsky J.L., 1982a. Astrophys. J., **258**, 740.

Giampapa M.S., Africano J.L., Klimke A. et al., 1982b. Astrophys. J., **252**, L39.

Giampapa M.S., Golub L., Peres G. et al., 1985. Astrophys. J., **289**, 203.

Giampapa M.S., Cram L.E., and Wild W.J., 1989. Astrophys. J., **345**, 536.

Giampapa M.S., Rosner R., Kashyap V. et al., 1996. Astrophys. J., **463**, 707.

Giampapa M.S., Prosser C.F., and Fleming T.A., 1998. Astrophys. J., **501**, 624.

Gibson D.M., 1983. In: S.L. Baliunas and L. Hartmann (eds). Cool Stars, Stellar Systems, and the Sun. Lecture Notes in Physics, **193**, 197.

Gibson D.M. and Fisher P.L., 1981. Proc. Southwest Reg. Conf., **6**, 33.

Gilliland R.L. and Baliunas S.L., 1987. Astrophys. J., **314**, 766.

Gilman P.A., 1983. In: J.O. Stenflo (ed). Solar and Stellar Magnetic Fields: Origins and Coronal Effects. Reidel, Dordrecht. p. 247; Astrophys. J. Suppl. Ser., **53**, 243.

Gizis J.E., 1998. Astron. J., **115**, 2053.

Gizis J.E., Monet D.G., Reid I.N., and Kirkpatrick J.D., 2000a. Monthly Not. Roy. Astron. Soc., **311**, 385.

Gizis J.E., Monet D.G., Reid I.N. et al., 2000b. Astron. J., **120**, 1085.

Glebocki R., Musielak G., and Stawikowski A., 1980. Acta Astron. **30**, 453.

Gliese W. and Jahreiss H., 1991. In: The ADC CD-ROM: Selected Astronomical Catalogs, 1, NASAADC, Greenbelt MD.

Golub L., 1983. In: P.B. Byrne and M. Rodonò (eds). Activity in Red Dwarf Stars. Reidel, Dordrecht., 83.

Golub L., Harnden F.R., Pallavicini R. et al., 1982. Astrophys. J. **253**, 242.

Gondoin Ph., Giampapa M.S., and Bookbinder J.A., 1985. Astrophys. J., **297**, 710.

Gordon K.C. and Kron G.E., 1949. Publ. Astron. Soc. Pacific, **61**, 210.

Gotthelf E.V., Jalota L., Mukai K., and White N.E., 1994. Astrophys. J., **436**, L91.

Grandpierre A., 1986. In: L.V. Mirzoyan (ed). Flare Stars and Related Objects. Publ. House Acad. Sci. Armenian SSR, Yerevan. p. 176.
Grandpierre A., 1988. In: O. Havnes, B.R. Pettersen, J.H.M.M. Schmitt, and J.E. Solheim (eds). Activity in Cool Star Envelopes. Kluwer, Dordrecht. p. 159.
Grandpierre A., 1991. In: B.R. Pettersen (ed). Stellar Flares. Mem. Soc. Astron. Ital., **62**, 401.
Grandpierre A. and Melikian N.D., 1985. Astrophys. Space Sci., **116**, 189.
Granzer Th., Schüssler M., Caligari P., and Strassmeier K.G., 2000. Astron. Astrophys., **355**, 1087.
Gray D.F., 1984. Astrophys. J., **277**, 640.
Gray D.F., Baliunas S.L., Lockwood G.W., and Skiff B.A., 1996. Astrophys. J., **465**, 945.
Greenstein J.L., 1950. Publ. Astron. Soc. Pacific, **62**, 156.
Greenstein J.L. and Arp H., 1969. Astrophys. Lett. 3, 149.
Grindlay J.E., 1970. Astrophys. J., **162**, 187.
Grinin V.P., 1973. Izv. Crimean Astrophys. Obs., **48**, 58.
Grinin V.P., 1976. Izv. Crimean Astrophys. Obs., **55**, 179.
Grinin V.P., 1979. Izv. Crimean Astrophys. Obs., **59**, 154.
Grinin V.P., 1980a. In: L.V. Mirzoyan (ed). Flare Stars, FUOrs, and Herbig-Haro Objects. Publ. House Acad. Sci. Armenian SSR, Yerevan. p. 23.
Grinin V.P., 1980b. Izv. Crimean Astrophys. Obs., **62**, 54.
Grinin V.P., 1991. In: B.R. Pettersen (ed). Stellar Flares. Mem. Soc. Astron. Ital., **62**, 389.
Grinin V.P. and Sobolev V.V., 1977. Astrofizika, **13**, 587.
Grinin V.P. and Sobolev V.V., 1988. Astrofizika, **28**, 355.
Grinin V.P. and Sobolev V.V., 1989. Astrofizika, **31**, 527.
Grinin V.P., Loskutov V.M., and Sobolev V.V., 1993. Astron. Zh., **70**, 350 (= Astron. Rep., **37**, 182).
Grossmann-Doerth U. and Solanki S.K., 1990. Astron. Astrophys., **238**, 279.
Güdel M., 1992. Astron. Astrophys., **264**, L31.
Güdel M., 1994. Astrophys. J. Suppl. Ser., **90**, 743.
Güdel M., 1997. Astrophys. J., **480**, L121.
Güdel M. and Audard M., 2001. In: R.J. Garcia-López, R. Rebolo, and M.R. Zapatero Osorio (eds). Cool Stars, Stellar Systems, and the Sun. Astron. Soc. Pacific Conf. Ser., **223**, 961.
Güdel M. and Benz A.O., 1989. Astron. Astrophys., **211**, L5.
Güdel M. and Benz A.O., 1993. Astrophys. J., **405**, L63.
Güdel M. and Benz A.O., 1996. In: A.R. Taylor and J.M. Paredes (eds). Radio Emission from the Stars and the Sun. Astron. Soc. Pacific Conf. Ser., **93**, 303.
Güdel M. and Gaidos E.J., 2001. In: R.J. Garcia López, R. Rebolo, and M.R. Zapatero Osorio (eds). Cool Stars, Stellar Systems, and the Sun. Astron. Soc. Pacific Conf. Ser., **223**, CD 662.

Güdel M. and Schmitt J.H.M.M., 1996. In: H.U. Zimmermann, J. Trümper, and H. Yorke (eds). Roentgenstrahlung from the Universe. MPE Report, **263**, 37.
Güdel M. and Zucker A., 2001. In: P.C.H. Martens and S. Tsuruta (eds). Highly Energetic Physical Processes and Mechanisms for Emission from Astrophysical Plasmas. Proc. IAU Symp. No 195., 393.
Güdel M., Benz A.O., Bastian T.S. et al., 1989. Astron. Astrophys., **220**, L5.
Güdel M., Schmitt J.H.M.M., Bookbinder J.A., and Fleming T.A., 1993. Astrophys. J., **415**, 236.
Güdel M., Schmitt J.H.M.M., and Benz A.O., 1994a. Science, **265**, 933.
Güdel M., Schmitt J.H.M.M., Kürster M., and Benz A.O., 1994b. In: J.-P. Caillault (ed). Cool Stars, Stellar Systems, and the Sun. Astron. Soc. Pacific Conf. Ser., **64**, 86.
Güdel M., Schmitt J.H.M.M., Benz A.O., and Elias N.M., 1995a. Astron. Astrophys., **301**, 201.
Güdel M., Schmitt J.H.M.M., and Benz A.O., 1995b. Astron. Astrophys., **293**, L49.
Güdel M., Schmitt J.H.M.M., and Benz A.O., 1995c. Astron. Astrophys., **302**, 775.
Güdel M., Benz A.O., Schmitt J.H.M.M., and Skinner S.L., 1996. Astrophys. J., **471**, 1002.
Güdel M., Guinan E.F., Mewe R. et al., 1997a. Astrophys. J., **479**, 416.
Güdel M., Guinan E.F., and Skinner S.L., 1997b. Astrophys. J., **483**, 947.
Güdel M., Guinan E.F., and Skinner S.L., 1998. In: R.A. Donahue and J.A. Bookbinder (eds). Cool Stars, Stellar Systems, and the Sun. Astron. Soc. Pacific Conf. Ser., **154**, CD 1041.
Güdel M., Audard M., Guinan E.F. et al., 2001a. In: R. Garcia López, R. Rebolo, and M.R. Zapatero Osorio (eds). Cool Stars, Stellar Systems, and the Sun. Astron. Soc. Pacific Conf. Ser., **223**, CD 1085.
Güdel M., Audard M., Magee H. et al., 2001b. Astron. Astrophys., **365**, L344.
Güdel M., Audard M., Skinner S.L., and Horvath M.I., 2002. Astrophys. J., **580**, L73.
Güdel M., Arzner K., Audard M., and Mewe R., 2003a. Astron. Astrophys., **403**, 155.
Güdel M., Audard M., Kashyap V.L. et al., 2003b. Astrophys. J., **582**, 423.
Güdel M., Audard M., Reale F. et al., 2004. Astron. Astrophys., **416**, 713.
Guinan E.F., McCook G.P., Fragola J.L. et al., 1982. Astron. J., **87**, 893.
Gunn A.G., Doyle J.G., Mathioudakis M., and Avgoloupis S., 1994a. Astron. Astrophys., **285**, 157.
Gunn A.G., Doyle J.G., Mathioudakis M. et al., 1994b. Astron. Astrophys., **285**, 489.
Gurzadian G.A., 1965. Astrofizika, **1**, 319.
Gurzadian G.A., 1969. Astrofizika, **5**, 383.
Gurzadian G.A., 1970. Tonantzintla y Tacubaya Bol., **5**, 255.
Gurzadian G.A., 1971a. Tonantzintla y Tacubaya Bol., **6**, 39.

Gurzadian G.A., 1971b. Astron. Astrophys., **13**, 348.
Gurzadian G.A., 1977. Astrophys. Space Sci., **48**, 313.
Gurzadian G.A., 1984. Astrophys. Space Sci. **106**, 1.
Gurzadian G.A., 1986. Astrophys. Space Sci. **125**, 127.
Haisch B.M., 1983. In: P.P. Byrne and M. Rodonò (eds). Activityin Red-Dwarf Stars. Reidel, Dordrecht. p. 255.
Haisch B.M., 1989. Astron. Astrophys., **219**, 317.
Haisch B.M. and Basri G., 1985. Astrophys. J., Suppl. Ser., **58**, 179.
Haisch B.M. and Linsky J.L., 1980. Astrophys. J., **236**, L33.
Haisch B.M., Linsky J.L., Lampton M. et al., 1977. Astrophys. J., **213**, L119. H.
Haisch B.M., Linsky J.L., Slee O.B. et al., 1978. Astrophys. J., **225**, L35.
Haisch B.M., Linsky J.L., Harnden F.R. et al., 1980. Astrophys. J., **242**, L99.
Haisch B.M., Linsky J.L., Slee O.B. et al., 1981. Astrophys. J., **245**, 1009.
Haisch B.M., Linsky J.L., Bornmann P.L. et al., 1983. Astrophys. J., **267**, 280.
Haisch B.M., Butler C.J., Doyle J.G., and Rodonò M., 1987. Astron. Astrophys., **181**, 96.
Haisch B.M., Schmitt J.H.M.M., Rodonò M., and Gibson D.M., 1990a. Astron. Astrophys., **230**, 419.
Haisch B.M., Butler C.J., Foing B., et al., 1990b. Astron. Astrophys., **232**, 387.
Haisch B., Strong K.T., and Rodonò M., 1991. Ann. Rev. Astron. Astrophys., **29**, 275.
Haisch B., Antunes A., and Schmitt J.H.M.M., 1995. Science, **268**, 1327.
Hall D.S., 1972. Astron. J., **84**, 323.
Hall D.S., 1994. I. A. P. P. Communication No 54, 1.
Hall D.S., 1996. In: K.G. Strassmeier and J.L. Linsky (eds). Stellar Surface Structure. Kluwer, Dordrecht. p. 217.
Hall D.S., Henry G.W., and Sowell J.R., 1989. Astron. J., **99**, 396.
Hall P.B., 2002. Astrophys. J., **564**, L89.
Hallam K.L. and Wolff C.L., 1981. Astrophys. J., **248**, L73.
Hallam K.L., Altner B., and Endal A.S., 1991. Astrophys. J., **372**, 610.
Hambarian V.V., 1982. Astrofizika, **18**, 654.
Harnden F.G., Adams N.R., Damiani F. et al., 2001. Astrophys. J., **547**, L141.
Haro G. and Chavira E., 1955. Bol. Observ. Tonantzintla, **12**, 3.
Haro G. and Parsamian E., 1969. Bol. Observ. Tonantzintla, **5**, 45.
Haro G., Chavira E., and Gonzalez G., 1982. Bol. Inst.Tonantzintla, **3**, 3.
Harris D.E. and Johnson H.M., 1985. Astrophys. J., **294**, 649.
Hartmann L., 1987. In: J.L. Linsky and R.E. Stencel (eds). Cool Stars, Stellar Systems, and the Sun. Lecture Notes in Physics, **291**, 3.
Hartmann L. and Anderson C.M., 1977. Astrophys. J., **215**, 188.
Hartmann L.W. and Noyes R.W., 1987. Ann. Rev. Astron. Astrophys., **25**, 271.
Hartmann L. and Rosner R., 1979. Astrophys. J., **230**, 802.

Hartmann L., Davis R., Dupree A.K. et al., 1979. Astrophys. J., **233**, L69.
Hartmann L., Bopp B.W., Dussault M. et al., 1981. Astrophys. J., **249**, 662.
Hartmann L., Baliunas S.L., Duncan D.K., and Noyes R.W., 1984a. Astrophys. J., **279**, 778.
Hartmann L., Soderblom D.R., Noyes R.W. et al., 1984b. Astrophys. J., **276**, 254.
Haruthyunian H.A., Krikorian R.A., and Nikoghossian A.G., 1979. Astrofizika, **15**, 431.
Hatzes A.P., 1995. In: K.G. Strassmeier (ed). Poster Proceedings of the IAU Symp., No 176, Stellar Surface Structure. Vienne. p. 90.
Haupt W. and Schlosser W., 1974. Astron. Astrophys., **37**, 219.
Hawley S.L., 1989. In: B.M. Haisch and M. Rodonò (eds). Solar and Stellar Flares. Catania Astrophysical Observatory Special Publication, p. 49.
Hawley S.L. and Feigelson E.D., 1994. In: J.P. Caillault (ed). Cool Stars, Stellar Systems, and the Sun. Astron. Soc. Pacific Conf. Ser., **64**, 89.
Hawley S.L. and Fisher G.H., 1992. Astrophys. J. Suppl. Ser., **78**, 565.
Hawley S.L. and Pettersen B.R., 1991. Astrophys. J., **378**, 725.
Hawley S.L., Fisher G.H., Simon T. et al., 1995. Astrophys. J., **453**, 464.
Hawley S.L., Gizis J.E., and Reid I.N., 1996. Astron. J., **112**, 2799.
Hawley S.L., Allred J.C., Johns-Krull Ch.M. et al., 2003. Astrophys. J., **597**, 535.
Hayrapetian V.S. and Nikoghossian A.G., 1989. Astrofizika, **30**, 534.
Hayrapetian V.S., Vikhrev V.V., Ivanov V.V., and Rozanov G.A., 1990. Astrofizika, **32**, 405.
Heise J., Brinkman A.C., Schrijver J., et al., 1975. Astrophys. J., **202**, L73.
Helfand D.J. and Caillault J.-P., 1982. Astrophys. J., **253**, 760.
Hempelmann A., Schmitt J.H.M.M., Schultz M. et al., 1995. Astron. Astrophys., **294**, 515.
Hempelmann A., Schmitt J.H.M.M., and Stępień K., 1996. Astron. Astrophys., **305**, 284.
Hempelmann A., Schmitt J.H.M.M., Baliunas S.L., and Donahue R.A., 2003. Astron. Astrophys., **406**, L39.
Henoux J.-C., Aboudarham J., Brown J.C. et al., 1990. Astron. Astrophys., **233**, 577.
Herbig G.H., 1956. Publ. Astron. Soc. Pacific, 68., 531.
Herbig G.H., 1962. Proc. Symp. on Stellar Evolution. La Plata Obs. Argentina. Contr. Lick Obs., No 134.
Herbig G.H., 1965. Astrophys. J., **141**, 588.
Herbig G.H., 1985. Astrophys. J., **289**, 269.
Herbst W. and Miller J.R., 1989. Astron. J., **97**, 891.
Higgins C.S., Solomon L.H., and Bateson F.M., 1968. Austr. J. Phys., **21**, 725.
Hodgkin S.T., Jameson R.F., and Steele, 1995. Monthly Not. Roy. Astron. Soc., **274**, 869.
Hoffleit D., 1952. Harvard Bull. No 921.

Holman G.D., 1986. In: M. Zeilik and D.M. Gibson (eds). Cool Stars, Stellar Systems, and the Sun. Lecture Notes in Physics, **254**, 271.
Horne J.H. and Baliunas S.L., 1986. Astrophys. J., **302**, 757.
Houdebine E.R., 1992. Irish Astron. J., **20**, 213.
Houdebine E.R., 2003. Astron. Astrophys., **397**, 1019.
Houdebine E.R. and Doyle J.G., 1994a. Astron. Astrophys., **289**, 169 and 185.
Houdebine E.R. and Doyle J.G., 1994b. In: J.P. Caillault (ed). Cool Stars, Stellar Systems, and the Sun. Astron. Soc. Pacific Conf. Ser., **64**, 285.
Houdebine E.R., Foing B.H., and Rodonò M., 1990. Astron. Astrophys., **238**, 249.
Houdebine E.R., Butler C.J., Panagi P.M. et al., 1991. Astron. Astrophys., Suppl. Ser., **87**, 33.
Houdebine E.R., Foing B.H., Doyle J.G., and Rodonò M., 1993a. Astron. Astrophys., **274**, 245.
Houdebine E.R., Foing B.H., Doyle J.G., and Rodonò M., 1993b. Astron. Astrophys., **278**, 109.
Houdebine E.R., Doyle J.G., and Koscielecki M., 1995. Astron. Astrophys., **294**, 773.
Houdebine E.R., Mathioudakis M., Doyle J.G., and Foing B.H., 1996. Astron. Astrophys., **305**, 209.
Hubrig S., Plachinda S.I., Hünsch M., and Schroder K.-P., 1994. Astron. Astrophys., **291**, 890.
Hudson H.S., 1978. Solar Phys., **57**, 237.
Hünsch M. and Reimers D., 1995. Astron. Astrophys., **296**, 509.
Hünsch M., Schmitt J.H.M.M., and Voges W., 1998. Astron. Astrophys. Suppl. Ser., **132**, 155.
Hünsch M., Schmitt J.H.M.M., Sterzik M.F., and Voges W., 1999. Astron. Astrophys., Suppl. Ser., **135**, 319.
Hünsch M., Weidner C., and Schmitt J.H.M.M., 2003. Astron. Astrophys., **402**, 571.
Huovelin J. and Saar S.H., 1991. In: I. Tuominen, D. Moss, and G. Rüdiger (eds). Lecture Notes in Physics, **380**, 420.
Huovelin J., Saar S.H., and Tuominen I., 1988. Astrophys. J., **329**, 882.
Inda M., Makishima K., Kohmura Y. et al., 1994. Astrophys. J., **420**, 143.
Ishida K., Mahasenaputra, Ichimura K., and Shimizu Y., 1991. Astrophys. Space Sci., **182**, 227.
Jackson P.D., Kundu M.R., and White S.M., 1987a. Astrophys. J., **316**, L85.
Jackson P.D., Kundu M.R., and White S.M., 1987b. In: J.L. Linsky and R.E. Stencel (eds). Cool Stars, Stellar Systems, and the Sun. Lecture Notes in Physics, **291**, 103.
Jackson P.D., Kundu M.R., and White S.M., 1989. Astron. Astrophys., **210**, 284.
Jackson P.D., Kundu M.R., and Kassim N., 1990. Solar Phys., **130**, 391.
James D.J., Jardine M.M., Jeffries R.D. et al., 2000. Monthly Not. Roy. Astron. Soc., **318**, 1217.

Jardine M., 2004. Astron. Astrophys., **414**, L5.
Jardine M. and Unruh Y.C., 1999. Astron. Astrophys., **346**, 883.
Jarrett A.H. and Eksteen J.P., 1972. Mon. Notes Astron. Soc. South Africa, **31**, 37.
Jarrett A.H. and Grabner G., 1976. Info. Bull. Var. Stars, No 1221.
Jeffers S.V., Barnes J.R., and Collier Cameron A., 2002. Monthly Not. Roy. Astron. Soc., **331**, 666.
Jeffries R.D., 1999. In: C.J. Butler and J.G. Doyle (eds). Solar and Stellar Activity: Similarities and Differences. Astron. Soc. Pacific Conf. Ser., **158**, 75.
Jeffries R.D. and Bromage G.E., 1993. Monthly Not. Roy. Astron. Soc., **260**, 132.
Jeffries R.D. and Tolley A.J., 1998. Monthly Not. Roy. Astron. Soc., **300**, 331.
Jeffries R.D., Elliott K.H., Kellett B.J., and Bromage G.E., 1993.Monthly Not. Roy. Astron. Soc., **265**, 81.
Jeffries R.D., Thurston M.R., and Pye J.P., 1997. Monthly Not. Roy. Astron. Soc., **287**, 350.
Jetsu L., 1993. Astron. Astrophys., **276**, 345.
Jevremovic D., Butler C.J., Drake S.A. et al., 1998a. Astron. Astrophys., **338**, 1057.
Jevremovic D., Houdebine E.R., and Butler C.J., 1998b. In: R.A. Donahue and J.A. Bookbinder (eds). Cool Stars, Stellar Systems, and the Sun. Astron. Soc. Pacific Conf. Ser., **154**, CD1500.
Jevremovic D., Doyle J.G., and Short C.I., 2000. Astron. Astrophys., **358**, 575.
Jevremovic D., Doyle J.G., and Butler C.J., 2001. In: R.J. Garcia López, R. Rebolo, and M.R. Zapatero Osorio (eds). Cool Stars, Stellar Systems, and the Sun. Astron. Soc. Pacific Conf. Ser., **223**, CD 815.
Johns-Krull C.M. and Valenti J.A., 1996. Astrophys. J., **459**, L95.
Johnson H.M., 1981. Astrophys. J., **243**, 234.
Johnson H.M., 1983a. In: P.B. Byrne and M. Rodonò (eds). Activity in Red Dwarf Stars. Reidel, Dordrecht. p. 109.
Johnson H.M., 1983b. Astrophys. J., **273**, 702.
Johnson H.M., 1986. Astrophys. J., **303**, 470.
Johnson H.M., 1987. Astrophys. J., **316**, 458.
Johnson H.L. and Mitchell R.I., 1958. Astrophys. J., **128**, 31.
Jones H.R.A. and Tsuji T., 1997. Astrophys. J., **480**, L39.
Jordan C. and Montesinos B., 1991. Monthly Not. Roy. Astron. Soc., **252**, L21.
Jordan C., Ayres T.R., Brown A. et al., 1987. Monthly Not. Roy. Astron. Soc., **225**, 903.
Jordan C., McMurry A.D., Sim S.A., and Arulvel M., 2001. Monthly Not. Roy. Astron. Soc., **322**, L5.
Joy A.H., 1958. Publ. Astron. Soc. Pacific, **70**, 505.

Joy A.H., 1960. In: J.L. Greenstein (ed). Stellar Atmospheres. University Chicago Press. Chicago. p. 653.
Joy A.H. and Abt H.A., 1974. Astrophys. J. Suppl. Ser., **28**, 1.
Joy A.H. and Humason M.L., 1949. Publ. Astron. Soc. Pacific, **61**, 133.
Joy A.H. and Wilson R.E., 1949. Astrophys. J., **109**, 231.
Kaastra J.S., 1992. An X-ray Spectral Code for Optically Thin Plasmas (SRON-Leiden report, updated version 2.0).
Kahler S. and Shulman S., 1972. Nature, Phys. Sci., **237**, 101.
Kahler S., Golub L., Harnden F.R. et al., 1982. Astrophys. J., **252**, 239.
Kahn F.D., 1969. Nature, **222**, 1130.
Kahn F.D., 1974. Nature, **250**, 125.
Kahn S.M., Linsky J.L., Mason K.O. et al., 1979. Astrophys. J., **234**, L107.
Kandel R., 1967. Ann. d'Astrophys., **30**, 999.
Kang Y.W. and Wilson R.E., 1988. Astron. J., **97**, 848.
Karpen J.T., Crannell C.J., Hobbs R.W. et al., 1977. Astrophys. J., **216**, 479.
Kashyap V.L., Drake J.J., Güdel M., and Audard M., 2002. Astrophys. J., **580**, 1118.
Kasinsky V.V. and Sotnikova R.T., 1988. Issledovaniya po Geomagnetismu, Aeronomii i Fizike Solntsa. Issue 83, 99.
Kasinsky V.V. and Sotnikova R.T., 1989. Issledovaniya po Geomagnetismu, Aeronomii i Fizike Solntsa. Issue 87, 43.
Kasinsky V.V. and Sotnikova R.T., 1997. Astron. Astrophys., Trans. **12**, 313.
Kato T., 1976. Astrophys. J. Suppl. Ser., **30**, 397.
Katsova M.M., 1981a. Astron Tsirk. No 1154, 1.
Katsova M.M., 1981b. Astron. Zh., **58**, 350 (= Sov. Astron., **25**, 197).
Katsova M.M., 1990. Astron. Zh., **67**, 1219 (= Sov. Astron., **34**, 614).
Katsova M.M., 1999. Activity of stars of late spectral types. Thesis. Moscow.
Katsova M.M. and Livshits M.A., 1986. In: L.V. Mirzoyan (ed). Flare Stars and Related Objects. Publ. House Acad. Sci. Armenian SSR, Yerevan. p. 183.
Katsova M.M. and Livshits M.A., 1988. In: O. Havnes, B.R. Pettersen et al. (eds). Activity in Cool Star Envelopes. Kluwer, Dordrecht. p. 143.
Katsova M.M. and Livshits M.A., 1989. Astron. Zh., **66**, 307 (= Sov. Astron., **33**, 155).
Katsova M.M. and Livshits M.A., 1991. Astron. Zh., **68**, 131 (= Sov. Astron., **35**, 65).
Katsova M.M. and Livshits M.A., 1992. Astron. Astrophys., Trans. **3** No 1, 67.
Katsova M.M. and Tsikoudi V., 1992. Astron. Zh., **69**, 821 (= Sov. Astron., **36**, 421).
Katsova M.M. and Tsikoudi V., 1993. Astrophys. J., **402**, L9.
Katsova M.M., Kosovichev A.G., and Livshits M.A., 1980. P. Astron. Zh., **6**, 498 (= Sov. Astron. Let. **6**, 275).
Katsova M.M., Badalyan O.G., and Livshits M.A., 1987. Astron. Zh., **64**, 1243 (= Sov. Astron., **31**, 652).

Katsova M.M., Livshits M.A., Butler C.J., and Doyle J.G., 1991. Monthly Not. Roy. Astron. Soc., **250**, 402.
Katsova M., Tsikoudi V., and Livshits M., 1993. In: J.L. Linsky and S. Serio (eds). Physics of Solar and Stellar Coronae. Kluwer, Dordrecht. p. 483.
Katsova M.M., Drake J.J., and Livshits M.A., 1995. In: J. Greiner, H.W. Duerbeck, and R.E. Gershberg (eds). Flare and Flashes. Lecture Notes in Physics, 454. Springer, Berlin. p. 146.
Katsova M.M., Boiko A.Ya., and Livshits M.A., 1997. Astron. Astrophys., **321**, 549.
Katsova M.M., Drake J.J., and Livshits M.A., 1999a. Astrophys. J., **510**, 986.
Katsova M.M., Hawley S.L., Abbett W.P., and Livshits M.A., 1999b. Preprint No 4 (1120). IZMIRAN. Moscow.
Katsova M.M., Livshits M.A., and Schmitt J.H.M.M., 2002. In: F. Favata and J.J. Drake (eds). Stellar Coronae in the Chandra and XMM-Newton Era. Proc. ESTEC Symp., Noordwijk, the Netherlands, June 25–29, 2001. Astron. Soc. Pacific Conf. Ser., **277**, 515.
Katsova M.M., Livshits M.A., and Belvedere G., 2003. Solar Phys., **216**, 353.
Kelch W.L., 1978. Astrophys. J., **222**, 931.
Kelch W.L., Linsky J.L., and Worden S.P., 1979. Astrophys. J., **229**, 700.
Kellett B.J. and Tsikoudi V., 1997. Monthly Not. Roy. Astron. Soc., **288**, 411.
Kellett B.J., Bromage G.E., Brown A. et al., 1995. Astrophys. J., **438**, 364.
Kiljachkov N.N. and Shevchenko V.S., 1980. In: L.V. Mirzoyan (ed). Flare Stars, FUOrs, and Herbig-Haro Objects. Publ. House Acad.Sci. Armenian SSR, Yerevan. p. 31.
Kiljachkov N.N., Melikian N.D., Mirzoyan L.V., and Shevchenko V.S., 1979. Astrofizika, **15**, 605.
Kim Y.-C. and Demarque P., 1996. Astrophys. J., **457**, 340.
Kirkpatrick J.D., Kelly D.M., Rieke G.H. et al., 1993. Astrophys. J., **402**, 643.
Kitchatinov L.L., Jardine M., and Donati J.-F., 2000. Astron. Astrophys., **318**, 1171.
Kitchatinov L.L., Jardine M., and Collier Cameron A., 2001. Astron. Astrophys., **374**, 250.
Kjeldsen H., Bedding T.R., Frandsen S., and Dall T.H., 1999. Monthly Not. Roy. Astron. Soc., **303**, 579.
Kjurkchieva D.P., 1987. Astrophys. Space Sci., **138**, 141.
Kjurkchieva D.P., 1989. Astrophys. Space Sci., **155**, 203; 159, 333.
Kjurkchieva D.P., 1990. Astrophys. Space Sci., **172**, 255.
Kjurkchieva D.P. and Shkodrov V.G., 1986. Astrophys. Space Sci., **124**, 27.
Klimishin I.A., 1969. Tsirk. Astron. Obs. Lvov Univ., No 43, 8.
Knobloch E., Rosner R., and Weiss N.O., 1981. Monthly Not. Roy. Astron. Soc., **197**, 45.
Kochukhov O.P., Piskunov N.E., Valenti J.A., and Johns-Krull C.M., 2001. In: R.J. Garcia López, R. Rebolo, and M.R. Zapatero Osorio (eds). Cool Stars, Stellar Systems, and the Sun. Astron. Soc. Pacific Conf. Ser., **223**, CD 985.

Kodaira K., 1977. Astron. Astrophys., **61**, 625.
Kodaira K., 1986. In: L.V. Mirzoyan (ed). Flare Stars and Related Objects. Publ. House Acad. Sci. Armenian SSR, Yerevan. p. 43.
Kodaira K. and Ichimura K., 1980. Publ. Astron. Soc. Japan, **32**, 451.
Kodaira K. and Ichimura K., 1982. Publ. Astron. Soc. Japan, **34**, 21.
Kodaira K., Ichimura K., and Nishimura S., 1976. Publ. Astron. Soc. Japan, **28**, 665.
Kolesov A.K. and Sobolev V.V., 1990. Astron. Zh., **67**, 357 (= Sov. Astron., **34**, 179).
Kopp R.A. and Pneumann G.W., 1976. Solar Phys., **50**, 85.
Kopp R.A. and Poletto G., 1984. Solar Phys., **93**, 351.
Kopp R.A. and Poletto G., 1990. In: G. Wallerstein (ed). Cool Stars, Stellar Systems, and the Sun. Astron. Soc. Pacific Conf. Ser., **9**, 119.
Kopp R.A. and Poletto G., 1992. In: P.B. Byrne and D.J. Mullan (eds). Surface Inhomogeneities on Late-Type Stars. Lecture Notes in Physics, **397**, 295.
Korotin S.A. and Krasnobabtsev V.I., 1985. Izv. Crimean Astrophys. Obs., **73**, 131.
Korovyakovskaya A.A., 1972. Astrofizika, **8**, 247.
Kostjuk N.D. and Pikel'ner S.B., 1974. Astron. Zh., **51**, 1002 (= Sov. Astron., **18**, 590).
Kövári Z., 1999. In: C.J. Butler and J.G. Doyle (eds). Solar and Stellar Activity: Similarities and Differences. Astron. Soc. Pacific Conf. Ser., **158**, 166.
Kövári Z. and Bartus J., 1997. Astron. Astrophys., **323**, 801.
Kövári Z., Strassmeier K.G., Granzer T. et al., 2004. Astron. Astrophys., **417**, 1047.
Kraft R., 1967. Astrophys. J., **150**, 551.
Kraft R.P. and Greenstein J.L., 1969. In: S.S. Kumar (ed). Low-Luminosity Stars. Gordon and Breach Sci. Publ., New-York. p. 65.
Krasnobabtsev V.I. and Gershberg R.E., 1975. Izv. Crimean Astrophys. Obs., **53**, 154.
Krishnamurthi A., Leto G., and Linsky J.L., 1999. Astron. J., **118**, 1369.
Krishnamurthi A., Reynolds C.S., Linsky J.L. et al., 2001a. Astron. J., **121**, 337.
Krishnamurthi A., Terndrup D.M., Linsky J.L., and Leto G., 2001b. In: R.J. Garcia López, R. Rebolo, and M.R. Zapatero Osorio (eds). Cool Stars, Stellar Systems, and the Sun. Astron. Soc. Pacific Conf. Ser., **223**, CD 1538.
Kron G.E., 1947. Publ. Astron. Soc. Pacific, **59**, 261.
Kron G.E., 1952. Astrophys. J., **115**, 301.
Krzeminski W., 1969. In: S.S. Kumar (ed). Low-Luminosity Stars. Gordon and Breach Sci. Publ. p. 57.
Krzeminski W. and Kraft R.P., 1967. Astron. J., **72**, 307.
Kuiper G.P., 1942. Astrophys. J., **95**, 201.
Kulapova A.N. and Shakhovskaya N.I., 1973. Izv. Crimean Astrophys. Obs., **48**, 31.

Kundu M.R. and Shevgaonkar R.K., 1985. Astrophys. J., **297**, 644.
Kundu M.R., Jackson P.D., White S.M., and Melozzi M., 1987. Astrophys. J., **312**, 822.
Kundu M.R., Pallavicini R., White S.M., and Jackson P.D., 1988a. Astron. Astrophys., **195**, 159.
Kundu M.R., White S.M., and Agrawal P.C., 1988b. Bull. Am. Astron. Soc., **20**, 696.
Kunkel W., 1967. An optical study of stellar flare. Thesis. Austin, Texas.
Kunkel W., 1968. Info. Bull. Var. Stars, No 315.
Kunkel W.E., 1969a. Nature, **222**, 1129.
Kunkel W.E., 1969b. In: S.S. Kumar (ed). Low-Luminosity Stars. Gordon and Breach Sci. Publ. p. 195.
Kunkel W.E., 1970. Astrophys. J., **161**, 503.
Kunkel W.E., 1971. Bull. Am. Astron. Soc., **3**, 13.
Kunkel W.E., 1972. Info. Bull. Var. Stars, No 748.
Kunkel W.E., 1973. Astrophys. J. Suppl. Ser., **25**, 1.
Kunkel W.E., 1974. Nature, **248**, 571.
Kunkel W.E., 1975a. In: V.E. Sherwood and L. Plaut (eds). Variable Stars and Stellar Evolution. Reidel, Dordrecht. p. 15.
Kunkel W.E., 1975b. In: V.E. Sherwood and L. Plaut (eds). Variable Stars and Stellar Evolution. Reidel, Dordrecht. p. 67.
Kurochka L.N., 1987. Astron. Zh., **64**, 443 (= Sov. Astron., **31**, 231).
Kurochka L.N. and Rossada V.M., 1981. Solar Data, No 7, 95.
Kurochka L.N. and Stasyuk L.A., 1981. Solar Data, No 5, 83.
La Fauci G. and Rodonò M., 1983. In: P.B. Byrne and M. Rodonò (eds). Activity in Red-Dwarf Stars. Reidel, Dordrecht. 1983. p. 185.
Lacy C.H., Evans D.S., Quigley R.J., and Sandmann W.H., 1978. Astrophys. J. Suppl. Ser., **37**, 313.
Lacy C.H., Moffett T.J., and Evans D.S., 1976. Astrophys. J. Suppl. Ser., **30**, 85.
Laming J.M. and Drake J.J., 1999. Astrophys. J., **516**, 324.
Laming J.M., Drake J.J., and Widing K.G., 1996. Astrophys. J., **462**, 948.
Landi E., Landini M., and Del Zanna G., 1997. Astron. Astrophys., **324**, 1027.
Landi E., Landini M., Dere K., and Risaliti G., 2001. In: R.J. Garcia López, R. Rebolo, and M.R. Zapatero Osorio (eds). Cool Stars, Stellar Systems, and the Sun. Astron. Soc. Pacific Conf. Ser., **223**, CD 991.
Landini M. and Monsignori Fossi B.C., 1984. Phys. Scri. **7**, 53.
Landini M., Monsignori Fossi B.C., Pallavicini R., and Piro L., 1986. Astron. Astrophys., **157**, 217.
Landsman W. and Simon T., 1991. Astrophys. J., **366**, L79.
Landsman W. and Simon T., 1993. Astrophys. J., **408**, 305.
Lang K.R. and Willson R.F., 1986a. Astrophys. J., **302**, L17.
Lang K.R. and Willson R.F., 1986b. Astrophys. J., **305**, 363.
Lang K.R. and Willson R.F., 1988. Astrophys. J., **326**, 300.

Lang K.R., Bookbinder J., Golub L., and Davis M.M., 1983. Astrophys. J., **272**, L15.
Lanza A.F., Rodonò M., and Zappala R.A., 1992. In: P.B. Byrne and D.J. Mullan (eds). Surface Inhomogeneities on Late-Type Stars. Lecture Notes in Physics, **397**, 297.
Lanza A.F., Catalano S., Cutispoto G. et al., 1998. Astron. Astrophys., **332**, 541.
Lanzafame A.C. and Byrne P.B., 1995. Astron. Astrophys., **303**, 155.
Larson A.M., Irwin A.W., Yang S.L.S. et al., 1993. Publ. Astron. Soc. Pacific, **105**, 332.
Latorre A., Montes D., and Fernandez-Figueroa M.J., 2001. In: R.J. Garcia López, R. Rebolo, and M.R. Zapatero Osorio (eds). Cool Stars, Stellar Systems, and the Sun. Astron. Soc. Pacific Conf. Ser., **223**, CD 997.
Lean J., 1992. In: P.B. Byrne and D.J. Mullan (eds). Surface Inhomogeneities on Late-Type Stars. Lecture Notes in Physics, **397**, 167.
Lecacheux A., Rosolen C., Davis M. et al., 1993. Astron. Astrophys., **275**, 670.
Lee T.A. and Hoxie D.T., 1972. Info. Bull. Var. Stars, No 707.
Leggett S.K., 1992. Astrophys. J. Suppl. Ser., **82**, 351.
Leggett S.K., Harris H.C., and Dahn C.C., 1994. Astron. J., **108**, 944.
Leinert Ch., Allard F., Richichi A., and Hauschildt P.H., 2000. Astron. Astrophys., **353**, 691.
Leto G., Pagano I., Buemi C.S., and Rodonò M., 1997. Astron. Astrophys., **327**, 1114.
Leto G., Pagano I., Linsky J.L. et al., 2000. Astron. Astrophys., **359**, 1035.
Liebert J., Saffer R.A., Norsworthy J. et al., 1992. In: M.S. Giampapa and J.A. Bookbinder (eds). Cool Stars, Stellar Systems, and the Sun. Astron. Soc. Pacific Conf. Ser., **26**, 282.
Liebert J., Kirkpatrick J.D., Reid J.N., and Fisher M.D., 1999. Astrophys. J., **519**, 345.
Liller W., 1952. Publ. Astron. Soc. Pacific, **64**, 129.
Lim J., 1993. Astrophys. J., **405**, L33.
Lim J. and White S.M., 1995. Astrophys. J., **453**, 207.
Lim J., White S.W., and Slee O.B., 1996. Astrophys. J., **460**, 976.
Lin R.P. and Hudson H.S., 1976. Solar Phys., **50**, 153.
Lingenfelter R.E. and Hudson H.S., 1980. In: R.O. Pepin, J.A. Eddy, and R.B. Merrill (eds). The Ancient Sun. Pergamon Press, New York. p. 69.
Linsky J.L., 1988. In: F. Cordova (ed). Multi-Wavelength Astrophysics. Cambridge University Press. p. 49.
Linsky J.L., 1991. In: I. Tuominen, D. Moss, and G. Rüdiger (eds). The Sun and Cool Stars: Activity, Magnetism, Dynamos. Lecture Notes in Physics, **380**, 452.
Linsky J.L., 1999. In: C.J. Butler and J.G. Doyle (eds). Solar and Stellar Activity: Similarities and Differences. Astron. Soc. Pacific Conf. Ser., **158**, 401.

Linsky J.L. and Ayres T.R., 1978. Astrophys. J., **220**, 619.
Linsky J.L. and Gary D.E., 1983. Astrophys. J., **274**, 776.
Linsky J.L. and Saar S.H., 1987. In: J.L. Linsky and R.E. Stencel (eds). Cool Stars, Stellar Systems, and the Sun. Lecture Notes in Physics, **291**, 44.
Linsky J.L. and Wood B.E., 1994. Astrophys. J., **430**, 342.
Linsky J.L. and Wood B.E., 1996. Astrophys. J., **463**, 254.
Linsky J.L., Worden S.P., McClintock W., and Robertson R.M., 1979a. Astrophys. J. Suppl. Ser., **41**, 47.
Linsky J.L., Hunten D.M., Sowell R. et al., 1979b. Astrophys. J. Suppl. Ser., **41**, 481.
Linsky J.L., Bornmann P.L., Carpenter K.G. et al., 1982. Astrophys. J., **260**, 670.
Linsky J.L., Wood B.E., and Andrulis C., 1994a. In: J.P. Caillault (ed). Cool Stars, Stellar Systems, and the Sun. Astron. Soc. Pacific Conf. Ser., **64**, 59.
Linsky J.L., Andrulis C., Saar S.H. et al., 1994b. In: J.P. Caillault (ed). Cool Stars, Stellar Systems, and the Sun. Astron. Soc. Pacific Conf. Ser., **64**, 438.
Linsky J.L., Wood B.E., Brown A. et al., 1995. Astrophys. J., **455**, 670.
Lippincott S.L., 1952. Astrophys. J., **115**, 582.
Lippincott S.L., 1953. Publ. Astron. Soc. Pacific, **65**, 248.
Lister T.A., Collier Cameron A., and Bartus J., 1999. Monthly Not. Roy. Astron. Soc., **307**, 685.
Lister T.A., Collier Cameron A., and Bartus J., 2001. In: R.J. Garcia López, R. Rebolo, and M.R. Zapatero Osorio (eds). Cool Stars, Stellar Systems, and the Sun. Astron. Soc. Pacific Conf. Ser., **223**, CD 1268.
Litvinenko Yu.E., 1994. Solar Physics, **151**, 195.
Liu M.C., Matthews B.C., Williams J.P., and Kalas P.G., 2004. Astrophys. J., **608**, 526.
Livshits M.A. and Katsova M.M., 1996. In: S. Bowyer and R.F. Malina (eds). Astrophysics in the Extreme Uultraviolet. Kluwer, Dordrecht. p. 171.
Livshits M.A., Badalyan O.G., Kosovichev A.G., and Katsova M.M., 1981. Solar Phys., **73**, 269.
Livshits M.A., Alekseev I. Yu., and Katsova M.M., 2003. Astron. Zh., **80**, 613 (= Astron. Rep., **47**, 562).
Lodenquai J. and McTavish J., 1988. Astron. J., **96**, 741.
López-Santiago J., Montes D., Fernandez-Figueroa M.J., and Ramsey L.W., 2003. Astron. Astrophys., **411**, 489.
Lovell A.C.B., Whipple F.L., and Solomon L.H., 1964. Nature, **201**, 1013.
Lovell B., 1964. Observatory, **84**, 191.
Lovell B., 1969. Nature, **222**, 1126.
Lovell B., 1971. Quart. J. Roy. Astron. Soc., **12**, 98.
Lovell B. and Chugainov P.F., 1964. Nature, **203**, 1213.
Lovell B. and Solomon L.H., 1966. Observatory, **86**, 16.
Lovell B., Whipple F.L., and Solomon L.H., 1963. Nature, **198**, 228.
Lovell B., Mavridis L.N., and Contadakis M.E., 1974. Nature, **250**, 124.

Lu E.T. and Hamilton R.J., 1991. Astrophys. J., **380**, L89.
Ludwig H.-G., Allard F., and Hauschildt P.H., 2002. Astron. Astrophys., **395**, 99.
Lukatskaya F.I., 1972. Astron. Astrophys., **17**, 97.
Lukatskaya F.I., 1976. P. Astron. Zh., **2**, 155 (= Sov. Astron. Lett. **2**, 61).
Lukatskaya F.I., 1977. Brightness and Color Variations of Non-Stationary Stars. Naukova Dumka, Kiev. 236 p.
Luyten W.J., 1949a. Astrophys. J., **109**, 532.
Luyten W.J., 1949b. Publ. Astron. Soc. Pacific, **61**, 179.
Luyten W.J. and Hoffleit D., 1954. Astron. J., **59**, 136.
MacConnell D.J., 1968. Astrophys. J., **153**, 313.
Maceroni C., Bianchini A., Rodonò M. et al., 1990. Astron. Astrophys., **237**, 395.
Maggio A. and Peres G., 1997. Astron. Astrophys., **325**, 237.
Maggio A., Sciortino S., Vaiana G.S. et al., 1987. Astrophys. J., **315**, 687.
Maggio A., Micela G., and Peres G., 1997. Mem. Soc. Astron. Ital., **68**, No 4, 1095.
Maggio A., Drake J.J., Kashyap V. et al., 2002. In: F. Favata and J.J. Drake (eds). Stellar Coronae in the Chandra and XMM-Newton Era. Proc. ESTEC Symp., Noordwijk, the Netherlands, June 25–29, 2001. Astron. Soc. Pacific Conf. Ser., **277**, 57.
Mahmoud F.M., 1993a. Astrophys. Space Sci., **208**, 217.
Mahmoud F.M., 1993b. Astrophys. Space Sci., **208**, 205.
Mahmoud F.M. and Soliman M.A., 1980. Info. Bull. Var. Stars, No 1866.
Majer P., Schmitt J.H.M.M., Golub L. et al., 1986. Astrophys. J., **300**, 360.
Mangeney A. and Praderie F., 1984. Astron. Astrophys., **130**, 143.
Maran S.P., Robinson R.D., Shore S.N. et al., 1994. Astrophys. J., **421**, 800.
Marcy G.W., 1981. Astrophys. J., **245**, 624.
Marcy G.W., 1983. In: J.O. Stenflo (ed). Solar and Stellar Magnetic Fields: Origins and Coronal Effects. Reidel, Dordrecht. p. 3.
Marcy G.W., 1984. Astrophys. J., **276**, 286.
Marcy G.W. and Basri G., 1989. Astrophys. J., **345**, 480.
Marcy G.W. and Chen G.H., 1992. Astrophys. J., **390**, 550.
Marilli E. and Catalano S., 1984. Astron. Astrophys., **133**, 57.
Marilli E., Catalano S., and Trigilio C., 1986. Astron. Astrophys., **167**, 297.
Marino A., Micela G., and Peres G., 1999. Astron. Astrophys., **353**, 177.
Marino A., Micela G., Peres G., and Sciortino S., 2002. Astron. Astrophys., **383**, 210.
Marino A., Micela G., Peres G., and Sciortino S., 2003a. Astron. Astrophys., **406**, 629.
Marino A., Micela G., Peres G., and Sciortino S., 2003b. Astron. Astrophys., **407**, L63.
Mariska J.T., 1987. In: J.L. Linsky and R.E. Stencel (eds). Cool Stars, Stellar Systems, and the Sun. Lecture Notes in Physics, **291**, 21.
Martin E.L., 1999. Monthly Not. Roy. Astron. Soc., **302**, 59.

Martin E.L. and Ardila D.R., 2001. Astron. J., **121**, 2758.
Martins D.H., 1975. Publ. Astron. Soc. Pacific, **87**, 163.
Mathioudakis M. and Doyle J.G., 1989a. Astron. Astrophys., **224**, 179.
Mathioudakis M. and Doyle J.G., 1989b. Astron. Astrophys., **232**, 114.
Mathioudakis M. and Doyle J.G., 1990. Astron. Astrophys., **240**, 357.
Mathioudakis M. and Doyle J.G., 1991a. Astron. Astrophys., **244**, 409.
Mathioudakis M. and Doyle J.G., 1991b. Astron. Astrophys., **244**, 433.
Mathioudakis M. and Doyle J.G., 1992. Astron. Astrophys., **262**, 523.
Mathioudakis M. and Doyle J.G., 1993. Astron. Astrophys., **280**, 181.
Mathioudakis M., Doyle J.G., Rodonò M. et al., 1991. Astron. Astrophys., **244**, 155.
Mathioudakis M., Drake J.J., Vedder P.W. et al., 1994. Astron. Astrophys., **291**, 517.
Mathioudakis M., Fruscione A., Drake J.J. et al., 1995a. Astron. Astrophys., **300**, 775.
Mathioudakis M., Drake J.J., Craig N. et al., 1995b. Astron. Astrophys., **302**, 422.
Mathioudakis M., McKenny J., Keenan F.P. et al., 1999. Astron. Astrophys., **351**, L23.
Mathys G. and Solanki S.K., 1987. In: G. de Strobel and M. Spite (eds). The Impact of Very High SN-Spectroscopy on Stellar Physics. Kluwer, Dordrecht. p. 325.
Mathys G. and Solanki S.K., 1989. Astron. Astrophys., **208**, 189.
Matthews L., Bromage G.E., Kellett B.J. et al., 1994. Monthly Not. Roy. Astron. Soc., **266**, 757.
Mauas P.J.D. and Falchi A., 1994. Astron. Astrophys., **281**, 129.
Mauas P.J.D. and Falchi A., 1996. Astron. Astrophys., **310**, 245.
Mauas P.J.D., Falchi A., Pasquini L., and Pallavicini R., 1997. Astron. Astrophys., **326**, 249.
Mavridis L.N. and Avgoloupis S., 1987. Astron. Astrophys., **188**, 95.
Mavridis L.N. and Avgoloupis S., 1993. Astron. Astrophys., **280**, L5.
Mavridis L.N., Asteriadis G., and Mahmoud F.M., 1982. In: E.G. Mariolopoulos, P.S. Theocaris, and L.N. Mavridis (eds). Compendium in Astronomy. Reidel, Dordrecht. p. 253.
Mavridis L.N., Avgoloupis S., Seiradakis J.H., and Varvolis P.P., 1995. Astron. Astrophys., **296**, 705.
McGale P.A., Pye J.P., and Hodgkin S.T., 1994. In: J.P. Caillault (ed). Cool Stars, Stellar Systems, and the Sun. Astron. Soc. Pacific Conf. Ser., **64**, 107.
McMillan J.D. and Herbst W., 1991. Astron. J., **101**, 1788.
Mekkaden M.V., 1985. Astrophys. Space Sci., **117**, 381.
Melikian N.D. and Grandpierre A., 1984. Info. Bull. Var. Stars, No 2638.
Melnik V.N., 1994. Kinemat. Phys. Celest. Bodies, **10**, No 5, 77.
Melnik V.N., 2000. Propagation and radiation of electron beams in cosmic plasma. Thesis. Kharkov.

Messina S. and Guinan E.F., 2002. Astron. Astrophys., **393**, 225.
Messina S. and Guinan E.F., 2003. Astron. Astrophys., **409**, 1017.
Messina S., Guinan E.F., Lanza A.F., and Ambruster C., 1999a. Astron. Astrophys., **347**, 249.
Messina S., Guinan E.F., and Lanza A.F., 1999b. Astrophys. Space Sci., **260**, 493.
Messina S., Rodonò M., and Guinan E.F., 2001. Astron. Astrophys., **366**, 215.
Messina S., Pizzolato N., Guinan E.F., and Rodonò M., 2003. Astron. Astrophys., **410**, 671.
Mestel L., 1984. In: S.L. Baliunas and L. Hartmann (eds). Cool Stars, Stellar Systems, and the Sun. Lecture Notes in Physics, **193**, 49.
Mewe R., Heise J., Gronenschild E.H.B.M. et al., 1975. Astrophys. J., **202**, L67.
Mewe R., Schrijver C.J., and Zwaan C., 1981. Space Sci. Rev., **30**, 191.
Mewe R., Gronenschild E.H.B.M., and van den Oord G.H.J., 1985. Astron. Astrophys., **62**, 197.
Mewe R., Lemen J.R., Schrijver C.J., and Fludra A., 1987. In: J.L. Linsky and R.E. Stencel (eds). Cool Stars, Stellar Systems, and the Sun. Lecture Notes in Physics, **291**, 60.
Mewe R., Kaastra J.S., Schrijver C.J., van den Oord G.H.J., and Alkemade F.J.M., 1995a. Astron. Astrophys., **296**, 477.
Mewe R., Kaastra J.S., and Liedahl D.A., 1995b. Legacy, **6**, 16.
Mewe R., Drake S.A., Kaastra J.S. et al., 1998a. Astron. Astrophys., **339**, 545.
Mewe R., Güdel M., Favata F., and Kaastra J.S., 1998b. Astron. Astrophys., **340**, 216.
Micela G., Sciortino S., Serio S. et al., 1985. Astrophys. J., **292**, 172.
Micela G., Sciortino S., Vaiana G.S. et al., 1988. Astrophys. J., **325**, 798.
Micela G., Sciortino S., Vaiana G.S. et al., 1990. Astrophys. J., **348**, 557.
Micela G., Favata F., Pye J., and Sciortino S., 1995. Astron. Astrophys., **298**, 505.
Micela G., Sciortino S., Kashyap V. et al., 1996. Astrophys. J. Suppl. Ser., **102**, 75.
Micela G., Pye J., and Sciortino S., 1997. Astron. Astrophys., **320**, 865.
Micela G., Sciortino S., Harnden F.R. et al., 1999a. Astron. Astrophys., **341**, 751.
Micela G., Sciortino S., Favata F. et al., 1999b. Astron. Astrophys., **344**, 83.
Micela G., Sciortino S., Jeffries R.D. et al., 2000. Astron. Astrophys., **357**, 909.
Middelkoop F., 1982. Astron. Astrophys., **107**, 31.
Mirzoyan L.V., 1981. Non-Stationarity and Evolution of Stars. Publ. House Acad. Sci. Armenian SSR, Yerevan. 380 p.
Mirzoyan L.V., 1986. Commun. Konkoly Obs., No 86., 409.

Mirzoyan L.V., 1990. In: L.V. Mirzoyan, B.R. Pettersen, and M.K. Tsvetkov (eds). Flare Stars in Star Clusters, Associations and the Solar Vicinity. Kluwer, Dordrecht. p. 1.

Mirzoyan L.V. and Ohanian G.B., 1977. Astrofizika, **13**, 561.

Mirzoyan L.V., Hambarian V.V., Garibjanian A.T., and Mirzoyan A.L., 1988. Astrofizika, **29**, 44.

Mochnacki S.W. and Schommer R.A., 1979. Astrophys. J., **231**, L77.

Mochnacki S.W. and Zirin H., 1980. Astrophys. J., **239**, L27.

Moffett T.J., 1972. Nature Phys. Sci., **240**, 41.

Moffett T.J., 1973. Monthly Not. Roy. Astron. Soc., **164**, 11.

Moffett T.J., 1974. Astrophys. J. Suppl. Ser., **29**, 1.

Moffett T.J., 1975. Info. Bull. Var. Stars, No 997.

Moffett T.J. and Bopp B.W., 1976. Astrophys. J. Suppl. Ser., **31**, 61.

Moffett T.J., Evans D.S., and Ferland G., 1977. Monthly Not. Roy. Astron. Soc., **178**, 149.

Moffett T.J., Helmken H.F., and Spangler S.R., 1978. Publ. Astron. Soc. Pacific, **90**, 93.

Mogilevsky E.I., 1980. In: Physics of Solar Activity. Moscow. IZMIRAN. p. 3.

Mogilevsky E.I., 2001. Fractals on the Sun. Fizmatlit. Moscow, 154p.

Mohanty S., Basri G., Shu F. et al., 2002. Astrophys. J., **571**, 469.

Moisseev I.G., Patston G.E., Solomon L.H. et al., 1975. Izv. Crimean Astrophys. Obs., **53**, 150.

Monsignori Fossi B.C. and Landini M., 1994. Astron. Astrophys., **284**, 900.

Monsignori Fossi B.C. and Landini M., 1996. In: S. Bowyer and R.F. Malina (eds). Astrophysics in the Extreme Ultraviolet. Kluwer, Dordrecht. p. 543.

Monsignori Fossi B., Landini M., Del Zanna G., and Drake J.J., 1994a. In: J.P. Caillault (ed). Cool Stars, Stellar Systems, and the Sun. Astron. Soc. Pacific Conf. Ser., **64**, 44.

Monsignori Fossi B., Landini M., Del Zanna G., and Bowyer S., 1994b. In: J.P. Caillault (ed). Cool Stars, Stellar Systems, and the Sun. Astron. Soc. Pacific Conf. Ser., **64**, 47.

Monsignori Fossi B.C., Landini M., Drake J.J., and Cully S.L., 1995a. Astron. Astrophys., **302**, 193.

Monsignori Fossi B.C., Landini M., Fruscione A., and Dupuis J., 1995b. Astrophys. J., **449**, 376.

Monsignori Fossi B.C., Landini M., Del Zanna G., and Bowyer S., 1996. Astrophys. J., **466**, 427.

Montes D., Fernandez-Figueroa M.J., de Castro E., and Cornide M., 1994. Astron. Astrophys., **285**, 609.

Montes D., Fernandez-Figueroa M.J., de Castro E., and Cornide M., 1995a. Astron. Astrophys., **294**, 165.

Montes D., de Castro E., Fernandez-Figueroa M.J., and Cornide M., 1995b. Astron. Astrophys. Suppl. Ser., **114**, 287.

Montes D., Fernandez-Figueroa M.J., Cornide M., and de Castro E., 1996. Astron. Astrophys., **312**, 221.

Montes D., Saar S.H., Collier Cameron A., and Unruh Y.C., 1999. Monthly Not. Roy. Astron. Soc., **305**, 45.
Montes D., Fernandez-Figueroa M.J., de Castro E. et al., 2000. Astron. Astrophys. Suppl. Ser., **146**, 103.
Montesinos B., 2001. In: R.J. Garcia López, R. Rebolo, and M.R. Zapatero Osorio (eds). Cool Stars, Stellar Systems, and the Sun. Astron. Soc. Pacific Conf. Ser., **223**, 284.
Montesinos B. and Jordan C., 1993. In: J.L. Linsky and S. Serio (eds). Physics of Solar and Stellar Coronae. Kluwer, Dordrecht. p. 579.
Montesinos B., Fernandez-Figueroa M.J., and de Castro E., 1987. Monthly Not. Roy. Astron. Soc., **229**, 627.
Moore R., McKenzie D.L., Švestka Z. et al., 1980. In: P.A. Sturrock (ed). Solar Flares. Colorado Ass. Univ. Press, Boulder. P. 341.
Mould J.R., 1978. Astrophys. J., **226**, 923.
Mullan D.J., 1973. Astrophys. J., **186**, 1059.
Mullan D.J., 1974. Astrophys. J., **192**, 149.
Mullan D.J., 1975a. Astron. Astrophys., **40**, 41.
Mullan D.J., 1975b. Astrophys. J., **200**, 641.
Mullan D.J., 1975c. Publ. Astron. Soc. Pacific, **87**, 455.
Mullan D.J., 1976a. Astrophys. J., **207**, 289.
Mullan D.J., 1976b. Astrophys. J., **210**, 702.
Mullan D.J., 1976c. Astrophys. J., **204**, 530.
Mullan D.J., 1977. Astrophys. J., **212**, 171.
Mullan D.J., 1983. In: P.P. Byrne and M. Rodonò (eds). Activity in Red-Dwarf Stars. Reidel, Dordrecht. p. 527.
Mullan D.J., 1984. Astrophys. J., **282**, 603.
Mullan D.J., 1989. Solar Phys., **121**, 239.
Mullan D.J., 1990. Astrophys. J., **361**, 215.
Mullan D.J., 1992. In: P.B. Byrne and D.J. Mullan (eds). Surface Inhomogeneities on Late-Type Stars. Lecture Notes in Physics, **397**, 233.
Mullan D.J. and Bell R.A., 1976. Astrophys. J., **204**, 818.
Mullan D.J. and Cheng Q.Q., 1993. Astrophys. J., **412**, 312.
Mullan D.J. and Cheng Q.Q., 1994. Astrophys. J., **420**, 392.
Mullan D.J. and Tarter C.B., 1977. Astrophys. J., **212**, 179.
Mullan D.J., Sion E.M., Bruhweiler F.C., and Carpenter K.G., 1989a. Astrophys. J., **339**, L33.
Mullan D.J., Stencel R.E., and Backman D.E., 1989b. Astrophys. J., **343**, 400.
Mullan D.J., Doyle J.G., Redman R.O., and Mathioudakis M., 1992a. Astrophys. J., **397**, 225.
Mullan D.J., Herr R.B., and Bhattacharyya S., 1992b. In: M.S. Giampapa and J.A. Bookbinder (eds). Cool Stars, Stellar Systems, and the Sun. Astron. Soc. Pacific Conf. Ser., **26**, 306.

Mullan D.J., Fleming T.A., and Schmitt J.H.M.M., 1995. In: K.G. Strassmeier (ed). Stellar Surface Structure. Poster Proceedings. Inst. Astron. Univ. Wien. p. 210.

Münch L. and Münch G., 1955. Tonantzintla y Tacubaya Bol., No 13, 36.

Musielak Z.E., Rosner R., and Ulmschneider P., 1990. In: G. Wallerstein (ed). Cool Stars, Stellar Systems, and the Sun. Astron. Soc. Pacific Conf. Ser., **9**, 79.

Natanzon A.M., 1981. Astron. Zh., **58**, 576 (= Sov. Astron., **25**, 328).

Neff D.H., Bookbinder J.A., and Linsky J.L., 1987. In: J.L. Linsky and R.E. Stencel (eds). Cool Stars, Stellar Systems, and the Sun. Lecture Notes in Physics, **291**, 161.

Neidig D.F. and Kane S.R., 1993. Solar Phys. **143**, 201.

Nelson G.J., Robinson R.D., Slee O.B. et al., 1979. Monthly Not. Roy. Astron. Soc., **187**, 405.

Nelson G.J., Robinson R.D., Slee O.B. et al., 1986. Monthly Not. Roy. Astron. Soc., **220**, 91.

Ness J.-U., Schmitt J.H.M.M., Burwitz V. et al., 2002. Astron. Astrophys., **394**, 911.

Ness J.-U., Schmitt J.H.M.M., Audard M. et al., 2003. Astron. Astrophys., **407**, 347.

Neuhauser R. and Comeron F., 2001. In: R.J. Garcia López, R. Rebolo, and M.R. Zapatero Osorio (eds). Cool Stars, Stellar Systems, and the Sun. Astron. Soc. Pacific Conf. Ser., **223**, CD 1097.

Newmark J.S., Buzasi D.L., Huenemoerder D.P. et al., 1990. Astron. J., **100**, 560.

Noyes R.W., 1983. In: J.O. Stenflo (ed). Solar and Stellar Magnetic fields: Origin and Coronal Effects. Reidel, Dordrecht., 133.

Noyes R.W., Hartmann L.W., Baliunas S.L. et al., 1984a. Astrophys. J., **279**, 763.

Noyes R.W., Weiss N.O., and Vaughan A.H., 1984b. Astrophys. J., 287, 769.

Nugent J. and Garmire G., 1978. Astrophys. J., **226**, L83.

Oláh K., 1986. In: L. Szabados (ed). Eruptive Phenomena in Stars. Commun. Konkoly Obs. Hung. Acad. Sci., Budapest. No 86, p. 393.

Oláh K. and Pettersen B.R., 1991. Astron. Astrophys., **242**, 443.

Oláh K., Kolláth Z., and Strassmeier K.G., 2000. Astron. Astrophys., **356**, 643.

Oláh K., Strassmeier K.G., Kövári Zs., and Guinan E.F., 2001. Astron. Astrophys., **372**, 119.

Oranje B.J., 1986. Astron. Astrophys., **154**, 185.

Oranje B.J., Zwaan C., and Middelkoop F., 1982. Astron. Astrophys., **110**, 30.

Oranje B.J. and Zwaan C., 1985. Astron. Astrophys., **147**, 265.

Oreshina A.V. and Somov B.V., 1997. Astron. Astrophys., **320**, L53.

Osawa K., Ichimura K., Noguchi T., and Watanabe E., 1968. Tokyo Astron. Bull., No 180.

Oskanjan V., 1953. Bull. Obs. Astron. Belgr., **18**, 24.
Oskanjan V., 1964. Publ. Obs. Astron. Belgr., No 10.
Oskanian V., 1969. In: L. Detre (ed). Non-Periodic Phenomena in Variable Stars. Proc. IAU Coll. No 4. Academic Press, Budapest. p. 131.
Oskanian V.S. and Terebizh V. Yu., 1971a. Astrofizika, **7**, 281.
Oskanian V.S. and Terebizh V. Yu., 1971b. Astrofizika, **7**, 83.
Oskanian V.S. and Terebizh V. Yu., 1971c. Info. Bull. Var. Stars, No 535.
Oskanyan V.S., Evans D.S., Lacy C., and McMillan R.S., 1977. Astrophys. J., **214**, 430.
Ossendrijver A.J.H., 1997. Astron. Astrophys., **323**, 151.
Pagano I., Ventura R., Rodonò M. et al., 1995. In: J. Greiner, H.W. Duerbeck, and R.E. Gershberg (eds). Flares and Flashes. Lecture Notes in Physics, **454**, 95.
Pagano I., Linsky J.L., Carkner L. et al., 2000. Astrophys. J., 532, 497.
Pagano I., Linsky J.L., Valenti J., and Duncan D.K., 2004. Astron. Astrophys., **415**, 331.
Pallavicini R., 1993. In: J.L. Linsky and S. Serio (eds). Physics of Solar and Stellar Coronae. Kluwer, Dordrecht. p. 237.
Pallavicini R., Golub L., Rosner R. et al., 1981. Astrophys. J., 248, 279.
Pallavicini R., Golub L., Rosner R., and Vaiana G.S., 1982. In: M.S. Giampapa and L. Golub (eds). Cool Stars, Stellar Systems, and the Sun. SAO Special report No 392/2, 77.
Pallavicini R., Peres G., Serio S. et al., 1983. Astrophys. J., 270, 270.
Pallavicini R., Willson R.F., and Lang K.R., 1985. Astron. Astrophys., **149**, 95.
Pallavicini R., Kundu M.R., and Jackson P.D., 1986. In: M. Zeilik and D.M. Gibson (eds). Cool Stars, Stellar Systems, and the Sun. Lecture Notes in Physics., **254**, 225.
Pallavicini R., Monsignori Fossi B.C., Landini M., and Schmitt J.H.M.M., 1988. Astron. Astrophys., **191**, 109.
Pallavicini R., Tagliaferri G., and Stella L., 1990a. Astron. Astrophys., **228**, 403.
Pallavicini R., Tagliaferri G., Pollock A.M.T. et al., 1990b. Astron. Astrophys., **227**, 483.
Pan H.C. and Jordan C., 1995. Monthly Not. Roy. Astron. Soc., **272**, 11.
Pan H.C., Jordan C., Makishima K. et al., 1995. In: J. Greiner, H.W. Duerbeck, and R.E. Gershberg (eds). Flares and Flashes. Lecture Notes in Physics, **454**, 171.
Pan H.C., Jordan C., Makishima K. et al., 1997. Monthly Not. Roy. Astron. Soc., **285**, 735.
Panagi P.M. and Mathioudakis M., 1993. Astron. Astrophys. Suppl. Ser., **100**, 343.
Panagi P.M., Byrne P.B., and Houdebine E.R., 1991. Astron. Astrophys. Suppl. Ser., **90**, 437.
Panov K.P. and Ivanova M.S., 1993. Astrophys. Space Sci., **199**, 265.

Panov K.P., Ivanova M.S., and Stegert J.S.W., 1995. In: J. Greiner, H.W. Duerbeck, and R.E. Gershberg (eds). Flares and Flashes. Lecture Notes in Physics, **454**, 103.
Panov K.P., Piirola V., and Korhonen T., 1988. Astron. Astrophys. Suppl. Ser., **75**, 53.
Papas C.H., 1977. In: L.V. Mirzoyan (ed). Flare Stars. Publ. House Acad. Sci. Armenian SSR. Yerevan., 175.
Papas C.H., 1990. In: L.V. Mirzoyan, B.P. Pettersen, and M.K. Tsvetkov (eds). Flare Stars in Star Clusters, Associations and Solar Vicinity. Kluwer, Dordrecht. p. 337.
Parker E., 1955a. Astrophys. J., **121**, 491.
Parker E., 1955b. Astrophys. J., **122**, 293.
Parker E., 1958. Astrophys. J., **128**, 664, 677.
Parker E., 1963. Interplanetary Dynamical Processes. Interscience, New York.
Parker E.N., 1970. Ann. Rev. Astron. Astrophys., **8**, 1.
Parker E.N., 1971. Astrophys. J., **165**, 139.
Parker E.N., 1988. Astrophys. J., **330**, 474.
Parsamian E.S., 1976. Astrofizika, **12**, 235.
Parsamian E.S., 1980. Astrofizika, **16**, 87.
Pasquini L., 1992. Astron. Astrophys., **266**, 347.
Pasquini L. and Pallavicini R., 1991. Astron. Astrophys., **251**, 199.
Pasquini L., Pallavicini R., and Pakull M., 1988. Astron. Astrophys., **191**, 253.
Patten B.M., 1994. Info. Bull. Var. Stars, No 4048.
Patten B.M. and Simon T., 1993. Astrophys. J., **415**, L123.
Patten B.M. and Simon T., 1996. Astrophys. J. Suppl. Ser., **106**, 489.
Pazzani V. and Rodonò M., 1981. Astrophys. Space Sci., **77**, 347.
Peres G., Ventura R., Pagano I., and Rodonò M., 1993. Astron. Astrophys., **278**, 179.
Peres G., Orlando S., Reale F. et al., 2000. Astrophys. J., **528**, 537.
Pestalozzi M.R., Benz A.O., Conway J.E., and Güdel M., 2000. Astron. Astrophys., **353**, 569.
Peterson R.C. and Schrijver C.J., 2001. In: R.J. Garcia López, R. Rebolo, and M.R. Zapatero Osorio (eds). Cool Stars, Stellar Systems, and the Sun. Astron. Soc. Pacific Conf. Ser., **223**, 300.
Petit M., 1954. Ciel et Terre, No 11–12., 3.
Petit M., 1955. J. Obs., **38**, 354.
Petit M., 1957. Astron. Zh., **34**, 805. (= Sov. Astron., **1**, 783).
Petit M., 1958. Contr. Osserv. Astrofisico Univ. Padova in Asiago, No 95, 29.
Petit M., 1959. Variable Stars, **12**, 4.
Petit M., 1961. J. Obs., **44**, 11.
Petit M., 1970. Info. Bull. Var. Stars, No 430.
Petit M. and Weber R., 1956. J. Obs., **39**, 51.
Petrov P.P., Chugainov P.F., and Shcherbakov A.G., 1984. Izv. Crimean Astrophys. Obs., **69**, 3.

Pettersen B.R., 1976. Theor. Astrophys. Institute Report, No 46. Blindern-Oslo.
Pettersen B.R., 1980. Astron. Astrophys., **82**, 53.
Pettersen B.R., 1987. Vistas Astron., **30**, 41.
Pettersen B.R., 1988. In: O. Havnes, B.R. Pettersen, J.H.M.M. Schmitt, and J.E. Solheim (eds). Activity in Cool Star Envelopes. Kluwer, Dordrecht. p. 49.
Pettersen B.R., 1989a. Solar Phys., **121**, 299.
Pettersen B.R., 1989b. Astron. Astrophys., **209**, 279.
Pettersen B.R., 1991. Mem. Soc. Astron. Ital., **62**, N2, 217.
Pettersen B.R. and Coleman L.A., 1981. Astrophys. J., **251**, 571.
Pettersen B.R. and Hawley S.L., 1987. Inst. Theor. Astrophys. Univ. Oslo Publ. Ser., No 2.
Pettersen B.R. and Hawley S.L., 1989. Astron. Astrophys., **217**, 187.
Pettersen B.R. and Hsu J.-Ch., 1981. Astrophys. J., **247**, 1013.
Pettersen B.R. and Sundland S.R., 1991. Astron. Astrophys. Suppl. Ser., **87**, 303.
Pettersen B.R., Coleman L.A., and Evans D.S., 1984a. Astrophys. J. Suppl. Ser., **54**, 375.
Pettersen B.R., Evans D.S., and Coleman L.A., 1984b. Astrophys. J., **282**, 214.
Pettersen B.R., Panov K.P., Sandmann W.H., and Ivanova M.S., 1986a. Astron. Astrophys. Suppl. Ser., **66**, 235.
Pettersen B.R., Hawley S.L., and Andersen B.N., 1986b. New Insight in Astrophysics. ESA SP-263, 157.
Pettersen B.R., Lambert D.L., Tomkin J. et al., 1987. Astron. Astrophys., **183**, 66.
Pettersen B.R., Panov K.P., Ivanova M.S. et al., 1990a. In: L.V. Mirzoyan, B.R. Pettersen, and M.K. Tsvetkov (eds). Flare Stars in Star Clusters, Associations and the Solar Vicinity. Kluwer, Dordrecht. p. 15.
Pettersen B.R., Sundland S.R., Hawley S.L., and Coleman L.A., 1990b. In: G. Wallerstein (ed). Cool Stars, Stellar Systems, and the Sun. Astron. Soc. Pacific Conf. Ser., **9**, 177.
Pettersen B.R., Hawley S.L., and Fisher G.H., 1992a. Solar Phys., **142**, 197.
Pettersen B.R., Oláh K., and Sandmann W.H., 1992b. Astron. Astrophys. Suppl. Ser., **96**, 497.
Phillips M.J. and Hartmann L., 1978. Astrophys. J., **224**, 182.
Phillips K.J.H., Bromage G.E., Dufton P.L. et al., 1988. Monthly Not. Roy. Astron. Soc. **235**, 573.
Phillips R.B., Hewitt J.N., Corey B.E. et al., 1989. Bull. Am. Astron. Soc., **21**, 710.
Phillips K.J.H., Bromage G.E., and Doyle J.G., 1992. Astrophys. J., **385**, 731.
Pickering E.C., 1880. Proc. Am. Acad. Arts Sci., **16**, 257.
Piirola V., 1975. Ann. Acad. Scien. Fennicae Ser. A, Physica, No 418.

Pillitteri I., Micela G., Sciortino S., and Favata F., 2003. Astron. Astrophys., **399**, 919.
Piters A.J.M., Schrijver C.J., Schmitt J.H.M.M. et al., 1997. Astron. Astrophys., **325**, 1115.
Pizzolato N., Maggio A., Micela G., and Sciortino S., 2002. In: F. Favata and J.J. Drake (eds). Stellar Coronae in the Chandra and XMM-Newton Era. Proc. ESTEC Symp., Noordwijk, the Netherlands, June 25–29, 2001. Astron. Soc. Pacific Conf. Ser., **277**, 557.
Plachinda S.I. and Tarasova T.N., 1999. Astrophys. J., **514**, 402.
Plachinda S.I. and Tarasova T.N., 2000. Astrophys. J., **533**, 1016.
Plachinda S.I., Johns-Krull C.M., and Tarasova T.N., 2001. Odessa Astron. Publ., **14**, 219.
Podlazov A.V. and Osokin A.R., 2002. Astrophys. Space Sci., **282**, 221.
Poe C.H. and Eaton J.A., 1985. Astrophys. J., **289**, 644.
Poletto G., 1989. Solar Phys., **121**, 313.
Poletto G., Pallavicini R., and Kopp R.A., 1988. Astron. Astrophys., **201**, 93.
Pollock A.M.T., Tagliaferri G., and Pallavicini R., 1991. Astron. Astrophys., **241**, 451.
Ponomarev E.A. and Rubo G.A., 1965. In: S.K. Vsekhsvyatsky (ed). Solar Corona and Corpuscular Radiation in Interplanetary Space. Publ. House Kiev Univ. p. 159.
Popper D.M., 1953. Publ. Astron. Soc. Pacific, **65**, 278.
Poveda A., Allen C., and Herrera M.A., 1995. In: J. Greiner, H.W. Duerbeck, and R.E. Gershberg (eds). Flares and Flashes. Lecture Notes in Physics, **454**, 57.
Preibisch Th. and Zinnecker H., 2002. In: F. Favata and J.J. Drake (eds). Stellar Coronae in the Chandra and XMM-Newton era. Proc. ESTEC Symp., Noordwijk, the Netherlands, June 25–29, 2001. Astron. Soc. Pacific Conf. Ser., **277**, 185.
Prosser C.F., Stauffer J.R., Caillault J.-P. et al., 1995. Astron. J., **110**, 1229.
Prosser C.F., Randich S., Stauffer J. et al., 1996. Astron. J., **112**, 1570.
Pustil'nik L.A., 1988. P. Astron. Zh., **14**, 940 (= Sov. Astron. Lett. **14**, 398).
Pustil'nik L.A., 1997. Astrophys. Space Sci. **252**, 325.
Pustil'nik L.A., 1999. Astrophys. Space Sci. **264**, 171.
Pye J.P., Hodgkin S.T., Morley J.E. et al., 1994. In: J.P. Caillault (ed). Cool Stars, Stellar Systems, and the Sun. Astron. Soc. Pacific Conf. Ser., **64**, 128.
Quast G.R. and Torres C.A., 1986. Rev. Mex. Astron. Astrofis. **12**, 209.
Quin D.A., Doyle J.G., Butler C.J. et al., 1993. Astron. Astrophys., **272**, 477.
Raassen A.J.J., Ness J.-U., Mewe R. et al., 2003. Astron. Astrophys., **400**, 671.
Radick R.R., Wilkerson M.S., Worden S.P. et al., 1983. Publ. Astron. Soc. Pacific, **95**, 300.
Radick R.R., Thompson D.T., Lockwood G.W. et al., 1987. Astrophys. J., **321**, 459.

Radick R.R., Lockwood G.W., Skiff B.A., and Baliunas S.L., 1998. Astrophys. J. Suppl. Ser., **118**, 239.
Rammacher W. and Cuntz M., 2003. Astrophys. J., **594**, L51.
Ramseyer T.F., Hatzes A.P., and Jablonski F., 1995. Astron. J., **110**, 1364.
Randich S., 1997. In: G. Micela, R. Pallavicini, and S. Sciortino (eds). Cool Stars in Clusters and Associations: Magnetic Activity and Age Indicators. Mem. Soc. Astron. Ital., **68**, 971.
Randich S. and Schmitt J.H.M.M., 1995. Astron. Astrophys., **298**, 115.
Randich S., Schmitt J.H.M.M., Prosser G., and Stauffer J.R., 1996a. Astron. Astrophys., **305**, 785.
Randich S., Schmitt J.H.M.M., and Prosser G., 1996b. Astron. Astrophys., **313**, 815.
Rao A.R. and Vahia M.N., 1987. Astron. Astrophys., **188**, 109.
Rao A.R. and Singh K.P., 1990. Astrophys. J., **352**, 303.
Rao A.R., Singh K.P., and Vahia M.N., 1990. Astrophys. J., **365**, 332.
Raymond J.C., 1988. In: R. Pallavicini (ed). Hot Thin Plasmas in Astrophysics. Kluwer, Dordrecht. p. 3.
Raymond J.C. and Smith B.W., 1977. Astrophys. J. Suppl. Ser., **35**, 419.
Raymond J.C., Cox D.P., and Smith B.W., 1976. Astrophys. J., **204**, 290.
Reale F., 2002. In: F. Favata and J.J. Drake (eds). Stellar Coronae in the Chandra and XMM-Newton Era. Proc. ESTEC Symp., Noordwijk, the Netherlands, June 25–29, 2001. Astron. Soc. Pacific Conf. Ser., **277**, 103.
Reale F. and Micela G., 1998. Astron. Astrophys., **334**, 1028.
Reale F., Peres G., Serio S. et al., 1988. Astrophys. J., **328**, 256.
Reale F., Serio S., and Peres G., 1993. Astron. Astrophys., **272**, 486.
Reale F., Betta R., Peres G. et al., 1997. Astron. Astrophys., **325**, 782.
Reale F., Güdel M., Peres G., and Audard M., 2004. Astron. Astrophys., **416**, 733.
Rebolo R., Garcia López R., Beckman J.E. et al., 1989. Astron. Astrophys. Suppl. Ser., **80**, 135.
Redfield S., Linsky J.L., Ake T.B. et al., 2002. Astrophys. J., **581**, 626.
Redfield S., Ayres T.R., Linsky J.L. et al., 2003. Astrophys. J., **585**, 993.
Reglero V., Fabregat J., and de Castro A., 1986. Info. Bull. Var. Stars, No 2904.
Rego M., Gonzalez-Riestra R., and Fernandez-Figueroa M.J., 1983. Astron. Astrophys., **119**, 227.
Reid I.N., Hawley S.L., and Gizis J.E., 1995. Astron. J., **110**, 1838.
Reiners A. and Schmitt J.H.M.M., 2002a. Astron. Astrophys., **384**, 155.
Reiners A. and Schmitt J.H.M.M., 2002b. Astron. Astrophys., **388**, 1120.
Reiners A. and Schmitt J.H.M.M., 2003. Astron. Astrophys., **398**, 647.
Rice J.B. and Strassmeier K.G., 1996. Astron. Astrophys., **316**, 164.
Rice J.B. and Strassmeier K.G., 1998. Astron. Astrophys., **336**, 972.
Richter G.A., Braeuer H.-J., and Greiner J., 1995. In: J. Greiner, H.W. Duerbeck, and R.E. Gershberg (eds). Flares and Flashes. Lecture Notes in Physics, **454**, 69.

Ripodas P., Sanchez Almeida J., Garcia López R.J., and Collados M., 1992. In: I. Tuominen, D. Moss, and G. Rüdiger (eds). The Sun and Cool Stars: Activity, Magnetism, Dynamos. Lecture Notes in Physics, **380**, 417.
Robb R.M. and Cardinal R.D., 1995. Info. Bull. Var. Stars, No 4221.
Robinson E.L. and Kraft R.P., 1974. Astron. J., **79**, 698.
Robinson R.D., 1980. Astrophys. J., **239**, 961.
Robinson R.D., 1989. In: B.M. Haisch and M. Rodonò (eds). Solar and Stellar Flares. Special publication Catania Astrophys. Obs., p. 83.
Robinson R.D. and Durney B.R., 1982. Astron. Astrophys., **108**, 322.
Robinson R.D., Worden S.P., and Harvey J.W., 1980. Astrophys. J., **236**, L155.
Robinson R.D., Cram L.E., and Giampapa M.S., 1990. Astrophys. J. Suppl. Ser., **74**, 891.
Robinson R.D., Carpenter K.G., Woodgate B.E., and Maran S.P., 1993. Astrophys. J., **414**, 872.
Robinson R.D., Carpenter K.G., Slee O.B. et al., 1994. Monthly Not. Roy. Astron. Soc., **267**, 918.
Robinson R.D., Carpenter K.G., Percival J.W., and Bookbinder J.A., 1995. Astrophys. J., **451**, 795.
Robinson R.D., Woodgate B.E., and Carpenter K.G., 1996. In: R. Pallavicini and A.K. Dupree (eds). Cool Stars, Stellar Systems, and the Sun. Astron. Soc. Pacific Conf. Ser., **109** p. 285.
Robinson R.D., Airapetian V., Slee O.B. et al., 2001. In: R.J. Garcia López, R. Rebolo, and M.R. Zapatero Osorio (eds). Cool Stars, Stellar Systems, and the Sun. Astron. Soc. Pacific Conf. Ser., **223**, CD 1151.
Robrade J., Ness J.-U., and Schmitt J.H.M.M., 2004. Astron. Astrophys., **413**, 317.
Rodonò M., 1974. Astron. Astrophys., **32**, 337.
Rodonò M., 1978. Astron. Astrophys., **66**, 175.
Rodonò M., 1980. Mem. Soc. Astron. Ital., **51**, 623.
Rodonò M., 1986a. In: L.V. Mirzoyan (ed). Flare Stars and Related Object. Publ. House Armenian Acad. Sci., Yerevan. p. 19.
Rodonò M., 1986b. In: M. Zeilik and D.M. Gibson (eds). Cool Stars, Stellar Systems, and the Sun. Lecture Notes in Physics, **254**, 475.
Rodonò M., 1986c. In: J.-P. Swings (ed). Highlight in Astronomy. Reidel, Dordrecht. **7**, 429.
Rodonò M., 1987. In: E.-H. Schroter and M. Schüssler (eds). Solar and Stellar Physics. Lecture Notes in Physics, **292**, 39.
Rodonò M., 1992. In: P.B. Byrne and D.J. Mullan (eds). Surface Inhomogeneities on Late-Type Stars. Lecture Notes in Physics, **397**, 201.
Rodonò M. and Cutispoto G., 1988. In: O. Havnes, B.R. Pettersen et al. (eds). Activity in Cool Star Envelopes. Kluwer, Dordrecht. p. 163.
Rodonò M., Pucillo M., Sedmak G., and de Biase G.A., 1979. Astron. Astrophys., **76**, 242.
Rodonò M., Cutispoto G., Catalano S. et al., 1984. ESA SP-218, 247.

Rodonò M., Cutispoto G., Pazzani V. et al., 1986. Astron. Astrophys., **165**, 135.
Rodonò M., Houdebine E.R., Catalano S. et al., 1989. In: B.M. Haisch and M. Rodonò (eds). Solar and Stellar Flares. Catania Astrophysical Observatory Special Publication. p. 53.
Roizman G. Sh., 1983. P. Astron. Zh., **9**, 41 (= Sov. Astron. Lett. **9**, 21).
Rojzman G. Sh., 1984. Astron. Zh., **61**, 500 (= Sov. Astron., **28**, 293).
Roizman G. Sh. and Lorents L.P., 1991. Astron. Tsirk. No 1549.
Rojzman G. Sh. and Kabichev G.I., 1985. Astron. Zh., **62**, 1095 (= Sov. Astron., **29**, 639).
Rojzman G. Sh. and Shevchenko V.S., 1982. P. Astron. Zh., **8**, 163 (= Sov. Astron. Lett. **8**, 85).
Roques P.E., 1953. Publ. Astron. Soc. Pacific, **65**, 19.
Roques P.E., 1954. Publ. Astron. Soc. Pacific, **66**, 256.
Roques P.E., 1958. Publ. Astron. Soc. Pacific, **70**, 310.
Roques P.E., 1961. Astrophys. J., **133**, 914.
Rosner R., 1980. In: A.K. Dupree (ed). Cool Stars, Stellar Systems, and the Sun. SAO Special Rep. No 389, 79.
Rosner R. and Vaiana G.S., 1978. Astrophys. J., **222**, 1104.
Rosner R., Tucker W.H., and Vaiana G.S., 1978. Astrophys. J., **220**, 643.
Rucinski S.M., 1979. Acta Astron., **29**, 203.
Rucinski S.M., 1994. Acta Astron., **44**, 75.
Rüedi I., Solanki S.K., Mathys G., and Saar S.H., 1997. Astron. Astrophys., **318**, 429.
Rutledge R.E., Basri G., Martin E.L., and Bildsten L., 2000. Astrophys. J., **538**, L141.
Rutten R.G.M., 1984. Astron. Astrophys., **130**, 353.
Rutten R.G.M., 1986. Astron. Astrophys., **159**, 291.
Rutten R.G.M., 1987. Astron. Astrophys., **177**, 131.
Rutten R.G.M., Schrijver C.J., Zwaan C. et al., 1989. Astron. Astrophys., **219**, 239.
Rutten R.G.M., Schrijver C.J., Lemmens A.F.P., and Zwaan C., 1991. Astron. Astrophys., **252**, 203.
Saar S.H., 1987. In: J.L. Linsky and R.E. Stencel (eds). Cool Stars, Stellar Systems, and the Sun. Lecture Notes in Physics, **291**, 10.
Saar S.H., 1988. Astrophys. J., **324**, 441.
Saar S.H., 1991. In: P. Ulmschneider, E. Priest, and R. Rosner (eds). Mechanisms of Chromospheric and Coronal Heating. Springer, Heidelberg. p. 273.
Saar S.H., 1992. In: M.S. Giampapa and J.A. Bookbinder (eds). Cool Stars, Stellar Systems, and the Sun. Astron. Soc. Pacific Conf. Ser., **26**, 252.
Saar S.H., 1994. In: D.M. Rabin, J.T. Jefferies, and C. Lindsey (eds). Infrared Solar Physics. Kluwer, Dordrecht. p. 493.
Saar S.H., 1996a. In: K.G. Strassmeier and J.L. Linsky (eds). Stellar Surface Structure. Kluwer, Dordrecht. p. 237.

Saar S.H., 1996b. In: Y. Uchida, T. Kosugi, and H.S. Hudson (eds). Magnetodynamic Phenomena in the Solar Atmosphere – Prototype of Stellar Magnetic Activity. Kluwer, Dordrecht. p. 367.

Saar S.H., 1998. In: R.A. Donahue and J.A. Bookbinder (eds). Cool Stars, Stellar Systems, and the Sun. Astron. Soc. Pacific Conf. Ser., **154**, 211.

Saar S.H., 2001. In: R.J. Garcia López, R. Rebolo, and M.R. Zapatero Osorio (eds). Cool Stars, Stellar Systems, and the Sun. Astron. Soc. Pacific Conf. Ser., **223**, 292.

Saar S.H. and Baliunas S.L., 1992a. In: K.L. Harvey (ed). The Solar Cycle. Astron. Soc. Pacific Conf. Ser., **27**, 150.

Saar S.H. and Baliunas S.L., 1992b. In: K.L. Harvey (ed). The Solar Cycle. Astron. Soc. Pacific Conf. Ser., **27**, 197.

Saar S.H. and Bookbinder J.A., 1998a. In: R.A. Donahue and J.A. Bookbinder (eds). Cool Stars, Stellar Systems, and the Sun. Astron. Soc. Pacific Conf. Ser., **154**, CD 1560.

Saar S.H. and Bookbinder J.A., 1998b. In: R.A. Donahue and J.A. Bookbinder (eds). Cool Stars, Stellar Systems, and the Sun. Astron. Soc. Pacific Conf. Ser., **154**, CD 2042.

Saar S.H. and Bopp B.W., 1992. In: M.S. Giampapa and J.A. Bookbinder (eds). Cool Stars, Stellar Systems, and the Sun. Astron. Soc. Pacific Conf. Ser., **26**, 288.

Saar S.H. and Brandenburg A., 1999. Astrophys. J., **524**, 295.

Saar S.H. and Huovelin J., 1993. Astrophys. J., **404**, 739.

Saar S.H. and Linsky J.L., 1985. Astrophys. J., **299**, L47.

Saar S.H. and Linsky J.L., 1986. In: M. Zeilik and D.M. Gibson (eds). Cool Stars, Stellar Systems, and the Sun. Lecture Notes in Physics, **254**, 278.

Saar S.H. and Neff J.E., 1990. In: G. Wallerstein (ed). Cool Stars, Stellar Systems, and the Sun. Astron. Soc. Pacific Conf. Ser., **9**, 171.

Saar S.H. and Schrijver C.J., 1987. In: J.L. Linsky and R.E. Stencel (eds). Cool Stars, Stellar Systems, and the Sun. Lecture Notes in Physics, **291**, 38.

Saar S.H., Linsky J.L., and Beckers J.M., 1986a. Astrophys. J., **302**, 777.

Saar S.H., Linsky J.L., and Duncan D.K., 1986b. In: M. Zeilik and D.M. Gibson (eds). Cool Stars, Stellar Systems, and the Sun. Lecture Notes in Physics, **254**, 275.

Saar S.H., Linsky J.L., and Giampapa M.S., 1987. In: Observational Astrophysics with High Precision Data. Proc. 27th Liege Int. Astrophys. Coll. Univ. de Liege Cointe–Ougree. p. 103.

Saar S.H., Huovelin J., Giampapa M.S. et al., 1988. In: O. Havnes, B.R. Pettersen, J.H.M.M. Schmitt and J.E. Solheim (eds). Activity in Cool Star Envelopes. Kluwer, Dordrecht. p. 45.

Saar S.H., Piskunov N.E., and Tuominen I., 1992. In: M.S. Giampapa and J.A. Bookbinder (eds). Cool Stars, Stellar Systems, and the Sun. Astron. Soc. Pacific Conf. Ser., **26**, 255.

Saar S.H., Brandenburg A., Donahue R.A., and Baliunas S.L., 1994a. In: J.p. Caillault (ed). Cool Stars, Stellar Systems, and the Sun. Astron. Soc. Pacific Conf. Ser., **64**, 468.

Saar S.H., Martens P.C.H., Huovelin J., and Linnaluoto S., 1994b. Astron. Astrophys., **286**, 194.

Saar S.H., Morgan M.R., Bookbinder J.A. et al., 1994c. In: J.P. Caillault (ed). Cool Stars, Stellar Systems, and the Sun. Astron. Soc. Pacific Conf. Ser., **64**, 471.

Saar S.H., Huovelin J., Osten R.A., and Shcherbakov A.G., 1997. Astron. Astrophys., **326**, 741.

Saar S.H., Peterchev A., O'Neal D., and Neff J.E., 2001. In: R.J. Garcia López, R. Rebolo, and M.R. Zapatero Osorio (eds). Cool Stars, Stellar Systems, and the Sun. Astron. Soc. Pacific Conf. Ser., **223**, CD 1057.

Sanamian V.A., Venugopal V.P., Chavushian H.S., 1978. Astrofizika, **14**, 283.

Sanwal B.B., 1995. In: J. Greiner, H.W. Duerbeck, and R.E. Gershberg (eds). Flares and Flashes. Lecture Notes in Physics, **454**, 115.

Sanz-Forcada J. and Micela G., 2002. Astron. Astrophys., **394**, 653.

Sanz-Forcada J., Favata F., and Micela G., 2004. Astron. Astrophys., **416**, 281.

Savanov I.S. and Savel'eva Yu. Yu., 1997. Astron. Zh., **74**, 919 (= Astron. Rep., **41**, 821).

Schachter J.F., Remillard R., Saar S.H. et al., 1996. Astrophys. J., **463**, 747.

Schaefer B.E., King J.R., and Deliyannis C.P., 2000. Astrophys. J., **529**, 1026.

Schatzman E., 1959. In: J.L. Greenstein (ed). The Hertzsprung – Russell Diagram. Proc. IAU Symp., No 10, 129.

Schatzman E., 1962. Ann. d'Astrophys. **25**, 18.

Scheible M. and Guinan E.F., 1994. Info. Bull. Var. Stars No 4110.

Schmitt D., Schüssler M., and Ferriz-Mas A., 1998. In: R.A. Donahue and J.A. Bookbinder (eds). Cool Stars, Stellar Systems, and the Sun. Astron. Soc. Pacific Conf. Ser., **154**, CD 1324.

Schmitt J.H.M.M., 1993. In: J.L. Linsky and S. Serio (eds). Physics of Solar and Stellar Coronae. Kluwer, Dordrecht. p. 327.

Schmitt J.H.M.M., 1994. Astrophys. J. Suppl. Ser., **90**, 735.

Schmitt J.H.M.M., 1996. In: K.G. Strassmeier and J.L. Linsky (eds). Stellar Surface Structure. Kluwer, Dordrecht. p. 85.

Schmitt J.H.M.M., 1997. Astron. Astrophys., **318**, 215.

Schmitt J.H.M.M. and Rosso C., 1988. Astron. Astrophys., **191**, 99.

Schmitt J.H.M.M. and Wichmann R., 2001. Nature, **412**, 508.

Schmitt J.H.M.M. and Liefke C., 2002. Astron. Astrophys., **382**, L9.

Schmitt J.H.M.M., Golub L., Harnden F.R. et al., 1985. Astrophys. J. **290**, 307.

Schmitt J.H.M.M., Pallavicini R., Monsignori Fossi B.C., and Harnden F.R., 1987. Astron. Astrophys., **179**, 193.

Schmitt J.H.M.M., Collura A., Sciortino S. et al., 1990. Astrophys. J., **365**, 704.

Schmitt J.H.M.M., Haisch B., and Barwig H., 1993a. Astrophys. J., **419**, L81.
Schmitt J.H.M.M., Kahabka P., Stauffer J., and Piters A.J.M., 1993b. Astron. Astrophys., **277**, 114.
Schmitt J.H.M.M., Güdel M., and Predehl P., 1994. Astron. Astrophys., **287**, 843.
Schmitt J.H.M.M., Fleming T.A., and Giampapa M.S., 1995. Astrophys. J., **450**, 392.
Schmitt J.H.M.M., Krautter J., Appenzeller I. et al., 1996a. In: R. Pallavicini and A.K. Dupree (eds). Cool Stars, Stellar Systems, and the Sun. Astron. Soc. Pacific Conf. Ser., **109**, 287.
Schmitt J.H.M.M., Drake J.J., Stern R.A., and Haisch B.M., 1996b. Astrophys. J., **457**, 882.
Schneeberger T.J., Linsky J.L., McClintock W., and Worden S.P., 1979. Astrophys. J., **231**, 148.
Schrijver C.J., 1983. Astron. Astrophys., **127**, 289.
Schrijver C.J., 1987. Astron. Astrophys., **172**, 111.
Schrijver C.J., 1990. Astron. Astrophys., **234**, 315.
Schrijver C.J. and Aschwanden M.J., 2002. Astrophys. J., **566**, 1147.
Schrijver C.J. and Mewe R., 1986. In: M. Zeilik and D.M. Gibson (eds). Cool Stars, Stellar Systems, and the Sun. Lecture Notes in Physics, **254**, 300.
Schrijver C.J. and Rutten R.G.M., 1987. Astron. Astrophys., J., **177**, 143.
Schrijver C.J. and Title A.M., 2001. Astrophys. J., **551**, 1099.
Schrijver C.J., Mewe R., and Walter F.M., 1984. Astron. Astrophys., **138**, 258.
Schrijver C.J., Dobson A.K., and Radick R.R., 1989a. Astrophys. J., **341**, 1035.
Schrijver C.J., Cote J., Zwaan C., and Saar S.H., 1989b. Astrophys. J., **337**, 964.
Schrijver C.J., Dobson A.K., and Radick R.R., 1992. Astron. Astrophys., J., **258**, 432.
Schrijver C.J., van den Oord G.H.J., Mewe R., and Kaastra J.S., 1996. In: S. Bowyer and R.F. Malina (eds). Astrophysics in the Extreme Ultraviolet. Kluwer, Dordrecht. p. 121.
Schütz O., Nielbock M., Wolf S. et al., 2004. Astron. Astrophys., **414**, L9.
Schüssler M. and Solanki S.K., 1992. Astron. Astrophys., **264**, L13.
Sciortino S., Micela G., Kashyap V. et al., 1994. In: J.P. Caillault (ed). Cool Stars, Stellar Systems, and the Sun. Astron. Soc. Pacific Conf. Ser., **64**, 140.
Sciortino S., Maggio A., Favata F., and Orlando S., 1999. Astron. Astrophys., **342**, 502.
Serio S., Peres G., Vaiana G.S. et al., 1981. Astrophys. J., **243**, 288.
Shakhovskaya N.I., 1974a. Izv. Crimean Astrophys. Obs., **50**, 84.
Shakhovskaya N.I., 1974b. Izv. Crimean Astrophys. Obs., **51**, 92.
Shakhovskaya N.I., 1975. Izv. Crimean Astrophys. Obs., **53**, 165.
Shakhovskaya N.I., 1979. Izv. Crimean Astrophys. Obs., **60**, 14.

Shakhovskaya N.I., 1989. Solar Phys., **121**, 375.
Shakhovskaya N.I., 1995. In: J. Greiner, H.W. Duerbeck, and R.E. Gershberg (eds). Flares and Flashes. Lecture Notes in Physics, **454**, 61.
Shakhovskoy N.M., 1993. In: O. Regev and G. Shaviv (eds). Cataclysmic Variables and Related Physics. Ann. Israel Phys. Soc., **10**, 237.
Shapley H., 1951. Harvard Reprint No 344.
Shapley H., 1954. Astron. J., **59**, 118.
Shcherbakov A.G., 1979. P. Astron. Zh., **5**, 360 (= Sov. Astron. Let., **5**, 193).
Shevchenko G.G., 1973. Astron. Tsirk., No 792.
Shmeleva O.P. and Syrovatsky S.I., 1973. Solar Phys., **33**, 341.
Short C.I. and Doyle J.G., 1998. Astron. Astrophys., **336**, 613.
Sim S.A. and Jordan C., 2002. Monthly Not. Roy. Astron. Soc., **346**, 846.
Sim S.A. and Jordan C., 2003. Monthly Not. Roy. Astron. Soc., **341**, 517.
Simon T., 1986. Astrophys. Space Sci., **118**, 209.
Simon T., 1990. Astrophys. J., **359**, L51.
Simon T., 2001. In: R.J. Garcia López, R. Rebolo, and M.R. Zapatero Osorio (eds). Cool Stars, Stellar Systems, and the Sun. Astron. Soc. Pacific Conf. Ser., **223**, 235.
Simon T. and Fekel F.C., 1987. Astrophys. J., **316**, 434.
Simon T. and Landsman W., 1991. Astrophys. J., **380**, 200.
Simon T., Kelch W., and Linsky J.L., 1980. Astrophys. J., **237**, 72.
Simon T., Herbig G., and Boesgaard A.M., 1985. Astrophys. J., **293**, 551.
Simon T., Ayres T.R., Redfield S., Linsky J.L., 2002. Astrophys. J., **579**, 800.
Singh K.P., Drake S.A., Gotthelf E.V., and White N.E., 1999. Astrophys. J., **512**, 874.
Sinvhal S.D. and Sanwal B.B., 1977. Info. Bull. Var. Stars, No 1263.
Skumanich A., 1972. Astrophys. J., **171**, 565.
Skumanich A., 1985. Austr. J. Phys., **38**, 971.
Skumanich A., 1986. Astrophys. J., **309**, 858.
Skumanich A. and McGregor K., 1986. Adv. Space Phys., **6**, No 8, 151.
Skumanich A., Smythe C., and Frazier E.N., 1975. Astrophys. J., **200**, 747.
Slee O.B., Higgins C.S., and Patston G.E., 1963a. Sky Telescope, **25**, 83.
Slee O.B., Solomon L.H., and Patston G.E., 1963b. Nature, **199**, 67.
Slee O.B., Higgins C.S., Roslund C., and Lynga G., 1969. Nature, **224**, 1087.
Slee O.B., Tuohy I.R., Nelson G.J., and Rennie C.J., 1981. Nature, **292**, 220.
Slee O.B., Stewart R.T., Nelson G.J. et al., 1988. Astrophys. Lett. Commun., **27**, No 4, 247.
Slish V.I., 1964. Astron. Zh., **41**, 1038.
Smale A.P., Charles P.A., Corbet R.H.D. et al., 1986. Monthly Not. Roy. Astron. Soc., **221**, 77.
Smith M.A., 1979. Publ. Astron. Soc. Pacific, **91**, 737.
Sobolev V.V. and Grinin V.P., 1995. Astrofizika, **38**, 33.
Soderblom D.R., 1983. Astrophys. J. Suppl. Ser., **53**, 1.
Soderblom D.R., 1985. Astron. J., **90**, 2103.
Soderblom D.R. and Clements S.D., 1987. Astron. J., **93**, 920.

Soderblom D.R., Duncan D.K., and Johnson D.R.H., 1991. Astrophys. J., **375**, 722.
Solanki S.K. and Mathys G., 1987. In: O. Havnes, B.R. Pettersen et al. (eds). Activity in Cool Star Envelopes. Kluwer, Dordrecht. p. 39.
Solov'ev A.A. and Kiritchek E.A., 2004. The Diffuse Theory of Solar Magnetic Cycle. Elista-S.-Petersburg, Kalmyk State University. 181 p.
Soon W.H., Baliunas S.L., and Zhang Q., 1994. Solar Phys., **154**, 385.
Soon W., Frick P., and Baliunas S.L., 1999. Astrophys. J., **510**, L135.
Spangler S.R., Shawhan S.D., and Rankin J.M., 1974a. Astrophys. J., **190**, L129.
Spangler S.R., Rankin J.M., and Shawhan S.D., 1974b. Astrophys. J., **194**, L43.
Spangler S.R., Shawhan S.D., and Rankin J.M., 1975. Bull. Am. Astron. Soc., **7**, 235.
Spangler S.R. and Moffett T.J., 1976. Astrophys. J., **203**, 497.
Spencer R.E., Davis R.J., Zafiropoulos B., and Nelson R.F., 1993. Monthly Not. Roy. Astron. Soc., **265**, 231.
Springfellow G.S., 1996. In: R. Pallavicini and A.K. Dupree (eds). Cool Stars, Stellar Systems, and the Sun. Astron. Soc. Pacific Conf. Ser., **109**, 293.
Spruit H.C., 1982. Astron. Astrophys., **108**, 348 and 356.
Spruit H.C., 1992. In: P.B. Byrne and D.J. Mullan (eds). Surface Inhomogeneities on Late-Type Stars. Lecture Notes in Physics, **397**, 78.
Stauffer J., 1984. Astrophys. J., **280**, 189.
Stauffer J.R. and Hartmann L.W., 1986. Astrophys. J. Suppl. Ser., **61**, 531.
Stauffer J.R. and Hartmann L.W., 1987. Astrophys. J., **318**, 337.
Stauffer J.R., Giampapa M.S., Herbst W. et al., 1991. Astrophys. J., **374**, 142.
Stauffer J.R., Caillault J.-P., Gagné M. et al., 1994. Astrophys. J., Suppl. Ser., **91**, 625.
Stauffer J.R., Liebert J., and Giampapa M., 1995. Astron. J., **109**, 298.
Stauffer J., Jones B., Fisher D. et al., 1998. In: R.A. Donahue and J.A. Bookbinder (eds). Cool Stars, Stellar Systems, and the Sun. Astron. Soc. Pacific Conf. Ser., **154**, CD 1331.
Stein R.F., 1981. Astrophys. J., **246**, 966.
Stelzer B. and Burwitz V., 2003. Astron. Astrophys., **402**, 719.
Stelzer B., Burwitz V., Audard M. et al., 2002. Astron. Astrophys., **392**, 585.
Stenflo J.O. and Lindegren L., 1977. Astron. Astrophys., **59**, 367.
Stepanov A.V., 2001. Phys. Usp., **46**, 106.
Stepanov A.V., Fürst E., Krüger A. et al., 1995. Astron. Astrophys., **299**, 739.
Stepanov A.V., Kliem B., Krüger A. et al., 1999. Astrophys. J., **524**, 961.
Stepanov A.V., Kliem B., Zaitsev V.V. et al., 2001. Astron. Astrophys., **374**, 1072.
Stępień K., 1989. Acta Astron. **39**, 209.
Stępień K., 1994. Astron. Astrophys., **292**, 191.

Stępień K. and Geyer E., 1996. Astron. Astrophys. Suppl. Ser., **117**, 83.
Stern R.A., 1984. In: S.L. Baliunas and L. Hartmann (eds). Cool Stars, Stellar Systems, and the Sun. Lecture Notes in Physics, **193**, 150.
Stern R.A., 1999. In: C.J. Butler and J.G. Doyle (eds). Solar and Stellar Activity: Similarities and Differences. Astron. Soc. Pacific Conf. Ser., **158**, 47.
Stern R.A. and Drake J.J., 1996. In: S. Bowyer and R.F. Malina (eds). Astrophysics in the Extreme Ultraviolet. Kluwer, Dordrecht. p. 135.
Stern R.A. and Zolcinski M.-C., 1983. In: P.P. Byrne and M. Rodonò (eds). Activity in Red Dwarf Stars. Reidel, Dordrecht. p. 131.
Stern R.A., Zolcinski M.-C., Antiochos S.K., and Underwood J.H., 1981. Astrophys. J., **249**, 647.
Stern R.A., Underwood J.H., and Antiochos S.K., 1983. Astrophys. J., **264**, L55.
Stern R.A., Antiochos S.K., and Harnden F.R., 1986. Astrophys. J., **305**, 417.
Stern R.A., Schmitt J.H.M.M., and Kahabka P.T., 1995. Astrophys. J., **448**, 683.
Stimets R.W. and Giles R.H., 1980. Astrophys. J., **242**, L37.
Stout-Batalha N.M. and Vogt S.S., 1999. Astrophys. J. Suppl. Ser., **123**, 251.
Strassmeier K.G., 1988. Astrophys. Space Sci., **140**, 223.
Strassmeier K.G. and Bopp B.W., 1992. Astron. Astrophys., **259**, 183.
Strassmeier K.G. and Rice J.B., 1998. Astron. Astrophys., **330**, 685.
Strassmeier K.G., Hall D.S., Zeilik M. et al., 1988. Astron. Astrophys. Suppl. Ser., **72**, 291.
Strassmeier K.G., Hooten J.T., Hall D.S., and Fekel F.C., 1989. Publ. Astron. Soc. Pacific, **101**, 107.
Strassmeier K.G., Rice J.B., Wehlau W.H. et al., 1993. Astron. Astrophys., **268**, 671.
Strassmeier K.G., Bartus J., Cutispoto G., and Rodonò M., 1997. Astron. Astrophys. Suppl. Ser., **125**, 11.
Sundland S.R., Pettersen B.R., Hawley S.L. et al., 1988. In: O. Havnes, B.R. Pettersen, J.H.M.M. Schmitt, and J.E. Solheim (eds). Activity in Cool Star Envelopes. Kluwer, Dordrecht. p. 61.
Švestke Z., 1967. Trans. IAU 13A, 141.
Swank J.H. and Johnson H.M., 1982. Astrophys. J., **259**, L67.
Syrovatskii S.I., 1976. P. Astron. Zh., **2**, 35 (= Sov. Astron. Lett. **2**, 13).
Szecsnyi-Nagy G., 1986. In: L. Szabados (ed). Eruptive Phenomena in Stars. Comm. Konkoly Obs., No 86, 425.
Szecsnyi-Nagy G., 1990. Astrophys. Space Sci., **170**, 63.
Tagliaferri G., White N.E., Giommi P., and Doyle J.G., 1987. In: J.L. Linsky and R.E. Stencel (eds). Cool Stars, Stellar Systems, and the Sun. Lecture Notes in Physics, **291**, 176.
Tagliaferri G., Doyle J.G., and Giommi P., 1990. Astron. Astrophys., **231**, 131.

Tagliaferri G., Covino S., Panzera M.R. et al., 2001. In: R.J. Garcia López, R. Rebolo, and M.R. Zapatero Osorio (eds). Cool Stars, Stellar Systems, and the Sun. Astron. Soc. Pacific Conf. Ser., **223**, CD 1177.

Tarasova T.N., Plachinda S.I., and Rumyantsev V.V., 2001. Astron. Zh., **78**, 550 (= Astron. Rep., **45**, 475).

Teplitskaya R.B. and Skochilov V.G., 1990. Astron. Zh., **67**, 1261 (= Sov. Astron., **34**, 636).

Terndrup D.M., Krishnamurthi A., Pinsonneault M.H., and Stauffer J.R., 1999. Astron. J., **118**, 1814.

Terndrup D.M., Stauffer J.R., Pinsonneault M.H. et al., 2000. Astron. J., **119**, 1303.

Thackeray A.D., 1950. Monthly Not. Roy. Astron. Soc., **110**, 45.

Thatcher J.D. and Robinson R.D., 1993. Monthly Not. Roy. Astron. Soc., **262**, 1.

Thatcher J.D., Robinson R.D., and Rees D.E., 1991. Monthly Not. Roy. Astron. Soc., **250**, 14.

Tinney C., 1995. Mem. Soc. Astron. Ital., **66**, No 3, 611.

Tinney C.G., Delfosse X., and Forveille T., 1997. Astrophys. J., **490**, L95.

Toner C.G. and Gray D.F, 1988. Astrophys. J., **334**, 1008.

Toner C.G. and LaBonte B.J., 1991. Astrophys. J., **368**, 633.

Toner C.G. and LaBonte B.J., 1992. In: P.B. Byrne and D.J. Mullan (eds). Surface Inhomogeneities on Late-Type Stars. Lecture Notes in Physics, **397**, 192.

Topka K. and Marsh K.A., 1982. Astrophys. J., **254**, 641.

Torres C.A.O. and Ferraz Mello S., 1973. Astron. Astrophys., **27**, 231.

Torres C.A.O., Ferraz Mello S., and Quast G.R., 1972. Astrophys. Lett., **11**, 13.

Torres C.A.O., Busko I.C., and Quast G.R., 1983. In: P.P. Byrne and M. Rodonò (eds). Activity in Red-Dwarf Stars. Reidel, Dordrecht. p. 175.

Torres C.A.O., Busko I.C., and Quast G.R., 1985. Rev. Mex. Astron. Astrofis., **10**, 329.

Tovmassian H.M., Haro G., Webber J.C. et al., 1974. Astrofizika, **10**, 339.

Tovmassian H.M., Recillas E., Cardona O., and Zalinian V.P., 1997. Rev. Mex. Astron. Astrofis., **33**, 107

Tsikoudi V., 1982. Astrophys. J., **262**, 263.

Tsikoudi V., 1989. Astron. J., **98**, 290.

Tsikoudi V., 1990. Astrophys. Space Sci., **179**, 69.

Tsikoudi V. and Hudson H., 1975. Astron. Astrophys., **44**, 273.

Tsikoudi V. and Kellett B.J., 1997. Monthly Not. Roy. Astron. Soc., **285**, 759.

Tsuboi Y., Chartas G., Feigelson E.D. et al., 2002. In: F. Favata and J.J. Drake (eds). Stellar Coronae in the Chandra and XMM-Newton Era. Proc. ESTEC Symp., Noordwijk, the Netherlands, June 25–29, 2001. Astron. Soc. Pacific Conf. Ser., **277**, 255.

Tsuji T., Ohnaka K., and Aoki W., 1996. Astron. Astrophys., **305**, L1.

Tuominen I., Huovelin J., Efimov Yu.S. et al., 1989. Solar Phys., **121**, 419.

Turner N.J., Cram L.E., and Robinson R.D., 1991. Monthly Not. Roy. Astron. Soc., **253**, 575.
Vaiana G.S., 1980. In: A.K. Dupree (ed). Cool Stars, Stellar Systems, and the Sun. SAO Special Rep. No 389, 195.
Vaiana G.S. and 15 coauthors, 1981. Astrophys. J., **245**, 163.
Valenti J.A., Marcy G.W, and Basri G., 1995. Astrophys. J., **439**, 939.
Valenti J.A., Johns-Krull C.M., and Piskunov N.E., 2001. In: R.J. Garcia López, R. Rebolo, and M.R. Zapatero (eds). Cool Stars, Stellar Systems, and the Sun. Astron. Soc. Pacific Conf. Ser., **223**, CD 1579.
van de Kamp P., 1969. In: S.S. Kumar (ed). Low-Luminosity Stars. Gordon and Breach Sci. Publ., New York. p. 199.
van de Kamp P. and Lippincott S.L., 1951. Publ. Astron. Soc. Pacific, **63**, 141.
van den Oord G.H.J., 1988. Astron. Astrophys., **207**, 101.
van den Oord G.H.J. and Doyle J.G., 1997. Astron. Astrophys., **319**, 578.
van den Oord G.H.J. and Mewe R., 1989. Astron. Astrophys., **213**, 245.
van den Oord G.H.J., Doyle J.G., Rodonò M. et al., 1996. Astron. Astrophys., **310**, 908.
van den Oord G.H.J., Schrijver C.J., Camphens M. et al., 1997. Astron. Astrophys., **326**, 1090.
van der Woerd H., Tagliaferri G., Thomas H.C., and Beuermann K., 1989. Astron. Astrophys., **220**, 221.
van Hamme M., 1993. Astron. J., **106**, 2096.
van Maanen A., 1940. Astrophys. J., **91**, 503.
van Maanen A., 1945. Publ. Astron. Soc. Pacific, **57**, 216.
Vaughan A.E. and Large M.I., 1986a. Proc. Astron. Soc. Austr., **6**, No 3, 319.
Vaughan A.E. and Large M.I., 1986b. Monthly Not. Roy. Astron. Soc., **223**, 399.
Vaughan A.H., 1980. Publ. Astron. Soc. Pacific, **92**, 392.
Vaughan A.H., 1983. In: J.O. Stenflo (ed). Solar and Stellar Magnetic Fields: Origins and Coronal Effects. Reidel, Dordrecht. p. 113.
Vaughan A.H. and Preston R.W., 1980. Publ. Astron. Soc. Pacific, **92**, 385.
Vaughan A.H., Baliunas S.L., Middelkoop F. et al., 1981. Astrophys. J., **250**, 276.
Veeder G.J., 1974. Astron. J., **79**, 702.
Vedder P.W., Patterer R.J., Jelinsky P. et al., 1993. Astrophys. J., **414**, L61.
Vedder P.W., Brown A., Drake J.J. et al., 1994. In: J.P. Caillault (ed). Cool Stars, Stellar Systems, and the Sun. Astron. Soc. Pacific Conf. Ser., **64**, 13.
Ventura R., Peres G., Pagano I., and Rodonò M., 1995. Astron. Astrophys., **303**, 509.
Ventura R., Maggio A., and Peres G., 1998. Astron. Astrophys., **334**, 188.
Vernazza J.E., Avrett E.H., and Loeser R., 1973. Astrophys. J., **184**, 605.
Vernazza J.E., Avrett E.H., and Loeser R., 1981. Astrophys. J. Suppl. Ser., **45**, 635.

Vetesnik M. and Esghafa M.M., 1989. Publ. Astron. Inst. Brno Univ., 34–36, 7.
Vikram Singh R., 1995. Astrophys. Space Sci., **228**, 135.
Vilhu O., 1984. Astron. Astrophys., **133**, 117.
Vilhu O., 1987. In: J.L. Linsky and R.E. Stencel (eds). Cool Stars, Stellar Systems, and the Sun. Lecture Notes in Physics, **291**, 110.
Vilhu O. and Rucinski S.M., 1983. Astron. Astrophys., **127**, 5.
Vilhu O., Neff J.E., and Walter F.M., 1986. In: E.J. Rolfe (ed). New Insight in Astrophysics. ESA SP-263, 113.
Vilhu O., Ambruster C.W., Neff J.E. et al., 1989. Astron. Astrophys., **222**, 179.
Viti S., Jones H.R.A., Maxted P., and Tennyson J., 2002. Monthly Not. Roy. Astron. Soc., **329**, 290.
Vogt S.S., 1975. Astrophys. J., **199**, 418.
Vogt S.S., 1980. Astrophys. J., **240**, 567.
Vogt S.S., 1981. Astrophys. J., **250**, 327.
Vogt S.S., 1983. In: P.P. Byrne and M. Rodonò (eds). Activity in Red Dwarf Stars. Reidel, Dordrecht. p. 137.
Wachmann A.A., 1939. B.Z. **21**, 25.
Wachmann A.A., 1968. Astron. Abhand. Bergedorf, **7**, No 8, 397.
Walker A.R., 1981. Monthly Not. Roy. Astron. Soc., **195**, 1029.
Walter F.M., 1981. Astrophys. J., **245**, 677.
Walter F.M., 1982. Astrophys. J., **253**, 745.
Walter F.M., 1990. In: M. Elvis (ed). Imaging X-ray Astronomy. A Decade of Einstein Observatory Achievements. Cambridge Univerity Press. p. 223.
Walter F.M., Linsky J.L., Bowyer S., and Garmire G., 1980. Astrophys. J., **236**, L137.
Webber J.C., Yoss K.M., Deming D. et al., 1973. Publ. Astron. Pacific Soc. **85**, 739.
Weis E.W., 1994. Astron. J., **107**, 1135.
White S.M., Kundu M.R., and Jackson P.D., 1986. Astrophys. J., **311**, 814.
White S.M., Jackson P.D., and Kundu M.R., 1989a. Astrophys. J. Suppl. Ser., **71**, 895.
White S.M., Kundu M.R., and Jackson P.D., 1989b. Astron. Astrophys., **225**, 112.
White S.M., Lim J., and Kundu M.R., 1994. Astrophys. J., **422**, 293.
Whitehouse D.R., 1985. Astron. Astrophys., **145**, 449.
Wild W.J., Rosner R., Harmon R., and Drish W.F., 1994. In: J.P. Caillault (ed). Cool Stars, Stellar Systems, and the Sun. Astron. Soc. Pacific Conf. Ser., **64**, 628.
Willson R.C., Gulkis S., Janssen M. et al., 1981. Science, **211**, 700.
Willson R.C., Hudson H.S., and Woodward M., 1984. Sky Telescope, **67**, 501.
Willson R.F., Lang K.R., and Foster P., 1988. Astron. Astrophys., **199**, 255.
Wilson O.C., 1961. Publ. Astron. Soc. Pacific, **73**, 15.
Wilson O.C., 1963. Astrophys. J., **138**, 832.

Wilson O.C., 1966. Astrophys. J., **144**, 695.
Wilson O.C., 1968. Astrophys. J., **153**, 221.
Wilson O.C., 1978. Astrophys. J., **226**, 379.
Wilson O. and Wooley R., 1970. Monthly Not. Roy. Astron. Soc., **148**, 463.
Wilson O.C. and Skumanich A., 1964. Astrophys. J., **140**, 1401.
Wood B.E. and Linsky J.L., 1998. Astrophys. J., **492**, 788.
Wood B.E., Brown A., Linsky J.L. et al., 1994. Astrophys. J. Suppl. Ser., **93**, 287.
Wood B.E., Linsky J.L., and Ayres T.R., 1997. Astrophys. J., **478**, 745.
Wood B.E., Linsky J.L., Müller H.-R., and Zank G.P., 2001. Astrophys. J., **547**, L49.
Wood B.E., Müller H.-R., Zank G.P., and Linsky J.L., 2002. Astrophys. J., **574**, 412.
Woodgate B.E., Robinson R.D., Carpenter K.G. et al., 1992. Astrophys. J., **397**, L95.
Worden S.P., 1974. Publ. Astron. Soc. Pacific, **86**, 595.
Worden S.P. and Peterson B.M., 1976. Astrophys. J., **206**, L145.
Worden S.P., Schneeberger T.J., and Giampapa M.S., 1981. Astrophys. J. Suppl. Ser., **46**, 159.
Worden S.P., Schneeberger T.J., Giampapa M.S. et al., 1984. Astrophys. J., **276**, 270.
Write J.T., Marcy G.W., Butler R.P., and Vogt S.S., 2004. Astrophys. J. Suppl. Ser., **152**, 261.
Young A., Sadjadi S., and Harlan E., 1987a. Astrophys. J., **314**, 272.
Young A., Mielbrecht R.A., and Abt H.A., 1987b. Astrophys. J., **317**, 787.
Young A., Skumanich A., and Harlan E., 1984. Astrophys. J., **282**, 683.
Young A., Skumanich A., Stauffer J.R. et al., 1989. Astrophys. J., **344**, 427.
Young A., Skumanich A., MacGregor K.B., and Temple S., 1990. Astrophys. J., **349**, 608.
Zaitsev V.V. and Stepanov A.V., 1991. Astron. Zh., **68**, 384 (= Sov. Astron., **35**, 384).
Zaitsev V.V., Stepanov A.V., and Tsap Yu.T., 1994. Kinemat. Phys. Celest. Bodies, **10**, No 6, 3.
Zaitsev V.V., Kislyakov A.G., Stepanov A.V. et al., 2004. P. Astron. Zh. **30**, 362 (= Sov. Astron. Lett. **30**, 319).
Zarro D.M., 1983. Astrophys. J., **267**, L61.
Zarro D.M. and Zirin H., 1985. Astron. Astrophys., **148**, 240.
Zarro D.M. and Zirin H., 1986. Astrophys. J., **304**, 365.
Zboril M., Byrne P.B., and Rolleston W.R.J.R., 1997. Monthly Not. Roy. Astron. Soc., **284**, 685.
Zheleznjakov V.V., 1967. Astron. Zh., **44**, 42. (= Sov. Astron., **11**, 33).
Zhilyaev B.E. and Verlyuk I.A., 1995. In: J. Greiner, H.W. Duerbeck, and R.E. Gershberg (eds). Flares and Flashes. Lecture Notes in Physics, **454**, 80.

Zhilyaev B.E., Verlyuk I.A., Romanyuk Ya.O. et al., 1998. Astron. Astrophys., **334**, 931.

Zhilyaev B.E., Romanyuk Ya.O., Svyatogorov O.A. et al., 2001. In: J. Seimenis (ed). Proc. 4th Astron. Conf. HEL. A.S. Samos. p. 87.

Zhilyaev B.E., Romanyuk Ya.O., Verlyuk I.A. et al., 2000. Astron. Astrophys., **364**, 641.

Zhilyaev B., Romanyuk Ya., Svyatogorov O. et al., 2003. Kinematics and Physics of Celestial Bodies, Suppl. No 4, 30.

Zirin H., 1976. Astrophys. J., **208**, 414.

Zirin H. and Ferland G.J., 1980. Big Bear Solar Observatory reprint No 0192.

Zirin H. and Liggett M.A., 1987. Solar Phys., **113**, 267.

Zolcinski M.-C., Antiochos S.K., Stern R.A., and Walker A.B.C., 1982. Astrophys. J., **258**, 177.

Index

α Cen (includes A and B components) 17, 64, 66, 83, 87, 92, 96, 99, 101, 102, 105, 114, 124, 132, 134, 152–157, 161, 180, 190, 335
α CrB   163
α Per   30, 57, 98, 124, 150–152, 270, 314, 396, 403, 404, 407
β Com   392
β Hyi   412, 417, 418
χ^1 Ori   43, 79, 87, 88, 99, 114, 125, 152, 154, 164, 166
δ Pav   108, 183
κ Cet   67, 79, 99, 101, 152, 154, 383, 389, 423
π^1 UMa   11, 156, 164, 261, 383, 423
σ Dra   52
τ Cet   99, 108, 183, 186
ε Eri   40, 43, 47, 49, 51, 52, 54–56, 62, 67, 68, 79, 82, 83, 87, 92, 99, 101, 102, 105, 108, 114, 116, 123, 124, 152, 153, 155, 158, 160, 162, 164, 178, 190, 351, 392
ε Ind   66, 180
ξ Boo A   39, 41–45, 47, 49, 53, 55, 79, 82, 83, 87, 94, 113–116, 155, 158, 392, 426, 427
ξ UMa   95
ζ Dor   99
111 Tau   114
12 Oph (= HD 149661)   113
1E0419.2+1908   137, 246, 261
2MASSI J1315309-264951   10

2MASSW J0149090+295613   10, 350, 351, 353
36 Dra   270
40 Eri (=Gl 166)   52, 125, 132–134, 180, 223
47 Cas   175, 188, 238, 248, 268
59 Vir   94
61 Cyg A (= HD 201091)   40, 42, 43, 46, 48, 49, 64, 80, 108, 113, 114, 123, 180, 183, 392
61 Cyg B   64, 83, 105, 106, 114
70 Oph   41, 42, 44, 45, 55, 67, 79, 99, 105, 123, 125, 178
71 Tau   135
78 UMa   88

AD Leo (=Gl 388)   45–48, 50, 52, 53, 71, 72, 78–80, 89, 90, 95, 99, 111, 112, 120, 121, 125, 152, 157–161, 168, 169, 171–173, 188, 190, 194–196, 202, 204, 206, 212, 213, 216, 218, 223, 225, 230, 235, 238, 240, 245, 248–252, 254, 268, 274–278, 281, 284–286, 289–300, 305, 306, 310–314, 317, 320, 321, 325, 329, 332, 334–336, 339–341, 343–346, 349, 351–358, 360, 369, 370, 374–377, 379, 383
AP 108   250
AP 20   250
AT Mic (=Gl 799)   81, 83, 86, 89, 92, 95, 152, 153, 167, 170, 218, 223, 239, 244, 245, 247, 252, 270,

286–288, 291, 303, 304, 310, 312, 319, 334, 353, 359, 360, 380
AU Mic (=Gl 803)   23, 47, 48, 72, 78–80, 82, 83, 86–89, 91–93, 95–99, 101, 102, 109, 115, 133, 152, 153, 166, 167, 202, 205, 209, 218, 250, 252, 278, 280, 285, 286, 291, 303–305, 307, 310, 312–315, 317–320, 322, 334, 376, 384
AZ Cnc   209, 248, 267

Barnard's star (= GJ 699)   145, 169, 404
BD+08°102   12
BD+14°690   246
BD+16°2708 (=Gl 569 B = CE Boo)   285
BD+16°577 (= HD 27130)   258
BD+19°5116   223
BD+19°5116 AB   281
BD+22°3406   328
BD+22°4409   20, 77
BD+22°669   322
BD+26°730 (= Gl 171.2 = V 833 Tau)   28
BD+44°2051   254, 255
BD+48°1958 A   139, 248
BD-8°4352 (= Gl 644 = V 1054 Oph = Wolf 630)   325
BE Cet (= HD 1835)   34, 39, 383, 423
BF Lyn (= HD 80715)   115, 124, 152
Blanco 1   151
BRI 0021-0214   10
BY Dra (= Gl 719)   19, 21–24, 26–28, 41, 47, 72, 73, 79, 86, 101, 112, 115, 125, 146, 169–171, 178, 193, 209, 218, 220, 223, 237, 240, 246, 261, 267, 304, 305, 307, 310, 311, 318, 319, 334, 335, 341, 374, 382–385

CC Eri   21, 22, 44, 113, 115, 144, 167, 170, 248, 267, 268, 308, 311, 384
CE Boo (= BD+16°2708 = Gl 1569 B)   193
CM Dra   15, 90, 97, 317
CN Leo (=Gl 406 =Wolf 359)   81, 206, 218, 248, 267, 288, 315, 317, 318, 374
Coma   150

CR Dra (= Gl 616.2 = BD+55°1823)   112

DH Leo   77, 101, 113, 115
DK Leo   72, 78, 79
DO Cep (=Gl 860 B)   218, 223, 295
DT Vir (= Gl 494)   52, 53, 218, 223, 295
DX Leo   382, 383, 423

EI Cnc (= LHS 2076)   384
EK Dra (=HD 129333)   27, 30, 102, 146, 157, 174, 175, 188, 190, 238, 248, 269, 298, 382, 383, 412, 414–417, 423
EQ 1839.6+8002   144, 248, 271–273
EQ Peg (= Gl 896 B)   72, 79, 82, 83, 86, 95, 113, 152, 153, 164–166, 168, 169, 171, 172, 187, 206, 209, 213, 214, 223, 225, 247, 250, 254, 262, 264, 265, 277, 278, 280, 282, 285, 291, 292, 298, 299, 303, 304, 306, 310, 311, 313, 314, 321, 322, 352, 370, 372
EQ Vir   28, 46, 47, 53, 56, 72, 79, 83, 86, 105, 106, 182, 245, 253, 322
ER Vul   161, 251, 280, 296
EUVE 2056   312
EUVE J0008+208   322
EUVE J0202+105   322
EUVE J0213+368   322
EUVE J0613-23.9 B   314, 323
EUVE J0725-004   322
EUVE J1147050   322
EUVE J1148-374   322
EUVE J1438-432   314, 322
EUVE J1808+297   322
EUVE J2056-17.1   320
EV Lac (=Gl 873)   27, 32, 33, 48, 50, 51, 53, 69, 72, 78, 79, 86, 92, 96, 99, 113, 119–121, 125, 152, 159, 172, 175, 179, 187, 190, 193, 197, 206, 208–210, 213, 216, 218, 223, 229, 234, 235, 237, 239, 245, 248, 249, 251, 253, 256, 264, 265, 268, 270, 271, 275–277, 281, 284, 295, 300, 303, 307, 310–313, 317, 318, 321, 322, 326, 328–331, 333–346,

Index   491

350, 353, 357–359, 362, 365, 368, 377–379, 382–384
EXO 040830-7134.7   246, 262
EXO 2041.8-3129   314

FF And   21, 334
FH Aqr   262
FK Aqr (= Gl 867 = L 717-22)   32, 115, 152, 188
FL Aqr (=Gl 867)   95, 382

G 102-21   249, 268, 340
G 208-45   322
G 25-16   285
G 3-33 (= TZ Ari)   285
G 32-6   322
GJ 1049   51
GJ 117   152
GJ 191   143
GJ 644   152
GJ 699 (= Barnard's star)   404
GJ 702   152
GJ 845   143
GJ 866   143
GJ 887   143
Gl 1   117
Gl 105 B   91, 92, 94, 109
Gl 113.1 (= VY Ari)   178
Gl 15 B (= GQ And)   73, 223
Gl 166 (= 40 Eri)   218, 223
Gl 171.2 (= BD+26°730 =V 833 Tau)   53, 56, 384
Gl 179   73
Gl 182 (= V 1005 Ori)   14, 107, 108, 170, 290, 305, 310, 311, 334
Gl 206   218
Gl 213   268, 383
Gl 229   47, 218
Gl 234 (= V 577 Mon)   80, 218, 223, 295
Gl 268 (= Ross 986)   223
Gl 278 C (= YY Gem)   218, 223
Gl 283 B   75
Gl 285 (= YZ CMi = Ross 882)   218, 223
Gl 33   100
Gl 338 A   246, 259
Gl 34 B   246, 259
Gl 375   80, 346

Gl 380 (=HD 88230)   108, 117
Gl 388 (= AD Leo)   218, 223
Gl 393   106
Gl 406 (= CN Leo = Wolf 359)   218, 223
Gl 410   73
Gl 411   73, 106, 117, 144
Gl 425   73
Gl 431   78, 80
Gl 447   91
Gl 455   10
Gl 461   170
Gl 473 (=Wolf 424)   218, 223
Gl 493.1 (= Wolf 461)   218
Gl 494 (= DT Vir)   218, 223
Gl 526   73
Gl 54.1 (= YZ Cet)   218
Gl 540.2   218
Gl 549   334
Gl 551 (= V 645 Cen =Proxima Cen)   218
Gl 569 B (=BD+16°2708 = CE Boo)   176
Gl 588   111
Gl 616.2 (= CR Dra = BD+55°1823)   106, 218, 223
Gl 628   111
Gl 644 (= V 1054 Oph = Wolf 630 = BD-8°4352)   178, 218, 223
Gl 65 B (= UV Cet)   178, 223
Gl 669 (includes A and B =Ross 867)   246, 259
Gl 687   307
Gl 719 (= BY Dra)   218, 223
Gl 725   50
Gl 729 (= V 1216 Sgr)   50, 53, 133, 218, 223
Gl 735 (= V 1285 Aql)   117, 133, 169, 178, 218
Gl 752 A   97
Gl 752 B (= VB 10)   97, 269, 313, 316
Gl 754   144
Gl 755   174
Gl 781   10
Gl 784   91
Gl 791.2 (=HU Del)   133
Gl 793   91
Gl 799 (= AT Mic)   218, 223
Gl 803 (= AU Mic)   218

Gl 815   218, 223
Gl 821   94
Gl 825   82, 91
Gl 841   95
Gl 860 B (= DO Cep)   218, 223
Gl 866   218, 308, 312, 345, 380
Gl 867 (includes A and B = L 717-22 = FK/FL Aqr)   82, 133, 139, 170, 223, 246, 262, 295, 302, 304, 310, 318
Gl 873 (= EV Lac)   178, 218, 223
Gl 875.1 (= GT Peg)   223
Gl 876   50, 383
Gl 887   117
Gl 890 (=HK Aqr)   12, 29, 76, 77, 114, 124, 139, 157, 170, 250, 273, 274, 307, 311
Gl 896 (= BD +19°5116, includes A and B =EQ Peg)   223
Gl 900   91, 117
Gl 97   174, 270
GQ And (=Gl 15 B)   206, 218, 223
GT Peg (=Gl 875.1)   223
GX And (= Gl 15 A)   72

HCG 143   249
HCG 144   270
HCG 181   249
HCG 307   270
HCG 97   249
HD 101501   63
HD 10476   391
HD 10700   391
HD 114710   63
HD 115383   53, 179
HD 129333 (= EK Dra)   96, 314, 387
HD 131511   53, 94
HD 131977   42
HD 134319   32, 36, 416
HD 147365   248, 269
HD 149661 (= 12 Oph)   116
HD 17925   49, 53, 56
HD 1835 (=BE Cet)   39, 116
HD 185114   53
HD 190007   116
HD 190406   63
HD 197890   157, 174, 250, 270, 273, 274, 312
HD 201091 (=61 Cyg A)   391

HD 2047   55
HD 20630   53
HD 206860 (=HN Peg)   63, 113, 114, 116
HD 216803   133
HD 218738   133, 170
HD 22049   53
HD 225239   174
HD 26965   53
HD 27130 (= BD+16°577)   246, 258, 259
HD 27836   135
HD 32147   49
HD 35850   157, 188
HD 3651   388, 391
HD 39587   179
HD 72905   179
HD 80715 (=BF Lyn)   113
HD 82558 (= LQ Hya)   308
HD 88230 (= Gl 380)   83
HDE 19139   114
HE 421   250
HII 1100   249
HII 1136   175, 298
HII 1306   335, 337
HII 1516   249, 270
HII 174   249
HII 1883   175
HII 191   249
HII 2034   249, 269
HII 212   249
HII 2147   249, 270
HII 2244   250
HII 2411   228, 229, 332–334, 384
HII 314   312, 315
HII 345   249
HII 625   175
HK Aqr (=Gl 890)   29, 77
HK Lac   27
HN Peg (=HD 206860)   383, 423
HR 3625   145
HR 6806   125
HU Del (=Gl 791.2)   206
Hyades   11, 12, 113, 134–136, 145, 148–152, 228, 258, 270, 384, 394–398, 401–407, 410, 416, 425
HZ 1136   258
HZ 1733   246, 258
Hz 892   270

Hz II 253   135

IC 2602   152
IC 318   161
IC 4665   149

Kelu 1   176

L 1113-55   318
L 717-22 (= FK/FL Aqr = Gl 867)
   223
L 726-8 (includes A and B =UV Cet)
   2, 3, 160, 168, 170, 172, 214, 278,
   288, 289, 291–293, 299
Lalande 21258 (= WX UMa=
   BD+44°2051)   1
LHS 1070   10
LHS 2   107
LHS 2065   10, 176, 353
LHS 2076 (= EI Cnc)   317
LHS 2924   16
LO Peg   30
LP 944-20   10, 251
LQ Hya (= HD 82558)   28–30, 34,
   44, 49, 52, 53, 56, 78, 113, 152, 190,
   277, 308, 311, 313, 314, 346, 350,
   353, 354, 360, 382, 383, 394, 423,
   424

Melotte 25 VA 334   322

NGC 2422   151
NGC 2451 AB   151
NGC 2516   150, 151, 163
NGC 2547   151
NGC 3532   151
NGC 6475   149

Orion   1, 5, 6, 136, 149, 209, 224, 227,
   240, 283, 405
OU Gem   80, 124

Pleiades   11, 12, 23, 30, 57, 98, 131, 135,
   136, 147–152, 163, 175, 213, 215,
   219, 221, 224, 227, 229, 230, 232,
   240, 258, 268–270, 284, 298, 314,
   315, 335, 383, 384, 393, 395–398,
   401–407, 414, 416, 425
Praesepe   150, 227–229, 395, 405

Proxima Cen (= Gl 551 = V 645 Cen)
   12, 20, 82, 83, 99, 102, 131, 132,
   140, 152, 176, 180, 187, 188, 194,
   199, 200, 202, 203, 243–248, 250,
   251, 254–257, 259, 260, 263, 265,
   267, 273, 279, 282, 286, 287, 299,
   302–304, 307, 308, 310, 311, 318,
   353, 355, 362, 372–374, 393
PW And   116
PZ Mon   384, 385

RE 0618+75   95
RE 1816+541   124
RE 1816541   77
RE J0241-53N   360
Ross 867 (= Gl 669)   285, 295
Ross 882 (= YZ CMi = Gl 285)   1
Ross 986 (= Gl 268)   223
Rossiter 137 B   173

T177   209
T48   209
TW Hya   102
TWA 5B   161
TZ Ari (=G 3-33)   245, 254, 255, 259

UV Cet (= L-726-8 B = Gl 65 B)
   VII–IX, 1, 3–7, 9–11, 14, 68, 72,
   82, 83, 86, 92, 100, 104, 136, 140,
   164–171, 173, 175, 176, 187, 188,
   191, 195, 198, 199, 204–206, 208,
   209, 211–214, 216–218, 222, 223,
   225, 229, 240–245, 247–251, 254,
   258, 259, 261–265, 267, 271, 278,
   281, 282, 284, 286, 288, 291–294,
   297, 299, 303, 304, 310, 325, 327–
   329, 337, 340, 342–352, 355–357,
   362–364, 374, 383, 384, 394, 395,
   419

V 1005 Ori (= Gl 182)   14, 72, 78, 79,
   86, 397
V 1054 Oph (= Wolf 630 = BD-8°4352
   = Gl 644)   72, 170, 188, 218, 282,
   286, 328, 334, 357
V 1216 Sgr (= Gl 729)   212, 218, 259,
   282
V 1285 Aql (=Gl 735)   218, 334, 335
V 371 Ori   282, 284, 302

V 471 Tau   30, 161, 177
V 478 Lyr   26
V 577 Mon (=Gl 234)   206, 208, 218, 318
V 645 Cen (= Gl 551 = Proxima Cen)   218, 334
V 775 Her   113
V 833 Tau (=BD+26°730 = Gl 171.2)   385
V 833 Tau (=BD+26°730 = Gl 171.2)   52, 79, 384
V 834 Tau   322
V 837 Tau   322
vA 288   246, 258
vA 500   246, 258
VB 10 (= Gl 752 B)   97, 160, 176, 250, 269, 278, 316
VB 141   270
VB 190   270
VB 8 (=GL 644 C)   101, 107, 139, 176, 247, 248, 263, 268, 354, 425
VW Cep   152
VXR45   416
VY Ari (=Gl 113.1)   27, 113, 158, 383

Wolf 1561   246, 261
Wolf 359 (= CN Leo =Gl 406)   352, 355
Wolf 424 (=Gl 473)   200, 206, 214, 223, 284, 286, 288, 295, 347, 353, 357
Wolf 461 (= Gl 493.1)   295
Wolf 47   295

Wolf 630 (=BD-8°4352 = Gl 644 AB = V 1054 Oph)   95, 133, 139, 165, 171, 172, 179, 212, 245, 247, 254, 255, 259, 260, 263, 291, 294, 307, 310
WT 486/487   322
WX UMa (=BD+44°2051 = Lalande 21258)   2
WX UMa (=BD+44°2051 = Lalande 21258)   1, 245

XRNumber 191   250

YY Gem (= Gl 278 C)   18–20, 22, 28, 29, 72, 80, 86, 92, 93, 113, 133, 139, 140, 152, 156, 157, 159, 160, 162, 165, 166, 172, 194, 214, 218, 223, 226, 246, 249, 251, 262, 273, 274, 277, 279, 291, 308, 311, 334, 341, 344, 357, 374
YZ Cet (=Gl 54.1)   218, 318
YZ CMi (=Ross 882 = Gl 285)   1, 23, 52, 53, 82, 83, 86, 91, 106, 120, 123, 129, 152, 164, 166, 168–173, 179, 201, 206, 209, 212, 213, 216, 218, 220, 223, 235, 242–245, 247, 248, 254, 256, 257, 259, 263, 265, 281, 283–288, 290, 292–296, 305, 310, 313, 316, 328, 332, 334, 339, 341, 343–348, 350, 352, 356, 357, 359, 362, 368, 375, 376, 384

ASTRONOMY AND ASTROPHYSICS LIBRARY

Series Editors: G. Börner · A. Burkert · W. B. Burton · M. A. Dopita
A. Eckart · T. Encrenaz · M. Harwit · R. Kippenhahn
B. Leibundgut · J. Lequeux · A. Maeder · V. Trimble

The Stars By E. L. Schatzman and F. Praderie

Modern Astrometry 2nd Edition
By J. Kovalevsky

The Physics and Dynamics of Planetary Nebulae By G. A. Gurzadyan

Galaxies and Cosmology By F. Combes, P. Boissé, A. Mazure and A. Blanchard

Observational Astrophysics 2nd Edition
By P. Léna, F. Lebrun and F. Mignard

Physics of Planetary Rings Celestial Mechanics of Continuous Media
By A. M. Fridman and N. N. Gorkavyi

Tools of Radio Astronomy 4th Edition
By K. Rohlfs and T. L. Wilson

Tools of Radio Astronomy Problems and Solutions 1st Edition, Corr. 2nd printing By T. L. Wilson and S. Hüttemeister

Astrophysical Formulae 3rd Edition (2 volumes)
Volume I: Radiation, Gas Processes and High Energy Astrophysics
Volume II: Space, Time, Matter and Cosmology
By K. R. Lang

Galaxy Formation By M. S. Longair

Astrophysical Concepts 2nd Edition
By M. Harwit

Astrometry of Fundamental Catalogues
The Evolution from Optical to Radio Reference Frames
By H. G. Walter and O. J. Sovers

Compact Stars. Nuclear Physics, Particle Physics and General Relativity 2nd Edition
By N. K. Glendenning

The Sun from Space By K. R. Lang

Stellar Physics (2 volumes)
Volume 1: Fundamental Concepts and Stellar Equilibrium
By G. S. Bisnovatyi-Kogan

Stellar Physics (2 volumes)
Volume 2: Stellar Evolution and Stability
By G. S. Bisnovatyi-Kogan

Theory of Orbits (2 volumes)
Volume 1: Integrable Systems and Non-perturbative Methods
Volume 2: Perturbative and Geometrical Methods
By D. Boccaletti and G. Pucacco

Black Hole Gravitohydromagnetics
By B. Punsly

Stellar Structure and Evolution
By R. Kippenhahn and A. Weigert

Gravitational Lenses By P. Schneider, J. Ehlers and E. E. Falco

Reflecting Telescope Optics (2 volumes)
Volume I: Basic Design Theory and its Historical Development. 2nd Edition
Volume II: Manufacture, Testing, Alignment, Modern Techniques
By R. N. Wilson

Interplanetary Dust
By E. Grün, B. Å. S. Gustafson, S. Dermott and H. Fechtig (Eds.)

The Universe in Gamma Rays
By V. Schönfelder

Astrophysics. A New Approach 2nd Edition
By W. Kundt

Cosmic Ray Astrophysics
By R. Schlickeiser

Astrophysics of the Diffuse Universe
By M. A. Dopita and R. S. Sutherland

The Sun An Introduction. 2nd Edition
By M. Stix

Order and Chaos in Dynamical Astronomy
By G. J. Contopoulos

Astronomical Image and Data Analysis
By J.-L. Starck and F. Murtagh

ASTRONOMY AND ASTROPHYSICS LIBRARY

Series Editors: G. Börner · A. Burkert · W. B. Burton · M. A. Dopita
A. Eckart · T. Encrenaz · M. Harwit · R. Kippenhahn
B. Leibundgut · J. Lequeux · A. Maeder · V. Trimble

The Early Universe Facts and Fiction
4th Edition By G. Börner

The Design and Construction of Large Optical Telescopes By P. Y. Bely

The Solar System 4th Edition
By T. Encrenaz, J.-P. Bibring, M. Blanc,
M. A. Barucci, F. Roques, Ph. Zarka

General Relativity, Astrophysics, and Cosmology By A. K. Raychaudhuri,
S. Banerji, and A. Banerjee

Stellar Interiors Physical Principles,
Structure, and Evolution 2nd Edition
By C. J. Hansen, S. D. Kawaler, and V. Trimble

Asymptotic Giant Branch Stars
By H. J. Habing and H. Olofsson

The Interstellar Medium
By J. Lequeux

Methods of Celestial Mechanics (2 volumes)
Volume I: Physical, Mathematical, and Numerical Principles
Volume II: Application to Planetary System, Geodynamics and Satellite Geodesy
By G. Beutler

Solar-Type Activity in Main-Sequence Stars
By R. E. Gershberg

Printing: Krips bv, Meppel
Binding: Stürtz, Würzburg